T5-AXN-083

HUMAN FACTORS IN ORGANIZATIONAL DESIGN AND MANAGEMENT

First Symposium on
Human Factors in Organizational Design and Management
held in Honolulu, Hawaii, 21–24 August, 1984

Co-sponsored by:
The Human Factors Society
The Organizational Design and Management Technical Group
The International Ergonomics Association

Symposium Chair:
Hal W. Hendrick

Program Chair:
Ogden Brown, Jr.

Symposium and Program Committee:
Ogden Brown, Jr., Sara J. Czaja, F. L. Ficks, Dorothy L. Finley,
Glen R. Gallaway, Peter A. Hancock, Martin G. Helander, Hal W. Hendrick,
Andrew S. Imada, Kageyu Noro, Gordon H. Robinson, Tapas K. Sen,
Arnold M. Small, Wanda J. Smith, and James W. Suzansky.

NORTH-HOLLAND
AMSTERDAM · NEW YORK · OXFORD

HUMAN FACTORS IN ORGANIZATIONAL DESIGN AND MANAGEMENT

Proceedings of First Symposium held in
Honolulu, Hawaii, 21–24 August, 1984

Edited by

H. W. Hendrick and O. Brown, Jr.
University of Southern California
Los Angeles
California
U.S.A.

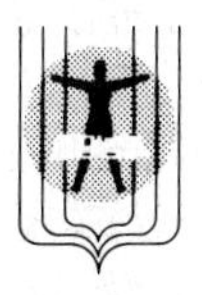

1984

NORTH-HOLLAND
AMSTERDAM · NEW YORK · OXFORD

ISBN: 0 444 87590 5

Published by:
ELSEVIER SCIENCE PUBLISHERS B.V.
P.O. Box 1991
1000 BZ Amsterdam
The Netherlands

Sole distributors for the U.S.A. and Canada:
ELSEVIER SCIENCE PUBLISHING COMPANY, INC.
52 Vanderbilt Avenue
New York, N.Y. 10017
U.S.A.

PRINTED IN THE NETHERLANDS

PREFACE

This book contains a series of papers which were presented during the First International Symposium on Human Factors in Organizational Design and Management held in Honolulu, Hawaii on 21–24 August, 1984. The symposium was sponsored jointly by the Human Factors Society, the Society's Technical Group on Organizational Design and Management, and the International Ergonomics Association. The symposium provided a unique opportunity for researchers, practitioners and students of human factors/ergonomics, organizational behavior and related fields to come together from around the world and share their expertise in this rapidly expanding interdisciplinary area.

New technology, the changing demographic composition, values and attitudes of work forces, and a renewed emphasis on both productivity and the quality of work life have created a need for a true macroergonomic systems approach to the design of organizational and managerial systems. These factors also have heightened the need to consider organizational design elements and managerial processes in the application of human factors/ergonomics to the design of specific subsystems, jobs and workstations. While a number of new methods have been developed and utilized to study and apply human factors on a macroergonomic level, there had been little communication of these methods, research and applications across continents prior to this symposium.

Included herein are a broad selection of papers on methodology, research findings, reviews and case studies from leading professionals throughout the world. These papers provide the reader with a good introduction into this aspect of organizational design and ergonomics, and a 'feel' for the existing knowledge base in this emerging interdisciplinary area.

Hal W. Hendrick and Ogden Brown, Jr.
Editors

THE HUMAN FACTORS SOCIETY

The Human Factors Society, established in 1957, is an interdisciplinary organization of professional people involved in the human factors field. The society promotes the discovery, exchange, and application of knowledge concerning the relationship of people to their machines and their environments. It furthers the assignment of appropriate roles to humans and machines in systems. It advocates the consideration of operators, maintainers, and users in the design of equipment and facilities. The Society supports the development of working and living environments that are comfortable and safe. It encourages the appropriate education and training of those who conceive, design, develop, manufacture, test, manage, and participate in systems.

Meetings and publications are sponsored by the Society. An annual meeting is held in the fall. It includes a technical program, consisting of research reports, panel discussions, and workshops.

Publications include the bimonthly journal *Human Factors* and the monthly *Human Factors Society Bulletin*.

The Human Factors Society has chapters throughout the United States which sponsor local meetings and publications.

The society's technical groups each focus on a particular aspect of the human factors field. The technical groups also hold meetings, publish newsletters, and sponsor technical sessions at the annual meeting.

For information write to:-

Human Factors Society
P.O. Box 1369
Santa Monica
CA 90406
U.S.A.
Phone: (213) 394-1811

TECHNICAL GROUP ON ORGANIZATIONAL DESIGN AND MANAGEMENT

HUMAN FACTORS SOCIETY

The Technical Group on Organizational Design and Management (ODAM) is concerned with improving organizational effectiveness and the quality of work life through an integration and joint optimization of psychosocial, cultural, and technological factors with human-machine performance interface factors in the design of jobs, work stations, organizations, and related management systems through a sociotechnical systems approach.

INTERNATIONAL ERGONOMICS ASSOCIATION

The International Ergonomics Association is an association of ergonomics and human factors societies around the world. Developing out of the Ergonomics Research Society, which was founded in 1949, the international organization has continued to flourish and now has a membership from some 15 federated ergonomics societies in both hemispheres, a central secretariat, and an executive council of international representation.

The International Ergonomics Association brings together organizations and persons concerned with ergonomics or human factors; i.e., the relations between man and his occupation, equipment, and environment in the widest sense, including work, play, leisure, home, and travel situations. To further these aims, it establishes international contacts among those acting in this field, promotes knowledge and practice of ergonomics on an international basis especially in areas where no Federated Society exists, cooperates with international associations to encourage the practical application of ergonomics in industry and other areas, and promotes scientific research by qualified persons in this field.

Apart from the international congress, held every three years, the IEA also sponsors joint conferences of an international nature which are organized by one or more of its member federated societies.

CONTENTS

ORGANIZATION AND MANAGEMENT OF ERGONOMICS

ORGANIZATIONAL DESIGN AND ORGANIZATIONAL DEVELOPMENT

QUALITY OF WORK LIFE, ATTITUDES, AND VALUES

SAFETY FACTORS, WORKLOAD AND PERFORMANCE FACTORS

THEORETICAL FACTORS

TRAINING SYSTEMS

LATE PAPERS*

*These papers arrived too late to be included under their appropriate session headings. The paper by K. Morooka et al. belongs to the "Human-Computer Interaction" section and that of W. Rohmert to the "Organizational Design and Organizational Development" section.

AGING

Human Factors in Organizational Design and Management
H.W. Hendrick and O. Brown, Jr. (Editors)

THE COMING OF AGE OF THE WORKER (PANEL)

Arnold M. Small, Chair
University of Southern California
Los Angeles, California, USA

Ogden Brown, Jr.
David B.D. Smith
University of Southern California
Los Angeles, California, USA

Gordon H. Robinson
University of Wisconsin, Madison
Madison, Wisconsin, USA

DEMOGRAPHIC CORRELATES OF THE FUTURE WORK FORCE

Ogden Brown, Jr.

Important and profound demographic changes are taking place in America's labor force. Increased participation by females has been extraordinary, and a large increase in the number of workers under 35 has also taken place. A coming shortage of youth and a vast increase in the number of prime age workers seem to be the most important changes ahead. Due to the low birth rate of the 1960s, a decline in the number of young workers will exceed one million. Along with that trend, an overabundance of prime age workers (or demographic bunching) will occur in the 25-44 age group. In 1975 there were 39 million in this group; by 1990 it will be over 60 million. As this bulge moves through the work force there will be fierce competition for advancement and great frustration for those who fail to do so. Another result should likely be the exertion of great pressure to accelerate withdrawal from the firm of older workers who block the opportunities for advancement. In turn, this should induce the proliferation of early retirement programs and entice from the work force those with time in middle and upper management. At the same time, increased availability of employment opportunities for older workers should occur, on a part-time basis or otherwise, because of the decline in youth within the work force. Several demographic correlates will be discussed with respect to the role and place of the aging worker and implications will be drawn about the demographic composition of the work force as it ages (the "greying of America").

ANTHROPOMETRY

Human Factors in Organizational Design and Management
H.W. Hendrick and O. Brown, Jr. (Editors)

META-ANALYSIS OF AGE-DEPENDENT DESIGN PARAMETERS: A MODIFIED RELAXATION TECHNIQUE

R.J. Marley
Department of Psychology
Wichita State University
Wichita, Kansas
U.S.A.

M. Rahimi, D.E. Malzahn, and D.L. Hommertzheim
Department of Industrial Engineering
Wichita State University
Wichita, Kansas
U.S.A.

A modified relaxation technique is introduced as a means of integrating anthropometric measures, particularly age-dependent data. The advantages of this method over regression models are discussed. This microcomputer based technique is a practical tool for pooling large groups of sample statistics with great effectiveness and at minimal costs. Implications for use in the design of work environments and equipment are drawn.

INTRODUCTION

As the proportion of the population classed as aged increases, the need to apply human factors principles and techniques to aging research increases. One area of current interest is the integration of basic anthropometric and performance data as a function of age. Improved methodologies for dealing with this subject will most certainly have positive implications for both the academic researcher and the practitioner alike.

In the design of office environments, consideration must now be given to the aging employee as these workers are now putting in years of service past traditional retirement ages. Often, employers extend the retirement dates because of specific talent demands (Kreps, 1977). Therefore, changes in design of the traditional work environments (office or otherwise) is called for in order to optimally utilize the capabilities of this growing population.

Traditionally, human factors researches on aging are concerned with some specified dependent design parameter as a function of an independent variable, age. However, in reviewing the literature on aging (Ruger and Stoessiger, 1927; Fisher and Birren, 1947; Kellor, Frost, Silberberg, Iverson, and Cummings, 1971; Lyddan, Caldwell, and Alluisi, 1971; Montoye and Lamphiear, 1977), several problems in methodology become apparent. First and foremost is the fact that researchers are often forced to draw conclusions and generalizations from inadequate sample sizes for sufficient statistical power and reliability.

Secondly, any conclusions drawn from such studies cannot be easily generalized because of their limited breadth of sampling on the target population. Often the range of the independent variable (age) has been limited to reduce sampling costs. This becomes a compounding problem since the demographics of these populations are continually changing thus hampering simple generalizations for the target population. Consequently, design specifications drawn from such generalizations fail to consider the entire user population. This design problem is a direct effect of both the narrow range of studies, and user demographics that are changing or poorly defined.

MODIFIED RELAXATION TECHNIQUE

There are a large number of aging-related anthropometric data available from previous research (e.g., Churchill, Laubach, McConville, and Tebbetts, 1978). Regression models appear to be a viable technique to indicate the relationship between any dependent variable and the independent variable, age. However, lack of sufficient data for each dependent variable group has forced adoption of linear regression models. On the other hand, these publications contain only the sample statistics (e.g., mean). In a majority of these reports, the original data points (raw data) are inaccessible. As a result, using the sample statistics in a regression analysis would significantly diminish the statistical power of the results generated. Thus, a technique is needed to integrate several groups of sample statistic data.

This paper proposes a new methodology, using an original microcomputer program, that allows the integration of age-dependent anthropometric data derived from previous studies. The previous studies used may or may not have been intended to perform age-dependent analysis. The procedure is based upon a modified relaxation technique (Jenson and Jefferys, 1972) that derives the relationship between groups of sample statistics as a function of an independent variable.

Specifically, the modified relaxation technique defines a relationship between an independent and dependent variable as a discrete function. This function can be defined to any desired interval size for the independent variable, approximating a continuous relationship. Also, this technique can be applied to sample statistics derived from samples taken over any possible set of intervals on the dependent data. The technique maintains the integrity of the original sample data. That is, from the final definition of the dependent-independent variable relationship, the original sample statistics can be regenerated. This procedure works for any sample statistic, i.e., mean, variance, skewness, or kurtosis. From a statistical point of view, this procedure may be considered as a technique for meta-analysis.

The advantages of the modified relaxation technique over regression models (in aging research) are: 1)Regression has been used to produce estimates of the expected value of the dependent variable. Generally, regression has not been used to predict other population statistics that may also be a function of the independent variable. The standard error of estimate in regression is derived for the entire data set, whereas, the relaxation technique derives the variance as a function of the independent variable. 2)The relaxation technique provides the ability to integrate large samples of data by simply gathering several samples of measures previously collected. Thus, eliminating the need for expensive and time-consuming data collection projects. 3)This microcomputer-based technique is capable of updating source data as often as necessary in order to keep abreast with changing population demographics. This would be as simple as adding current research data into the existing pool of data files.

AN APPLICATION

The proposed modified relaxation technique is a form of meta-analysis that allows the integration of any statistic as a function of its independent variable. Therefore, interpretation of the application is determined by the nature of the statistics used. As an application of this technique, the literature regarding grip strength (GS) measures which use age as an independent variable were searched. Four major studies (Fisher and Birren, 1947; Burke, Tuttle, and Thompson, 1952; Kjerland, 1960; Montoye, and Lamphiear, 1977) were found that had consistent methodologies. Reported within these four studies were a total of 41 separate sample statistics for the mean and standard deviation of grip strength. Each of the samples were specified by a different age interval. A graph of the mean grip strength versus age intervals and standard deviation of grip strength versus age intervals are presented in Figure 1 and Figure 2, respectively.

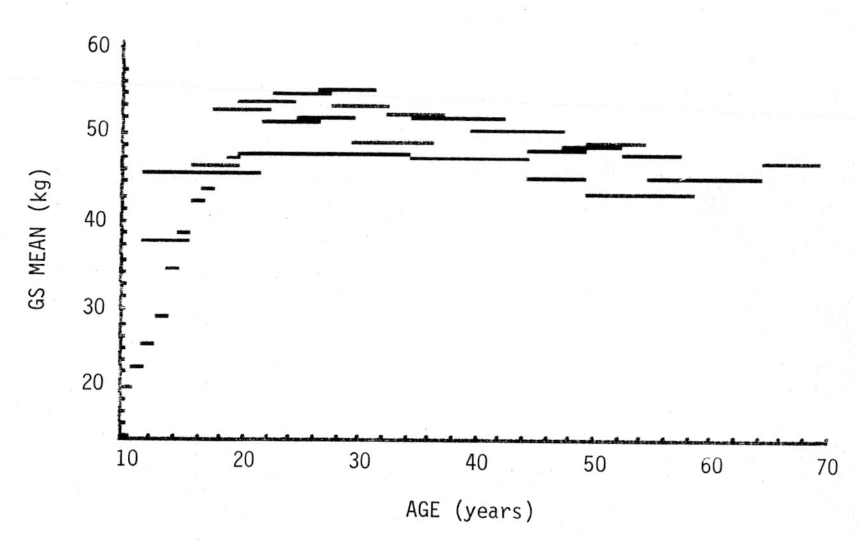

Figure 1
Grip Strength Means for 41 Age Intervals

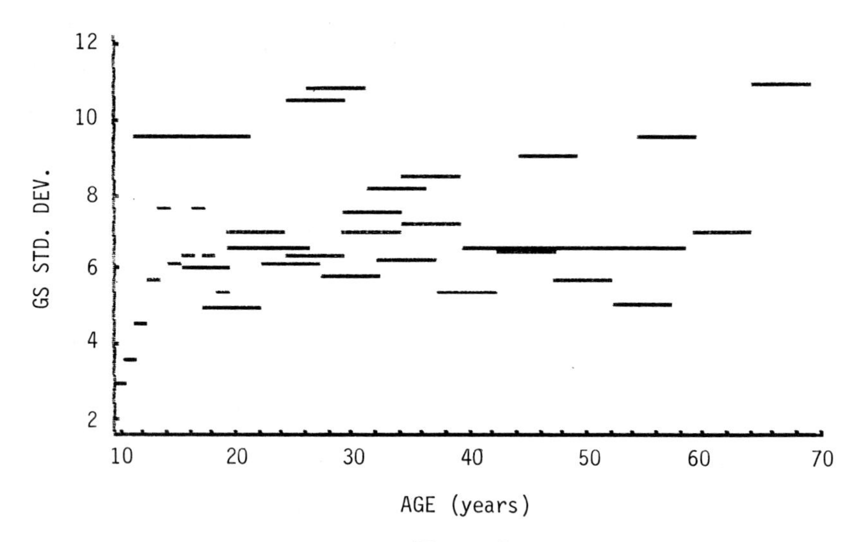

Figure 2
Grip Strength Standard Deviations for 41 Age Intervals

The sample data for both the mean and standard deviation were relaxed to yield the curves in Figure 3 and Figure 4, respectively.

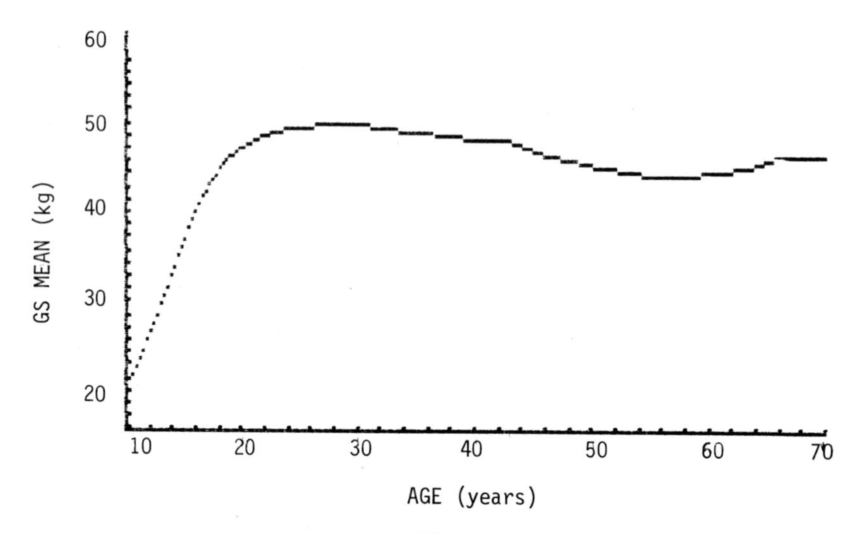

Figure 3
Relaxed Mean Grip Strength vs. Age

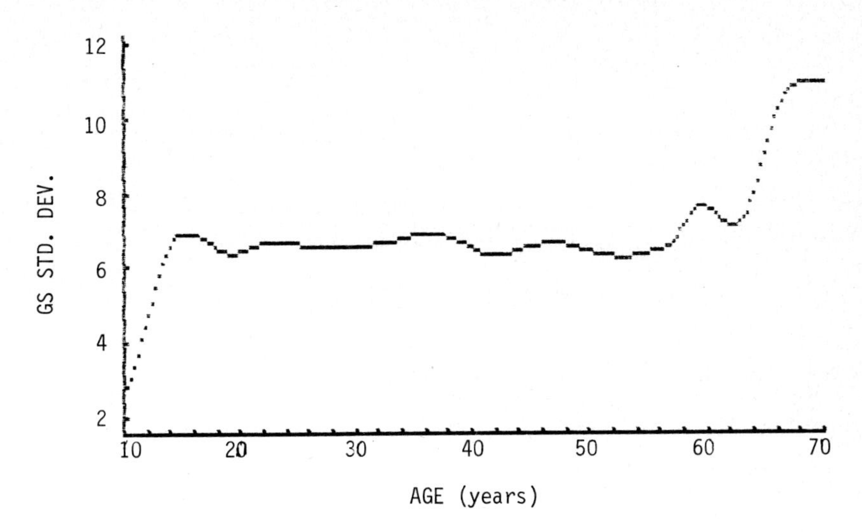

Figure 4
Relaxed Standard Deviation of Grip Strength vs. Age

The trend in these curves agree with other major studies (e.g., Kellor et al., 1971). However, the final integrated relationship was derived from a much larger effective sample size (N=3,200). This technique also allows the evaluation of the variability of a measure as a function of age (Figure 4). For example, most published data indicate that the variability of grip strength is constant across the entire age span. The results presented in Figure 4 indicate that the variability of grip strength may in fact increase significantly with age. The samples included in the higher age ranges were considerably smaller than the others, therefore, are less reliable. Additional data need be collected only for the under sampled age ranges to enhance the accuracy of this meta-analysis. As additional studies in the higher age groups are published the meta-analysis of grip strength can be easily updated.

REFERENCES

Burke, W.E., Tuttle, W.W., Thompson, C.W., Janney, C.D., and Weber, R.J. The Relation of Grip Strength and Grip-Strength Endurance To Age. Journal of Applied Physiology, 1952, 5, 628-630.

Churchill, E., Laubach, L.L., McConville, J.T., and Tebbetts, I. (Eds.). Anthropometric Source Book Volume 1: Anthropometry for Designers. NASA Reference Publication 1024, Scientific and Technical Information Office, 1978.

Fisher, M.B. and Birren, J.E. Age and Strength. Journal of Applied Psychology, 1947, 31 (5), 490-497.

Jenson, V.E. and Jefferys, G.V. Mathmatical Methods in Chemical Engineering. London: Academic Press, 1972.

Kellor, M., Frost, J., Silberberg, N., Iverson, I., and Cummings, R. Hand Strength and Dexterity. American Journal of Occupational Therapy, March 1971, 25 (2), 77-83.

Kjerland, R.N. Age and Sex Differences in Performance in Motility and Strength Tests. Proceedings of the Iowa Academy of Sciences, 1960, 60, 519-522.

Kreps, J.M. Age, Work, and Income. Southern Economic Journal, 1977, 43, 1423-1437.

Lyddan, M.J., Caldwell, L.S., and Alluisi, E.A. Measurements of Muscular Strength, Endurance, and Recovery Over Fifteen Successive Days. Journal of Motor Behavior, 1971, 3 (3), 213-223.

Montoye, H.J. and Lamphiear, D.E. Grip and Arm Strength in Males and Females Age 10 to 69. Research Quarterly, 1977, 48, 109-120.

Ruger, H.A. and Stoessiger, B. On the Growth Curves of Certain Characters in Man (Males). Annals of Eugenics, 1927, 2, 76-110.

COGNITIVE COMPLEXITY AND ORGANIZATIONAL DECISION MAKING

Human Factors in Organizational Design and Management
H.W. Hendrick and O. Brown, Jr. (Editors)

COGNITIVE COMPLEXITY, CONCEPTUAL SYSTEMS AND ORGANIZATIONAL DESIGN AND MANAGEMENT: REVIEW AND ERGONOMIC IMPLICATIONS

Hal W. Hendrick

Institute of Safety and Systems Management
University of Southern California
Los Angeles, California
U.S.A.

Empirical findings concerning the nature of the higher-order personality dimension of cognitive complexity and related conceptual systems are reviewed, including managerial behavior and organizational design requirements. Demographic characteristics of the U.S. population in terms of cognitive complexity are noted, including ongoing demographic shifts in the work force. The importance of the cognitive complexity variable to the ergonomic aspects of organizational and work systems design, and the design of related management support systems is discussed.

INTRODUCTION

During the past two decades there has been a growing body of research data, largely outside of the field of ergonomics, which I believe to be relevant to our understanding of human differences in values, attitudes, motives, and stylistic leadership and work behavior. To the extent that this is true, it carries direct implications for the ergonomics of job and systems design. This research concerns the higher-order predispositional structural cognitive style personality dimension of cognitive complexity or concreteness-abstractness of thinking. The fact that this research also indicates that the high technology societies of the world are undergoing significant yet understandable age-related changes in this higher-order personality dimension further enhances its potential value to those of us who are concerned with the science and practice of ergonomics.

This area of structured personality research has its roots in the classical work of Piaget (1948) on child development. Two groups of researchers have extended the study of concreteness-abstractness into the adult range in ways important to ergonomics. These are O.J. Harvey, D.E. Hunt, and H. Schroder (1961) in the area of conceptual functioning or conceptual systems and behavior, and Lawrence Kohlberg and his colleagues at Harvard with respect to moral reasoning (1969).

Concreteness-abstractness or cognitive complexity is reported to have two major structural aspects, differentiation and integration (Harvey, Hunt & Schroder, 1961). Differentiation can be operationally defined as the number of dimensions extracted from a set of data, and integration as the number of interconnections between rules for combining structured data (Bariff & Lusk, 1971). A concrete cognitive style is one in which relatively little differentiation is used in structuring concepts. That is, experiential data are categorized by the individual within relatively few conceptual dimensions, and within concepts there exists for the individual relatively few categories or shades of grey. In the extreme, a concept is divided into just two categories, characteristic of either/or, absolutist thinking. In addition, concrete thinkers are relatively poor at integrating conceptual data in assessing complex problems and developing unique or creative, insightful solutions. In contrast to concrete persons, abstract individuals tend to demonstrate high differentiation and effective integration in their conceptualizing (Harvey, 1966; Harvey et al., 196l).

Based on their initial research, Harvey, et al. (1961) concluded that there appear to be at least four fundimentally different ways in which people organize or structure and integrate their experiences of reality. Further, these four ways appear to lie along an invariant developmental continuum, with the underlying dimension being concreteness-abstractness. As persons develop greater differentiation and integration in there conceptual functioning, they may move on to a new, more cognitively complex conceptual way of viewing reality. These four systematic ways or stages, in order of their developmental occurance, are labled simply as conceptual systems 1,2,3, and 4 (Harvey, et al., 1961).

Because a given person's level of abstractness can fall at any point along the continuum, it is possible to function primarily out of one conceptual system, but also secondarily out of another conceptual orientation. Also, while the four conceptual systems represent points along the underlying dimension of cognitive complexity, the specific location of these points can vary somewhat from person to person.

COGNITIVE COMPLEXITY AND MORAL REASONING

When Lawrence Kohlberg set out to study moral reasoning, he was looking for structures, forms and relationships that are common to all societies and languages (Kohlberg, 1969). Over the years he gradually elaborated a topological scheme for describing the general structure and forms of moral reasoning that he and his colleagues have found throughout the world. Of particular importance was the finding that moral thought could be defined independently of the specific moral content of moral decisions and actions.

As with research on conceptual systems, Kohlberg found that there are stages of moral reasoning which lie along the concreteness-abstractness continuum. Kohlberg identified three developmental levels of moral thinking. Within each of these levels he identified two separate moral philosophies or conceptualizations of the social-moral world. In Kohlberg's words, "moral thought, then seems to behave like all other kinds of thought. Progress through moral levels and stages is characterized by increasing differentiation and integration." (Kohlberg, 1969, p.186) As with conceptual systems, each level of moral reasoning represents an invariant developmental sequence. Whether or not the specific stages or philosophies within each level are invariant remains questionable. All researchers working in this area appear agreed that about 50 percent of most people's reasoning will be at a single stage, regardless of the moral dilemma involved.

Preconventional or First Level of Moral Reasoning

This level is oriented around concepts of good and bad which are interpreted in terms of physical concepts (punishment and reward) or in terms of the physical power of those who make the rules (i.e., might is right). Within the preconventional level are the first two discernable stages. Stage 1 is an orientation toward punishment and unquestioning deference to superior power. Here, a good decision is one which leads to avoidance of punishment. Stage 2 is an orientation toward personal need satisfaction and occasionally, the needs of others. Elements of fairness, sharing and reciprocity are present, but it is a "you scratch my back and I'll scratch yours" kind, rather than a reciprocity based on loyalty or justice. This stage sometimes is referred to as the morality of the marketplace.

Conventional or Second Level of Moral Reasoning

This level can be described as conformist in the sense of maintaining the expectations and rules of one's family, group, culture or nation. The maintenance of the existing ways is perceived as a valuable end in itself. Stage 3 is referred to as the good-boy, good-girl orientation. Here the goodness of an action is based on whether it pleases or helps others and is approved by them. Stage 4 is an orientation toward authority, fixed rules, regulations, and the maintenance of the existing social order. The goodness or rightness of behavior is judged by the entent to which a person is doing one's duty, showing respect for authority, and maintaining the existing order as and end in itself.

Postconventional or Third Level of Moral Reasoning

The posconventional level is characterized by autonomous, universal moral principles. These principles are seen as existing independent of the authority of the particular groups or individuals who hold them, and apart from one's personal identification with these groups or persons. Whereas Conventional moral reasoning is an externally based value system, Postconventional reasoning is based on an internalized and more principled set of values. Stage 5 is a social contract orientation with legalistic overtones. The rightness of action tends to be evaluated in terms of respecting the general individual rights of persons, and the standards which have been critically examined and agreed upon by the whole society. Stage 5 is the "official" morality of the American government and the U.S. Constitution. Stage 6 is an orientation of universal moral and ethical principles. Morality is not defined by rules and laws of a given society but by one's own conscience in accordance with self-determined ethical principles. These are not concrete rules like the Ten Commandments; rather, they are broad and abstract and often include universal principles of justice and of the reciprocity and equality of human rights.

COGNITIVE COMPLEXITY AND CONCEPTUAL SYSTEMS

In their initial research, Harvey, et al. (1961) identified the following sets of characteristics with each conceptual level of functioning. Subsequent research consistently has confirmed these findings.

System 1 Functioning: Conventional (conformist) Thinking and Behavior

All persons, regardless of culture or nationality, appear to start out with a highly concrete System 1 orientation toward reality. Many, as adults, become more abstract in their functioning with experience, but maintain a System 1 perspective of reality. In general, in comparison with persons with other conceptual orientations, System 1 persons are characterized by conventional thinking and behavior. If highly concrete, the majority of their moral decisions are at Kohlberg's Preconventional level; if somewhat more abstract, their moral reasoning is focused at the Conventional level, and they tend to rely on rules, regulations, tradition, and other external sources as the basis of their decisions. The research consistently has shown System 1 persons to have a high need for (a) structure and order and (b) simplicity and consistency, to be (c) authoritarian, (d) absolutist rather than relativistic in their thinking, (e) closed in their belief system, (f) ethnocentric, (g) paternalistic, (h) low in creativity, (i) personally rigid, (j) highly accepting of prevailing rules, norms, procedures and social roles (and to see them as relatively static and unchanging), and to (k) have a high belief in external fate control.

Research to date indicates that approximately 60% of the American population is operating primarily from a System 1 orientation. However, it is age related. Among those in their mid 50's and older, 80% to 85% may be operating at the System 1 stage; among those in their middle thirties and younger, less than half have remained at the System 1 level. Those in the middle thirties to middle fifties represent the national average. Limited research in other industrialized nations indicates a similar pattern.

System 2 Functioning: General Negativism

For many persons, the System 1 reality eventually "breaks down" as they become more abstract. These individuals react by becoming focused on and sensitized to what is "wrong" with the "system", its institutions and those who exercise authority in one's life. As an essential part of this reaction, developmentally, the individual also appears to learn more about oneself as distinct from the generalized cultural standards which had been applied to both self and others during System 1 functioning. In their moral decision-making, System 2 persons often seem in a kind of psychological vacume. They tend to see the external norms, which heretofore they had relied upon, as having let them down and, thus, no longer reliable. As yet, they have not replaced this external basis for one's moral thinking with an internalized basis. About all System 2 persons can do is to react in a distrustful, negative manner--a kind of Preconventional Stage 2 hedonistic relativism which is confused with more principled moral reasoning. Like System 1 persons, System 2 conceptualizers tend to have a high need for structure and order and for simplicity and concistency. They also tend to be absolutist, closed in their belief system, not highly creative, and personally rigid. Unlike System 1 persons, they tend not to be authoritarian and to reject the prevailing rules, norms and social order, and to advocate change.

Movement into System 2 functioning often takes place during the later high school or college years. Behaviorally, a person's reaction may take either an "approach" or "avoidance" direction--the individual may become the campus left wing activist, determined to tear down that which has let him or her down or, alternately, simply "drop out" from the society which has disappointed them. In either case, it is likely to be a somewhat indiscriminate "throw the baby out with the bathwater" reaction. If asked what they believe should replace the existing social order, the answer invariably is some form of social anarchy--a simplistic belief that everyone should "do their own thing", for example.

For most persons who transition from a System 1 orientation towards reality the System 2 stage is a transitory one, and they eventually move on to a System 3 orientation. As a result, while perhaps as many as 20% of those in their late teens or early twenty's may be in this stage in the U.S., only several percent of older adults conceptualize at this level.

System 3 Functioning: The World is People

The System 2 breaking away from the norm and learning about how one is distinctly oneself provides the basis for empathically understanding and accepting differences (from both the norm and oneself) in others With development of this more abstract realization about others the individual moves into the third stage of conceptual functioning. As an overview of this orientation, we can note that it is characterized by a strong, empathic people orientation. Instead of seeing differences in values, culture, beliefs life styles and institutions as deviant or "less than", as do more concrete persons, System 3 individuals tend to value these differences and to see them as enriching their personal lives and the human condition. With this increase in empathy, valuing of differences, and strong people orientation comes a shift to Postconventional moral reasoning.

In marked contrast to System 1 individuals, System 3 functioning persons tend to be characterized by a low need for structure and order, and for simplicity and consistency(In fact, they often will express a preference for complexity and change) and to demonstrate a high tolerance for ambiguity, low absolutism, openness of beliefs (They expect beliefs to change with added experience), and low ethnocentrism. They tend to be creative, personally flexible, and to rely on internal fate control. They tend to be moderately high on authoritarainism (but do not hold authority figures in awe), accepting of the prevailing rules, mores, traditions and social roles, but also to see them as changing rather than static, and in need of periodic examination to determine if they still are functional in the light of societal change. They tend to have a high need for people and for helping others. They empathically are concerned with the human condition and with factors that affect the quality of life.

In the United States, approximately 25% of all adults appear to be operating primarily from a System 3 orientation, but there are significant age differences. Among those in their middle fifties or older, about 15% fall into the System 3 category; among those in their middle twenties to middle thirties, about one-third appear to be functioning at the System 3 stage. Based on only limited research, this same age related pattern appears to hold for other industrialized nations.

System 4 Functioning: Autonomous, Creative Behavior; Conceptual Maturity

The major developmental work at the fourth conceptual level is the integration of standards which apply to both self and others. This integration enables the individual to understand both self and others as occupying different positions on the same transcendent dimension rather than seeing self and others simply as being on different standards. In accomplishing this integration task, the individual develops greater autonomy in thought and action.

To an even greater extent than System 3 persons, System 4 thinkers rely on Postconventional moral reasoning as the basis of their actions, have a low need for structure and order, and for simplicity and consistency, are relativistic rather than absolutist, are creative, flexible and very open in their belief system. Like System 3 persons, System 4 individuals are people oriented, but they are not people dependent. Although others may have a high percieved self worth, this characteristic seems universal among

System 4 functioning persons.

System 4 persons seem to be the same individuals which, through very different research approaches, Maslow identified as self actualizers and Carl Rogers as fully functioning persons. All three approaches have identified eight to ten percent of the adult U.S. population as functioning at this level.

Characteristics Unrelated to Conceptual Systems

At this point it may be useful to note several important characteristics that might seem highly related to conceptual stage, but which are not. First, only a weak relationship has been found between conceptual systems and intelligence. Although some minimal level of general intelligence seems to be required for highly abstract functioning, some of the most brilliant persons in all walks of life have been, and are System 1 functioning individuals. Secondly, conceptual functioning does not appear related to generosity, friendliness, or numerous other valued personality characteristics which have not been mentioned.

ETIOLOGICAL DETERMINANTS OF LEVEL OF COGNITIVE COMPLEXITY

As Harvey has noted, "from the vast amount of work done on early stimulation and experience, there seems little doubt that exposure to diversity is probably the most central prerequisite of differentiation and integration". (1966, p. 63) The research on moral growth further indicates that passive exposure is not a sufficient condition, but that active exposure does result in children becoming more abstract in their functioning (Blatt & Kohlberg, 1971; Holstein, 1969). These studies also indicate that the nature of the child's training environment at home and school with respect to providing active exposure is critical to fostering or inhibiting abstractness and moral development. Harvey et al. (1961) also came to the conclusion that these training environments appeared to contain important determinents. In particular, the nature of the "trainer" role appeared critical. In general, the more absolutist and authoritarian the parent, teacher or other trainer, the greater is the liklihood that active exposure was inhibited, and that the child will pleateau at a relatively concrete level of conceptual functioning as an adult. The more relativistic, the less authoritarian, the more the "trainer" encourages the child to think things through and draw personal conclusions, and the more the trainer instills a strong positive sense of personal self-worth in the child, the more abstract or cognitively complex the child will become in his or her conceptual functioning as an adult.

Since World War II, there has been a dramatic shift in the childrearing patterns in the United States and other industrialized nations from the absolutist-authoritarian one to a more relativistic-permissive pattern. In addition, the dramatic increase in affluence, quantity and quality of transportation and communications systems, educational systems, and other technological advances--particularly that of television--has greatly increased exposure to diversity during the childhood and teenage years. for example, television has enabled an entire generation to grow up looking at the spaceship earth from the outside-in; it also has brought the reality of war, poverty and other events from around the world to our livingrooms. So dramatic has the impact of these changes been on those born since World War II that Yankelovich (1978), based on extensive longitudinal research on the U.S. work force, has labeled this new group of workers "the new breed". In summary, the dramatic changes in childrearing patterns, affluence, education level (which structures exposure to diversity) and technology have resulted in a work force in which the majority of those in their middle thirties and younger are operationg from beyond a System 1 orientation, and a middle aged and, especially, older segment which is operating from predominently a System 1 orientation. This dichotomy in the work force has great significance, in my opinion, for ergonomics because it is profoundly affecting work values, motives and expectancies with respect to how organizations and jobs are designed and managed.

COGNITIVE COMPLEXITY AND CLASSICAL ORGANIZATIONAL DESIGN AND MANAGEMENT

When we consider that approximately 80% of the older persons--those occupying most of the positions of power and influence in our complex organizations--are operating from a System 1 conception of reality, it is understandable that the classic Weberian (1946) bureaucratic, scientific management (Taylor, 1911) basis of organizational design has been dominant throughout the world; it is directly reflective of a System 1 conception of the world. The bureaucratic structure models the relatively unambiguous and hierarchical structuring of reality which forms the very core of System 1 conceptualizing. Any substantially different organizational form would be inconsistent with System 1 reality and, thus, seem unnatural and "unreal".

By the same token, the authoritarian, somewhat paternalistic managerial style with its emphasis on individual, centralized decision-making--as opposed to participative, decentralized decision-making--also is consistent with System 1 conceptualizing, and heretofore has complemented the classic bureaucratic form of organizational design.

In the past, American organizations were comprized of workers, as well as managers, who functioned predominently from a System 1 orientation. Thus, even though subordinates might not always have liked the effects of bureaucratic design and authoritarian, paternalistic management on employee life, both workers and managers accepted the bureaucratic, scientific management form of organizational functioning as natural. For both groups, it was consistent with their conception of reality. Ergonomic applications within this classical framework also seemed appropriate, and often were beneficial.

As Burns and Stalker's (1961) extensive research in England has shown--and as suggested by Nagandhi's cross cultural studies in Argentina, Brazil India and the Philippines of both American subsidiaries and local firms--this classical "mechanistic" form of organization can be highly effective under relatively simple and stable environmental conditions, such as exists when there is (a)comparatively little competition for raw materials and markets, (b) stable government and governmental policies towards businesses, (c)relatively stable and unchanging values in the population, and (d) only slow or predictable changes in technology. These conditions more-or-less characterized the environments of American industry until approximately the middle 1960's. It also was not until the middle 1960's that the children of the Post World War II "baby boom"--Yankelovitch's (1978) "new breed"--with their more abstract conceptual functioning begain entering the work force in progressively larger numbers.

COGNITIVE COMPLEXITY, AND CONTEMPORARY ORGANIZATIONAL DESIGN AND MANAGEMENT REQUIREMENTS

During the late 1960's the socio-cultural, economic, and technological environments in the United States (and other countries) begain to undergo dramatic change, and came to full fruition during the early and middle 1970's. Other industrial countries reached their stride and became highly competitive with American industry both for the U.S. and other world markets. The more cognitively complex post war "baby boom" children reached adulthood and became both a strong political force and a significant part of the labor market. Technology, as best represented by the invention and exploitation of the silicon chip, entered a period of rapid and unpredictable change, with no end yet in sight.

As the research by Burns and Stalker has shown, and that of Neghandi suggests, under these conditions of complex and unpredictable environments, mechanistic or classical bureaucratic organizational designs and related management systems tend _not_ to be effective or responsive. They also are _not_ consistent with a more cognitively complex conception of reality and, hence, do not fit the conceptions of the majority of the younger work force.

With their lower need for structure and order, and their comfort with ambiguity and change, to this "new breed" of worker the rigid hierarchical bureaucracy seems inappropriate, unnecessary, and stifling of individual initiative and worth. As Yankelovitch's (1978) research has shown, this "new breed" of worker particularly values "recogntion as an individual person" and "opportunity to be with pleasant people with whom I like to work". These values not only tend to be at odds with classical bureaucracy, but also with traditional authoritarian management. Instead, these values and their underlying more cognitively complex conceptions of reality are more consistent with what Burns and Stalker refer to as _organic_ organizational designs, and participative management. These are the very organizational and managerial system designs that they have found to be the most effective and responsive to complex and changing environments, such as characterize the industrialized world now, and will for the foreseeable future.

Organic Organizational Designs

This family of organizational designs can take a variety of forms, but all have a common set of characteristics. First, they tend to be less hierarchical, or flatter and broader in structure. Secondly, there tends to be broad spans of control; units tend to function relatively more autonomously than in classic organizational designs. Thirdly, decision-making--particularly _tactical_ decision-making--tends to be decentralized, and is delegated to the lowest level where competence exists. Fourthly, employees _participate_ in decision-making on issues of importance to them and/or to their jobs. Fifthly, prestige is based relatively more on personal competence than on position in the system. Sixth, the locus of authority is at whatever level the necessary skill or competence exists, rather than in a select group at the top of the system as in classically designed systems. Seventh, organic systems tend to be characterized by relatively few formal rules as compared with traditional mechanistic designs. Finally, information tends to be widely disciminated and forms the primary basis of communication, rather than direction and orders from above. In general, Organic organizational structures tend to be more ambiguous and fluid, and hence more responsive to relevant environmental changes.

Cognitive Complexity and "Natural" Hierarchical Organizational Design

In what Harry Levinson, Emeritous Professor of Management at Harvard and a leading world authority on organization and management, has called "the single most significant contribution to this field" (1977, p.97), Elliott Jaques in his book, A General Theory of Bureaucracy concludes that there is a systematic psychological structure which underlies the bureaucratic work system as it has evolved across the spectrum of economic and political systems throughout the world. He goes on to say that this underlying structure leads to systematic levels of organizational hierarchy which, in turn, are a function of the "time span of discression"--the largest period a person can work on any task without having to report to one's superior. Jacques has identified five levels of hierarchy and associated time spans which, in turn, are a function of the degree of cognitive complexity inherent in the positions at each level. Cognitive complexity, then, provides the underlying psychological structure for natural hierarchical arrangements in organizations.

Elliott Jacques, Gillian Stamp and their colleagues at Brunell University in England have continued research on, and have further refined the specific nature of the cognitive complexity requirements of each natural hierarchical level

I have noted that typical large bureaucratic organizations tend to be characterized by as many as 15 hierarchical levels, whereas the organically designed organizations with which I am familiar are characterized by only five levels which tend to be differentiated on the basis that Jacques and his colleagues suggest.

Cognitive Complexity and Management

We already have noted that the authoritàrian, somewhat paternalistic management style, characteristic of most classical bureaucratic organizations is highly consistent with the characteristics of System 1 functioning. We also have noted that this form of management was not consistent with the characteristics of more cognitively complex conceptions of reality. Since progressively more System 3 and 4 persons are not only moving into the work force, but also into management, what will be their style of operation that, collectively, will form their system of management?

Part of the answer to this question already has been suggested by Maccoby's (1976) psycho-social personality research on managers in high-technology (and somewhat more organically designed) organizations. Maccoby has noted that a new psycho-social personality type of manager has emerged and moved into the upper managerial positions in these organizations. Maccoby has labeled this personality type the "gamesman". From Maccoby's description, and from my own ongoing research, the gamesman most frequently exhibits a primary System 1, secondary System 3 conception of reality--a pattern that I have found among approximately one-third of my graduate management students. This pattern tends to be held by persons raised in a System 1 trainer environment, but through higher education, travel, active use of the media, and managing culturally diverse groups have become more abstract, less absolutist, less ethnocentric, somewhat more people oriented, and are more interpersonally skillful in their behavior than classic System 1 managers. Yet, the System 1 reality never has truly broken down for them. They remain fairly conformist in their thinking and behavior, and primarily rely on Conventional moral reasoning. As managers, they are very open to innovation, fairly comfortable with more ambiguous organizational structures, and readily utilize participative management.

During the next decade, the "gamesmen" should, in turn, begin being replaced in substantial numbers by true System 3 (and 4) conceptualizers. Given their low need for structure and order, relativism, open-mindedness, people orientation, and greater capacity for empathy, dramatic changes in organizational functioning should result. Included are likely to be the following.

Managerial value system. A change in the dominent interpersonal value system to one based on more complex and empathic assumptions regarding the nature of human beings. Included will be greater use of group problem-solving and a better integration of the emotional with the rational aspects of problems in the decision process. One of the most striking differences that I have been able to demonstrate experimentally has been in the teamwork of concrete, as compared with abstract managerial task groups (Hendrick, 1979). In contrast with the behavior of abstract teams, concrete teams took twice as long to successfully complete the task, worked at a slow pace, made poor use of available cues, focused on their own task rather than also on the work of others, and did minimal testing of the rules to determine their true limits. In short, more cognitively complex managers appear both more comfortable with and more effective at participative problem solving.

Structure and process. A religation of less routine aspects of organizational life which significantly affect employees or their work to more democratic and participative management practices, including looser, more ambiguous and flexible organizational structuring.

Organizational goals. Improving the quality of work life will become an internalized organizational goal. External pressure from a progressively more abstract public, as well as internal work force will further ensure the acceptance of this as a fundimantal managerial and corporate responsibility, and one ergonomists are likely to be called upon to help implement.

COGNITIVE COMPLEXITY AND CONTEMPORARY ORGANIZATIONAL DESIGN AND MANAGEMENT: ERGONOMIC IMPLICATIONS

From the above discussion I believe it is evident that we can not continue to ergonomically design systems based on classical organizational design and managerial practices. In the future a macroergonomics approach will be required which systematically considers alternative organizational designs and the complexity level of those whom will manage the system. Only then should the networking requirements for decision support systems, and the design of work stations and integrated environments be considered.

At present, much ergonomics research needs to be done before we can adequately respond to the above challenge. For example, what are the differing ergonomic requirements for the complexity differences that characterize the five "natural" levels of organizational hierarchy identified by Jacques and his colleagues? Once identified, how do we translate these into engineering design specifications and guides for office systems, decision support systems, networking requirements, and integrated office furniture design? How do we further modify these design guides to allow for the flexibility requirements of organic organizational designs and related practices of a more cognitively complex managerial work force than has heretofore characterized our organizational enterprizes? As ergonomists, I believe that the challenge is both a demanding and an exciting one. Our potential to contribute to society has never been greater.

REFERENCES

[1] Bariff, M.L. and Lusk, E.J., Cognitive and personality tests for the design of management information systems, Management Science, 23 (1977) 820-837.

[2] Blatt, M., The effects of classroom discussion upon children's moral judgement, in: Kohlberg, L., and Turiel, E. (eds.), Moral Research: The Cognitive-Developmental Approach (Holt, Rinehart and Winston, New York, 1971).

[3] Burns, T., and Stalker, G.M., The Management of Innovation (Tavistock, London, 1961).

[4] Harvey, O.J., System structure, flexibility and creativity, in: Harvey, O.J. (ed.), Experience, Structure, and Adaptability (Springer, New York, 1966).

[5] Harvey, O.J., Hunt, D.S., and Schroder, H.M., Conceptual Systems and Personality Organization (Wiley, New York, 1961).

[6] Hendrick, H.W., Differences in group problem-solving behavior and effectiveness as a function of abstractness, Journal of Applied Psychology, 64 (1979) 518-525.

[7] Holstein, C., The relation of children's moral judgement to that of their parents and to communication patterns in the family. Unpublished dissertation, University of California at Berkeley (1969).

[8] Jacques, E., A General Theory of Bureaucracy (Halstead Press, New York, 1976).

[9] Kohlberg, L., The child as a moral philosopher, in: Readings in Psychology Today (CRM Books, DelMar, California, 1969).

[10] Levinson, H., Review of Elliott Jacques, A General Theory of Bureaucracy, The Columbia Journal of World Business, (Fall 1977) 96-100.

[11] Maccoby, M., The Gamesman (Simon and Schuster, New York, 1976).

[12] Negandhi, A. R., A model for analysing organizations in cross-cultural settings: a conceptual scheme and some research findings, in: Negandhi A.R., England, G.W., and Wilpert, B., Modern Organizational Theory (Kent State Univ. Press, Kent, Ohio, 1977).

[13] Piaget, J. The Moral Judgement of the Child (Free Press, Glencoe, Illinois, 1948).

[14] Taylor, F.W., Principles of Scientific Management (Harper and Brothers, New York, 1911).

[15] Weber, M. Essays in Sociology, trans. Gerth, H.H. and Miller, C. W. (Oxford, New York, 1946).

[16] Yankelovich, D. The new psychological contracts at work, Psychology Today (May 1978) 46-50.

Human Factors in Organizational Design and Management
H.W. Hendrick and O. Brown, Jr. (Editors)

A COGNITIVE COMPLEXITY BASED SIMULATION FOR MANAGEMENT ASSESSMENT AND TRAINING[1]

Robert W. Swezey

Behavioral Sciences Research Center
Science Applications, Inc.
McLean, Virginia
U.S.A.

A microcomputer-based complex decision-making simulation was developed and pilot tested. The simulation is based upon Interactive Complexity Theory and involves a hypothetical political-military crisis in Eastern Europe. Participant information load was varied in a fashion that allows for assessment of individual decision style under differing load conditions.

An issue of concern in the domain of organizational behavior involves the use of simulation-based techniques as a medium for training and performance evaluation. This paper reports on the design and development of a computer-based decision-making simulation for use by managers. Its purpose was to provide a vehicle for use in relevant training and assessment oriented situations where complex, strategic decision making is required.

The simulation was designed to enable assessment of the decision-making performance of participants according to the parameters of a social psychological theory known as Interactive Complexity Theory (Streufert & Streufert, 1978). This theory describes cognitive information processing as an interactive function of the complexity ability and requirements of individual and environmental variables, and is concerned primarily with the "structure" of information processing. The term "structure" in this sense refers to patterns of relationships employed in information processing; whereas the term "content" is applied to the substance, or meaning, of the information being processed. In the area of decision making, interactive complexity theory is concerned with such structural variables as the number of sources of information requested, the categories of information employed in each decision, and the use of strategic planning. Content aspects of the situation in which the decisions are made are not considered by the theory.

Recent work in interactive complexity theory suggests that as the environmental complexity of a situation increases, a person's ability to demonstrate flexible differentiative and integrative performance in complex decision-making tasks follows a series of inverted U-shaped curves as shown in Figure 1. According to Streufert (1982), the potential for multidimensional (differentiative/integrative) behavior is considered to be optimal at some intermediate level of environmental load. However, differential maximum elevations of the U-shaped curves at that optimal point reflect differential categories of information processing activity. Individuals employing different decision-making categories presumably show different constructed inverted U-shaped curves.

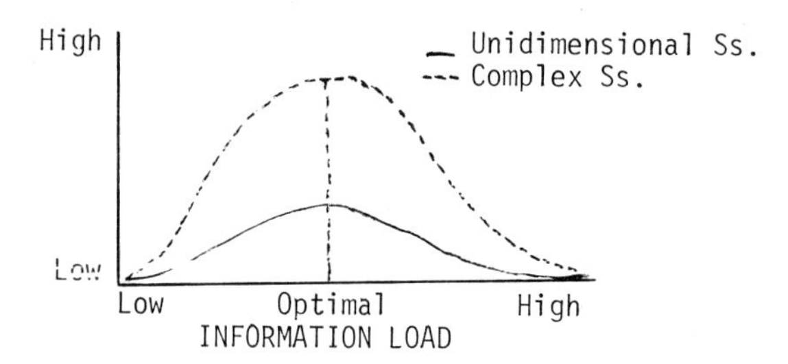

Figure 1. Theoretical relationship between decision complexity and information load.

Current interactive complexity theory suggests the existence of nine categories of decision making. Two categories pertain to unidimensional decision making and seven involve multidimensionality. Figure 2 presents the categories and their interrelationships. Basically, according to this theory, unidimensional decision making involves responses to nearly all situations in terms of a single evaluative dimension (e.g., good-bad), whereas multidimensionality enables simultaneous use of several dimensions. Differentiated decision making, while involving multiple dimensions, does not address the possible interrelationships among the dimensions; whereas integrated decision making does. These categories are discussed in detail in Swezey, Streufert, Criswell, Unger, and van Rijn (1984).

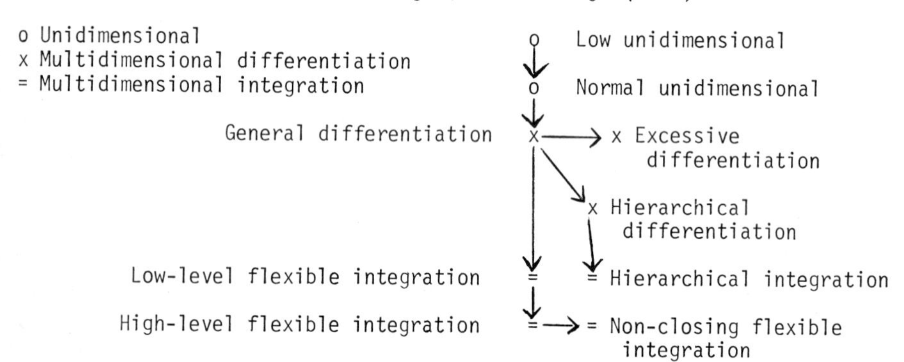

Figure 2. Categories of decision making predicted by interactive complexity theory.

Streufert and Streufert (1978) have made the important point that no one category is "better" than another since the utility of a decision-making style interacts with situational demands. Multidimensional decision making has value in situations where behavior should proceed in a flexible way and where a large number of stimuli must be taken into account. On the other hand, a unidimensional approach may be advantageous where decisions must be made according to a clear criterion, or where rapid action is required.

Although measurement of interactive complexity has until recently been primarily subjective using formats such as Sentence Completion, a large body of validity data exists for these measures (reviewed in Streufert & Streufert, 1978). Recent activity has involved development of various other measurement techniques including so-called "Time Event Matrices" where categories of decisions are plotted across time and their interactive and strategic relationships charted. Other work involves development of criterion-based categories of complexity, such as various categories of decisions made in complex environments. Since it is not the purpose of this paper to review measurement issues, refer to Streufert and Streufert (1978) and to Swezey et al. (1984) for detailed consideration of this topic.

Complexity theory as a basis for a complex decision-making simulation. A recent series of efforts (summarized in Swezey et al., 1984) have focused on the development of a complexity theory-based managerial decision-making simulation presented on a microcomputer. The simulation was designed to create an environment which supports use of multidimensional decision-making strategies. The environment was designed to be challenging (with no clearly correct plan of action) and to allow time to develop and execute plans. The primary scenario (a hypothetical political-military dilemma set in Eastern Europe) contains three periods, each having 30 minutes of simulation time (approximately 1-1/2 to 2 hours of real time). At predetermined intervals, the computer presents information concerning a hypothetical escalating military crisis, and the participant is allowed the option to select various choices of action from a list of alternatives or to make no selection. The code number of the action is then entered into the computer which collects and stores data on participant action during the three periods for the purpose of generating quantitative measures of decision-making strategy. The main independent variable of interest, information load, is varied across the three periods although an option exists to experimentally manipulate this condition as desired. In order, the three periods currently contain medium, low, and high information load conditions.

In the simulation, both the participant and the experimenter control simulation events. The participant controls events in the sense that the computer delivers specific responses to specific decisions made by the participant over the course of the simulation. It is highly unlikely that any two participants would encounter the same sequence of messages during the simulation. The experimenter controls events in the simulation in the sense that the outcome of the dilemma is not resolved by the end of the simulation no matter what the participant's responses. The simulation runs on an Apple II Plus microcomputer system supported by a hard disk, printer and color monitor, and is programmed in BASIC. Software for the simulation is stored both in the hard disk and on supplementary floppy disks. A list of the hundreds of files required and a description of the eight main simulation programs is documented in Swezey et al. (1984).

Each decision alternative has its own unique code number which is entered into the computer. The major political-military simulation has 411 decision alternatives. The alternatives are divided into six action areas: economic (64); political (106); military (88); covert operations (88); public opinion (8); and information request (57).

The simulation includes a scenario map with grid squares labelled by their x, y coordinates. The major simulation map is a color, commercially produced map of Eastern Europe on which a 32x45 grid has been overlayed. If, during the simulation, a participant needs to enter a location into the computer, the computer is programmed to accept x, y coordinates from the scenario map. The simulation sequence (including the practice session) usually takes between 6 and 8 hours, although this estimate varies depending on how much information an individual participant enters into the computer.

The simulation contains four manipulable variables: information load, number of success vs. failure messages presented, number of fixed vs. responsive messages presented, and the extent of the time compression.

Information load refers to the number of messages presented to the participant per 30-minute period in the simulation. Information load is of importance to complexity theory which proposes that perceptual and information processing activity vary as a function of the interaction between the decision-style category being employed by the decision maker and environmental load (Streufert & Streufert, 1978). The simulation was designed to enable experimental manipulation of load condition and to present low, medium, and high levels.

A success message is a response to a participant's decision that conveys that the action taken was successfully accomplished. A failure message indicates that action was not successfully accomplished. The ratio of success to failure messages in the simulation may be easily manipulated via simple software changes in the program.

A responsive message is one which specifically replies to a participant's decision. A fixed message is one which does not speak specifically to a participant's decision. The program administers either a fixed or a responsive message every x minutes according to programmed information load schedule. In the present simulation, a minimum of 40% of the messages in each period are fixed messages; these messages (the same for every participant) keep the scenario unfolding. These fixed messages are programmed to occur at specific times; responsive messages may not be delivered at those times.

Manipulable variables in the dilemma which concern time compression are ratio of real to simulation time, amount of simulation time advanced by each decision, length of session in simulation minutes, and simulation start time. Although this category of variables is not necessarily related to current interactive complexity theory, they may be modified as desired in the simulation.

The simulation has three periods, each with 30 minutes of simulation time. Again, actual elapsed time varies widely depending on how long the participant spends making decisions and entering plans. If the participant makes no decisions, the scenario will last 90 real minutes. In simulation time, 2-1/2 days elapse in each period and there are a total of 7-1/2 days in the hypothetical political-military dilemma. Nine measures of decision making in the simulation relate specifically to predictions based on interactive complexity theory. These measures include such topics as: number of general unintegrated decisions, number of respondent decisions, number of backward integrations, and number of forward integrations, for example. For detailed explanations and sample calculations of the nine measures, see Criswell, Unger, Swezey, and Streufert (1983).

As mentioned previously, interactive complexity theory postulates the existence of various categories of decisions based on the concepts of differentiation and integration. However, the present simulation concentrated on three of these categories only (i.e., unidimensional, multidimensional differentiative, and integrative decision making). This was done for two reasons: the measures used in the simulation were not developed on the basis of fine distinctions; and the existing data base, which provides at least partial support for some of the interactive complexity theory measures included in the simulation, is much too small to permit testing for more extensive individual categories of structural information processing.

In assessing the measurement scheme, seven simulation participants were also administered various other measures of cognitive complexity. This sample of seven participants consisted of two unidimensional persons, two multidimensional differentiators, two multidimensional low-level integrators, and one multidimensional high-level integrator. Consistent and high associations were found among the various complexity scores and the measurement profiles based upon the interactive complexity theory contained in the simulation. (For details of this pilot test, see Swezey et al., 1984.)

Conclusions. The future of the simulation appears to center around three activities: empirically establishing its validity, modifying its content for different applications, and employing it in organizational and training situations. Establishment of validity is of primary concern. This activity must precede the establishment of norms to assist in the interpretation of the cognitive style profiles. The simulation has the potential to become a valuable aid in several areas: (1) It may be useful in training personnel how to employ a multidimensional decision-making style and to teach them when to apply the style for best results; (2) it may help place people who typically employ a certain style into positions where that style has special usefulness; (3) it may be possible, using this simulation, to teach people how to regulate their own information load in order to minimize the likelihood of underload or overload; and (4) the simulation may help people better understand their own cognitive styles and to more appropriately adapt their styles to the requirements of the situation. Finally, the simulation may be useful in helping design environments which foster the multidimensional strategies that will be most useful for the particular information processing task at hand.

FOOTNOTE: Portions of the work described in this paper were developed under Contract No. MDA 903-79-C-0699 with the U.S. Army Research Institute. Other portions were supported by Science Applications, Inc. No endorsement by the U.S. Army is either intended or implied.

REFERENCES:

[1] Streufert, S. and Streufert, S. C., Behavior in the complex environment (John Wiley, New York, 1978).

[2] Streufert, S., Overview of Cognitive Complexity Theory Employed in a Managerial Assessment and Training Simulation System, paper presented at the 1982 Annual Convention of the American Psychological Association, Washington, D.C., 1982.

[3] Swezey, R. W., Streufert, S., Criswell, E. L., Unger, K. W. and van Rijn, P., Development of a Computer Simulation for Assessing Decision-making Style Using Cognitive Complexity Theory (SAI Report No. SAI-84-04-178), U.S. Army Research Institute for the Behavioral and Social Sciences, Alexandria, Virginia (1984).

[4] Criswell, E. L., Unger, K. W., Swezey, R. W., and Streufert, S., Researcher's Manual to Accompany the Yugoslav Dilemma (a computer simulation) (SAI Report No. SAI-83-06-178), U.S. Army Research Institute for the Behavioral and Social Sciences, Alexandria, Virginia (1983).

Human Factors in Organizational Design and Management
H.W. Hendrick and O. Brown, Jr. (Editors)

A SYSTEMS ANALYSIS APPROACH TO INTEGRATING OD AND ERGONOMIC SOLUTIONS TO ORGANIZATIONAL PROBLEMS

Michelle M. Robertson
Hal W. Hendrick

Institute of Safety and Systems Management
University of Southern California
Los Angeles, California
U.S.A.

A seven step systems analysis model is proposed for assessing complex organizational problems having Ergonomic implications. The model is applied to an illustrative problem concerning a decrease in productivity as a result of a dehumanized work system. A major benefit of the model, demonstrated through this application, is the integration of Ergonomic approaches to organizational problems with those of other disciplines.

INTRODUCTION

In today's highly advanced technologically oriented societies, many complex organizational problems arise. Some of these problems concern productivity and the quality of work life. Because these organizational problems are extremely complex and multi-causal, many of the underlying critical factors may not be identified and assessed. Consequently, only the more obvious symptoms of the problem may be analyzed, and thus result in a superficial, narrow and non-effective solution.

Traditionally, this scenario has occurred when an organizational problem in the area of quality of working conditions arises and a specialist in one or the other of the fields of Industrial Psychology, Organizational Development or Ergonomics is asked to analyze the proposed problem and specify a solution. Unfortunately, each expert holds a bias in approaching the problem and subsequent solution. For example, organizational behavior specialists may identify solutions in the areas of leadership, motivation or job enrichment. Ergonomists, on the other hand, would be likely to focus on solutions involving redesign of human-machine interfaces. In each case, not only is the approach going to be narrow in focus but also atomistic, rather than viewed from a true systems perspective.

In order to assure a more thorough interdisciplinary and comprehensive approach to these multi-causal organizational problems, an integrated approach that incorporates the methodological framework of systems analysis is needed. The present paper utilizes and demonstrates a generalized seven step systems analysis model. This model is a modification of one proposed by Mosard (1982) based on earlier work in systems engineering by A.D. Hall (1969).

To illustrate the model, a hypothetical organizational problem concerning a decrease in productivity as a result of a dehumanized work system was utilized. This problem was selected because of its relevance to our high technology organizations. Since the turn of the century, assembly line processes involving workpacing and repetitive tasks have become widely used in industry. It is estimated that in the U.S. alone over three million workers peform repetitive or paced jobs on assembly lines (Beith, 1981). The result often has been a dehumanization of work systems which has multiple effects, such as stress, a decrease in performance, reduced productivity, and job dissatification (Bulat, 1982).

STEPS OF THE PROPOSED SYSTEMS ANAYSIS MODEL

There are seven steps in the proposed systems analysis framework. These are: 1. Defining the problem 2. Setting the objectives (and developing evaluation criteria) 3. Developing alternatives 4. Modeling alternatives 5. Evaluating alternatives 6. Selecting an alternative and 7. Planning for implementation.

DEFINING THE PROBLEM

The first step in the systems analysis approach is defining the problem. Large-scale problems and issues display a complexity that necessitates a fairly formal, systematic analytical approach to problem definition (Mosard, 1982). An analytical tool that has proven useful in defining large, complex problems is a "problem factor tree" (Warfield, 1971). The problem factor tree is developed by identifying the major problem, subproblems, and causal factors including their interrelationships. The major problem would be sectionalized by identifying problem elements, including needs, constraints, and the persons/groups involved.

These problems and problem factors should be precisely stated and carefully linked together by an iterative process that produces a graphical "model" (Mosard, 1982). The lower level problem factors on the model contribute to or "cause" the sub-problems, which in turn, contribute to the major problem. Feedback loops also may be shown. The model then depicts a hierarchal, logical structure of the encompassing problem elements.

The following model, Figure 1, depicts a hypothetical problem involving a decrease in productivity of assembly line workers caused by dehumanization of the work system. Four sub-problems are defined along with the many causal problem factors.

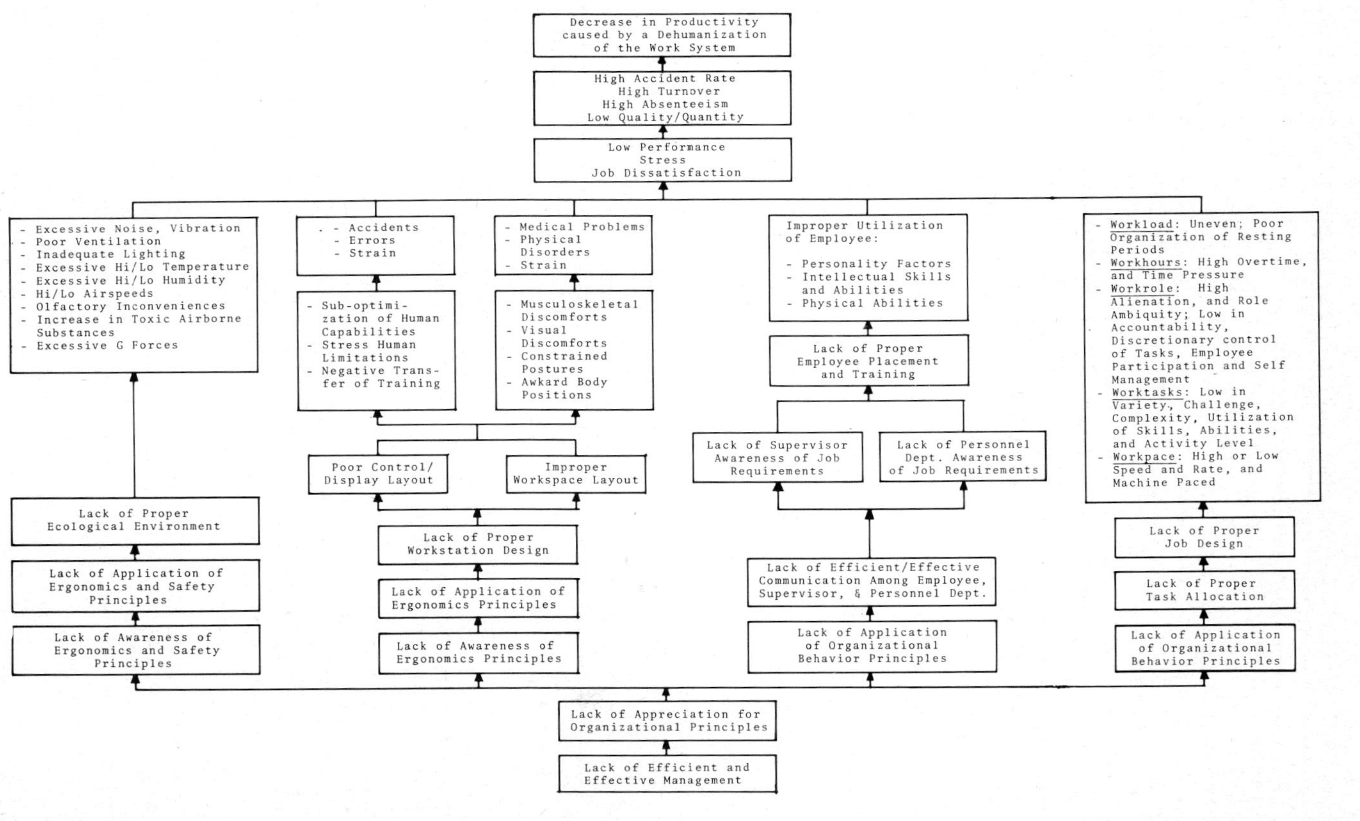

Problem Factor Tree
Figure 1

SETTING THE OBJECTIVES AND DEVELOPING EVALUATION CRITERIA

Step II of the systems analysis approach is setting the objectives and developing evaluation criteria for selecting the best alternative for achieving the objectives. This is accomplished by first identifying the major needs, goals, objectives and sub-objectives, and graphically may be depicted in the form of an objectives tree. An Objectives Tree is a hierarchal structure of objectives that addresses the problems that have been identified and structurally analyzed (Mosard, 1983). An Objectives Tree for our problem is depicted in the upper half of Figure 2. The logic of the objectives tree is that the lower level objectives contribute to the attainment of the middle level objectives which contribute to the upper level needs or goals (Warfield, 1973). The development of the Objectives Tree can be aided by an analysis of the degree of interaction between objectives, constraints, and the persons/groups that are involved (Mosard, 1982). After the objectives are set, and precisely defined in terms of specific objective measures, preliminary decision criteria can be developed. These criteria will be used to evaluate alternative ways to accomplish the objectives. Decision criteria usually will include various risks, costs, benefits, and measures of effectiveness (Mosard, 1983).

DEVELOPING ALTERNATIVES

The third step of the systems analysis methodology requires identifying alternative ways of attaining the objectives and subobjectives. An alternative can be defined as a specified set of activties or tasks designed to accomplish an objective (Mosard, 1983). Quade (1975), recommends that a wide range of alternatives be developed by first identifying major classes of alternatives and then developing distinct alternatives within each class. These alternative sets of activities are portrayed below the Objectives Tree in Figure 2, thus creating an Objectives/Activities Tree. Hybrid alternatives also may be created which incorporate the best features of several of the initially identified alternatives. The Objective/Activities Tree in Figure 2 depicts the major goal of increasing productivity by alleviating a dehumanized work system. Also shown are two objectives and two sub-objectives, two hybrid alternatives and one single alternative. For each alternative the activities also are exhibited. The two hybrid alternatives represent only two of the many possible combinations of common activities that could be specified. These three alternatives A, B, and C, will be the focus of the analysis.

Figure 3 presents preliminary decision criteria which will be utilized to evaluate alternatives later in this study. The criteria are based on a long term perspective. The cost criteria are based on both short and long term perspectives.

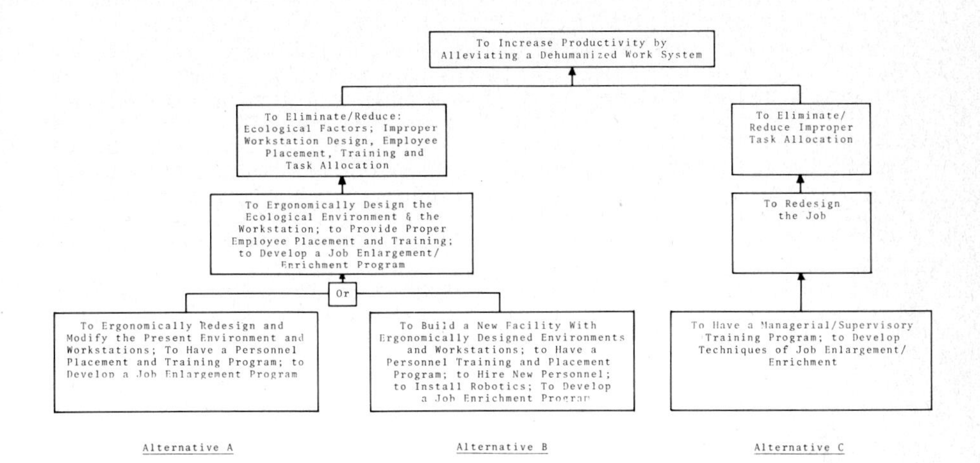

Objectives/Activities Tree
Figure 2

Scope	Risk of Failure	Costs (Short/Long Term)	Benefits/ Effectiveness
1. Will Help Entire Organization Internally/Externally? 2. Long Term Program Effectiveness?	1. Employees Resistance to Change 2. Employees Learning New Skills 3. Management Acceptance	1. Materials 2. Resources 3. Loss of Productivity 4. Training Programs 5. Decrease in Product Quality	1. Increase in Productivity 2. Increase in Morale 3. Decrease in Job Stress 4. Decrease in Absenteeism 5. Decrease in Turnover

Preliminary Decision Criteria
(Long Term)
Figure 3

MODELING THE ALTERNATIVES

Step IV of the systems analysis approach is modeling the alternatives. Modeling, in this context, is defined by Mosard (1982), as the development of either a descriptive or predictive model representing each alternative set of activities (or representing the entire system) in such a way as to allow alternative configurations of the system to be analyzed. In order to aid in determing the effectiveness and other consequences of each alternative set of activities, the system element interrelationships and/or gross resource requirements should be depicted (Mosard, 1983). There are many modeling techniques that may be utilized, such as flow charts, simulation and systems dynamics modeling.

The model utilized in this illustration was that of an input-ouput flow diagram. This is a simplified modeling technique that allows alternative configurations of the system to be analyzed.

In utilizing the input-output flow diagram, the inputs consist of people, resources and information, and the outputs are the results and products. These outputs can, in themselves, become the sources of inputs to other sub-systems, and thus extend the diagram to fully represent the entire system being analyzed.

Figure 4 exhibits only one alternative, A, in the simplified model of the input-output flow diagram. There are two phases of this model that are depicted, the redesign phase and the operations phase.

EVALUATING ALTERNATIVES

Step V of the systems analysis approach is evaluating alternatives. This is accomplished by evaluating and comparing each alternative set of activities utilizing the major decision criteria. These criteria include effectiveness, risk, cost, and benefit for all appropriate future conditions. An evaluation criteria scorecard can be developed for use in evaluating and comparing alternatives.

Figure 5 depicts an evaluation criteria scorecard for the proposed problem and its respective three alternatives. The scorecard is based on a long term perspective and incorporates the preliminary decision critieria table (Figure 3). Since this problem is not a specific case study, quantitative figures for each main decision criterion were not developed. Instead qualitative rankings were developed, using a five point scale. These rankings are based on job redesign studies and experiences at SAAB, Volvo, and Purina (Szilagyi & Wallace, 1983).

In actually applying this systems analysis approach within the context of a specified, well-defined organization and its encompassing environment, the individual "score" for each

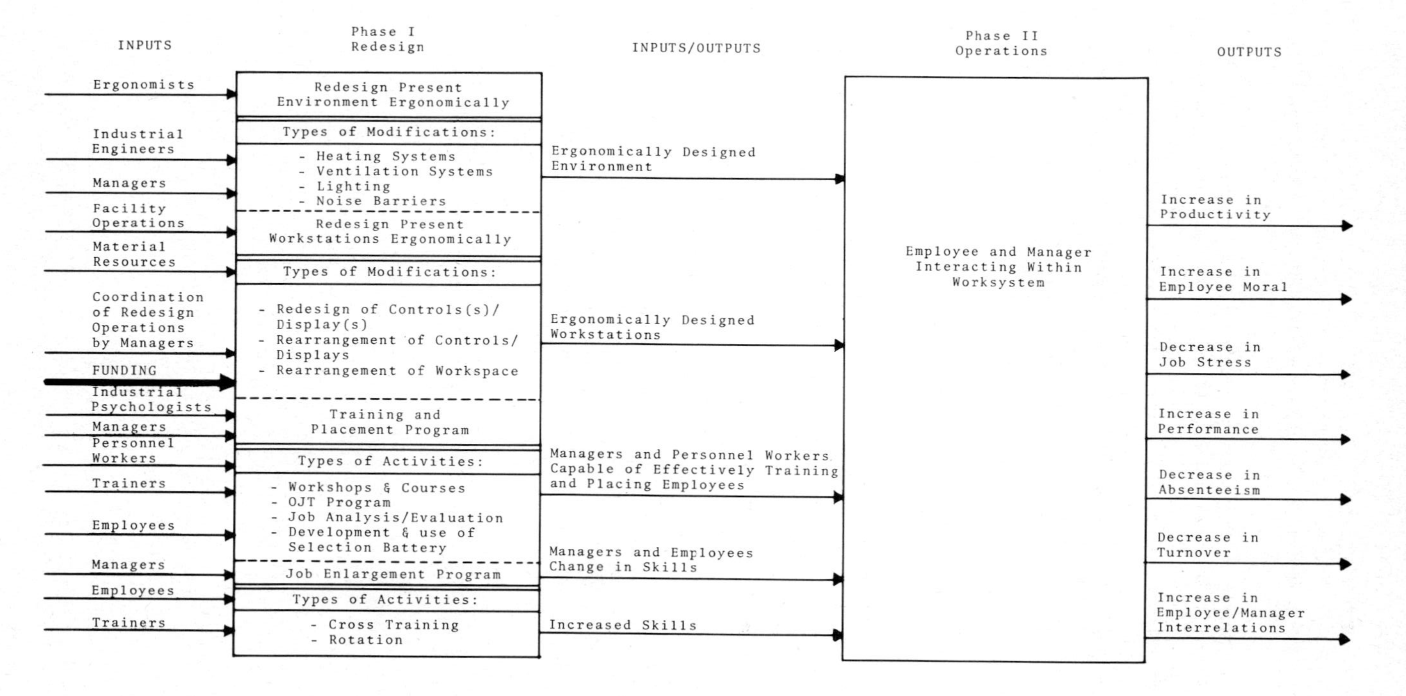

Input-Output Model
Alternative A
Figure 4

criteria for each alternative could be quantitative and/or qualitative in nature. These scores would be the result of considerable data gathering, analysis, and interpretation.

Alternatives	Effective-ness	Risk of Failure	Costs	Benefits
Alternative A Redesign/ Training and Placement Pro-grams/Job Enlargement	Moderate	Low	Moderate	Moderate
Alternative B Build/Training and Placement Programs/Hire Personnel/In-stall Robotics/ Job Enrichment	High	Moderate	High	High
Alternative C Managerial/ Supervisor Training Program	Low	Moderate	Low	Low

Evaluation Criteria "Scorecard" (Long Term)
Rating Scale: Very Low, Acceptable, Moderate, High, Very High
Figure 5

SELECTING AN ALTERNATIVE

The first step in this process is to establish the importance or weight of each evaluation criteria.This may be accomplished by either objective or subjective measures. Next, a decision table is utilized in the combining and structuring of the evaluation information. Lastly, there is the selection of the desired alternative set of activities. Figure 6 exhibits a decision table for the three alternatives of the proposed problem. The future conditions were based on the assumption that the proposed project has not yet been funded. Therefore, three levels of potential funding have been identified. The three alternatives thus can be evaluated in terms of each level of funding. For illustrative purposes, an arbitrary probability for each future condition was assigned. In this example, the alternative set of activities that most likely would be selected is alternative A, as judged by the best combination of effectiveness, low risk, low cost and ultimate benefit to the organization.

Future Conditions	High Level of Funding	Moderate Level of Funding	Low Level of Funding
Probability of Future Condition	.25	50	.25
Alternative A Redesign/Training and Placement Programs/Job Enlargement	2	1	2
Alternative B Build/Training and Placement Programs/Hire Personnel/Install Robotics/Job Enrichment	1	2	3
Alternative C Managerial/Supervisor Training Program	3	3	1

Decision Table
Figure 6

PLANNING FOR IMPLEMENTATION

Step VII in this systems analysis approach is planning for implementation. This is accomplished by developing network plans for initiating and implementing the selected alternative set of activities. Also, a schedule and a sequence of tasks, responsibilities and requirements should be developed for the implementation activities. This schedule might include a contingency plan with scheduled decision points and decision responsibilities.

There are several scheduling techniques that are available for implementation planning, including PERT/CPM charts, The Gantt chart and the DELTA chart. These may be used either alone or in conjunction with one another.

Figure 7 presents a DELTA chart for the implementation of alternative A. This chart provides an overview of plans for initiating and implementing the selected activities.

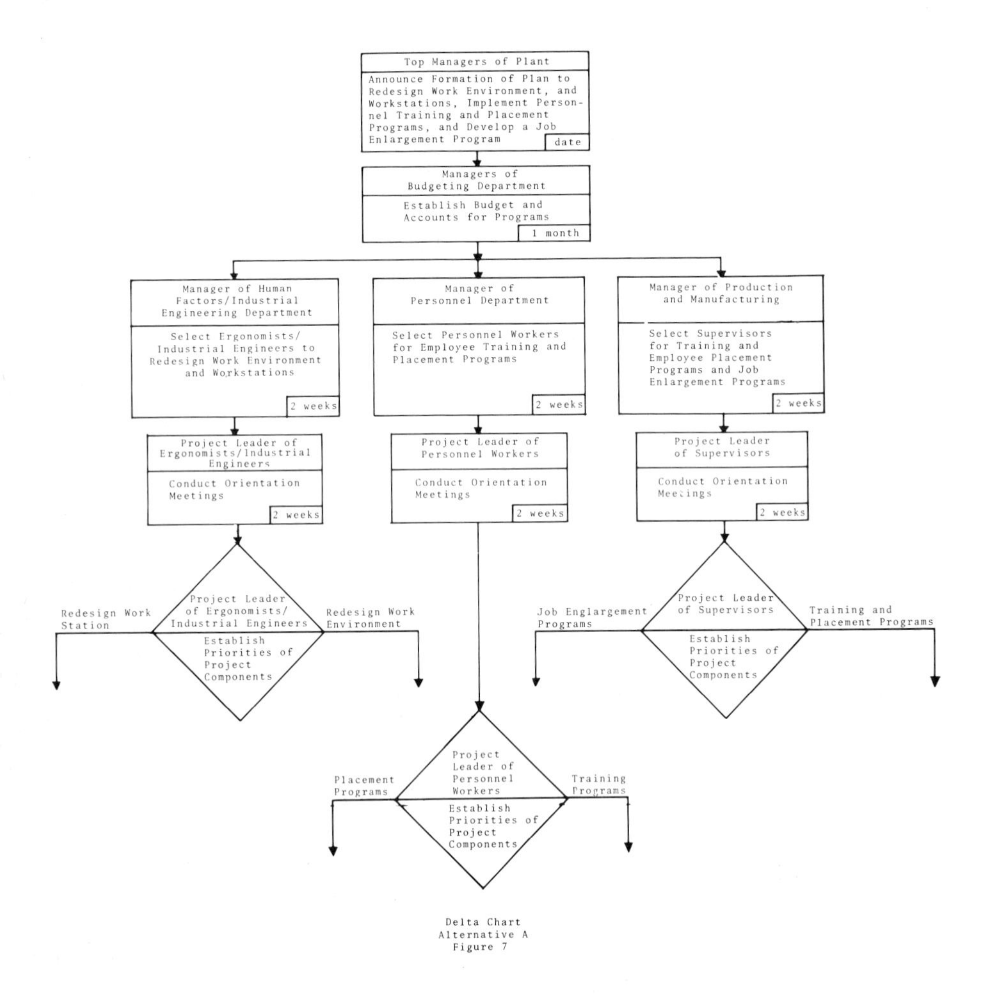

Delta Chart
Alternative A
Figure 7

CONCLUSION

This paper has presented and demonstrated a seven step systems analysis approach for integrating organizational development and ergonomic solutions for organizational problems. It was applied to an illustrative problem which was defined as a decrease in productivity of assembly line workers caused by a dehumanization of the work system.

This model represents a systems methodology which can be utilized to avoid atomistic, simplistic, and unidisciplined approaches to what, in reality, are multi-causal systems problems requiring multidisciplinary approaches and expertise to ensure sound solutions.

REFERENCES

[1] Beith, B., Work repetition and pacing as a source of occupational stress, in: Salvendy, G. and Smith, M.J. (eds.), Machine Pacing and Occupational Stress (Taylor and Francis Ltd., London,1981).
[2] Bulat, V., Jakov D., Vic, D., Parezanin, V., Parezanin, R., Rhythmical work in production lines and humanization of work, in: Noro, K. (Ed.) Proceedings of the 8th Congress of theInternational Ergonomics Association, (Tokyo, Intergroup, 1981).
[3] Hall, A. D., A three dimensional morphology of systems engineering, IEEE Transactionson Systems Science Cybernetics., SSC-5 (1969) 156-160.
[4] Mosard, G., A generalized framework and methodology for systems analysis, IEEE Transactions on Engineering Management, EM-29 (1982) 81-87.
[5] Mosard, G., Problem definition: Tasks and techniques, Journal of Systems Managment, 34 (1983) 16-21.
[6] Quade, E.S., Analysis for public decisions (American Elsevier, New York, 1975).
[7] Szilagyi, A.D. & Wallace M.J., Organizational behavior and performance (Scott, Foresman and Company, Glenview, Illinois, 1983).
[8] Warfield, J. N., Toward interpretation of complex structural models, IEEE Transaction on Systems Management Cybernetics, SMC-2 (1972) 610-621.
[9] Warfield, J. N., Intent structures, IEEE Transaction on Systems Management Cybernetics, 3 (1973).

Human Factors in Organizational Design and Management
H.W. Hendrick and O. Brown, Jr. (Editors)

OBSTACLES IN THE UTILIZATION OF DECISION-ANALYSIS BY MANAGERS

Dan Zakay

Department of Psychology
Tel-Aviv University

A survey was conducted among 40 midlevel managers who were graduates of Decision-Analysis (DA) workshops. The survey revealed eight major obstacles to optimal usage of DA. These obstacles were categorized as environmental, informational, emotional, and intuition loss. "Utilizers" of DA were differentiated from "non utilizers" mainly in regard to emotional obstacles while environmental obstacles were found to be major for both "utilizers" and "non utilizers". Implications for future development of DA as a practical decision aid were discussed.

INTRODUCTION

Decision makers and managers face decision problems which are made today in an environment more complex than ever before (Keeney, 1982). Nevertheless, recent research has revealed systematic divergences from optimal decision procedures by intuitive decision makers (e.g., Kahneman, Slovic & Tversky, 1981). As a result of this situation, attention is focused today on the training of decision makers in the use of optimal procedures (e.g., Nickerson & Feerer, 1975). Normative models like simple Multi-Attribute Utility models (e.g., Kepner & Tregoe, 1965), Decision trees (e.g., Davies, 1970) or Decision Analysis (DA) (e.g., Brown, Kahr & Peterson, 1974) are offered as aids to optimal decision making. However, it would seem that some obstacles and pitfalls lie in the way of effective utilization of such normative procedures by decision makers.

Decision Analysis is considered to be the most advanced normative decision aid. DA is defined by Keeney (1982) as "a philosophy, articulated by a set of logical axioms and a methodology and collection of systematic procedures, based upon those axioms, for responsibly analyzing the complexities inherent in decision problems". Keeney gives, in addition, an intuitive definition of DA which is: "a formalization of common sense for decision problems which are too complex for informal use of common sense" (Ibid, p. 806). However, it seems that despite the fact that DA is rooted in common sense, managers find its daily utilization to be problematic. Indeed, Brown (1970) reported, on the basis of a survey conducted among 20 U.S. companies, that "business executives have often found Decision Theory Analysis frustrating and unrewarding" (p. 78). Brown (1970) indicated the following reasons as responsible for that situation: Personal incompetence, not being sure about the practicality of DA; difficulties in picking up all of the variables in a complex decision making problem; difficulties in measuring uncertainty on the basis of weak information and inappropriate organizational arrangements for offering Decision Theory Analysis' services.

It would seem that the situation has not improved much in the past ten to fifteen years. Keeney (1982) claims that "most analyses of important decision problems have left the incorporation of judgments and values to informal procedures with unidentified assumptions and to the intuition of the decision makers" (p. 830), and Ulvila and Brown (1982) admit that although DA has gained acceptance in many corporations, it "has not become the dominant analytic discipline that some people expected" (p. 131).

The present study is aimed at defining the obstacles which stand in the way of practical effective utilization of DA, as felt by managers of the 1980s. The understanding of these obstacles is, no doubt, a necessary condition for improving the utilization of DA.

METHOD

SUBJECTS:

Forty-six midlevel managers, from three advanced electronic industries, agreed to participate in the study. All of the respondents had graduated from DA workshops between 6-12 months prior to the survey.

PROCEDURE:

In the first stage of the study, six managers, randomly selected from the sample, were individually interviewed in order to explore possible obstacles which prevent the utilization of decision analysis. On the basis of these interviews, a list of 8 major obstacles was constructed. In the second stage, the remaining 40 managers were asked to rate each possible obstacle on a 10-point scale, where 0 denoted that this was not an obstacle at all, and 10 denoted that this was a very serious obstacle. Each manager was also asked whether or not he actually used DA in his daily decisions, and whether he thought the quality of his decisions was improved as a result. The list of obstacles was found to be exhaustive since no other obstacle was suggested by the managers. The following is the list of the 8 obstacles:

OB-1 - The use of DA takes too much time; OB-2 - The use of DA forces me to treat qualitative information quantitatively, and this is very difficult or unwise to do; OB-3 - The use of DA requires unavailable information; OB-4 - The use of DA requires too much investment of effort and energy; OB-5 - It is impossible to take feelings and hunches into account using DA; OB-6 - By using DA one loses the advantages of intuition; OB-7 - My decisions are effective enough without the use of DA; OB-8 - Formal responsibility of the decision maker is too high because the decision process is open and can be checked easily.

RESULTS

Only 3 managers reported that they frequently utilized DA. Twenty-three managers reported that they utilized DA some of the time, and 14 reported not using DA at all. Hence, 26 managers were categorized as "utilizers" and 14 as "non utilizers". Only 4 of the "utilizers" thought that the use of DA did not cause any improvement in the quality of their decisions. A two-way ANOVA (organizations x type of obstacle), with repeated measures, revealed that the organizations did not differ significantly in regard to the rating of obstacles ($F(2,37) = 1.42$), and also did not have any significant interaction with type of obstacles. The frequencies of "utilizers" and "non utilizers" did not differ significantly among the three organizations. Hence, organization was not included as a factor in further analyses. The ratings obtained for each obstacle were subjected to another two-way (utilization x type of obstacle) ANOVA with repeated measures. The main effect of utilization was found to be significant ($F(1, 38) = 4.62$; $p \; 0.05$). "Non utilizers" rated each obstacle as more severe as compared to "utilizers". "Utilizers" were compared to "non utilizers" in regard to the ratings of each obstacle by means of Scheffe tests. Mean ratings of each obstacle and the results of the Schéffé tests are presented in Table 1.

Table 1

Schéffé tests between "utilizers" and "non utilizers"

Obstacle	"Utilizers" (n=26)		"Non Utilizers" (n=14)		Total		F
	Mean	S.D.	Mean	S.D.	Mean	S.D.	df=7;313
OB-1	6.44	2.80	6.80	2.75	6.56	2.74	0.45
OB-3	5.46	3.63	6.08	2.52	5.67	3.24	0.77
OB-2	4.60	2.93	6.50	2.45	5.26	2.88	2.37**
OB-4	4.73	2.87	5.00	3.16	4.82	3.03	0.33
OB-7	3.10	2.61	5.32	3.36	3.87	3.06	2.77***
OB-5	3.17	2.78	4.83	2.74	3.75	2.72	2.07**
OB-6	2.99	2.90	4.58	3.36	3.54	3.12	1.99*
OB-8	1.06	2.66	3.25	3.47	1.82	3.12	2.74***

* $P < 0.1$; ** $P < 0.05$; *** $P < 0.01$

A stepwise discriminant analysis using Rao's method was conducted between "utilizers" and "non utilizers". It was found that obstacles 7, 8, 5 and 2 contributed to a significant discrimination between the two groups.

DISCUSSION

The data obtained in the survey indicated that only a negligible proportion of the managers (6.73%) utilize DA frequently, while most of them utilize it some of the time or never. However, most of the managers who utilize DA did think that the quality of their decisions was improved by it. Only 15.38% of the "utilizers" thought that it did not. These managers, of course, utilize DA only some of the time, and the reason given for their utilization was to check their own decisions. However, both "utilizers" and"non utilizers" expressed the feeling that obstacles did interfere with optimal utilization of DA. The most critical one was that the use of DA takes too much time. The second obstacle in importance was found to be that the use of DA requires unavailable information. No difference was found between "utilizers" and "non utilizers" regarding these two obstacles. The third obstacle, on which overall agreement was found, was that the use of DA requires too much investment of effort and energy. This obstacle belongs to the same category as that of time-consuming. "Non utilizers" were discriminated from "utilizers" with regard to the following obstacles:

(1) They thought that their own intuitive decisions were effective

enough without the use of DA; (2) they thought that formal responsibility involved in the use of DA was too high; (3) they thought that it was not possible to take feelings and hunches into account while using DA; (4) they found it relatively more difficult and unwise to treat qualitative information quantitatively - all of these to a higher extent than "utilizers". "Non utilizers" also felt more strongly than "utilizers", although the difference reached only minor significance, that by using DA one lost the advantages of intuition.

The eight obstacles might be divided on the basis of content into the following four categories: (a) Environmental obstacles (OB-1; OB-4); (b) Informational obstacles (OB-2; OB-3); (c) Intuition loss (OB-5;OB-6); and (d) Emotional obstacles (OB-7; OB-8). The environmental obstacles might be caused by the fact that all three organizations included in the survey did not supply their managers with effective DA services. Thus, Brown's (1970) conclusion about the importance of appropriate organizational arrangements for DA utilization are supported here. An important category of obstacles not mentioned by Brown is the "emotional" one in which fear of responsibility and a feeling of self-satisfaction with one's own decisions were included. This category might represent specific personality traits, on the one hand, and specific organizational climate, on the other, however this hypothesis must be tested empirically. What can be concluded is that in order to cause an effective utilization of DA within an organization, a process of education and organizational development must accompany it in order to create an appropriate climate of support and of high risk tolerance (Sanford, Hunt, Bracey, 1976), and in order to provide the needed technical arrangement and resources. This might facilitate the accessibility of information as well, by facilitating the flow and sharing of information within the organizations and by creating reliable computerized data banks (Zakay, 1982). Nevertheless, the two categories of informational problems and of intuitive utilization present real and essential problems of DA as well as for other existing normative decision strategies. As for training, it seems that managers must be trained so that they will gain insight regarding the danger of relying too heavily on intuitive decision making (Goodman, Fischhoff, Lichtenstein & Slovic, 1976).

It might be concluded that DA offers decision makers many potential advantages and benefits. However, in order to enable its utilization, appropriate educational and developmental processes must first take place on both personal and organizational levels. In addition, the model itself should be adjusted to the demands posed by real life conditions on decision makers. In doing this, the "action-science" approach suggested by Argyris (1980), according to which the primary purpose is the production of knowledge that can be implemented, seems appropriate.

REFERENCES

(1) Argyris, C., Inner contradiction of rigorous research. (Academic Press, New York, 1980).

(2) Brown, R.V., Do managers find decision theory useful? Harvard Business Review. 48 (1970) 78-89.

(3) Brown, R.V., Kahr, A.A., & Peterson, C., Decision analysis for the manager. (Holt, Rinehart, & Winston, New York, 1974).

(4) Davies, I., Get immediate relief with an algorithm. Psychology Today. April (1970) 53-69.

(5) Goodman, B., Fischhoff, B., Lichtenstein, S., & Slovic, P., The training of decision makers. Oregon Research Institute (July, 1976).

(6) Kahneman, D., Slovic, P., & Tversky, A., Judgment under uncertainty: Heuristics and biases. (Cambridge University Press, New York, 1981).

(7) Keeney, R.L., Decision analysis: An overview. Operational Research. 30 (1982) 803-838.

(8) Kepner, C.H., & Tregoe, B.B., The rational manager. (McGraw-Hill, New York, 1965).

(9) Nickerson, R.S., & Feerer, C.E., Decision making and training. (Bolt Beranek and Newman, Inc., Cambridge, Mass., 1975).

(10) Sanford, A.C., Hunt, G.T., & Bracey, H.J., Communication behavior in organizations. (Merrill, Columbus, Ohio, 1976).

(11) Zakay, D., Reliability of information as a potential threat to the acceptability of Decision Support Systems. IEEE Transactions on Systems, Man and Cybernetics. SMC-12 (1982) 518-520.

(12) Ulvila, J.W., & Brown, R.V., Decision Analysis comes of ages. Harvard Business Review. 60 (1982) 130-141.

Human Factors in Organizational Design and Management
H.W. Hendrick and O. Brown, Jr. (Editors)

HUMAN FACTORS ISSUES IN EVALUATING THE QUALITY OF ORGANIZATIONAL DECISION MAKING

Jeffrey P. Schwartz Louis J. Blazy

University of Maryland
College Park, Maryland
U.S.A.

A question of major concern in decision research is how to assess the overall quality or success of decision making. Two general approaches to the problem are (1) examining the consequences of choosing an alternative and (2) evaluating the thoroughness of the procedures (e.g., correctly assimilates new information, intensely searches for new alternatives) used by the decision maker. Both approaches ignore aspects which are critical to evaluating decision quality. In general, the sole use of consequences as quality criteria can allow fortuitous results to be dubbed "good decision making"; similarly, using the decision maker's procedures alone could allow negatively-valued consequences to be considered "successful decision making" provided the procedures themselves were judged vigilant. The evaluation of organizational decision making is not only hampered by the lack of a quality metric but also by other human factors issues which remain to be examined. One particularly important human factors procedure, typically overlooked, is needs assessment. Needs assessment (entailing task analysis, job analysis, selection of criteria, and evaluation goals) is a traditional tool of the human factors analyst, yet is often bypassed when decision making is the behavior in question. For example, an important factor is the specific types of validity that one intends to establish when training organizational members to be "good" decision makers (e.g., performance validity, intra-organizational validity, or inter-organizational validity). The validity that one chooses is a function of the criteria selected which in turn are a function of the goals of the organization. These goals are identified through the needs assessment process. Any evaluation of the overall quality of decision making is multidimensional, and must proceed in a hierarchical, step-down process. This will ensure that (a) the organizational goals are identified and (c) the environmental constraints are identified. Once these goals are established, the defining of a quality metric is a meaningful problem.

A quality metric is required for several reasons, including (a) many organizational decisions have both positive and negative consequences (b) many organizational decisions have no objectively correct answer and (c) a potential criterion for evaluating quality often does not appear until long after the decision is made. These constraints are bothersome because decision makers need a method for assessing quality at the moment a decision is made. A method which bypasses these constraints is to use the "lens method" of Brunswik, along with its associated concepts, as a rationale for assessing decision quality. According to a "lens model" analysis, a high quality decision is one in which the decision maker has both high knowledge of and cognitive control over the information present. These two conditions are likely to lead to successful decision making even when no objective criterion exists in the environment. To produce quality decisions, organizational decision makers need to be trained to (a) correctly determine all relevant information within situations (knowledge) and (b) consistently implement an optimal rule for integrating the information (cognitive control).

For any organization to be effective, organizational members must perform at a satisfactory or superior level. Evaluation of decisions made by organizational members is one of the primary ways in which organizations analyze member performance (Gannon, 1979). Organizational members who make high-quality decisions are viewed as superior performers, whereas members whose decision making is of lesser quality are rated as less valuable to the organization. Therefore, it is incumbent upon organizational members, be they entry level or managerial, to learn both how to make high-quality decisions and what methodology (if any) will be used to evaluate their decision making.

As noted by Carroll and Tosi (1977), decision making is an essential aspect of any manager's job, and managers are continually evaluated on their ability to make effective decisions; "the manager is usually evaluated on decision making to the degree that the desired end result was achieved and to the degree that the resources employed were less than those that would have been required by other alternatives" (p. 332). These two general criteria -- goal achievement and resource utilization -- coincide well with the concepts of decision effectiveness and decision efficiency, especially when resource utilization is defined to include both physical and mental resources.

EFFECTIVENESS AND EFFICIENCY

A decision is considered effective if it is performed as well as possible in relation to some predefined performance criterion. A decision is considered efficient if information gathering is vigilant and if the chosen criterion is the relevant one (Keen & Scott Morton, 1978). It is possible for any decision to be both efficient and effective, neither, or one or the other. Some decisions, however, need to be made in spite of inherent conflict between efficiency and effectiveness. For example, time pressure often disallows sufficient criterion evaluation. In any case, the aim in most organizational decision situations is to choose an alternative that provides the greatest effectiveness relative to the objectives of the decision maker (Radford, 1981). If the alternative chosen also provides the greatest efficiency in the use of resources, this is well and good, but primary emphasis is on effectiveness. The twin goals of decision efficiency and decision effectiveness form the basis for assessing the overall quality of any organizational decision.

Effective organizational decision making is hampered by several constraints (Hogarth, 1982). A major constraint is that no single index of decision quality exists, preventing both decision makers and organizational observers from knowing when good decisions have been made. The source of this difficulty is threefold:

(1) Decision making can be likened to ill-structured problem solving (Simon, 1976). This means, among other things, that decisions do not often have objectively right and wrong answers.

(2) Decisions typically have multiple consequences, some of which are positively-valued and some of which are negatively-valued (Janis & Mann, 1977). If consequences are to be used as the basis of a quality metric, the ambiguity surrounding how many and which consequences to use needs to be resolved. Evaluators could use a weighting scheme to combine the consequences into a quality metric, but, as with any decision, it is not clear which weighting scheme to use (Einhorn & Hogarth, 1981). In addition, the use of a weighting scheme to combine multiple decision consequences could lead to the absurdity in which evaluating the decision is more resource-consuming than the original decision.

(3) Decision quality has been discussed as if it had to be evaluated against criteria external to the decision itself. Several problems occur here. First, as with any criterion-related situation, criterion relevance and validity themselves are open to question. Second, appropriate criteria may never appear, or may only appear long after the decision has been made, rendering any assessment of decision quality fruitless. Third, multiple criteria often conflict with one another. Fourth, criteria often have differential validity for the multiple organizational subunits affected by a decision (Ebert & Mitchell, 1975). If criteria are used to assess decision quality, they must be sensitive to not only the needs of the decision maker, but also to others affected by his decisions.

APPROACHES TO DEFINING DECISION QUALITY

One attempt at developing a general quality metric was provided by Maier (1963), although he refers to it as a metric of decision effectiveness. According to Maier, an effective decision is one that is high on both Quality (the degree to which technical and rational factors govern in the selection of alternatives), and Acceptance (the extent to which those who are ultimately going to carry out the decision are willing to implement it in a way which minimizes problems). In other words, Effective Decision= Quality x Acceptance. When either quality or acceptance is zero, then an organizational decision will be highly ineffective. Maier seems to be saying that the evaluation of organizational decisions involves both an information processing component (Quality) and a social component

(Acceptance). However, recent evidence (Heller & Wilpert, 1981) suggests that the importance of acceptability in determining overall decision effectiveness is equivocal.

Hoffman and his associates (1979, 1982) have extended Maier's approach in a way which distinguishes among the Adoption of a decision solution, its Quality, and its Acceptance, where Adoption is a function of the solution's attractiveness for the entire group, and Acceptance is a function of each individual member's valence for the solution. Hoffman and O'Day (1979) show that the valence-Adoption and valence-Acceptance relationships occur for solutions of both high and low quality, corroborating an earlier finding by Maier himself (1963). Hoffman has been primarily concerned with delineating the antecedents of solution adoption and acceptance, rather than determining when a decision is of high quality.

> To produce a solution of high quality, a group must define the problem accurately, have relevant information available, use that information to support the adoption of a best solution (1982, p. 104).

A high quality decision would appear to be one that produces the "best" solution, but Hoffman provides no definition of what "best" means within his framework.

A different approach to defining decision quality has been taken by Sutherland (1977), who advocates measuring the quality of a decision maker by the aggregate cost of error (or loss) associated with the decisions made during some period of time; that is, as a product of the frequency of errors, their severity, and their duration. The smaller this product is, the higher the quality of the decision maker. Sutherland also states that any attempt to measure the quality of decision making performance must consider the category of problems confronted by the decision maker. For example, the decisions faced by organizational planners are often ill-defined and intractable, whereas those faced by lower-level organizational members are more easily formulated. Accordingly, the function of errors stated above would be most appropriate for lower-level organizational members. Sutherland recognizes the problem with using an error function for planning-type decisions, and offers suggestions for modelling these decisions in a way that approaches the use of an error function.

Several approaches to developing a quality metric involve specifying a quantified model of organizational goal achievement. Goal programming (Lee & Clayton, 1972) has been used to analyze decisions consisting of multiple goals, where alternatives are not readily expressed on a single scale of measurement (such as dollars or cost). The goal-consistency model (Vesper & Sayeki, 1973) has been developed to permit the decision maker to assess the consistency among his perceptions of organizational goals, policies, and alternatives. All such approaches insist that "managerial decisions must be judged ultimately in terms of their impact upon organizational goal achievement" (Shull, Delbecq & Cummings, 1970, p. 13); yet, they also acknowledge that the evaluation of decisions is "at best, a difficult task, especially in the short run" (Shull et al., 1970, p. 12).

Since it is the short-run evaluation of decision making which most interests organizations (Gannon, 1979), the use of goal achievement as the criterion for assessing quality is difficult, as goals are loosely defined (Hogarth, 1982). It is also unclear whether organizational goals actually direct the future decision making of individual organizational members (Weick, 1979). Furthermore, within organizational decision problems, a single goal rarely dominates to the exclusion of others. The resulting comparison among multiple and often conflicting goals is extremely problematic (Ebert & Mitchell, 1975). The use of goal achievement as the basis for defining a quality metric, therefore, does not seem promising.

The quality of the information-processing procedures used by the decision maker represents a major alterantive to using either the consequences of choosing an alternative or global goal achievement as the basis for defining a quality metric. Janis and Mann (1977) list seven "ideal" procedural criteria that decision makers should meet in order to be deemed successful.

> The decision maker, to the best of his ability and within his information-processing capabilities
> 1. thoroughly canvasses a wide range of alternative courses of action;
> 2. surveys the full range of objectives to be fulfilled and the values implicated by the choice;
> 3. carefully weighs whatever he knows about the costs and risks of negative consequences, as well as the positive consequences, that could flow from each alternative;
> 4. intensively searches for new information relevant to further evaluation of the alternatives;
> 5. correctly assimilates and takes account of any new information or expert judgment to which he is exposed, even when the information or judgment does not support the course of action he initially prefers;
> 6. reexamines the positive and negative consequences of all known alternatives, including those originally regarded as unacceptable, before making a final choice;
> 7. makes detailed provisions for implementing or executing the chosen course of action, with special attention to contingency plans that might be required if various known risks were to materialize (p. 11).

When a decision maker meets all seven criteria, his orientation at arriving at a choice is characterized as _vigilant information processing_. Note that, using Janis and Mann's (1977) taxonomy, the use of procedures as overall decision quality criteria entails a cognitive utilization of consequences, as criteria 3 and 6 involve the decision maker thinking through possible consequences of alternatives.

The notion of using decision-making procedures, rather than environmental criteria, as the basis for assessing quality seems as if it is a necessary development. However, there are general problems with using procedures to the complete exclusion of any external criteria. First, decision makers are limited in their ability to predict the future; perhaps they cannot conceive of every possible consequence of choosing every possible alternative; perhaps they mistakenly assign a consequence to an alternative. This limitation may be as much a result of environmental complexity as due to information-processing laxness. Second, the move to use the cognitive utilization of consequences, rather than the consequences themselves, as one type of quality criteria does not remove the problems inherent with consequences-as-criteria. Most importantly, decision makers within organizational settings will never be praised merely for "vigilant information processing" if either unforeseen or negative consequences result from their judgments. In short, the careful organizational decision maker must be more than careful; he must produce results. Vigilant information processing remains a necessary, but insufficient, standard for evaluating decision quality.

Janis and Mann's criteria should not be thought of as the basis of a metric of decision quality but rather as a list of what good decision makers (ought to) do. If a decision maker does these seven things, then perhaps he will produce some good decisions, but the quality of said decisions must be assessed in some way external to his cognitive procedures.

The evaluation of organizational decision making is not only hampered by the lack of a quality metric but also by other human factors issues which remain to be examined.

Four human factors issues are important in evaluating the quality of organizational decision making. They are needs assessment, criterion selection, selection of an evaluation model, and establishment of performance validity. Typically, they are not given sufficient priority or importance within the decision evaluation schemes of organizations.

Most needs assessment procedures include three major elements: organizational analysis, task analysis, and person analysis. The primary advantage to using a needs assessment model for evaluating the quality of organizational decision making is that it provides a systematic approach to (1) detailing how organizational and environmental factors could impact upon decisions, (2) selecting criterion measures for evaluating the effects of decisions on the organization, and on target groups within the organization, (3) determining what evaluation model should be used, and (4) determining internal and external validity of the decision.

Organizational analysis is generally concerned with (1) determining the short and long-term goals of the organization, (2) tracking trends that affect these goals, (3) evaluating resources needed for implementing the decision, and (4) examining internal and external constraints such as political, cultural, and economic factors which may adversely (or positively) affect the quality of decisions (Goldstein, 1974, 1980; Moore & Dutton, 1978).

The specific components of an organizational analysis depend upon the type of decision being considered and the characteristics of the organization. For example, if an organization is considering the implementation of a training program, then the needs assessment process would likely identify

(1) long and short-term goals, (2) assessing whether training is needed, (3) the time, money, and facilities the organization is willing to commit. On the other hand, if the organization is considering a merger with another company, then the specific components of the needs assessment process would probably change, but the same needs assessment model would still be used.

Unfortunately, selection of a needs assessment strategy is more of an art than a science (Steadham, 1980). It is learned through experience and application. Nevertheless, Steadham (1980) has listed nine methods for selecting a needs assessment strategy. These include observation, questionnaires, key consultation, media usage, interviews, group discussion, tests, record reports, and work samples. Application of these 'tools of the trade' is limited by organizational constraints (e.g., time, resources), the availability of personnel to supply inputs to the needs assessment process, and specific constraints upon the evaluators themselves.

Although task and person analyses are critical components of the needs assessment process, particularly in the context of instructional design and training programs, their relevancy to evaluating the quality of organizational decision making is limited, and will not be discussed here.

The importance of a properly conducted needs assessment is highlighted by the observation that the criteria one selects for evaluating decision quality are dependent upon the quality of the needs assessment process. Information from the needs assessment is used to describe the criterion measures needed to properly evaluate decisions. Criterion measures are derived from the organizational objectives identified during needs assessment. One extreme manifestation of evaluation via organizational objectives is management by objectives (MBO). MBO defines exactly what objectives the organization should be striving to achieve. Goldstein (1980) has criticized this approach because it depends too closely on who designs the objectives, and on how the objectives are to be used.

Another important issue in selecting criteria is their reliability and acceptability. Chosen criteria must have real-world meaningfulness to the organizational decisions they purport to measure, and they must measure decision quality in a consistent fashion.

Once criterion measures have been specified, the next step is to determine what evaluation model to use to achieve the objective of decision quality evaluation. Evaluation is the systematic collection of information necessary to make effective decisions. In this context, evaluation is value-free, stressing neither good nor bad aspects of decision making solely. Traditional experimental designs (e.g., randomized block designs) are generally inadequate models for evaluating decision making because they fail to incorporate time as an explanatory variable; organizational decisions typically occur across a wide period of time. Cook and Campbell (1979) have outlined a number of quasi-experimental designs which, in principle, can be used to evaluate the effectiveness of decisions over time.

Closely allied with issues of criterion evaluation are issues of criterion validity. Broadly defined, validity is concerned with potential sources of error within the criteria themselves. First, criteria must be sensitive enough to detect changes within (to) the organization that are due to specific decisions (i.e., internal validity). Second, will decisions of this kind always have the same kind of effect (i.e., external validity)?

Within the context of evaluating the quality of organizational decision making, three related types of validity must be addressed: performance validity, intra-organizational validity, and inter-organizational validity. Performance validity refers to evaluation of the effects of the decision on a target group or organization. For example, if an organizational objective is to train individual members to be 'good' decision makers, then the needs assessment would focus on this one target group without concern for transferring 'good decision making' techniques to other groups within the organization. Where training is concerned, performance validity establishes who the organization wishes to become better decision makers.

Performance validity is a prerequisite for predicting whether other groups can be trained to be good decision makers, both within (intra-organizational validity) and without (inter-organizational validity) the organization that developed the training program. Which validity to emphasize is a function of the organizational objectives identified during the needs assessment process.

Once the needs assessment process is completed, the defining of a quality metric is a meaningful problem. The needs assessment process places boundaries on what criteria should be used to evaluate decision quality, but it does not remove the problems with criteria generally.

The metric used to evaluate decision quality must be sensitive to problems with both consequences-as-criteria and procedures-as-criteria. What is needed, then, is a technology which (1) models cognitive processes at a level different from simple information search, (2) has readily interpretable constructs which reflect aspects of decision quality, (3) can operate in the absence of an external criterion, yet (4) can incorporate relevant criteria if they are identified.

One possibly relevant technology is Brunswik's (1956) "lens model", along with its associated constructs. The "lens model" has been used to implement regression-based modelling of human judgment (Slovic & Lichtenstein, 1971); potentially, its constructs can be used to model the key components of effective and efficient decision making.

According to a "lens model" analysis, a high quality decision is one in which the decision maker has both high <u>knowledge</u> of and <u>cognitive control</u> over the information relevant to making the decision. Hammond and Summers (1972) demonstrate that knowledge and cognitive control can be disentangled empirically, as well as statistically, and therefore contribute independently to the assessment of decision quality. Other authors have also argued that knowledge and control are important to the decision maker (e.g., Oxenfeldt, Miller, and Dickinson, 1981).

The basic "lens model" equation (Tucker, 1964) states that a decision maker's effectiveness (achievement) is a function of (1) environmental complexity-inherent uncertainty surrounding the decision (2) knowledge - the extent to which the decision maker correctly represents this complexity in his mind and (3) cognitive control - the extent to which the decision maker uses his knowledge in a consistent manner. In other words, Effectiveness= Environmental Complexity x Knowledge x Cognitive Control. In practice, each of these terms is replaced by an appropriate correlation coefficient.

The usage of the "lens model" as a way of assessing organizational decision quality would proceed as follows. If a relevant criterion is identified through needs assessment, then the version of the "lens model" given above would be used. If no criterion can be identified, or if any of the aforementioned problems concerning criteria exist, then a modified version of measuring the "lens model" constructs would be undertaken.

CRITERION PRESENT

When a criterion is present, the terms within the "lens model" equation are measured in the following way (Hammond & Summers, 1972).

Environmental Complexity - the multiple correlation between the informational cues present and the criterion.

Knowledge - the correlation between the linear prediction of the criterion and the decision maker's judgment from the cues.

Cognitive Control - the multiple correlation between the cues and the decision maker's judgment.

The correlation between the criterion and the decision maker's judgment -- his effectiveness -- can be decomposed into the product of these three terms.

It should be obvious that the "lens model" technology, being regression-based, requires many judgments of the decision maker. This is done by varying the values of the informational cues, one judgment being made to each possible combination.

As mentioned earlier, one reason why simple information-processing criteria are inappropriate quality standards is that they cannot reflect poor decision-making performance caused by <u>environmental complexity</u>. A "lens model" analysis overcomes this constraint by noting that environmental complexity sets an upper limit on decision quality. If there is much complexity within the environment, then decision quality will be less than perfect, regardless of the values of knowledge and control. Thus, use of the "lens model" does not inappropriately penalize decision makers for poor decisions that result from inherently complex situations.

In spite of how much complexity there is within the environment, a decision is said to be of perfect quality if both knowledge and control equal 1.00. Variations from this circumstance yield decisions of correspondingly lesser quality. Use of the "lens model" would allow organizations to evaluate whether judgmental deviations from criteria are due mainly to environmental complexity, faulty knowledge, or lack of cognitive control.

CRITERION ABSENT

It has been argued that, for short-term evaluation of decisions, appropriate criteria are rarely present. The "lens model" can still be used to evaluate quality for decisions in which criteria are absent. Cognitive control would be measured the same way as when a criterion is present. However, knowledge and control must be measured differently, since normally they are measured relative to a criterion. Knowledge could be assessed with models of standard decisions faced by organizations (e.g., investments, hiring, promotions).

Here, knowledge would reflect the decision maker's ability to gather information concerning all relevant cues, including what they are and their interrelationship. Environmental complexity could be measured in a number of ways. Beach and Mitchell (1978) suggest using the unavailability, unreliability, and imprecision of relevant information as possible measures.

Quality, or achievement, would still be the product of environmental complexity, knowledge, and control, but without a criterion, no separate index of achievement could be calculated. To produce successful decision makers, organizations would need to devise methods for training individuals to have both high knowledge of and cognitive control over the information relevant to any decision.

REFERENCES

Beach, L.R., and Mitchell, T.R. (1978). A contingency model for the selection of decision strategies. Academy of Management Review, 3, 439-449.

Brunswik, E. (1956). Perception and the representative design of psychological experiments. Berkeley: University of California Press.

Carroll, S.J., and Tosi, H.L. (1977). Organizational behavior. Chicago: St. Clair Press.

Cook, T.D., and Campbell, D.T. (1979). Quasi Experimentation: design and analysis issues for field settings. Chicago: Rand McNally.

Ebert, R.J., and Mitchell, T.R. (1975). Organizational decision processes. New York: Crane, Russak, and Company.

Einhorn, H.J., and Hogarth, R.M. (1981). Behavioral decision theory: Processes of judgment and choice. Annual Review of Psychology, 32, 52-88.

Goldstein, I.L. (1974). Training: Program Development and Evaluation. Monterey, Calif.: Brooks/Cole.

Goldstein, I.L. (1978). The pursuit of validity in the evaluation of training programs. Human Factors, 20, 131-144.

Goldstein, I.L. (1980). Training in work organizations. Annual Review of Psychology, 31, 229-272.

Gannon, M.J. (1979). Organizational behavior. Boston: Little, Brown and Company.

Hammond, K.R., and Summers, D.A. (1972). Cognitive control. Psychological Review, 79, 58-67.

Heller, F.A., and Wilpert, B. (1981). Competence and power in managerial decision-making. New York: AMACOM.

Hoffman, L.R. (1979). The group problem-solving process: Studies of a valence model. New York: Praeger.

Hoffman, L.R. (1982). Improving the problem-solving process in managerial groups. In R.A. Guzzo (Ed.) Improving group decision making in organizations. New York: Academic Press.

Hoffman, L.R. and O'Day, R. (1979). The process of solving reasoning and value problems. In L.R. Hoffman (Ed.) The group problem-solving process: Studies of a valence model. New York: Praeger.

Hogarth, R.M. (1982). Decision making organizations and the organization of decision making. (Unpublished manuscript). Chicago: University of Chicago Center for Decision Research.

Janis, I.L., and Mann, L. Decision making: A psychological analysis of conflict, choice, and commitment. New York: Free Press.

Keen, P.G.W., and Scott Morton, M.S. (1978). Decision support systems: An organizational perspective. Reading, Mass.: Addison-Wesley.

Lee, S.M., and Clayton, E.R. (1972). A good programming method for academic resource allocation. Management Science, 18, B-395-398.

Maier, N.R.F. (1963). Problem-solving discussions and conferences: Leadership methods and skills. New York: McGraw-Hill.

Moore, M., and Dutton, P. (1978). Training needs analysis: Review and critique. Academy of Management Review, July, 532-545.

Oxenfeldt, A.R., Miller, D.W., and Dickinson, R.A. (1978). A basic approach to executive decision making. New York: AMACOM.

Radford, K.J. (1981). Modern managerial decision making. Reston, Va.: Reston Publishing.

Shull, F.A., Jr., Delbert, A.L., and Cummings, L.L. (1970). Organizational decision making. New York: McGraw-Hill.

Simon, H.A. (1976). Administrative behavior: A study of decision-making processes in administrative organization. New York: Macmillan.

Slovic, P., and Lichtenstein, S. (1971). Comparison of Bayesian and regression approaches to the study of information processing in judgment. Organizational Behavior and Human Performance, 6, 649-744.

Steadham, S.U. (1980). Learning to select a needs assessment strategy. Training and Development Journal, January, 56-61.

Sutherland, J.W. (1977). Administrative decision-making. New York: Van Nostrand.

Tucker, L.R. (1964). A suggested alternative formulation in the developments by Hursch, Hammond, and Hursch, and by Hammond, Hursch, and Todd. Psychological Review, 71, 528-530.

Vesper, K.H., and Sayeki, Y. (1973). A quantitative approach for policy analysis. California Management Review, 15, 119-126.

Weick, K.E. (1979). The social psychology of organizing (2nd edition). Reading, Mass.: Addison-Wesley.

Human Factors in Organizational Design and Management
H.W. Hendrick and O. Brown, Jr. (Editors)

THE IMPACT OF COGNITIVE PROCESSING LIMITATIONS ON DECISIONS MADE USING A COMPUTER BASED MANAGERIAL WORKSTATION

William Remus
Professor of Decision Sciences
University of Hawaii
2404 Maile Way
Honolulu, Hawaii 96822

When managers make decisions they are subject to numerous cognitive processing limitations. The limitations result largely from the way in which they human brain is organized. This paper first reviews and ties those two literatures together.

Given the integrated view presented of the sources and symptoms of the managerial decision biases, the paper then discusses how the interface design for workstations can either exaggerate or compensate for the biases. Suggestions are made for good interface designs for workstations.

1. INTRODUCTION

When managers make decisions they are subject to numerous decision making errors (biases). Most of these errors appear not to result from stupidity but instead to the limitations resulting from the way in which the brain is organized.

2. BIASES AND HOW THE BRAIN IS ORGANIZED

Human decision makers receive data and analyze that data prior to reaching a decision. In processing that input stream of data, they are prone to certain common errors. In this section, a simple model of the brains' organization (Kent, 1982) is presented and then used to suggest reasons the processing errors occur.

The visual input processing can be viewed as the following sequence of actions:

(1) The light reflected or emitted by an object strikes the retina.

(2) The light is changed into neural coding.

(3) The image of the object and the object's location as separated and processed independently.

(4) The image is reduced to geometric features (e.g. arcs, lines, corners, edges) and fourier representations by simple feature extractors.

(5) The small geometric features are reconstructed into a global rendering of the object (but not unlike a stick drawing of the object).

(6) The object is held in short term memory as it is referred to associative memory for identification. Short term memory is quite limited being able to hold seven plus or minus two "chunks" of information.

(7) The identity of the object is found using a "fuzzy key" to access memory.

(8) The object's identity allows associative memory to look-up the relationships the object has to other objects.

At this point, the information is ready to be passed on for higher levels of processing.

In Table 1, many of the human input errors are summarized. From that list one gets the impression that humans are prone to numerous errors which they could easily avoid. Yet even training often fails to eliminate these errors. The reason that those errors persist is that most of the errors result from the neurophysiological limitations of the human mind. And training can't do much about those kind of limitations.

Consider the impact of dealing with the image only as it is coded by the simple and global feature extractors (that is, composed of linked lines, arcs, corners, edges, and fourier signatures). The automatic global object composing mechanism can fit lines to data points so that the decision maker sees patterns even in random data. Also what is irrelevant information is not a determination made by higher level processes; it is a determination made by the feature extractors and the memory look-up system. What the eyes focus on is what the feature extractors process in detail for memory look-up. The focus point of the eyes not only determines what is relevant but also what we selectively process and look-up in memory.

The memory look-up uses a fuzzy key arrangement to identify the objects and determine it's relationship with the rest of the world. But the key system is a function of our life experience. Thus objects are filtered and identified as a function of our life experience. Also higher processes can fine tune the look-up process. To the extent that we have expectations, the look-up may reflect those expectations (and we often find what we expect -- especially given the flexible nature of the fuzzy keys).

Since memory keying is based on our experiences, information which is consistent with out experiences is easily filed -- and thereby reinforces our predispositions. Contradictory information requires different keying than naturally occurs given our life experience. In this way sets of similar information is gathered. The more that is organized under one key, the easier the information is to access and use. It is not only easier to use but becomes a basis for causal attributions.

The preceding and other human input errors are related to our brain's wiring. But the hard-wiring effects not only the input stream but also the decisions made. First note the many human decision making errors shown in Table 2. Again one gets the impression that these obvious errors should be easy to correct through proper training. Unfortunately, these errors persist because they are a function of our brain's organization. Consider the way in which the frontal lobe processes a command to grasp an object:

(1) The command is given

(2) The current and final positions of the hand is identified

(3) The task to reach the goal is conservatively fractionated and passed to several parallel processors (it is much better to be conservative because overshooting the goal can injure the hand but undershooting will not)

(4) Each parallel processor repeats the conservative fractionation and passes that task on to sub-processors and onward to muscle groups

(5) The net impact of the conservative fractionation is movement toward the goal (but not totally reaching the goal)

(6) Dynamic feedback allows the fractionation of the remaining distance until the hand is correctly positioned

Human planning/decision making piggybacks on the frontal lobe's control system just discussed. And this has great implications for the limitations human decision makers have.

The working storage for human decision makers is called short term memory. The short term memory can usually hold no more than seven plus or minus two "chunks" of information. Not only does this limitation cause input problems -- for example, the primacy and recency effects described in Table 1 are caused by exceeding the limits of short term memory and dumping either the earliest or latest received data -- but it also affects the number of factors which a human decision maker can use. Heuristics are a necessary way to overcome short term memory limitations since short term memory is the holding place for decision factors. Short term memory also allows only limited working storage for search strategies -- hence we use limited search strategies.

The way in which we forecast the future parallels the processes used in the frontal lobe. We anchor based on past history (i.e. current position of our hand) and then adjust to reach the goal (i.e. the final position). Human decision makers underadjust just as the conservative fractionation processes underadjust. The motor system can correct using dynamic feedback -- but decision makers don't have dynamic feedback on their decisions.

Thus decision making errors could result from our hardware limitations. And the hardware problems are difficult to solve with software fixes (i.e. cognitive overrides on messages derived from the hardware).

3. IMPROVING DECISION MAKING WITH PROPER WORKSTATION DESIGN

To improve decision making, both the input and processing biases need to be addressed. Some of the earlier discussed biases can be improved upon by (1) alerting decision makers to their limitations, (2) adding working memory (e.g. sheets of paper) to supplement the limited short term memory, (3) adding formal models to provide analytical solutions, (4) providing graphs to display relationships and other simple remedies. But with the advent of the managerial workstations, it is now possible to have intelligent decision support which can better compensate for the previously discussed problems. In the remaining pages of this paper, examples of software are presented to make several key points about the approach necessary to build workstations which compensate for the manager's weaknesses.

Consider a typical graphics display of time series data presented on a managerial workstation. A good design would make the default display start at zero on the Y axis. This would avoid the distortion in the "perceived data variability" if the Y axis were shortened to zoom-in on the data. The trend line projection tools should be prompted to try to get a better prediction than the conservative "anchor and adjust" heuristic. Automatic testing for statistical significance of the trend and 4 (quarterly) and 12 (monthly) period seasonals should also be displayed. This might reduce the decision makers tendency to "see patterns in random data."

When dealing with multivariate data, decision makers may use heuristics which are oversimplified. When a historical data base is available, regression can be used to capture the policy of the decision makers for future use. Heuristics of this sort consistently outperform the decision makers themselves (Goldberg, 1970; Moskowitz and Miller, 1975; Remus, Carter, and Jenicke, 1979). At minimum these heuristics will reduce erratic decision making. The keys to successful use of the regression boot stopping model are (1) embedding a user transparent regression in the workstation, (2) keep the crucial details of its operation out of the users way, (3) automatically determining when its use might be appropriate and (4) prompting its use. Of particular note here is the need to competently perform the regression without involving the user in the details. This calls for an mini-expert system to automatically take care of outliers, transformations, hetroscedascity, and other related tasks. The user gets the finished product.

When the user has a natural language interface into a database, many of the tasks become simplified. Consider the query, "Compare the sales commissions earned in Boston and Chicago." The data could be retrieved and displayed but additionally statistical comparisons could be automatically made to avoid the decision maker's biases. With a mini-expert system, the proper test would be chose, variances pooled, outliers handled, etc. The manager could then receive a statement such as "There is a greater than 95% chance that the Commission differ between Boston and Chicago" without having called for the test.

4. Summary

The themes that emerge from all the preceding are that:

(1) even simple things like the length of the Y axis can bias the decision maker.

(2) many of the hard-wired biases can be detected automatically and the manager guided to avoid the pitfall.

(3) the workstation software should go beyond the query to provide the compensating information without prompting.

(4) to accomplish the latter, small expert systems will need to be constructed so that the naive user will not get involved in the details (the user, however, should be able to query the expert system for its logic and the details).

(5) the use of natural language queries to data bases will allow the workstation to be friendlier, more intelligent, and more compensatory.

(6) to accomplish this, statistical packages, data base systems, graphics packages, natural language dialog systems, and expert systems need to be combined.

Table 1

Biases associated with presenting data to a decision maker

Irrelevant information

Irrelevant information can influence decision makers and may reduce the quality of their decisions (Ebert, 1972; Slovic and MacPhillamy, 1974).

Data presentation biases

(a) The type of information. Data acquired through human interaction has more impact on the decision maker than just the data itself (Borgida and Nisbett, 1977).

(b) The format of the data display. The display format of the data crucially affects the decision made. Generally speaking, summarized data presentations (e.g., statistics, tables, graphs) are preferred to raw data (Chervany and Dickson, 1974). The choice of whether to use graphical summaries or tabular summaries depends on the level of environmental stability (Remus, 1981).

(c) Logical data displays. When a set of alternatives is presented which seems to capture all the possibilities, then the decision maker may not detect other alternatives that may exist (Fischhoff, Slovic, and Lichtenstein, 1978).

(d) Order effects. The order in which data are presented can affect the data retention (Slovic and Lichtenstein, 1971; Moskowitz, Schaefer, and Borcherding, 1976). In particular, the first piece of data may assume undue importance (primacy effect) or the last piece of data may become overvalued (recency effect).

(e) The context may affect perceived data variability. When decision makers assess the variability in a series of data points, their assessment will be affected by the absolute (i.e., mean) value of the data points and the sequence in which they are presented (Lathrop, 1967).

Selective perception

Decision makers selectively perceive and remember only certain portions of the data (Egeth, 1967).

(a) People filter data in ways that reflect their experience. When data from a problem is presented to decision makers, they will particularly notice the data that relates to areas in which they have expertise (Dearborn and Simon, 1958; Egeth, 1967). This leads to differing and often limited problem formulations.

(b) Expectations bias perceptions. When decision makers are reviewing data which is contrary to their expectations, they may remember incongruent pieces of data inaccurately (Bruner and Postman, 1949; Makridakis and Hibon, 1979).

(c) People seek information consistent with their own views. If decision makers bring to a problem some prejudices about the problem, they will seek data confirming their prejudices (Batson, 1975; Bruner, Goodnow, and Austin, 1956, pp. 88-89, 92-93; Wason and Johnson-Laird, 1972).

(d) People downplay data which contradicts their views. If the decision makers have an expectation about a problem or if they think they have arrived at the solution, they will downplay or ignore conflicting evidence (Anderson and Jacobson, 1965; Slovic, 1966b). Their expectations will often persist even in the light of continued conflicting evidence (Bach, 1971, p. 174; Wason and Johnson-Laird, 1972, pp. 180-181).

Frequency

(a) Ease of recall. The ease with which certain data points can be recalled will affect not only the use of this data but the decision maker's perception of the likelihood of similar events occurring (Kunreuther, 1976; Lichtenstein, Slovic, Fischhoff, Layman, Combs, 1978; Tversky and Kahneman, 1973, 1974).

(b) Base rate error. Often the decision maker determines the likelihood of two events by comparing the number of times each of the two events has occurred. However, the base rate is the crucial measure and is often ignored (Borgida and Nisbett, 1977; Kahneman and Tversky, 1973). This is a particular problem when the decision makers have concrete experience with one problem but only statistical information about another. They will generally be biased toward thinking the concrete problem to be more troublesome. Decision makers particularly overestimate the frequency of catastrophic events (Lichtenstein, Slovic, Fischhoff, Layman, Combs, 1973).

(c) Frequency is used to imply strength of relationship. The more pairs of co-varying data points that decision makers have, the stronger they think the relationship between variables is (Estes, 1976; Jenkins and Ward, 1963; Ward and Jenkins, 1965).

(d) Illusory correlation. Decision makers may erroneously believe certain variables to be correlated. From plots of data they can often "recognize" patterns, even when the points displayed here have no true correlation (Chapman, 1967; Smedslund, 1963; Tversky and Kahneman, 1973; Ward and Jenkins, 1965).

Table 2

Biases in information processing

Heuristics

As Ackoff (1967) pointed out, the decision maker's major problem is often not the lack of information upon which to base a decision but too much information. Generally the decision maker reduces the problem to a problem involving only 3 or 4 crucial factors. The decision is made using a heuristic based on those factors. Unfortunately, these heuristics often have built-in biases (Tversky and Kahneman, 1974). The following are some of the problems with heuristics.

(a) People structure problems based on their own experiences. Decision makers may try to find the best fit between the new problems they have and old problems they have had. When a match is found, the decision maker then alters the old decision slightly to reflect the new circumstances (Kahneman and Tversky, 1972, 1973; Tversky and Kahneman, 1973).

(b) Rules of thumb. If the decision maker has prior experience in solving a problem, he may again use the same rule of thumb since it proved satisfactory last time (Knafl and Burkett, 1975).

(c) Anchoring and adjustment. Prediction is often made by selecting an "anchoring" (e.g. on a mean) value for a set of data and then adjusting the value for the circumstances in the present case. Generally, the adjustments are insufficient (Slovic and Lichtenstein, 1971; Tversky and Kahneman, 1974).

(d) Inconsistency in the use of the heuristic. Bowman (1963) theorized that the major economic consequences of decisions did not result from a manager's use of poor heuristics but from the manager's inconsistent use of the heuristics. Numerous studies have shown that heuristics based on regression outperform the decision maker himself (Goldberg, 1970; Moskowitz and Miller, 1975; Remus, Carter, and Jenicke, 1979) and the performance difference, as theorized, has been shown to result from inconsistency (Remus, 1977).

Misunderstanding of the statistical properties of data.

(a) Mistaking random variation as a persisting change. Data available to a decision maker often has statistical properties; that is, it has a mean value around which the data points are randomly distributed. When the next observation is a high or low value data point, the decision maker may believe an upward or downward trend is occurring. In fact, it is just a random variation and not a persisting change. Often decision makers infer causes for these random variations (Langer, 1977; Smedslund, 1963; Tversky and Kahneman, 1972; Ward and Jenkins, 1965).

(b) Inferring from small samples. The characteristics of small samples are often believed to be representative of the population from which they are drawn. Thus, in data inference too much weight is given to small sample results (Tversky and Kahneman, 1971, 1974).

(c) Gambler's fallacy. In a probabilistic process where each event is independent of the ones preceding it (i.e., it is random), decision makers may erroneously make inferences about future events based on the occurrence of past events (Kahneman and Tversky, 1972; Langer, 1977).

(d) Ignoring uncertainty. When a decision maker is faced with several sources of uncertainty simultaneously, he may simplify the problem by ignoring or discounting some of the uncertainty. The decision maker may then choose the most likely alternative for his decision (Gettys, Kelly, and Peterson, 1973).

Limited search strategies.

Decision makers have to decide when to stop gathering data and begin the analysis. They often use truncated search strategies which prematurely exclude potentially relevant information (Shaklee and Fischhoff, 1979). The use of truncated search strategies increases as task complexity increases (Payne, 1976).

Conservatism in decision making.

When new information arrives, the decision maker will tend to revise his probability estimates in the direction prescribed by Bayes' theorem, but usually the revision is too small (Moskowitz, Schaefer, and Borcherding, 1976). This conservatism increases with message informativeness (Slovic and Lichtenstein, 1971) and is subject to the primacy effect (Mason and Moskowitz, 1972).

Inability to extrapolate growth processes.

When exponential growth is occurring, the decision maker may underestimate the outcomes of the growth process (Timmers and Wagenaar, 1977; Wagenaar and Sagaria, 1975). This underestimation is not improved by presenting more data points (Wagenaar and Timmers, 1978).

REFERENCES

1. Ackoff, R. L., "Management Misinformation Systems," Management Science, Vol. 14, No. 2 (1967), B147-B156.

2. Anderson, N. H. and A. Jacobson, "Effect of Stimulus Inconsistency and Discounting Instructions in Personality Impression Formation," Journal of Personality and Social Psychology, Vol. 2 (1965), 531-539.

3. Bach, G. R. and R. M. Deutsch, Pairing (Avon: New York City, 1978).

4. Batson, C. D., "Rational Processing or Rationalization: The Effect of Disconfirming Information on Stated Religious Belief," Journal of Personality and Social Psychology, Vol. 32 (1975), 176-184.

5. Borgida, E. and R. E. Nisbett, "The Differential Impact of Abstract vs. Concrete Information on Decisions," Journal of Applied Social Psychology, Vol. 7 (1977), 258-271.

6. Bowman, E. H., "Consistency and Optimality in Management Decision Making," Management Science, Vol. 10, No. 1 (1963), 310-321.

7. Bruner, J. S., J. J. Goodnow, and G. A. Austin, A Study of Thinking (New York: Wiley, 1956).

8. __________ and L. J. Postman, "On the Perception of Incongruity," Journal of Personality, Vol. 18 (1949), 206-223.

9. Chapman, L. J., "Illusory Correlation in Observational Report," Journal of Verbal Learning and Verbal Behavior, Vol. 6 (1967), 151-155.

10. Chervany, N. L. and G. W. Dickson, "An Experimental Evaluation of Information Overload in a Production Environment," Management Science, Vol. 20, No. 10 (1974), 1335-1344.

11. Dearborn, D. C. and H. A. Simon, "Selective Perception: A Note of the Departmental Identification of Executives," Sociometry, Vol. 21 (1958), 140-144.

12. Ebert, R. J., "Environmental Structure and Programmed Decision Effectiveness," Management Science, Vol. 19 (1972), 435-445.

13. Egeth, R. J., "Selective Attention," Psychological Bulletin, Vol. 67 (1967), 41-57.

14. Estes, W. K., "The Cognitive Side of Probability Learning," Psychological Review, Vol. 83 (1976), 37-64.

15. __________, "Fault Trees: Sensitivity of Estimated Failure Probabilities to Problem Presentation," Journal of Experimental Psychology: Human Perception and Performance, Vol. 4 (1978), 330-344.

16. Gettys, C. F., C. W. Kelly, III, and C. R. Petterson, "The Best Guess Hypothesis in Multistage Inference, Organizational Behavior and Human Performance, Vol. 10 (1973), 364-373.

17. Goldberg, L. R., "Man Versus Model of Man: A Rationale, Plus Some Evidence for a Method of Improving on Clinical Inferences," Psychological Bulletin, Vol. 73 (1970), 422-432.

18. Jenkins, H. M. and W. C. Ward, "Judgment of Contingency Between Responses and Outcomes," Psychological Monographs: General and Applied, Vol. 79, No. 1 (1965), Whole No. 594.

19. Kahneman, D. and A. Tversky, "Subjective Probability: A Judgment of Representativeness," Cognitive Psychology, Vol. 3 (1972), 430-454.

20. __________, "On the Psychology of Prediction," Psychological Review, Vol. 80 (1973), 237-251.

21. Kent, G. The Brains of Men and Machines. (Bytes Books: New York, 1982)

22. Knafl, K. and G. Burkett, "Professional Socialization in a Surgical Speciality: Acquiring Medical Judgment," Social Science and Medicine, Vol. 9 (1975), 397-404.

23. Kunreuther, H., "Limited Knowledge and Insurance Protection," Public Policy, Vol. 24 (1976), 227-261.

24. Langer, E. J., "The Psychology of Chance," Journal for the Theory of Social Behavior, Vol. 7, No. 2 (1977), 185-207.

25. Lathrop, R. G., "Perceived Variability," Journal of Experimental Psychology, Vol. 73 (1967), 498-502.

26. Lichtenstein, S. C. and Slovic, P., B. Fischhoff, M. Layman and B. Combs, "Judged Frequency of Lethal Events," Journal of Experimental Psychology Human Learning and Memory, Vol. 4 (1978), 551.578.

27. Makridakis, S. and M. Hibon, "Accuracy of Forecasting: An Empirical Investigation," Journal of the Royal Statistical Society A, Vol. 142 (1979), (Part II), 97-145.

28. Mason, R. O. and H. Moskowitz, "Conservatism in Information Processing: Implications for Management Information Systems," Decision Sciences, Vol. 3 (1972), 35-95.

29. Moskowitz, M. and J. Miller, "Information and Decision Systems for Production Planning," Management Science, Vol. 22, No. 3 (1975), 359-370.

30. __________, R. E. Schaefer, and K. Borcherding, "Irrationality of Managerial Judgments: Implications for Information Systems," Omega, Vol. 4, No. 2 (1976), 125-140.

31. Payne, J. W., "Task Complexity and Contingent Processing in Decision Making: An Information Search and Protocol Analysis," Organization Behavior and Human Performance, Vol. 16 (1976), 366-387.

32. Remus, W., "Bias and Variance in Bowman's Managerial Coefficient Theory," Omega, Vol. 5, No. 3 (1977), 349-351.

33. __________, P. Carter, and L. Jenicke, "Regression Models of Decision Rules in Unstable Environments," Journal of Business Research, Vol. 7, No. 2 (1979), 187-196.

34. Shaklee, H. and B. Fischhoff, "Strategies of Information Search in Causal Analysis," Decision Research Report 79-1 (Eugene, Oregon - Unpublished, 1979).

35. Slovic, P., "Value as a Determiner of Subjective Probability," Transactions of the Institute of Electronic Engineers: Human Factors Issue, HFE-7 (1966b), 22-28.

36. __________ and S. Lichtenstein, "Comparison of Bayesian and Regression Approaches to the Study of Information Processing in Judgment," Organization Behavior and Human Performance, Vol. 6 (1971), 649-744.

37. __________, and D. MacPhillamy, "Dimensional Commensurability and Cue Utilization in Comparative Judgment," Organizational Behavior and Human Performance, Vol. 11 (1974), 172-194.

38. Smedslund, J., "The Concept of correlation in Adults," Scandinavian Journal of Psychology, Vol. 4 (1963), 165-173.

39. Timmers, H. and W. A. Wagenaar, "Inverse Statistics and Misperception of Exponential Growth," Perception and Psychophysics, Vol. 21, No. 6 (1977), 558-562.

40. Tversky, A. and D. Kahneman, "The Belief in the 'Law of Small Numbers'," Psychological Bulletin, Vol. 76 (1971), 105-110.

41. __________, "Availability: A Heuristic for Judging Frequency and Probability," Cognitive Psychology, Vol. 5 (1973), 207-232.

42. __________, "Judgment Under Uncertainty: Heuristics and Biases," Science, Vol. 185 (1974), 1124-1131.

43. Wagenaar, W. A. and S. D. Sagaria, "Misperception of Exponential Growth," Perception and Psychophysics, Vol. 18, No. 6 (1975), 416-422.

44. __________ and H. Timmers, "Extrapolation of Exponential Time Series is Not Enhanced by Having More Data Points," Perception and Psychophysics, Vol. 24, No. 2 (1978), 182-184.

45. Ward, W. C. and H. M. Jenkins, "The Display of Information and the Judgment of Contingency," Canadian Journal of Psychology, Vol. 19 (1965), 231-241.

46. Wason, P. C. and P. N. Johnson-Laird, Psychology of Reasoning: Structure and Content (Cambridge, Massachusetts: Harvard University Press) 1972.

CULTURAL AND CROSS-CULTURAL FACTORS

Human Factors in Organizational Design and Management
H.W. Hendrick and O. Brown, Jr. (Editors)

WITH THE TREND OF THE TIME

Sadao Sugiyama

Faculty of Sociology
Kwansei Gakuin University
Nishinomiya, Hyogo
Japan

It is needless to say that, whether we like it or not, all of us, ergonomists and human factors engineers, are exposed to some influence or other of our academic environment which is crammed with various unfamiliar disciplines, scientific knowledges, theories and methods which are at variance with one another.
The fact is, we, as scientists and engineers, tend to believe almost unconditionally our scientific knowledge as one accepted worldwidely, because this was given by Dr. X who is an expert specialist, even if it is an unfamiliar discipline for us.
Our attitudes toward scientific knowledge is neither more nor less than a feeling closely akin to the one toward an absolute truth if there is such a thing at all. Because of the fact that we tend to feel like this, we can easily fall into belief what I believe is the fact that should be accepted by others. Is ergonomics like an absolute thing? Aren't there some other disciplines which help human welfare? Do we know enough about what other scientists are doing for human welfare? Those questions recur to my mind.

Since International Ergonomics Association was established in 1949, after many turns and twists, the present IEA came to exist. Naturally any organization like IEA should experience countless turns and twists during the course of its development. As I recall the early days of IEA, our activities were, so to speak, like an international extention of domestic activities. It was, I suspect, because of the fact that ergonomists in the past wished to establish a concrete framework for ergonomic sciences in the broad range of the scientific world and that they believed what they could do for human society was absolutely necessary for human welfare.
During the course of our development, ergonomics has exchanged experiences, knowledges and techniques with other scientific disciplines and, as a result, the range of our scopes and objects have been widened and our techniques have been highly sophisticated.
However, more importantly our social objects have been and will be more highly sophisticating as the progress of our society and our social objects to be solved are taking up many divergences.

In the past, if you remember, ergonomists attacked various problems, such as equipment design, physical and mental work load and fatigue, design of work environment, man-machine interface, man-computer interface, etc. Those topics were, of course, necessary ones to be solved in order to maintain human welfare and thus were presented to us by the human society. Those trends show that social demands have been always presented under certain social circumstances. This brings to my mind that our activities have been just like those of a medical doctor who operates on his patients. In our case, patients are our society, the indication of illness is the social demands and we ergonomists perform the role of a medical doctor to cure illness.

However, I think, still several other aspects are necessary to cure our patients. They are, for example, about self-curative properties which human organism originally has in order to maintain his health and also about necessary health guidance by the doctor.
In our case, matters related with the result of self-curative properties might be shown by the fact that the society never complains again about the same demand which was presented long time ago, as the society forgets about it. Matters related with health guidance by a doctor will be interpreted as follows: in order to accomplish it, any individual must realize by himself what his physical ability was in the past and is at present, also in what situation he was involved and how he has been behaving in the situation. Knowledges about health of his body may provide an answer about what to do by himself, if he is properly guided by an able doctor. In our case, we must provide a prescription or a guideline of how society can maintain its health condition from the viewpoint of human beings living in such a rapidly changing technologically oriented society. In this connection, I think ergonomics must perform as an advisor for healthy industrial society.

I suspect that our industrial society will continue almost forever to present us numerous social problems depending upon every social circumstances. I believe that these problems are waiting for our solution in the future. If we consider that this ever-changing human society is the source of problems,we must provide a series of bright ideas one after another as the needs of the case demand.

What is needed for the industrial society might be wisdom for living and thus resourceful men who can solve a series of problems with a series of bright ideas. In addition to the fact that those abilities are hard to be cultivated by mere training, we must prepare for establishing the roles and systems in the organization for making use of those man power. In this sense, an organization or the society is not like mere hardware systems which can easily be designed and operated just like a machine.
In the long history of mankind, it seems to me that man had lost some of important abilities instantaneous with the time when man acquired technical skills. Among them is the wisdom for living and instinct to cooperate with others. For example, an animal is equipped with fangs and claws which function as weapon, if necessary, to protect itself from a sudden attack

made by other species. However,such a weapon is dangerous for an animal of the same kind; thus, for example, a monkey controls his behavior by stopping to fight against another monkey if the latter turns his back to him. The human race has already lost such an inherited pattern of behavior. In this sense, modern man, that is ourselves, is considered as an animal without instinct to protect his fellow human species. However, a human being is equipped with rational faculty instead of instinct.
If you look back more closely to the history of the human race, you may notice the fact that human wisdom, not instinct, came to attach much importance to the question of survival, and then that scientific knowledge has taken place of a human wisdom.

In order to get over the critical situation induced by the ever-changing human society, I think we should learn more about the history of the human race and make more proper use of knowledges about a human being from wider points of view. Of course, our lost instinct never return, but I think at least human wisdom should possibly be cultivated, as we systematically produce scientific knowledges and make use of them. I believe that making use of human scientific knowledges by human wisdom is most significant for the future development of human organization, because almost all of the components in the organization is human.
To design such an organization is most difficult, if you compare it with any other hardware design. It is not mere combination of different kinds of human roles in an organizational system. Organization, after it is established, has to do with political circumstances, economical situations and scientific and technological advancement which are always floating. It is said that there is no axiom which can be applied to all the time in such a complicated situation. Therefore, I think planning of human organization has to be focused on how to maintain softness and flexibility of the behavior of an organization so that it can cope with a kaleidoscopic change of society, because we know, from our daily experiences, almost any organization would become rigid after some time.
If you think this way, you may realize that the design of an organization is almost identical with the design of an organic existence such as human body. Despite this difficulty, modern industrial society in any countries requests us to solve the problem.

It comes to my mind that either scientists or engineers would be banished by sciences and technology themselves, if they stop to pursue the eternal goals of sciences and technology which are basically for human welfare requested by the ever-changing human society. And I also think that they are destined to be banished, if they failed to be with this ever-changing society and the age.

Human Factors in Organizational Design and Management
H.W. Hendrick and O. Brown, Jr. (Editors)

ORGANISATION TRANSFER TOWARDS INDUSTRIALLY DEVELOPING COUNTRIES

Alain Wisner

Laboratoire d'Ergonomie
Conservatoire National des Arts et Metiers
41 rue Gay-Lussac
75005 PARIS
FRANCE

When a transfer of technology has been made to an industrially developing country, disapointing results have often been noted with regard to working conditions, production, and financial returns in the company.
These negative effects can be forestalled by different means, particularly by having a proper transfer of work organisation. If it is to be successful, such transfer must be exhaustive, effective and properly adapted. 5 studies undertaken in Brazil, the Philippines, the Central African Republic and in Tunisia have provided examples of this in the fields of maintenence and communications within the company. Increased efforts need to be made in order for organisation transfers to be successful. Ergonomic work analysis will enable the research for efficient solutions to be properly directed.

INTRODUCTION

In spite of present difficulties in international trade and in spite of the tendency to repatriate towards industrially developed countries, the sales of machinery and production systems has continued in considerable volume, and will undoubtedly increase strongly when the world economy has an upswing. It is, however fairly clear that the selling countries will not just be the old industrially developed nations like the United States, Japan, Western Europe, U.S.S.R.; but the newly industrialized countries as well: one tenth of India's export product is machinery and the increasing proportion of countries like Brazil or South Korea in the world engineering market is well known.

With such conditions, buyer countries are and will increasingly be in a position to choose. Such a choice will be obviously very much governed by the political and economic situation of the buyer country as well as by the financial conditions of the contract. Nevertheless, the quality of the product itself is being increasingly taken into account because of the frequent failures of industrial transfer. Too many transferred factories have had to be either shut down or else are operating in disastrous conditions.

UNDESIRABLE EFFECTS OF TECHNOLOGICAL TRANSFER

These aspects have already been related elsewhere (A. Wisner, 1982, 1984). However it is possible to identify negative effects in the area of health:

-Large amount of accidents at work,
-Greater frequency and more categories of work-related illnesses than in the country of origin,
-Specific development-related pathologies (for instance, spread of parasitic diseases because of increased amount of stagnant water used for irrigation, psychopathology linked with urban life in shantytowns etc...). All of which justify a "development hygiene".

Other negative aspects can be described from the production viewpoint (Krishna, 1980).
-Reduced volume of production due to low operation rate of machines. -Low quality of products being produced which limits them to national markets which then require protection from outside.
-Deterioration of production material because it has been put to bad use. It is known that on an average, one tenth of such incidents involve a worker, thus becoming accidents.

In terms of the financial aspects, such poor production results can have a disastrous effect:

-The company becomes unable to provide an adequate level of salary, social benefits and proper working conditions to their staff.
-The government of the industrially developing country is unable to obtain the financial returns it was counting on and may even have to go looking for further investment in order to maintain the plant's activity, thus involving further debts and interest payments. This will then increase their

dependency on funding sources and/or force them to increase the requirements demanded of the agricultural workers who are producing most of the exportable goods.
-It becomes impossible to repay bank loans, and the banks then become unable to make further loans which are needed for development.
It can thus be seen that breakdowns in technological transfer provoke very serious effects in many highly important areas.

. SUCCESSES OF INTEGRAL TRANSFER : ANTHROPOTECHNOLOGICAL ISLANDS

One can only be surprised at the fact that most of the aforementioned difficulties hardly ever or never arise when the particular conditions of "total transfer" are applied in certain cases by multinational or transnational companies. Since 1975 we have been using the term "anthropotechnological islands" to describe these highly interesting situations.

These establishments produce similar technical, human and financial results in industrially developing countries and industrially developed countries. This generally involves companies which are selling a uniform product throughout the world and who therefore must arrive at an interchangable quality level in all production centres. In order to make the same product, these companies have not only only transferred the same technical infrastructure and machinery they had elsewhere, but the same work organisation and training system as well. Nevertheless, since this proved to be insufficient, these multinational companies also applied very severe criteria in their hiring policy, and provided housing, transport, and even schools and hospitals, following the example of some European and American companies in the 18th and 19th centuries.

In this way anthropotechnological islands have grown up where the industrial entity is so close to that of the country of origin that one finds the same pathological symptoms (for example, nervous breakdowns in the electronics industry) as well as the same advantages (low accident rate, low staff turnover, low absenteeism etc). The term of islands is properly used here, because the workers in these companies are in the position of being deeply cut off from the kind of life their compatriots are leading although they are still living geographically in their own country.

This extreme situation is both theoretically and practically very important, and it can help us to analyse the usual technological transfer situations in the national enterprises of the buying country.

EQUIVALENCE OF STAFF COGNITIVE CAPACITY AND DIVERSITY OF INDUSTRIAL RESULTS

The first basic consequence to be derived from the aformentioned facts is that there exists no difference in basic cognitive capacity between men belonging to different peoples and cultures because these "anthropotechnological islands" have been set up with equal success in the most widely differing countries. Recent research by K. Meckassoua (1984) has shown that a Central African citizen who has spent his whole childhood and adolescence in a village far from any modern technical civilization is quite able to elaborate a highly complex operational image without adequate training if his job is to monitor a very complicated production unit (beer bottling production line including labling and packing into crates). By means of this complex image, the operator is able to undertake a true teaching process of this in progressive learning stages.

These are facts that are by no means new in terms of basic neuropsychology, but since the way this is expressed sometimes takes on a philosophical or even ideological mode of communication, it may be a good thing to provide proof of this in an albeit modest industrial domain.

The question that we wish to deal with is therefore the following: why will workers from the same country, the same town and the same population give perfectly acceptable results in an "anthropotechnological island" and unsatisfactory results in a national factory? The usual reply to this is in socio-cultural terms. This doesn't take into account any failures of organisational transfer and it leaves scant place for ergonomics, particularly the most modern ergonomics of cognitive and organisational activities. This approach is best exemplified in the book by A. Chapanis (1975) entitled "Ethnic Variables in Human Factor Engineering". In it, for instance, can be found a text by C.N. Daftuar on the squatting position in India, a study by C.H. Wyndham about the effect of food scarcity and parasite-related diseases on the working capacity of South African miners and a text by P. Verhagen on the stereotypes of visual movements. Although these biological and socio-cultural aspects are important, they are not relevant to the present text. This also holds true for the capital question of technological choice, or even for the need to industrialize the Third World, because these problems have already had a considerable amount of literature devoted to them, albeit more political and ideological in nature than purely scientific writing (Schumacher, Miske, 1981; Lenoir, 1984; etc.).

INDUSTRIAL SOCIETIES AND WORK ORGANISATION

Ever since Emery and Thorsrud's work on sociotechnics in 1964, it is known that the state of a society has to be taken into account in order to design a production system (Davis and Taylor 1972) and that solutions which may be valid at a given time in an industrially developed country may turn out to be inadequate at a later period. In this way taylorism corresponded to a certain state of technology and human resources. We feel that one can extend this reasoning geographically to consider that the production system will, of course, have to take into account the technology being transferred, but also the local human resources available. However, there is another important element that has to be taken into consideration: the density of the industrial network. When judging the success and productivity of large factories in industrially developing countries, what very often happens is that one underestimates the importance of the surrounding industrial context, i.e.the many small and medium-sized businesses that can provide specialized material, and what is even more important, that can provide highly qualified staff if difficulties arise, as well as the fact that outlets for foreign-made measuring and control apparatus will be nearby, and spare parts will be delivered quickly. A breakdown which could be repaired in a day in Paris or New York will require 2 or 3 days in a small town in France or in the USA, 2 or 3 weeks in Algiers or Manila, 2 or 3 months in Africa below the Sahara or in some regions of the Andes, and this simply because of the differences in industrial density. In order to afford protection against such difficulties, two types of solution have been designed:

-Large stockpiles of spare parts have been set up. This can represent up to ten times more spares for an industrially developing country than for an industrially developed one, and the organisation of extensive maintenance departments in the companies too. This kind of organisation requires large outlays of immobilized funds, and very often foreign technicians will be needed, who will also be very expensive.

-A subsidiary of the selling company is set up alongside of the buying company and it is then given a lucrative exclusive maintenance contract for the plant which has just been sold (this is very frequently practised for computers or automatic systems manufacturing). The outcome of this will therefore be a durable situation of technological subordination, i.e. an institutionalized incomplete transfer with the know-how still in the hands of the seller. It is unfortunate that all too often in transfer contracts the question of maintenance will frequently not be taken into consideration. The results will then be highly negative for the buyer, whose factory will deteriorate very quickly, determining high human and financial costs.

ERGONOMIC WORK ANALYSIS AND MODES OF ORGANISATIONAL TRANSFER

5 theses have now been or are due for completion in our laboratory's anthropothechnological group. All have been undertaken in their native countries by research workers who have come for ergonomic training in Paris. In all cases, a methodology was adopted with the objective of finding out why analagous technical infrastructures gave different results in the country of transfer from those observed in the country of origin. This method of comparison was used in the phosphate mines at Gafsa, in Tunisia, and in coal mines in France and Belgium by N. Sahbi; in the control rooms of the Paris metro and the Rio de Janeiro subway by N. dos Santos; for a beer-bottling plant at a brewery in Bangui, Central African Republic, and at Armentieres, France, by K. Meckassoua; for a telephone exchange in Manila, the Philippines, and in Munich, Federal Republic of Germany, by C. Rubio; and in the case of Brazilian sugar-cane alcohol distilleries, the comparison was made by J. Abrahao between the model plant at Ribeiro- Preto in the state of Sao Paulo and other distilleries working in a less industrialized state of Brazil, the state of Goias.

The fact that these studies were undertaken by research workers originating from the buying countries enabled us to go very far in studying the effects of work organisation anomalies on the operators` modes of action. A knowledge of the language and its attendant cultural references is indispensable for any success in analyzing the cognitive activities of the operators in ergonomic terms.

The anomalies in organisational transfer can be classified according to three main categories: incomplete, imperfect and inadequate transfer, and this for the main areas of organisational transfer. We wll be studying two of these areas : maintenance organisation and communications.

MAINTENANCE ORGANISATION

Incomplete Transfer

In spite of the fact that the concept of maintenance is an old idea and has been formalised for a long time (Goldman, Slattery, 1964) and in spite of the participation of Human Factor specialists (Cunningham, Cox, 1972), maintenance organisation is very often left right out of the transfer as if a maintenance market simply didn't exist. The buyer appears to consider that this is a secondary problem which will be overcome with a little ingenious handywork and the seller will frequently not even have a systematic description of the maintenance organisation that will have to be transferred. Even in very large-scale enterprises, it can often be seen that maintenance will often be the responsibility of relatively autonomous professional groups. An exception to this rule is the safety and security provisions in air control, nuclear energy plants or data processing equipment. In the latter case there will often be a subsidiary of the selling company dealing with all maintenance (Manila telephone exchange, C. Rubio).

In most of the transferred technical units, the absence of any accompanying maintenance transfer has resulted in a relatively rapid degradation of the initial unit. This degradation is sometimes attributed to the poor quality of the staff or to the cultural habits in the buying country : such explanations are very often inaccurate and, in any case, cannot be used as a basis for proposing improvements.

Imperfect Transfer

There are other cases in which the maintenance organisation is given an imperfect transfer. Generally speaking, the imperfection will be in the transfer language. In this way the maintenance system for hydraulic pit-props for Tunisia was transferred with an instruction booklet and parts list written in German, when only French and Arabic are spoken in Gafsa.

Quite recently a large company in an industrially developed country sold a consignment of diesel engines to an industrially developing country for a hundred million dollars. A good translation of the forty-page instruction handbook was provided to the buyers in their own language, which was portuguese. But the maintenance handbook was 800 pages long and wasn't translated! This probably represented a savings of 100,000 dollars at most out of an overall sum which would have been a thousand times greater.

Sometimes imperfect transfer takes a more subtle form : a translation will be made into the language of the buyer country, but it will be of poor quality. This applies to nearly all the translations into French of instructions for electronic and data processing material originating in South East Asia. And yet with regard to a well-documented historical situation, that of repairing American military equipment by Vietnamese workers (Sinaiko and colls., 1972), a number of fundamentally important facts were revealed : there was a significant relationship between the number and degree of importance of worker errors and the quality of the maintenance handbook's translation into vietnamese. If the translation was a good one, there was little difference between american and vietnamese workers. If the translation was poor, it was better for the vietnamese to use a handbook in english even if they didn't know the language very well. If the translation was really bad, the result was an absolute catastrophe. These differences became more accentuated as the technical text became more difficult. Because of the cost of a translation by a technical expert, H.W. Sinaiko studied the effect of a translation by computer. The result was only acceptable if the computerized translation was revised by an expert, but in any case it was very much inferior to a direct translation by an expert. And again, these differences increased with the level of difficulty of the texts involved.

A countless number of machines are sold with dials, labels and operating instructions drafted in either an unknown language or else in an absolutely incomprehensible translation. It is often alleged that the workers, or their supervisors or at least their trainers, have a working knowledge of the seller country's language. This assumption is generally mistaken when one is considering the degree of in-depth understanding of technical data. It is no longer possible to consider that there is a french-speaking and english-speaking Africa, that all the Americas south of the Rio Grande are either spanish or portuguese-speaking, or that southern and south-eastern Asia is english-speaking. In many cases the knowledge of the main vehicular languages is highly superficial, and the workers as well as many others in the company will go on with their reasoning in their vernacular language, which is the only one they can really understand properly. Nevertheless, these kinds of problems can sometimes be highly complex: some berber-speaking algerians will understand french better than arabic, there are filipinos that know english better than tagalog. In India, english was maintained as the official language because although hindi was the most important vernacular language, it was inacceptable for many citizens who spoke tamil, urdu or bengali.

The same kind of comments can be made about graphic representations. In the book by A. Chapanis, C.H. Wyndham and H.W. Sinaiko have documented examples of graphic misunderstandings, covering the area of work previously treated by W. Hudson.

Inadequate Transfer

Operational difficulties that require close attention to maintenance are not the same everywhere. For instance there can be other sources of differences, such as: climate, state of the roads, variation in the electricity voltage.

-Hot and humid climates which are very frequent in industrially developing countries will entail considerable problems in corrosion, fluidity of lubricants, and degree of stickiness in bonding materials, which means that monitoring operations must be undertaken more often on an increased list of indications, and this can lead to setting up maintenance stategies that will be quite different to those practised in the countries of origin of the technology.

-The very poor condition of roads and other transport surfaces mean that there will often be a fast and highly specific deterioration of rolling stock. Ways and means of monitoring this need to be set up as well as repair facilities that are properly adapted to the situation.

-Extensive and sudden variations in the grid voltage level will often provoke considerable and complex ajustment changes in automatic systems. In such cases, overcoming these problem incidents will have to be done more often and the incidents themselves may be more serious.

An accurate approach to the inadequate characteristics of the maintenance organisation being transferred will explain some of the breakdowns that have been observed in buyer countries. It can be seen that the diagnosis of breakdowns will often be more complex, and that the different pattern of problem incidents will mean that the spares requirement will be different to the one needed in the selling country, and that because of these situations, the maintenance department will often be forced to adopt "wire and hairpin" solutions for its repairs, which in turn will provoke further incidents.

It is indispensable to undertake an ergonomic analysis of maintenance activities in the plant being transferred, and this will lead to the formulation of very accurate recommendations for completing and improving the efficiency of the organisational transfer.

ORGANISATION OF COMMUNICATIONS

As in the case of maintenance, the organisation of communications is a highly important aspect of the satisfactory operation of technical systems. Because its characteristics are non-material and because it involves exchanges having cultural connotations, its transfer is even more neglected and difficult than maintenance.

Incomplete Transfer

This characteristic of incomplete transfer seem to be quite prevalent. An example of this can be seen in the case of the hydraulic pit-props used in phosphate mines. Here it would seem that the transfer of the pit-props and of the maintenance workshop was done purely technically without anything being done about transferring the organisation. This led to the maintenance workers not repairing the pit-prop handles and not putting the base coverings back on the bottom since they didn't know how important these things were for the miners. They would limit themselves to repairing or replacing the input valves, which was quite insufficient. Thus the existence of a communication system between the pithead and the workshop is a highly important aspect of the work organisation and it is often very much neglected when modernizing a mine (Sahbi, 1984).

The incomplete nature of a communications transfer is often due to the fact that the situation is really different in the buying country and that this is not perceived in the selling country. For instance at a brewery, the operator in charge of drawing the beer has a most important function, since in fact he is governing the whole adjustment of the bottling line. In Bangui, this worker has to take into account three more elements of information than his counterpart in Armentieres: the bottles in Bangui are unequal in size, the labels don't always stick properly onto the bottles because the paste used is poorly adapted to the high surrounding temperature, and there are periodical backups in the packing area because the bottles are put into cartons by hand in Bangui whereas this is done automatically in Armentieres. If he is to regulate the whole system successfully, the operator has to set up an original communications network, since the problems he has to deal with in association with the other workers hardly ever exist in Armentieres.

Imperfect Transfer

Once again, language problems arise when dealing with the characteristic of imperfect transfer of communications. Hence there is a communications scenario in case of incidents on the Rio de Janeiro subway lines which really corresponds to the communications history on the Paris metro rather than to the realities of the situation in Rio de Janeiro.

In fact the basic question is that of the relationship between the work that has been prescribed by the methods department and the work that really has to be done to make the system function. The difference between the prescribed and real work already exists in the repetitive actions of man production tasks, and this in a system where everything has been done to bring the two together. The more the work becomes complex, the more it becomes neccessary for the operator to be able to conceive a personal image of his unit's operation, in order to be able to follow up the relevant indications and to take the proper decisions for the ajustment of the unit and to overcome any problem incidents.

What is most often transferred is the prescribed work, since it is the only one the engineers know about and the only one they can communicate to their counterpart engineers when the technological and organisational transfer is taking place. Now, it is well known that in most cases the formal statements defining control and action procedures will not alone enable an efficient operation to be implemented within the required time limitations. If nothing is done to communicate the way the real work is done, there will be great difficulties when the time comes to start the transferred unit up. Sometimes these difficulties turn out to be insurmountable and this explains why some costly plants have had to be shut down permanently. In other cases, the operators in the buyer country have set up their own efficient communications network, although this can only happen when the formal operating instructions don't prevent them from doing so. It can be considered that the quality circle movement that originated in Japan should help people to realize that is is neccessary to take the real and efficient communication networks into account when transferring technology.

Inadequate Transfer

Organisational transfer is very often inadequate insofar, as in the above cases, such a transfer requires that the reality of the technology and the local situation both be taken into account. An extreme example can be given of an automatic control unit for a cigarette-packing system that didn't work. The organisation for communication between the operators was still transferred to the buyer country. Such a case as this may appear to be a caricature, and yet it is unfortunately not an isolated instance. One can give examples of the innumerable non-operational dials in control rooms which go on flashing completely meaningless signals and are no more than a visual noise in the background. In the most modern control rooms, one now finds symbolic representations of circuits that simply don't correspond to any real indication, and which only serve to introduce errors since the operators have not yet learned to ignore them.

FINAL COMMENTS

Research under way by the anthropotechnological team at the Ergonomics Laboratory of the C.N.A.M. shows that over and above the serious problems of transferring hardware, there is a huge area of organisational transfer in which many mistakes and problems arise, and these to a certain extent explain the operational breakdowns in the systems that have been transferred. The inherent interest which cannot be denied for the ergonomics of organisational transfer, is that such an analysis can lead towards operational solutions rather than towards sterile and unjustified considerations about the quality of the labour or the technical culture of the buyer country.

REFERENCES

-CHAPANIS A., (1975) Ethnic Variables in Human Factors Engineering.
John Hopkins University Press, Baltimore.

-CUNNINGHAM C.E., COX W., (1972) Applied Maitainability Engineering.
Wiley, New York.

-DAVIS L.E., TAYLOR J.C. (1972) Design of Jobs. Penguin, London.

-EMERY F.E., THORSRUD E. (1964) Form and Content in Industrial Democracy. Oslo University Press, Oslo.

-GOLDMAN A.S., SLATTERY T.B. (1964) Maintainability. Wiley, New York.

-KRISHNA R. (1980), The Economic Development of India. Scientific American 243 3 p. 118 - 133.

- LENOIR R. (1984) Le tiers-monde peut se nourrir. Fayard, Paris.

-MECKASSOUA K. (1984) Etude comparee des activites de regulation dans le cadre d'un transfert de technologie. These Ergonomie, C.N.A.M. Paris.

-MISKE A.B., (1981) Lettre ouverte aux elites du tiers-monde. Sycomore, Paris.

-SAHBI N., (1984) Anthropometric Measurement and Work Analysis of Modern Technology in the Tunisian Phosphate Mines, in: SHANAVAZ H., BABRI M., Ergonomics in Developing Countries. C.E.D.C. Lulea University Press, Sweden.

-SCHUMACHER E.F. (1973) Small is Beautiful. Blond and Briggs, London.

-SINAIKO H.W. (1975) Verbal Factors in Human Engineering : Some Cultural and Psychological Data, in: CHAPANIS A., Ethnic Variables in Human Factors Engineering. John Hopkins University Press, Baltimore.

-WISNER A., (1982) Ergonomics, Mental Load, Anthropotechnology. Conservatoire National des Arts et Metiers, Paris. 192 p.

-WISNER A. (1984) Ergonomics or Anthropotechnology : A Limited or Broad Approach in Technology Transfer, in: SHAHNAVAZ H., BABRI H., Ergonomics in Developing Countries. C.E.D.C. Lulea University Press, Sweden.

Human Factors in Organizational Design and Management
H.W. Hendrick and O. Brown, Jr. (Editors)

A CROSS-CULTURAL STUDY OF INTERGROUP CONFLICTS IN INDUSTRY AND ITS IMPLICATIONS FOR JOB DESIGN AND QUALITY OF WORKING LIFE

M.R. ALI, R. HUMBULO
Department of Psychology, University of Zambia, Lusaka, Zambia,

A. KHALEQUE and A. RAHMAN
Department of Psychology, University of Dhaka, Dhaka-2, Bangladesh

The present study was designed to identify: the areas and the causes of intergroup conflicts in industry, and the relative effectiveness of conflict resolution techniques. The study was conducted in Bangladesh on 87 managers and 60 workers, and in Zambia on 25 managers and 30 workers. A questionnaire consisting of 3 sub-scales: the areas and the causes of conflicts, and techniques for resolving conflict was used. The subjects of both the countries considered the basic needs as the import areas; political pressure, regionalism, communication gap as the important causes; and the third party approach, and participatory leadership as the effective techniques for resolving conflicts.

INTRODUCTION

Research literature on industrial conflict shows that most of the studies were carried out in the Western developed countries (Coser, 1956; Boulding 1962). But the industrial scene in the developing countries of the Third World is not free from it. Industrial conflict is rather one of the major causes of low productivity in the Third World countries. A cross-cultural study was jointly undertaken during 1980-81 by a group of industrial psychologists at the University of Dhaka, Bangladesh and at the University of Zambia to investigate the management and workers' perception of the areas and causes of conflict, and the effectiveness of the techniques used in resolving the industrial conflicts in these two countries.

This was an exploratory study and we did not have any specific hypothesis to test. Nevertheless, our aims were to identify: (1) the areas of frequent industrial conflict; (2) the important causes of industrial conflict; and (3) the effectiveness of the techniques used in resolving industrial conflict as perceived by the workers and managers in these two countries.

METHOD

Research setting and sample

As already mentioned, this study was carried out in Bangladesh and Zambia. The Bangladesh sample consisted 87 managerial employees and 60 workers selected randomly from a jute mill. The Zambian sample consisted of 25 managers and 30 workers selected randomly from five industries.

Measuring instruments

A questionnaire consisting of 62 items was used to measure the managements' perception of the areas and the causes of conflicts and the effectiveness of the techniques used for resolving such conflicts. The questionnaire consisted of three sub-scales:

(1) The sub-scale "areas of conflict", consisted of 22 items, dealing with the possible areas of conflict.

(2) The sub-scale "causes of conflict," consisted of 27 items, relating to the possible causes of conflict.

(3) The sub-scale "techniques of resolving conflict" consisted of 13 items describing the possible techniques for resolving conflict.

Every item of each of the three sub-scales could be replied by checking either "Yes" or "No" response to indicate whether that item describes a particular area or a cause of conflict, or a technique for resolving conflict. The respondent would indicate the frequency of occurrence of conflict in an area on a 5-point scale, ranging from least frequent (1) to most frequent (5); the importance of a cause of conflict on a 7-point scale, ranging from least important (1) to most important (7); and the effectiveness of a technique in resolving conflict also on a 7-point scale, ranging from least effective (1) to most effective (7).

The questionnaire was developed partially on the basis of items taken from the relevant literature on industrial conflict (Stagner, 1956; Blake et al., 1964; McCormic and Tiffin, 1974) and partially on the basis of a pilot survey of opinions of the workers and management concerning different aspects of industrial conflict. Before final selection of the items, the questionnaire was pretested with a number of managerial employees and workers. Those questions which appeared difficult and confusing to them were dropped. This questionnaire before using in Zambia was modified and adapted to suit the condition in Zambia.

RESULTS

The results of the present study have been presented in the following tables (Bangladesh data: Table 1, and Zambian data: Table 2).

Table 1. Mean ranks and the rank orders of the areas of conflict (in terms of frequency), causes of conflict (in terms of importance), and the techniques for resolving conflict (in terms of effectiveness) as perceived by the managers and workers of Bangladesh.

Items of the three subscales	Managers (N=87)		Workers (N=60)	
1) Area of conflict	Mean ranks	Rank orders	Mean ranks	Rank orders
Allotment of house	2.89	1	3.03	1
Waste	2.49	2	2.79	2
Absenteeism	2.44	3	2.38	4
Discipline	2.35	4	2.15	10
Promotion	2.17	5	2.11	11
Maintenance	2.09	6	2.20	8.5
Purchase and ware housing	2.07	7	2.39	3
Financial matters	2.00	8	2.31	5
Selection of employees	1.96	9	2.20	8.5
Salary and increment	1.84	10	2.23	7
Training	1.81	11.5	2.03	13
Accident prevention	1.81	11.5	1.74	20
Fulfilment of production target	1.79	13	1.78	19
Labour welfare	1.78	14	2.24	6
Quality control	1.73	15	1.93	15
Fringe benefits	1.72	16	2.09	12
Bonus	1.60	17.5	1.88	17
Compensation	1.60	17.5	1.89	16
Developmental planning	1.57	19	1.82	18
Sales	1.50	20	1.97	14
Recreational facilities	1.47	21	1.68	22
Assignment of shift duty	1.37	22	1.74	21

Spearman's rank correlation (rho): p = .80 (Significant at .01 level).

2) Causes of conflict	Mean ranks	Rank orders	Mean ranks	Rank orders
Basic needs (food and shelter)	5.34	1	5.55	1
Nepotism	5.04	2	5.02	3
Rumour	4.81	3	5.21	2
Regionalism	4.78	4	4.46	5.5
Centralisation of power	4.76	5	4.19	13
Job-security	4.60	6	4.72	4
Misunderstanding	4.50	7	4.46	5.5
Political pressure	4.49	8	3.68	21
Personality clash	4.30	9	4.21	11.5
Leadership style	4.26	10	3.70	20
Managements policy	4.15	11	3.63	22
Non-satisfaction of status need	4.07	12.5	3.98	16
Lack of effective communication	4.07	12.5	4.30	8
Sense of deprivation	4.06	14	4.24	10
Feeling of injustice	4.04	15	3.94	17
Need for autonomy	4.00	16	3.56	24
Excessive competition	3.92	17.5	4.21	11.5
Power inequality	3.92	17.5	4.04	15
Lack of fellow feeling	3.84	19	4.36	7
Personal anxiety & mental pressure	3.78	20	4.05	14

Table 1 (continued)

Items of the subscales Causes of conflict	Managers (N=87) Mean ranks	Managers (N=87) Rank orders	Workers (N=60) Mean ranks	Workers (N=60) Rank orders
Not getting desired job	3.77	21	3.43	26
"We-they feeling"	3.75	22	4.27	9
Unrealistic production target	3.70	23	3.81	18
Ignorance about pay rules	3.66	24	3.59	23
Ignorance about human relations	3.65	25	3.04	27
Undefined jurisdiction	3.64	26	3.78	19
Job dissatisfaction	3.59	27	3.48	25

Spearman's rank correlation (rho): p = .67 (significant at .01 level).

3) Techniques for resolving conflict

Problem solving technique	4.59	1.5	4.29	6
Effective leadership	4.59	1.5	4.42	3
Third party approach	4.57	3	4.70	1
Participation in decision making	4.55	4	4.57	2
Compromise	4.48	5	4.38	4
Decentralisation of power	4.12	6	3.87	8
Clinical approach	4.05	7	4.37	5
Peaceful co-existance	4.04	8	3.92	7
Bargaining	3.76	9	2.83	9
Divide and rule policy	2.68	10	2.35	11
Confrontation	2.34	11	2.17	12
Coercive method	2.25	12	2.16	13
Avoidance or withdrawal	2.19	13	2.60	10

Spearman's rank correlation (rho): p = .86 (significant at .01 level).

Table 2. Mean ranks and rank orders of the areas of conflict (in terms of the frequency of conflict), the causes of conflict (in terms of importance), and the techniques for resolving conflict (in terms of effectiveness) as perceived by the managers and workers of Zambia.

Items of the three subscales 1) Areas of conflict	Managers (N=25) Mean ranks	Managers (N=25) Rank orders	Workers (N=30) Mean ranks	Workers (N=30) Rank orders
Accommodation	2.76	1	3.53	1
Wage or salary increment	2.72	2	3.27	2
Promotion	2.57	3	2.50	3
Finance and accounting	2.56	4	2.43	4
Selection of personnel	2.15	5	1.63	5
Execution of production planning	1.68	6	1.13	9
Absenteeism	1.64	7	1.43	7
Discipline	1.56	8.5	1.50	6
Assignment of shift work	1.56	8.5	1.30	8

Spearman's rank correlation (rho): p = .87 (significant at .01 level).

Table 2 (continued)

Items of the subscales	Managers (N=25)		Workers (N=30)	
2) Causes of conflict	Mean ranks	Rank orders	Mean ranks	Rank orders
Communication breakdown	2.64	1	2.93	1
Political pressure	2.40	2	1.67	6
Tribalism	2.04	3	2.10	3
Misunderstanding	1.96	4	1.60	7
Misuse or damage to property	1.91	5	2.33	2
Poor supervision	1.60	6	1.97	4
Negligence of duty	1.52	7	1.53	8
Lack of recreational facilities	1.48	8	1.43	9
Excessive competition	1.44	9	1.68	5

Spearman's rank correlation (rho): p = .53 (not significant)

3) Techniques for resolving conflict				
Third party approach	2.61	1	2.70	1
Problem solving technique	2.60	2	2.60	2
Effective leadership	2.36	3	2.20	4
Splitting the differences	2.24	4	2.30	3
Smoothing-over technique	2.16	5	2.10	5
Coercive method	1.80	6	1.70	6
Participatory decision making	1.60	7	1.60	7

Spearman's rank correlation (rho): p = .96 (significant at .01 level).

DISCUSSION

After analysing the results of the present study, we have found significant rank order correlations between ratings of the managers and the workers of Bangladesh about the perceived importance of some specific areas of conflict, causes of conflict and techniques for resolving conflicts (see Table 1). Similar results have been obtained from the Zambian samples, excepting that there is no significant correlation between ratings of the managers and workers concerning the importance of some specific causes of conflict (see Table 2).

Thus the results show that although there are some differences in the perception of intergroup conflict situations by the workers and managers of Bangladesh and Zambia, yet there are good number of general agreements within and between the samples of the two countries.

The results show that both the managers and workers of Bangladesh and Zambia perceive allotment of houses or accommodation problem as an area where conflict occurs most frequently and assignment of shift duty as an area where conflict occurs least frequently. However, the managers and the workers in Bangladesh reported that waste, absenteeism and discipline are also important areas of conflict but the workers and the managers in Zambia considered wage, promotion, finance and accounting as some other important areas of conflict.

The managers and the workers of Bangladesh consider unfulfilment of the basic needs, nepotism, rumours and regionalism as important causes of conflict. Whereas, the managers and workers of Zambia think that communication breakdown, tribalism as important causes of conflict. However, the managers in Zambia consider political pressure as the second important cause of conflict but the workers from that country consider misuse of or damage to property as the second important cause of conflict.

Third party approach, effective leadership and problems solving techniques were mentioned by the samples in both the countries as effective techniques for resolving intergroup conflicts in industry. Participation in decision making was also considered as another important technique for preventing industrial conflict by the Bangladesh respondents. However, it is surprising to note that neither the managers nor the workers of Zambia (where the government has introduced industrial participatory democracy) consider participation in decision making as an effective preventive measure for dealing with industrial conflict.

The results of the present study indicate that wage, dwelling, supervision, leadership style and communication patterns are considered by the workers and the managers as important sources of industrial conflict. So any job design approach should take these factors into account. Whether a job is horizontally or vertically loaded, it should bring more satisfaction to the basic needs. For example, if more tasks are added to the present job, it should bring more wage. Similarly, additional responsibilities should bring additional financial benefits and better accommodation.

The results also show that nepotism, regionalism and trabalism are considered by the managers and workers in both the countries as important causes of industrial conflict. In order to alleviate the feelings of nepotism, regionalism and tribalism, effective selection and promotion procedures should be developed to ensure fair practices in these matters.

The results also indicate that unlike the Western countries, the quality of working life in the Third World is more linked with the satisfaction of the basic needs. So any attempt to improve the quality of working life in these countries must ensure that the basic human needs, such as food and shelter, are fulfilled.

REFERENCES

1. Blake, R.R. and Mouton, J.S., The Managerial Grid (Gulf, Houston, Texas, 1964).
2. Boulding, K.E., Conflict and Defense (Harper and Brothers, New York, 1962).
3. Coser, L.A., The Functions of Social Conflict (Free Press, Glencoe, Illinois, 1956).
4. McCormic, E.J. and Tiffin, J., Industrial Psychology (George Allen and Unwin, London, 1979).
5. Stagner, R., Psychology of Industrial Conflict (John Wiley, New York, 1956).

Human Factors in Organizational Design and Management
H.W. Hendrick and O. Brown, Jr. (Editors)

PRODUCTIVITY AND JAPANESE MANAGEMENT PRACTICES IN AN AMERICAN SUBSIDIARY

Ogden Brown, Jr.
University of Southern California
Los Angeles, California, USA

Subsidiaries of Japanese organizations operating in the United States and employing Japanese management systems and techniques appear to be successful in terms of productivity and worker involvement. One such company, Kyocera International, Inc., is examined in this case study and the management systems and techniques are discussed. It is concluded that the concern for human assets is both practical and desirable in the United States as well as in Japan.

Productivity, here defined as output per worker, is a current topic and one of national concern. When it rises, the economy grows as do living standards. When it goes down, growth slows or becomes dormant. Productivity itself results from many variables such as employee motivation, improvements in technology, capital investment, and a myriad of other factors. During the 1970s, 19 countries surpassed the United States average annual productivity growth rate (less than 2.5%). Japan was the leader with an annual increase of nearly 10%, and worker output per hour was one of the world's highest (Business Week, 1980). It is held by some that rather than cultural or personal factors which are related to performance, the principal reason for this growth in productivity may be the superiority of the management systems and practices in many Japanese companies (Hatvany and Pucik, 1981: Hayes and Abernathy, 1980).

In this context, there are subsidiaries of Japanese firms operating in the United States and Europe. These organizations typically are staffed by a small number of Japanese managers and enjoy various levels of autonomy with respect to the parent firm. Even though the prevailing culture may differ, many of these subsidiaries are quite successful. In point of fact, some even enjoy the same rate of productivity as their parent firms in Japan (Hatvany and Pucik, 1981). It would thus appear that a closer examination of the management systems and techniques in an American subsidiary might prove useful in identifying those systems and techniques which seem to be related to productivity and worker involvement.

The Kyocera Corporation was founded in 1959 by Kazuo Inamori with $10,000 in borrowed capital and a line of credit of $100,000. Total assets today are about $830 million and shareholders equity is about $675 million. Net world wide sales in fiscal year 1983 were about $723 million and net income was $87 million (Kyocera Corporation Annual Report, 1983). Projected sales for 1984 are $1 billion.

The principal product is high technology ceramics for semiconductor packaging (about 70% of the world's market), but the company has embarked on a large scale expansion and diversification program (Interview, 1984). Other products now include industrial ceramics, recrystallized gemstones and jewelry, bio-materials such as dental implants and artificial bones and joints, solar systems, and electronic equipment such as portable computers, plain paper copy machines, CB transceivers, and cordless telephones. In 1983, Kyocera acquired the Yashica, Co. Ltd. to add another dimension to the diversification. This year Kyocera successfully built and tested the world's first ceramic automobile engine (Interview, 1984).

The wholly owned American subsidiary, Kyocera International, Inc. (KII), was established in 1969 and corporate headquarters are in San Diego, CA. Assets of KII are about $120 million. Net sales for fiscal year 1983 were about $256 million and net profit was $184 million (Kyocera Corporation Annual Report, 1983). There are sales offices located in nine states as well as three main manufacturing plants in San Diego.

As is the case in most Japanese companies, KII considers its human assets to be the most profitable and most important in the long term. Accordingly, secure and stable employment is a management practice. Over 300 workers who would have been terminated by a typical American company were retained by KII during the recent flat semiconductor market in 1981-1982 (Interview, 1984). Research reflects that commitment to the organization is positively related to long tenure which, in turn, may reduce turnover (less than 2% at KII). Security and secure employment appears to be the foundation for the use of other management strategies.

The motto or ideals of KII are expressed in a calligraphy of four Chinese letters which says: "Respect the Divine and Love People". Workers adhere to Inamori's sayings such as the "spirit, philosophy, vision, humanism, and management policy" that he has created and which are known as "Inamorisms". No task is too menial, and the firm has a "job description covering all jobs" (Kyocera Guidelines, 1978). Respecting the divine infers that some things happen which are beyond one's control and workers must accept these things (such as changes in the economy). Respecting people (love is a literal translation) is discussed and lectured about on a daily basis.

Workers arrive at 7 a.m. five days a week for a five to ten minute meeting held outdoors in the parking lot. The speech of the day may be given by a young supervisor or a high-level manager and may cover anything from an appeal to adhere to the spirit of Kyocera to safety matters to profit and loss reports. Employees then engage in a designated exercise program which is followed by a short group or departmental meeting (also held outdoors). All of these activities are done on company time and are a part of the job.

At KII, everyone wears a light blue or yellow windbreaker-style jacket all day long. The philosophy here is that all people are exactly alike, so if everyone looks alike then they will feel alike (Kyocera Guidelines, 1978). People who apply for employment are evaluated not only on skills, but also on enthusiasm and a positive mental attitude. The company philosophy and policies are also explained, and applicants are told that if they cannot or will not accept them, KII is not the place for them.

Kyocera's Guidelines consist of four pages of company policy, vision, some humanism and philosophy, "Inamorisms", and a two page description of the "Kyocera Spirit". The Kyocera Spirit stresses the following attitudes:

- o Loyalty: Faithful to the Kyocera cause, ideal, philosophy, and customers.
- o Dedication: A devoting or setting aside for Kyocera purposes.
- o Zeal: Eagerness and ardent interest in the pursuit of Kyocera goals.
- o Spirit: Full of energy, animation, and courage.
- o Management by Consensus/Cooperation: Group solidarity in sentiment and belief; a judgment arrived at by most of those concerned. (Kyocera Spirit, 1978)

Pay is a secondary matter even though KII attempts to pay prevailing wages for the area. The person who is accepted at Kyocera has to prove himself through hard work and dedication to the company. In return, the company offers secure employment and will take care of the worker. Their policy is not to seek out the "superstar", but rather people who will accept and work under a team approach. The consensus management practiced at KII is not necessarily democratic in nature, however. Each manager is responsible for decisions, but he also has the responsibility to listen to all viewpoints before reaching decisions. Once the decision has been made, the manager is then responsible for convincing those people on his team that he reached the right decision.

Each manager is in charge of a self-sufficient entity which produces an "amoeba" effect and approach within the firm, where each unit attempts to be an independent profit center. Upper management monitors and controls the various units, and they must consider Kyocera a "24 hour job" which might even have to come before one's family (Interview, 1984).

The belief that the company is there to support people leads to the belief that people are the only real asset in an organization. Equipment and other machinery may be bought and sold, but people are the prime asset in Kyocera's judgment. It also follows that in order to support people the company needs to realize a profit. It would appear that Kyocera is meeting that goal, for international net sales rose 9.8% and net income rose 38% in fiscal year 1983 as compared with the previous year.

At KII the Vice-President level of management is about equally divided between Japanese and American managers. In contrast, middle management is composed of engineers who are mostly from Japan. In the early years of the American operation, those who came from Japan were told they would not be rotated back to Japan but should integrate with the Americans. This policy has changed somewhat and some of the engineers and managers will be returning to Japan when the company so desires (Interview, 1984).

People do not have private offices, and in the KII operation of consensus management work is done in an open office because everyone is there to work for the company and thus privacy is not necessary. A conference room is available for personal matters and confidential company information. Each person is responsible for generating profit to eliminate layoffs. KII has an honor system for people who have not been late and who have done well so that no timecard is used by these individuals; this reinforces the notion of treating people with respect and dignity. The company pays a monthly full attendance bonus, birthday bonus, and has a percentage-of-pay bonus plan for performance. The company sponsors picnics, a Christmas party, and family recreational activities. There is an open door policy, a "family atmosphere" based on mutual respect. Employees may talk to executives on both business and personal matters.

Kyocera has a formula to measure achievement or performance:

Performance = Ability x Desire to Succeed x Method of Thinking

(Kyocera Guidelines, 1978). The desire to succeed is held to be the most important factor. Enthusiasm is energy and managers should have a desire to excel. One "Inamorism" holds that"management is not the result of what is achieved, but rather the realization of what has been desired". Another says "our hearts are bound together as colleagues with mutual trust, and we spare no efforts to accomplish our goals and help others". A final one, and indicative of Inamori's beliefs: "Our goal is to strive toward both the material and spiritual fulfillment of all employees in the Company, and through this successful fulfillment, serve mankind in its progress and prosperity". These beliefs, coupled with hard work, appear to be at least partially responsible for the company's success record. Indeed, if such practices and beliefs can be internalized by each employee, both worker involvement and productivity should be the result.

In summary, people in every culture value security, equitable treatment, and the opportunity to meet their needs in the work place. It would seem that Japanese managership has created organizational systems which do respond to these values and needs. As Hatvany and Pucik (1981) have noted, Japanese management practices are "highly congruent with the way in which tasks are structured, with individual members' goals, and with the organization's climate". This fit thus should result in a high productivity rate and increased organizational effectiveness. As Hatvany and Pucik (1981) conclude, and as this case study has supported, the key features of Japanese management practices can successfully be transferred to other places and other cultures: they are not unique. The concern for human assets is both practical and desirable in the United States as well as in Japan, and this fact can lead to both productivity and worker motivation.

REFERENCES:

(1) "The Decline of U.S. Industry", Business Week, June 30, 1980, p. 59.
(2) Hatvany, N. and Pucik, V., Japanese Management Practices and Productivity, Organizational Dynamics, Spring 1981.
(3) Hayes, R. and Abernathy, W., Managing Our Way to Economic Decline, Harvard Business Review, July-August 1980.
(4) Kyocera Corporation Annual Report, 1983 (Kyocera Corporation, 1983).
(5) Interview with Gene Lyons, Personnel Manager, San Diego Plant, Kyocera International, Inc., April 19, 1984.
(6) Kyocera Guidelines and Spirit (Kyocera Corporation, 1978).

Human Factors in Organizational Design and Management
H.W. Hendrick and O. Brown, Jr. (Editors)

DESIGNING AND IMPLEMENTING AN ERGONOMICS INVENTORY TO IMPROVE MANAGEMENT OF HUMAN FACTORS PROGRAMS

Kenneth A. Johnson

Cooperative Education Coordinator
Honolulu Community College
Honolulu, Hawaii
U.S.A.

The subject of human factors in organizational design and management is multidimensional and requires collecting and interpreting information from a variety of sources in order to maximize human efficiency. This paper analyzes the self report inventory in a dual role of providing valuable information to employers and as an intervention to improve employee attributes. Factors that affect performance are identified and translated into questionnaire items. A specific inventory is then constructed and a brief discussion of administration is included.

INTRODUCTION

Ergonomics is a science which studies the natural laws of work in order to maximize human efficiency in job performance. Areas of focus include people, the work they do, the things they work with, and the environment of the workplace. Through analysis of the interaction between areas of focus, people can be selected and trained to fit some jobs while other jobs can be modified to fit the people who are available to work. The information necessary to facilitate this analytical process can be acquired from a variety of sources including statistical statements, direct observation, and from the people themselves. One method of gathering information from people is called the self report inventory. Self report inventories seek to obtain precise and dependable information by placing the individual in a standardized situation to which he/she must respond.

Cronbach (1960) describes the various advantages and disadvantages involved in use of the self report inventory. Since each employee has had many opportunities to observe himself/herself in a variety of situations, a valuable report of typical behavior can be given. Not only can past experiences and unobservable aspects of a subjects life be shared, but far more incidents can be included than an observer could record and some of the difficulties of field observation can be avoided. However, if the self report is to be representative of typical behavior, responses to the questionnaire must be accurate and honest. Unfortunately, people tend to present a public self concept that is socially acceptable rather than one that is totally accurate. Even when the individual attempts to be truthful, one cannot assume that he/she is a completely accurate, detached and impartial observer. Other disadvantages involved in the use of self report inventories are that the items can be ambiguous, that subjects might interpret questions differently due to individual self concept or situational experiences, that individual response styles (Lorge, 1937) and interpretation of adverbs used for answers vary from subject to (Simpson, 1944), and that the intent of the tester may vary from the intent of the subject (Edwards, 1957).

Distortion in self report can be overcome to a certain extent, although it cannot be eliminated completely. The test and testing situation can be designed in such a way as to promote an atmosphere of cooperation rather than one of evaluation. The terms "questionnaire" or "inventory" should be used instead of "test". If the questionnaire is administered by an independent consultant, subjects can be assured that the results will not be self incriminating. The consultant could make a detailed report to the employer recommending improvements. Concordantly, during the testing session every attempt must be made to establish rapport by creating a friendly and relaxed atmosphere. Subjects will be more receptive and willing to participate if they can understand that the results will be used to help them perform their duties more efficiently, and, in fact, could actually make their job easier. It is also preferable to design test items in such a way so that possible responses range on a numeric continuum (seven point scale) between extremes rather than by merely using ambiguous adverbs (i.e., sometimes or often, like or dislike, etc.). In addition, the self report inventory sheet should be free of possibly incriminating details (name, age, etc.) since the top page could be seen by unqualified bystanders or other employees. If the front cover includes only the name and basic concept of the inventory, then all personal and specific information can be contained between the covers to insure confidentiality.

When using a self report inventory, one must realize that the information obtained is merely a statement of the subject's public self concept. Ambiguity of test items and inevitablie distortion in self observation reduce the correspondence between responses and actual behavior. Therefore, information obtained from a self report inventory can only be considered as one part of a complex whole and not as an entity in itself. However, in terms of an intervention designed to improve employee attributes, valuable information can be passed on to the subjects provided that at the end of the inventory there is a section detailing implications of questionnaire items. This section can be given to the subjects for later reference.

The entire process would also be facilitated by a review session following test administration where all items and related issues could be discussed. The appropriateness of test items in relation to factors that affect performance is a critical issue involved in determining the effectiveness of the inventory.

FACTORS THAT AFFECT PERFORMANCE

The interraction between job factors contributes to employee health, safety, wellbeing and productivity. For any specific situation, the employee, the tools, the task or job content, and the work station and environment each have specific factors which can be identified and translated into questionnaire items. If employees can gain significant understanding regarding the interraction between factors, then quality of experience and quantity of performance can be improved.

No two individuals are exactly alike and no two actions are performed in exactly the same manner. Consequently, when analyzing the relationship between employee attributes and performance of a task one must consider similarities and differences within specific areas. For the purposes of this paper, discussion of employee attributes will be limited to the areas of 1) physical characteristics, 2) attitudes and beliefs, and 3) skills and behavior.

Physical charactristics include structural (standing and sitting height, etc.) and dynamic dimensions (sideways and forward reach, etc.), sensory ability (sight, hearing, touch, smell and taste), age, sex, physical strength, posture, and whether one is right-handed, left handed or ambidextrous to name a few. The workspace design should be based on body dimensions. Since people vary greatly from the "average dimensions", the workplace needs to be adjustable. Questions concerning physical characteristics can provide information that can be used to evaluate work stations, and such items can increase employee awareness of the relationship between physical characteristics and workplace design (i.e., seating, physical dimensions of workspace, design of controls and displays, etc.).

Some of the implications of physical characteristics can be realized by analyzing statistics from the U.S. Department of Labor. Length of service and age affect the rate of injury reported in workman's compensation claims. Individuals who have been employed one year or less have twice the probability of being injured on the job as people with longer tenure. Employees between the age of 25 and 34 years of age also have a greater incidence of reported injury on the job (followed by those under 25 years, then gradually decreasing rates for individuals over 35 years old). The most frequently reported types of injuries involve situations where workers were struck by something, followed by overexertion and being struck against something. Highest reported nature of injury or illness involves sprains and strains (25%), cuts, lacerations and punctures (24%), and various other sources (i.e., contusions, scratches, burns, fractures and multiple injuries). These statistics support the creation and implementation of safety education programs. However, no program is effective unless the workers have suportive attitudes, beliefs and skills.

According to Barry Schlenker (1980) the impression people create is affected by attitudes and beliefs. Attitudes describe an individual's feelings about self or the world, and beliefs describe one's thoughts. Since birth, we have accumulated attitudes and beliefs (through social interaction) about who we are, what we can and can't do, what's right and wrong, etc. These feelings and thoughts can be self supportive or destructive, appropriate or inappropriate, and may require behavior change in order to improve performance. Techniques of behavior self modification involve: selecting a problem behavior then analyzing the antecedent (set of conditions occuring before), the behavior itself and the consequences of that behavior (Watson and Tharp, 1981). In the work environment, attitudes and beliefs affect performance. By analyzing one's thoughts and feelings toward self, job duties, tools and the workplace environment (including other people) the antecedents of performance can be discovered. One's ability to control thoughts and feelings can be viewed as a skill that can be developed (similar to job skills that improve with information, motivation and practice).

Personal characteristics, as well as various dimensions of the task, tools and workplace all affect the extent of human efficiency. Information from various sources can be compared with the results of the self report inventory in order to gain an accurate and realistic perspective.

DESIGNING THE INVENTORY

Questionnaire items have been selected in two general areas: Part I includes personal characteristics and Part II includes tasks, tools and workspace. An asterisk at the end of an item indicates that there is a footnote for that statement on the last page. By reviewing the footnotes individually as well as in the group session following administration of the questionnaire, subjects can gain a better understanding of ergonomic principles. Evaluation of the results would involve reviewing subject responses in comparison to work conditions, available training, and future performance. Because the major focus of this discussion involves construction of an ergonomics inventory, designed primarily for its value as an intervention rather than as an evaluation tool, the remainder of this paper will be devoted to the actual inventory.

INSTRUCTION PAGE

Date of Employment________________

Name______________________________ Height_______Weight_______ Sex____

Ethnicity___________________Date of Birth________Married: Yes__ No__

Right Handed__ Left Handed__ Ambidextrous__ Job Title________________

Name and Age of Dependent Children__________________________________

Disability that Affects Work (Describe)____________________________

____

____

In this inventory, you will find questions about you and your job. This information will help you gain a better understanding of the interaction between personal characteristics, job content, tools and the work station environment. Please read carefully and answer every item as accurately as possible.

Start with the two simple examples below, for practice. As you see, each inquiry is actually put in the form of a sentence. Indicate how each statement applies to you by circling one of the numbers on the seven point scale located on the right side of the page. MARK NOW.

	Rarely						Always
1. I enjoy my job	1	2	3	4	5	6	7
2. I have enough energy to do what I want	1	2	3	4	5	6	7

Now: 1. Make sure you have filled in all the blanks at the top.
2. Answer all items. Your responses will be entirely confidential.

Most people finish in five minutes, some in ten. After completing all items, take a few minutes to review the FOOTNOTES on the last page.

TURN THE PAGE AND BEGIN

PART I - PERSONAL CHARACTERISTICS	Rarely						Always
1. I feel tired or rundown (except after strenuous work)*	1	2	3	4	5	6	7
2. I can talk to family or friends about important matters	1	2	3	4	5	6	7
3. I experience aches and pains during or after work	1	2	3	4	5	6	7
4. I enjoy my work	1	2	3	4	5	6	7
5. I can say "no" without guilt	1	2	3	4	5	6	7
6. My supervisor listens to my comments and suggestions*	1	2	3	4	5	6	7
7. I meditate or center myself for 15-30 mins. at least once a day* . . .	1	2	3	4	5	6	7
8. I use a shoulder/lap seat belt while riding in a motor vehicle*	1	2	3	4	5	6	7
9. I smoke cigarettes	1	2	3	4	5	6	7
10. I do a vigorous type of exercise (like swimming, brisk walking, etc.) for 15-30 mins. at least 3 times a wk*	1	2	3	4	5	6	7
11. I do stretching exercises (like yoga or calisthenics) that enhance my muscle tone for 15-30 mins. at least 3 times a wk*	1	2	3	4	5	6	7
12. I eat a balanced diet every day (protein, legumes, seeds & nuts, whole grains, fresh fruits & vegetables)*	1	2	3	4	5	6	7
13. I don't deny my feelings of anger, joy, fear or sadness*	1	2	3	4	5	6	7
14. I have an avocational activity outside of work that I enjoy	1	2	3	4	5	6	7
15. I keep informed of local, national and world news	1	2	3	4	5	6	7
16. I enjoy spending some time without planned or structured activities* . .	1	2	3	4	5	6	7
17. I look forward to being at least 75* .	1	2	3	4	5	6	7
18. I limit the amount of salt, sugar and animal fat that I eat*	1	2	3	4	5	6	7
19. I limit the amount of alcohol and caffeine that I drink	1	2	3	4	5	6	7
20. I have less than 4 colds or other minor illness per year	1	2	3	4	5	6	7
21. I have a good appetite and maintain a weight within 15% of my ideal weight	1	2	3	4	5	6	7
22. My job is stressful	1	2	3	4	5	6	7
23. Men and women are better off when employed in jobs suitable for their social role*	1	2	3	4	5	6	7
24. I am able to "turn off" thoughts that make me feel bad*	1	2	3	4	5	6	7
25. I am able to relax whenever I want* .	1	2	3	4	5	6	7

PART II - TASKS, TOOLS AND WORKSPACE	Rarely						Always
1. Every day, I stand still in one place at work*	1	2	3	4	5	6	7
2. Throughout the day, I sit down to work*	1	2	3	4	5	6	7
3. On the job, I'm required to lift or move heavy objects*	1	2	3	4	5	6	7
4. My job duties are repetitive*	1	2	3	4	5	6	7
5. I'm required to keep pace with electronic or mechanical devices* . .	1	2	3	4	5	6	7
6. My job requires lots of responsibility and little recognition*	1	2	3	4	5	6	7
7. I get along well with people at work	1	2	3	4	5	6	7
8. I use tools with short handles that press into the palm of my hand* . . .	1	2	3	4	5	6	7
9. I use tools that weigh more than 30 pounds*	1	2	3	4	5	6	7
10. The controls I use are easy to reach and within my normal field of vision*	1	2	3	4	5	6	7
11. I receive adequate training to do the work required of me	1	2	3	4	5	6	7
12. I'm able to maintain a normal, comfortable body temperature at work .	1	2	3	4	5	6	7
13. The lighting at work is good (not too bright, or too dim, no glare, etc.) .	1	2	3	4	5	6	7
14. It is too noisy to concentrate at work (sharp, loud, or continuous) . .	1	2	3	4	5	6	7
15. I am informed of all potential hazards, and provided with necessary information/equipment to reduce them .	1	2	3	4	5	6	7
16. I smell gaseous fumes at work	1	2	3	4	5	6	7
17. I have been injured on the job	1	2	3	4	5	6	7
18. I fit comfortably in my workspace . .	1	2	3	4	5	6	7
19. The visual displays that I work with are easy to see without straining . .	1	2	3	4	5	6	7
20. The seating is adjustable*	1	2	3	4	5	6	7
21. I am required to hold a posture for long periods of time*	1	2	3	4	5	6	7
22. My job requires me to do a lot of reaching (above, below, sideways or behind)	1	2	3	4	5	6	7
23. The tools I use are balanced and easy to hold in one hand (if not mounted) .	1	2	3	4	5	6	7
24. I use tools that weigh between 5-30 pounds*	1	2	3	4	5	6	7
25. I understand exactly what is required of me at work, and how that fits into the entire operation of the company* .	1	2	3	4	5	6	7
26. I feel that no matter how well I perform, I could always do better* . .	1	2	3	4	5	6	7

FOOTNOTES

PART I - PERSONAL CHARACTERISTICS

1. Fatigue without apparent cause is not a normal condition and usually indicates illness, stress or denial of emotional expression.(Travis)
6. Intrinsic factors such as challenge and considerate, thoughtful employers affect job satisfaction. (Work in America)
7. Meditation or centering greatly enhances well-being. (Naranjo, Ornstein, McCamy, Pearce, Abrams & Siegel, Bloomfield & Kory, Vattano, Benson)
8. Shoulder/lap belts are much safer than lap belts alone.
10. Vigorous aerobic exercise that keeps the heart rate at 150 beats per minute for 15-30 minutes produces enormous physical and psychological benefits. (Bloomfield and Kory)
11. Such exercise produce a greater sense of well-being while preventing stiffness of joints and musculo-skeletal degeneration. (Mishra)
12. A daily balanced diet provides the necessary vitamins and minerals necessary to maintain health and energy. (Bloomfield and Kory)
13. Repressing emotions can cause disease, anxiety, depression and irrational behavior which interferes with performance. (Travis)
16. Spending some time without structured activities can be self-renewing. (Travis)
17. With proper care, most can reach this age in good health.
18. Salt added while cooking draws many vitamins out of the food and it is also harmful to the kidneys and contributes to high blood pressure. Sugar provides only "empty" calories and contributes to arteriosclerosis. Consumption of animal fat increases cholesterol in the system. (Kirschmann)
24. Social role stereotyping can limit an individual's choices regarding employment decisions and can be a source of stress. (Brislin, Schlenker, Goleman, Argyris, Carter and Patterson)
25. Overall performance can be improved through acquisition of self controlled behavior change techniques. (Watson and Tharp)
26. Systematic relaxation techniques have enormous physical and psychological benefits. (Selye, Jacobson)

PART II - TASKS, TOOLS AND WORKSPACE

1,2,20,21. Holding a posture for an extended period of time can cause aches and pains. The workspace should be adjustable with armrests, foot pads, etc., to reduce physical strain. (Morris, Burdick, Webb Singleton, IAPA)

3,9,24,25. Care should be used whenever one is required to do prolonged or heavy lifting. Twisting can cause strain, as can lifting from an unbalanced posture. Expectations or stereotypic reactions can be deceiving (i.e., very large objects or dark objects imply heaviness; heavy objects are expected to be at the bottom, etc.). (Woodson)

25. Employees who understand their job and the way their work fits into the company as a whole are generally more effective, less stressed and more successful. (Drucker, Ordione, Ouchi, Tarrant)

26. Type "A" personalities who continuously pressure themselves to achieve greater goals have higher rates of heart attack. However, Type "A" personalities who exercise regularly and see themselves as being in control have a lower rate of heart disease. (Bloomfield and Kory)

References

Abrams, Allan I., and Larry M. Siegel. "The Transcendental Meditation Program and Rehabilitation of Folsom State Prison." Criminal Justice and Behavior, 5(1) March 1978, pp. 3-19.

Argyris, Chris. "Theories of Action that Inhibit Individual Learning." American Psychologist, September 1976, pp. 638-654.

Benson, Herbert, and M.Z. Clipper. The Relaxation Response. New York: Avon Books, 1976.

Bloomfield, Harold H., and Robert B. Kory. The Holistic Way to Health and Happiness. New York: Simon & Schuster, 1971.

Brislin, Richard. "Culture and Background Experience." Unpublished study. Honolulu: East West Center, 1984.

Burdick, W.E. Ergonomics Handbook. New York: International Business Machines, Inhouse Publication.

Carter, Bruce, and Charlotte J. Patterson. "Sex Roles as Social Conventions: The Development of Children's Conceptions of Sex-Role Stereotypes." Developmental Psychology 1982, Vol. 18, No. 6, 812-824.

Cronbach, Lee. Essentials of Psychological Testing. 2nd ed. New York: Harper and Row Brothers, 1960.

Edwards, Allen L. The socially desirability variable in personality research. New York: Dryden, 1957.

Goleman, Daniel. "Special Abilities of the Sexes: Do they Begin in the Brain?" Psychology Today, November 1978, pp. 48-223.

Jacobson, E. Progressive Relaxation. Chicago: University of Chicago Press, 1983.

Kirschmann, John D., Director. Nutrition Almanac. Revised Edition. McGraw-Hill, 1973.

Lorge, Irving. "Gen-like: halo or reality." Psychology Bulletin, 1937, 34, pp. 545-546.

McCammy, J.C., and J. Presley. Human Life Styling. New York: Harper and Row, 1975.

Mishra, R.S. Fundamentals of Yoga. New York: Anchor Press, Doubleday, 1974.

Morris, Julius. Ergonomics Script. Unpublished. Honolulu, 1984.

Naranjo, Claudio, and Robert E. Ornstein. On The Psychology of Meditation. New York: Viking Press, 1971.

Ordione, George S. Management by Objectives. New York: Pitman Publishing Corp., 1965.

Ornstein, R.E. The Psychology of Consciousness. San Francisco: W.H. Freeman, 1972.

Ouchi, W. Theory Z, Reading, Addison Wesley, 1981.

Pearce, Joseph C. The Crack in the Cosmic Egg. New York: Washington Square Press, 1971.

Schlenker, Barry R. Impression Management. California: Brooks/Cole, 1980. pp. 201-226.

Selye, Hans. Stress Without Distress. Philadelphia: J.B. Lippencott, 1974.

Simpson, Ray N. "The specific meanings of certain terms indicating different degrees of frequency." Quarterly Journal of Speech, 1944, 30, pp. 328-330.

Singleton, W.T. Introduction to Ergonomic. Geneva: World Health Organization, 1972.

Special Task Force, Secretary of Health, Education and Welfare, Work in America. Cambridge Mass: MIT Press, 1971.

Tarrant, John T. Drucker. New York: Cahners Books, Inc., 1976.

Travis, J.W., M.D. Wellness Inventory. Mill Valley, California: Wellness Resource Center, 1977.

Vattano, Anthony. "Self Management Procedures for Coping with Stress." Social Work, March 1978, v. 203, N2, pp. 113-118.

Watson, David L., and Roland G. Tharp. Self-Directed Behavior, 3rd ed. California: Brooks/Cole, 1981.

Webb, R.D.G. Industrial Ergonomics. Ontario: Industrial Accident Prevention Association, 1982.

Woodson, W.E. Human Factors Design Handbook. New York: McGraw-Hill Book Company, 1981.

ENVIRONMENTAL AND OFFICE DESIGN

Human Factors in Organizational Design and Management
H.W. Hendrick and O. Brown, Jr. (Editors)

SHIFTING FROM TRADITIONAL TO OPEN OFFICES: PROBLEMS SUGGEST DESIGN PRINCIPLES

Harvey Wichman

Aviation Psychology Laboratory
Claremont McKenna College
Claremont, California
U.S.A.

This paper reports the lessons learned in two studies of office design and one of informal dormitory innovations made by students. One of the office design studies followed the move of a small (100 employees) high-tech service company employing white-collar workers from a traditional office building to an open plan office building nearby. The other study was an evaluation of an open plan office that had been in operation for one and one-half years. This office system, which occupied the entire top floor of a building was compared with a traditional office system in the same organization. Certain problems appeared to be characteristic of open plan offices (e.g. visual and auditory distractions) and were often difficult to deal with or not dealt with at all. This paper provides a set of guiding principles for dealing with such problems and gives illustrations of the application of the principles.

This paper draws upon the results of two studies of open plan offices and one study of student-created innovations in college dormitory rooms. Collectively, the lessons learned from these studies suggest several general considerations that would be relevant in designing open plan office systems or when moving from a traditional to an open plan system.

The first study to be considered was conducted at the Jet Propulsion Laboratory (JPL) in Pasadena, California (Bayley, Green, King, Lemmex, and Rogers, 1978). This was part of a self-initiated study by JPL to evaluate an open plan office that had been in operation for one and one-half years. The office system consisted of the entire top floor of a building. The group housed there had previously been scattered in several different locations and was consolidated in one area for the first time when moved to the open plan office.

The study attempted to identify and quantify behaviors that took place in the setting by using time-sampled observations in a variety of locations. In addition, questionnaires and interviews were utilized to obtain measures of the perceived importance of various elements of the environment and to measure satisfaction with the work environment.

In general, the results showed that the supportiveness of the individual open plan offices was rated as significantly ($p < .05$) less than that of the traditional offices. In addition, the supportiveness of the overall floor environment was rated significantly lower for the open plan offices. The major problems with the open plan offices revolved around lack of visual privacy, lack of auditory privacy, and the high ambient noise level. In all other respects, traditional and open plan offices were rated about the same in terms of their providing facilitative work environments.

The open plan offices were seen as facilitating a greater degree of interaction and brief consultations. Work-life in them was perceived as more spontaneous and informal. Responses were strongly affected by the type of office one had previously occupied. Persons who had worked in shared offices (such as secretaries) adjusted much better to the open plan offices than did those who were accustomed to private offices. Intragroup communication was enhanced and the climate of the place was pleasantly informal. The summary conclusion of the study was that the open plan offices were acceptable (the group was pleased to be all together in one place) but with some seriously irritating drawbacks.

The study just referred to examined an open plan office that had been in operation for one and one-half years and compared it to a traditional office system in the same location and organization. The second study examined the effects of a company's move from a modern but traditional office building to an ultramodern landscaped office building (Hanson, 1978). The company was a small (100 employees) high-tech service company employing white-collar workers. The company built the new building on a plot of land only about 300 yards from the old one, so the locations were essentially the same. Measurements similar to those in the JPL study were made twice before the move and twice afterwards over a period of a year. Ninety-six employees participated in the study. The leaders of this young corporation felt a great need to increase communications within the company both horizontally and vertically. Indeed, at the time that a decision was made to construct a new building the principal investigator in this study was working as a management consultant to the company, attempting to improve communications, esprit de corps, and a sense of community of purpose among the staff. The top executives of the company hoped that a building could be designed that would help to further develop this family-like sense of community, shared ideas, and enthusiasm. An architect was hired and a more spacious three-story building was designed in a landscaped format with open offices, a large atrium with birch trees and beautiful wood framing huge windows. The building, which was quite dramatic looking and beautiful on the inside, would provide more space, improve the company image to all who visited it, and help create the warm, friendly, communicative atmosphere that the executives felt the company needed if it were to thrive. The architect was frequently heard to say "physical openness leads to psychological openness," and so he designed the new building with vast vistas outside and no hidden areas inside. The old building had all of the executive offices along a hallway in one area known as **executive row.** In the new building, executives were distributed throughout the structure to bring them into more and closer contact with the rank-and-file in the company. The idea was to improve executive visibility and to discourage any cliquish tendencies fostered by close proximity to one another. As one walked about in the old building it was obvious that the company was getting too big for it. Things had a somewhat crowded look, with files and notebook binders stacked up on shelves, in corners, and on top of file cabinets. As construction on the new building neared completion, excitement mounted and everyone looked forward to moving into their beautiful and more spacious new quarters.

Six months after the move to the new building it was clear that the dream was dashed. In fact, the move was almost a corporate catastrophe calling for immediate and dramatic changes. The architect was bewildered and angry at the behavior of his clients. A beautiful conference room was never used, but spontaneous conferences were held on a landing of the stairway, which promptly disrupted traffic. Some people complained that their "cubicles" were too confining and others complained that they had insufficient privacy. The beautiful atrium was resented as squandered space by some who felt they were still as crowded as before. The beauty of the wood paneling was lost on those disappointed that they now couldn't pin up personal items.

> The physical changes concomitant to the disruption were the wide physical dispersion throughout the new setting, the introduction of the desk-top organizers, and the 163 centimeters high panels that provided some seclusion for executives, supervisors and some technical positions. Perceptions of the setting, work habits, and general behaviors were affected but the actual work and social relationships remained fairly stable. In order to accomplish this, people found themselves making conscious efforts to restore the original patterns of face-to-face contacts that were habitual in the old setting.
>
> The perceptions generated by the rating scales showed that employees did not see the new setting as more convenient or functional and that it appeared no less impractical than the original one. The behaviors necessary to accommodate to the changed surroundings no doubt were the basis of these perceptions. People reported traveling much more to see people and get information, making an effort to maintain face-to-face contact, and having to organize their trips in order not to waste work time. The isolation from work associates was revealed in the open-ended questions after the move in such statements as "We have lost the close family-like feeling."
>
> At the executive level the physical closeness experienced in the original setting contributed to effective decision-making because of the immediacy of face-to-face exchange of views. Comments on the difficulty of getting top executives together quickly and changes in the process of group decision-making were common in post-move interviews and questionnaires. The efficient flow of information among those at this level was perceived as having decreased with the move. At clerical levels, the desk top organizers separated people visually and disrupted cross-desk communication which was characteristic of the former setting and had added in a natural manner to the efficiency of work flow and information exchange. The change was reported by both clerical and supervisory people.
>
> Participants' comments on behavioral changes, and observations made by the researcher, indicated another effect of the new layout upon former interaction patterns that had grown out of the informal organizational climate characteristic of the centralized management format in the original setting. In that building an unbroken travelpath through the executive wing allowed people from all levels to pass by the offices and observe if an executive was free to accept "drop in" calls. These visits were a way of keeping the newer and usually younger members of the firm informed of current activities and allowed them to learn "the business" in an unstructured manner. Thus, an informal, unspoken, and somewhat unconscious "apprenticeship system" had evolved during the time the company occupied the former quarters. The system was disturbed with the advent of the new layout in which doorless openings to executive work spaces were located at the end of travelpaths which stopped directly at the office entrances. A psychological reaction to the new physical surroundings was reported by executives and non-executives alike. A typical statement was:
>
>> If you go around the corner of an "open" office, you feel committed to enter the office, even if you were just trying to

> check to see if the person is busy. There is not the easy "flow-by" access we used to have. You have to plan what you want to talk about--not just drop in.
>
> The casual drop-in visits decreased enough to be noticable to several of those who had been a part of the learning system. So it seemed that the more open, informal style of interior in actuality produced, for many, a more formal style of behavior and perceptions of a less "open" atmosphere at the social level (Hanson, 1978, pp. 89-90).

Two open plan office settings have been presented. Neither lived up to the expectations of its designers nor the hopes of the executives responsible for their installation. What is more, in one setting the open plan office seemed to enhance informal communication while in the other it inhibited it.

Before attempting to resolve these problems and paradoxes one more study will be referenced. Wichman and Healy (1979) studied student-built lofts in dormitory rooms at two of the Claremont Colleges. These were wood platforms set on wooden posts. The platform then either became a bed when a mattress was placed upon it or it became an elevated surface on which to place a desk. If the desk was raised then the bed was usually beneath it and vice versa. These lofts were a source of great pride to their designers and had a salutary effect upon the relationship between roommates. A correlational evaluation showed that roommates with lofts spent more time together, spoke to each other more, and got along better with each other than those in similar rooms but without lofts. The study spaces so designed were extremely small and crowded, resembling the navigator's station in an airplane. They were the kind of settings no designer in his right mind would try to sell. Yet the students with lofts spent many pleasant hours in the cramped study spaces they had created. The lofts seem to have given individuals considerable control over their environment, e.g., at night one student could study below without his light disturbing the other one sleeping above.

Indeed, the very act of designing, building and installing the lofts was in itself an act of control over the environment. Glass and Singer (1972) found that when an aversive stimulus was controlable, exposed subjects showed a dramatically reduced stress response, and Sherrod (1974) showed that people who could leave crowded rooms, even though they did not exercise that option, performed better on subsequent frustration tolerance tasks. Thus, perceiving that one has control (whether exercised or not) raises one's tolerance for otherwise stressful events.

The design principles for open plan offices that distill out of these three studies can be divided into two categories; physical design principles that involve material objects and their arrangements, and psychological design principles that refer to rules, management, and perceptions. Those principles are presented next.

Physical Design Principles

1. **Use sound absorbing material on all major surfaces wherever possible.** Noise is always more of a problem then expected.

2. **Leave some elements of design for the workstation user.** Workstations are usually overly designed and inflexible. People need to have control over their environments. Leave some opportunities for changing or rearranging things. Uniformity can be efficient but psychologically deadly because people want to personalize their spaces--we are all designers to some extent.

3. **Provide both vertical and horizontal surfaces for the display of personal belongings**, e.g., knickknack shelves, bulletin boards (cork) and blackboards. We make spaces our own by marking them as ours.

4. **Install telephones that ring "silently".** Phones produce unpredictable noise and are designed to be heard thirty feet away. In small workstations phones that flash a light for the first two "rings" before sounding an auditory signal would dramatically reduce unpredictable noise.

5. **Provide all private work areas with a system to signal the willingness of the occupant to be disturbed.** Normally, doors serve this function, e.g., closed--do not disturb; somewhat open--I am busy, important intrusions only; open--drop in. A primary aspect of effective environmental design is to allow expression of changing needs for privacy and community.

6. **Provide several easily accessable islands of privacy**, such as small (e.g., four person) rooms with full walls and doors (visual and auditory privacy) that can be used for conferences, private telephone calls and long distance calls. These must not require reservations and should have in-use signals easily visible.

7. **Have clearly marked flow paths for visitors**, e.g., signs hanging from the ceiling showing where secretaries are located, department boundaries, etc. The visual stimulus configuration of an open plan office is too complex for visitors to quickly develop a cognitive map. They always need more help than they get from the environment so they interrupt people to ask directions. This was a serious problem at JPL.

8. **Design workstations so it is easy for drop-ins to sit down while speaking.** At JPL 92% of the personnel were at least 170 cm tall while the partitions were only 157 cm high.

9. **Ventilation air flow must be planned.** Traditional offices have ventilation ducting to them. Open plan cubicles usually do not and can become dead air cul-de-sacs that are maddeningly resistant to post hoc resolution. This situation is exacerbated by heat generating office machines and incandescent lamps.

10. **Overplan for storage space.** Open plan systems with their emphasis on tidiness seem to chronically underserve the storage needs of humans.

Psychological Design Principles

1. **Personnel must perceive that they can exercise control over their environment.** It is management's job to arrange this by such mechanisms as standing committees or regularly meeting groups that address problems that arise. These must be truly useful groups with the power to get things done.

2. **Social norms particular to the situation must be established, communicated and enforced.** Normally these evolve slowly. Often (e.g., the second study cited here) this process must be speeded up by social planning. Here are some norms appropriate to most open plan office systems.

   Speak softly
   Don't hold conferences while walking down aisles
   Sit down when you speak on the phone
   Sit down during drop-in conversations
   Do not be a Kilroy (peeking over partitions) 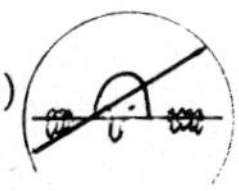 We found signs like this posted at JPL.

3. **Avoid cliche thinking when designing,** e.g., physical openness leads to psychological openness. Sometimes it may, but not necessarily.

4. **Find out in advance who needs to be together and then keep them together.** An environment that impedes communication is extremely frustrating.

5. **Plan certain locations as socializing areas.** These should be locations that naturally tend to be gathering places such as drinking fountain, food and drink machines, mail room, duplicating area, and lunch room. These should be noise insulated and encourage socializing and informal conversations.

6. **Do an observational study (or have one done) of behavior in the traditional office area before designing an open space office system.** If your design will require behavioral changes, plan for them in advance. If you (and/or your client) cannot develop a behavioral substitution for an established norm, change your design!

7. **Budget for a follow-up study and system changes based on its results.** Include this in your bid. No designer can anticipate all contingencies (that's why aircraft companies have test pilots) but if you don't do follow-ups you don't get corrective feedback. Plan funding in advance to allow some inevitable modifications.

REFERENCES

(1) Bayley, B., Green, T., King, E., Lemmex, D. and Rogers, M. Open Plan Office Evaluation Report, Bldg. 264: Eighth Floor, Human Environment Group, Claremont Graduate School, (June, 1978).

(2) Hanson, L.A., Effects of a Move to an Open Landscape Office, Ph.D. Thesis, Department of Psychology, Claremont Graduate School (April 1978).

(3) Wichman, H. and Healy, V., In their own spaces: Student built lofts in dormitory rooms, Presented Paper, American Psychological Association Convention, New York (September, 1979).

(4) Sherrod, D.R., Crowding, perceived control, and behavioral aftereffects, **Journal of Applied Social Psychology 4** (1974) 171-186.

(5) Glass, D. and Singer, J., **Urban stress,** (Academic Press, New York, 1972).

Human Factors in Organizational Design and Management
H.W. Hendrick and O. Brown, Jr. (Editors)

HUMANIZING THE UNDERGROUND WORKPLACE: ENVIRONMENTAL PROBLEMS AND DESIGN SOLUTIONS

James A. Wise and Barbara K. Wise

The University of Washington
Seattle, Washington
U.S.A.

Underground workplaces represent a major source of unused potential for American buildings. While they provide strong economic incentives, particularly in life-cycle costs, they have proven less than acceptable with the workers who inhabit them. This paper reviews the major human factors problems with underground workplaces and assesses how they may be resolved through appropriate strategies of environmental and organizational design. With the sorts of modifications outlined here, underground workplaces can become much more supportive, responsive and acceptable to their occupants.

INTRODUCTION

The United States and Japan lead the world in the volume of useable subsurface space. However, in proportion to population, Sweden and Norway are the world leaders in underground buildings. In a hundred Russian cities, thirty-five percent or more of new construction is underground (Dean, 1978), while overall, subsurface use is increasing at a rate that will double such space in ten years (Jansson & Winqvist, 1977).

Most of these underground projects are driven by considerations for energy conservation, environmental controls, or preservation of extant land use/ buildings on site. Energy savings can easily reach 50-60% over comparable aboveground structures, and the earth quickly damps out adverse weather, temperature and noise impacts on an underground building. When there are strong arguments for the preservation of downtown neighborhoods or campus open space, underground buildings such as the San Francisco Moscone Convention Center and the University of Michigan Law Library addition become imaginative and appropriate solutions to the dilemmas posed by their surface alternatives.

Many workplaces — particularly those incorporating high technology — would seem to benefit by going underground. There is greater building security, and the environment is less noisy, relatively vibration free, and easy to maintain at a constant temperature and humidity. In an information society the automated workplace would seem ostensibly well-served when it is below grade.

WHAT OCCUPANTS THINK OF UNDERGROUND WORKPLACES

Unfortunately, underground workplaces are often not proposed and developed because they have not yet "made the grade" in terms of habitability for their occupants. When the occupants of such buildings are surveyed, either carefully (Hollon, Kendall, Norsted & Watson, 1980) or informally (Sommer, 1974) their responses to working underground are consistently and overwhelmingly negative.

Sommer (1974) found that employees who worked underground said they "felt like moles", had "lost the meaning of time", and would make frequent trips to the bathroom upstairs just "to look outside".

When Hollon et al. (1980) surveyed underground workers on the perceived characteristics of their job settings, they found them significantly less satisfied than workers in two comparison settings: either aboveground without windows or aboveground with windows. The underground employees thought more negatively about their surroundings and were much less likely to choose such a space for other work or leisure activities. Their mood ratings on the MAACL (The Multiple Affect Adjective Checklist) also showed them to be more anxious, depressed and hostile.

In addition, the subjective ratings of underground workplaces were not related to awareness of the strong arguments for such workplaces such as energy conservation and economic savings. Paulus (1976) had hypothesized that underground workers may vicariously participate in the landscape benefits afforded to the aboveground users of an underground building, and thus come to positively regard their subterranean workplace. It seems, however, that just as with the ubiquitous "commons problems", one cannot expect people to prefer a social gain to compensate for an individually felt loss.

Such responses are not unexpected given the much more extensive survey literature on windowless aboveground buildings. In her review, Collins (1975) had also found that studies of windowless environments elicited similar negative evaluations from the settings' users. Without windows, workers complain of 'lack of daylight', 'poor ventilation', and 'inability to know weather' or 'have a view'. They also express feeling of claustrophobia, tension and depression(Ruys, 1970). More recently, Cuttle (1983) replicated these general findings with a survey of 471 English and New Zealand office workers, and concluded that it is important for offices to have large windows, and that workers should be able to sit close to them.

Wyon and Nilsson (1980) have produced an important conceptual analysis that helps determine what it is that office workers find so valuable about windows. Their thesis is that windows actually perform several functions in the indoor environment. These are dominated by access to daylight and view-out considerations, but also extend to providing visual and acoustic information from outside, affecting temperature and air exchange and even act as an emergency exit. The idea, then, is that windows are valued because of their singular fulfillment of collective functions, rather than their response to any basic biological need.

While the evidence is scanty, this hypothesis seems to be mildly confirmed by results from other studies. For example, children in windowless schools do not show measureable ill effects in terms of performance or learning achievement (Dean, Nov., 1978) and the school settings show reductions in vandalism, glare, and overheating (Collins, 1975) — all of which are detrimental to the learning experience.

In a study of an intensive care unit (Wilson, 1972) it was found that twice as many patients in a windowless care unit developed post-operative delirium as did patients in a windowed intensive care unit. But the windowless unit comparison in this study had no compensatory features for what the window might provide. In a similar well-controlled study (Ulrich, 1984) showed that just having a window is not sufficient in itself. He found significant differences in recovery time and medication needs dependent on the view through the window. When trees were viewed as opposed to a brick wall, the patients benefited.

Being able to manipulate windows also seems to affect peoples' thermal comfort. Davis and Szigeti (1983) report that when individuals are able to control the opening of windows, they accept wider ranges of variation in room temperature and drafts than when they are enclosed in windowed spaces without this capability.

These findings are consistent with a "collective functions" hypothesis of what it is windows provide, and they suggest that windows are an exceptionally convenient means of exercising a sense of personal control over one's environment. Windows become part of adaptive personal control loops for exercising curiosity, alleviating boredom, and sensually responding to different changing ambient conditions. Little wonder that they are sorely missed in windowless and underground workspaces.

But this conceptual model argues that windows are important for what they do, rather than what they are. So it should be possible to enhance the windowless, underground workspace by introducing, through other means, what windows fulfill. And if there are other negative aspects of underground settings that are not connected to windows per se, might not they be compensable too?

The results of Hollon et al. (1980) suggest that there are other sources of dissatisfaction since underground settings were still rated below their aboveground, windowless counterparts. Much of this may be due to the cultural conditioning we receive, with its rich symbolism of life's rewards and their directional contexts. Affective and evaluative connotations of "above and below" suffuse our language, and relative height is a strong indicant of status and social power. It is not unreasonable to hypothesize that social and cultural expectations, as well as sensory ones, need to be compensated for in other ways if they are violated in a move to an underground work setting.

HEALTH AND SAFETY CONCERNS OF UNDERGROUND ENVIRONMENTS

Some of the negative reaction to underground workspaces arises from the popular conception that they are unhealthy and unsafe. In particular, workers are concerned about the physical consequences of loss of sunlight and fresh air, and the building's response to earthquake, fire and (less frequently) flood.

Arguments about sunlight are exacerbated and confused by a lack of distinctions between what sunlight and artificial light produce as visual and non-visual physiological processes. As Cuttle (1983) observed, workers' preferences for windows were based not so much on their belief in the beneficial effects of daylight, but on their assessment of the poor quality of the artificial lighting.

Light not only enables us to see, it stimulates the functions of the hormones and endocrine system (Hollwich, 1980). Most of these effects do depend on the reception of light through the eye (Wurtman, 1968; Kuller, 1980). The biological effects on the human system, while acknowledged, are largely unknown. However, they seem to be more substantial in young children and elderly than other age groups (Kuller, 1980; Wurtman, 1975). Although many biological dangers are reputed to reside in artificial lighting (see Ott, 1965), the authors believe that many of these consequences are traceable to the poor design of lighting systems rather than to the power spectrum of different fluorescent lights themselves. Full spectrum fluorescents are available in the marketplace (at a considerable premium), but any adult who has access to the out-of-doors for fifteen minutes a day receives as much UV radiation as if they had worked under 500 fc of full-spectrum fluorescents all day. This artificial light level, of course, is over 10 x a reasonable ambient amount for interiors.

People who work next to windows do <u>not</u> receive their daily (beneficial?) dose of UV B — necessary for the production of vitamin D_3 — because this band is usually absorbed by window glass. So full-spectrum fluorescents, although advertised for their biological value, seem biologically substitutable by a noon-time walk. But full spectrum fluorescents are a good investment for their color-rendering properties, and with the skillful use of such lighting, an underground workplace can be made to appear as <u>if</u> it were illuminated by daylight. This sort of psychological enhancement will be treated in the next section.

Besides light, air is an important quality in underground settings. An underground building has no air infiltration at all, so as less 'fresh' (outside) air is brought in, the recirculated air accumulates odors and potentially hazardous pollutants.

Indoor air pollution has become a major health threat in all buildings (Budiansky, 1980), but it is exacerbated by the 'sealed' quality of an underground structure. The solution lies in an enhanced air intake and filtration system, but too often these are modeled after aboveground buildings, whose standards were set with an expected infusion rate that simply does not apply below grade. The occupants of even energy conserving, aboveground buildings benefit by air infiltration that persists despite the best efforts of thermally conscious engineers. Underground building occupants are totally reliant on the mechanical ventilation systems, which often has major circulation deficits (Budiansky, 1980; Rand, 1981).

Another potentially serious health hazard that is related to air quality and is unique to underground buildings is the infusion of radon-222 into the interior. This is a decay product of the naturally occurring uranium-238 that is a constituent of rock throughout the world. Radon-222 diffuses through porous building material such as concrete into a habitable interior space. Radon enters aboveground buildings through their foundations only. However, it enters underground buildings from all sides.

Breathing radon itself is not dangerous, however, as it decays, it emits by-products known as radon daughters that, as single ions, attach themselves to submicroscopic aerosols. When these are breathed in, they are retained in the lungs, where they may be responsible for 1,000 - 2,000 lung cancer deaths each year in the U.S. (Budiansky, 1980).

Radon infusion may be controlled by including a rather expensive filtration system into the building's HVAC, or by applying a polymeric sealant to the building floor and walls. Again, proper construction and engineering of underground buildings can protect their occupants from this largely unknown health hazard.

While sunlight and air quality are chronic concerns to the health of underground building occupants, there are also acute threats to safety due to hazards produced by fire, flood and earthquake. In all of these respects, underground buildings must be specially designed and engineered to meet unique conditions created in the subterranean environment. From a human factors view, they must also communicate to their occupants that such precautions have been respected.

Fire is the greatest hazard, again because of the lack of air filtration into an underground building. This creates a low pressure which will inhibit the exhaust fans to vent smoke while in "fire mode" (Fitzgerald, 1981). Because there are no windows to 'blow out', a positive pressure can also be created underground, which significantly lessens the fire retardancy of materials (Degenkolb, 1981). Both of these conditions exacerbate the problem of emergency egress for the building's occupants. The first by increasing the likelihood of obscuring smoke, and the second, by giving the occupants less time to escape.

Occupants of an underground building are already debilitated because they must travel <u>up</u> the stairs rather than down, to safety. In this respect, there seems to be a 10% decrease in flow rate from downward to upward movement (Grandjean, 1973), and fatigue will also occur at a rate five times faster for upward travel (McKinnon, 1981).

Design responses to emergency egress from underground buildings can include re-sized stairs (in terms of riser/tread ratio), isolated elevators, capacity stairwells (to hold the entire occupants of the building), and 'safe havens.' On a lower scale, fire safety design should include emergency lights and signs low on the walls, door hardware that can be operated from a crawling position, and interior layouts that do not allow dead end corridors.

An unresolved question is whether occupants of an underground building will be more liable to panic in an emergency situation. Guten and Allen (1972) point to expectations of entrapment, or a judged small likelihood of escape, as inducers of panic behavior. Underground environments seem liable to intensify these precursors, due to the closed nature of the building. Training of occupants, emergency signage, and voice instruction systems all can be sources of reassurance for the occupants of such structures. The latter are particularly important if a building is used by a high proportion of visitors — e.g. museums, libraries.

ENHANCING UNDERGROUND WORK ENVIRONMENTS

From both biological and safety perspectives, then, underground buildings impose special demands on their construction, environmental control systems, layout and finishing details. These do not pose insurmountable problems, but the solutions do not involve the sorts of ambient conditions that would overcome the general negative reactions to working underground. What this means is that aside from being made healthy and safe, underground work environments must be made psychologically habitable. Other ways must be employed to reintroduce what aboveground buildings provide as a matter of happenstance for their occupants.

For underground buildings, there are three basic approaches outside of other environmental controls (i.e. acoustics, thermal, air quality). These concern the spatial aspects, aesthetic lighting qualities and furnishings/finishing details of the interior. When adroitly combined into a 'design' that is appropriate for underground living, these elements can significantly influence perceived spaciousness, the perceived 'warmth' of a setting, and the overall sense of 'ambience' that a user perceives as a supportive environment or not.

Recall that Sommer (1974) found that workers in underground, windowless spaces complained of feeling cooped up in 'tight' spaces. The issue here is partially one of perceived spaciousness. This can be relieved by the shape of the interior space and the way it is subdivided. For example, interiors that have all boundary points visible from all inside points are judged as less spacious because everything to be seen always can be seen. The architect who wishes to make a space seem larger will not leave it totally open, but will subdivide it so that its parts open onto each other in unexpected ways (Benedikt, 1978).

There is also an effect of the mass/space ratio at work here. When a space has its perceived interior mass concentrated in one place, rather than dispersed, the space seems more filled (Titus, Dainoff, Hill, Oskamp, McClelland, & Riley, 1977). Thus, how one arranges furniture and partitions (as in an open plan office) affects its perceived spaciousness.

Lighting plays a crucial role, too. By using indirect lighting, the walls and ceiling can be made brighter than the floor and interior furnishings. This makes those enclosing surfaces appear to be farther away (Boyce, 1980). Aubree (1978) showed that such indirect lighting of the walls and ceiling periphery by fluorescent lamps was preferred over other systems.

While engaging such a lighting strategy to increase perceived interior space, even the choice of lamp is important. Generally, 'cooler' color temperature lamps aid an impression of largeness over warmer ones (Watson & Payne, 1968). Since full spectrum fluorescents have 'cooler' color temperatures, they aid the perception of spaciousness when properly employed in windowless settings (Aksugur, 1979).

Thus, the perceived spaciousness of an underground interior can be positively influenced by collaborative interior and lighting design. But underground office spaces have all too often been treated like their surface counterparts, and this potentially ameliorative strategy has been lost.

When light comes through a window to illuminate an interior, it "models" the interior elements. This means that minor variations of light and dark are produced on the surfaces of objects in a room. In short, they show what is called 'relief'. In experiments by Hopkinson (1970) occupants of a room were found not to like the modeling effects of light that came predominately from a homogeneous overhead lighting scheme. Instead, the subjects found such rooms to be 'flat' and 'boring'. To reduce this impression, light must be admitted from the side of an interior space to produce a 'horizontal modeling' component. Such horizontal modeling is particularly favored for lighting the faces of room occupants, with the preferred direction being 0-30 degrees above the horizontal and 45 degrees from either side (depending on the person's sex) (Barton, Spivack, & Powell, 1972; Canter, 1976). Curiously enough, these angles most closely simulate light entering through a window to illuminate a person sitting beside it.

So perceived interest of a space, and even the attractiveness with which people are rendered within it, are affected by the light distribution, as measured by its modeling quality — the vector/scalar ratio. Here is another powerful tool that could be used to design the lighting of underground spaces to make them appear to their occupants to be illuminated as aboveground spaces usually are.

Lighting techniques can be employed to carry various symbolic messages, or to create the illusion of amenities that are not really there. For example, 'washing' a wall with light — i.e. illuminating it at a low angle from either above or below — can be used in conjunction with a heavily textured wall surface to create a highly complex, topographical relief effect on the wall. This turns the wall into a source of visual interest, and lessens its perceived role as a spatial boundary. Lighting may also be recessed within a wall, above eye level, and made to illumiante the ceiling periphery, which suggests to the occupant a view of an overcast sky. This is called a "symbolic window" by Hopkinson (1970).

The illusion of a window in this arrangement is quite convincing, but it vanishes if the lights are brought to eye level, because the expected view is not there. The illusion can be enhanced if the recessed lights are fitted with daylight fluorescents, while the immediate interior lighting uses a warmer color temperature lamp.

A similar strategy can be employed to create the strong impression of an interior, translucent 'skylight' overhead. Since light levels through a window change with the passing of clouds overhead, or with diurnal sun passage, the lighting in a 'false' skylight could be programmed to reproduce such expected variations. Deliberate swings in temperature control have already been initiated to relieve "thermal boredom" in buildings (Gerlack, 1974). The analogous concept of "illumination boredom" would seem to merit investigation.

Still missing from these schemes is the actual view provided by windows in aboveground buildings. But this aspect seems also to be reproducible through an optical assist. The University of Minnesota uses a system of mirrors and lenses to reflect daylight and view down to a deep underground workspace (Carmody & Sterling, 1983). A more refined system is presently being developed at the University of New Mexico (Taschek, Thomas, & Lusk, 1983) whereby a fairly refined view is projected into an interior room. The images produced are colorful and surprisingly three-dimensional, and the view even changes as one moves back and forth in front of it, much like a real window. This interactive quality, of course, is exactly what is lost in attempts to bring a view into a windowless workspace through a closed video system.

Just as lighting design interacts with spatial arrangement to produce more psychologically habitable interiors, the furnishings and finishing details add another means whereby design may alter users' reactions to underground spaces. For example, underground interiors are oftentimes labeled 'dank' and 'cold', but the building's thermal controls are not the only way to affect this perception. Interiors can be made to appear "warmer" through the proper application of decor such as carpeting, paneling and wall color. This actually makes people feel warmer in a room (Rohles & Wells, 1977).

The use of finishing details such as wood handrails, fabric or wood arm rests and door handles also give occupants the chance to touch 'warm surfaces' during their daily activities, perhaps enhancing a sense of thermal comfort.

Rohles and Krohn (1982) found that thermal comfort was greater in cloth covered chairs than in vinyl upholstered ones, and that this effect was not mediated by actual differences in skin temperature.

Besides their effect on thermal comfort (and spaciousness, as discussed earlier), interior furnishings seem able to alleviate some negative psychological reactions simply by their ability to make certain associations with the outside world.

Janowski (1980) investigated an underground hospital unit where the deliberate use of green plants, murals, a skylight and the appropriate colors greatly improved environmental satisfaction. In her words, "The feeling of outdoors . . .is extended. . .through the use of materials associated with outdoors, and through the use of warm or earthy colors." Wall murals of regional significance seemed to be quite effective — again by turning what would otherwise be thought of as an enclosing barrier into a source of interest itself.

A more recent study by Laviana, Mattson & Rohles (1983) reaffirmed the importance of plants in indoor settings. Plants contributed positively to perceived quality and one's evaluation of an indoor space. Surprisingly, the plants used in this study did not influence thermal satisfaction or comfort. Plants have long been used as acoustical diffusers to enhance the quality of sound in interiors. It would seem that their positive effects on occupants extend to other domains as well.

Applications of color and pattern to interiors provide another set of opportunities to enhance the status and acceptability of workplaces. Stanwood (1983) reports again that natural earth tones were a good choice for windowless settings in their ability to invoke outdoors associations. Kleeman (1981) argues that "sharp-serrated patterns evoke a significantly greater (arousal) response than equally complicated rounded patterns." If this is a consistent result, it has implication for patterns in wall-paper, carpeting, and upholstery that might be applied in underground settings.

CONCLUSIONS

This paper has only been able to briefly sketch the rich complexity of the human factors of underground work environments[1]. The important message here is that underground workplaces do not have to be accepted and endured as the relatively uninhabitable settings that people have experienced. Throughout the human factors and building science literature, there are numerous examples of ways in which interior windowless spaces may be designed to offset the otherwise negative effects of working underground. Where architects and clients have taken the time, effort, and extra cost to design back in the "amenities of habitability", underground settings can be supportive and acceptable to their occupants.

To the professional manager, problems with personnel in the workplace are often taken for granted and treated as if the causes lie in the job or organizational design of the work environment. But the work environment is also the physical setting of the worker, and this deserves equal attention, especially in today's automated offices. Humanizing the underground workplace is a necessary and desirable part of utilizing the potential of such available space. It is also the one setting where a close cooperation of organizational and environmental design can produce the strongest results.

[1] A full report, The Human Factors of Underground Work Environments, edited by the authors is the basis for this article

REFERENCES

(1) Dean, A.O. Underground architecture, AIA Journal. 67 (April, 1978) 34-37.

(2) Jannson, R. & Winqvist, T. Planning of subsurface use (Pergamon Press, Inc., New York, 1977).

(3) Hollon, S., Kendall, P.C., Norsted, S. & Watson, D. Psychological responses to earth-sheltered, multilevel and aboveground structures with and without windows. Undgerground Space 5 (1980) 171-178.

(4) Sommer, R. Right spaces: hard architecture and how to humanize it. (Prentice Hall, New York, 1974).

(5) Paulus, P. On the psychology of earth covered buildings. Underground Space, 1 (1976), 127-130.

(6) Collins, B. Windows and people: a literature survey. NBS Building Science Series, Washington, D.C., 1975.

(7) Ruys, T. Windowless offices. Masters Thesis, Department of Architecture, University of Washington, 1970.

(8) Cuttle, K. People and windows in workplaces, in D. Joinier(Ed), Conference on People and Physical Environment Research (Wellington, New Zealand, 1983).

(9) Wyon, D. & Nilsson, I. Human experience in windowless environments in factories, offices, shops and colleges in Sweden. In the CIE

Proceedings of the Symposium on Daylight: Physical, Psychological, and Architectural Aspects. (Berlin, 1980).

(10) Dean, A.O. Two communities building subsurface schools. AIA Journal 67 (1978) 47-49.

(11) Wilson, L.M. Intensive care delirium: the effect of outside deprivation in a windowless unit. Arch. Int. Medicine 130 (1972) 225-226.

(12) Ulrich, R.S. View through a window may influence recovery from surgery. Science 224 (April 27, 1984) 420-421.

(13) Davis, G. & Szigeti, F. Development of a functional program for overall building performance, in D. Joinier (Ed) Conference on People and Physical Environment Research (Wellington, New Zealand, 1983).

(14) Hollowich, F. The effect of natural and artificial light via the eye on the hormonal and metabolic balance of animal and man. In the CIE Proceedings of the Symposium on Daylight: Physical, Psychological, and Architectural Aspects (Berlin, 1980).

(15) Wurtman, R. Biological implications of artificial illumination. Illuminating Engineering (October, 1968) 523-529.

(16) Kuller, R. Non-visual effects of daylight. In the CIE Proceedings of the Symposium on Daylight: Physical, Psychological & Architectural Aspects (Berlin, 1980).

(17) Wurtman, R. The effects of light on the human body. Scientific American (July, 1975) 68-77.

(18) Ott, J. Effects of unnatural light. New Scientist 429 (Feb. 4, 1965) 294-296.

(19) Budiansky, S. Indoor air pollution. Environmental Science and Technology 14 (September, 1980) 1023-1027.

(20) Rand, G. Thinking ecologically about indoor environments. Underground Space 6 (1981) 105-108.

(21) Fitzgerald, R. Smoke movement in buildings, in G.P. McKinnon (Ed), The Fire Protection Handbook Edition 15 (National Fire Protection Association, Mass., 1981).

(22) Degenkolb, J.G. Fire protection for underground buildings. Underground Space 6 (Sept. - Oct., 1981) 93-95.

(23) Grandjean, E. Ergonomics of the home (John Wiley, New York, 1973).

(24) McKinnon, G.P. (Ed). The fire protection handbook Edition 15 (National Fire Protection Association, Mass., 1981).

(25) Guten, S. & Allen, V. Likelihood of escape, likelihood of danger, and panic behavior. Jr. of Soc. Psychology 87 (1972), 29-36.

(26) Benedikt, M. An introduction to isovists. Technical Paper, School of Architecture, The University of Texas at Austin, 1978.

(27) Titus, W.C., Dainoff, M., Hill, M., Oskamp, R., McClelland, B. & Riley, R. The psychophysics of mass-space. Man-Environment Systems 6 (1977) 370-371.

(28) Boyce, P.R. Human factors in lighting.(Applied Science Publishers, London, 1981).

(29) Aubree, A. Artificial lighting during the day of a deep room. Paper presented at Illumianting Engineering Society Conf., 1978.

(30) Watson, N. & Payne, I. The influence of fluorescnet lamps of different color on the perception of interior volume. Environmental Research Group, University, London, 1968.

(31) Aksugur. The effect of hues of walls on the perceived magnitude of space in a room under two different light sources having different spectral distributions. Architectural Bulletin 4 (1979) 22-47.

(32) Hopkinson, R.G. The ergonomics of lighting.(MacDonald Technical and Scientific, London, 1970).

(33) Barton, M., Spivack, M. & Powell, P. The effects of angle of light on the recognition and evaluation of faces. Journal of IES (April, 1972) 231-234.

(34) Canter, D. Environmental interaction. (Int. University Press, New York, 1976).

(35) Gerlack, K.A. Environmental design to counter occupational boredom. Jr. of Arch. Research, (Sept. 3, 1974) 15-19.

(36) Carmody, J. & Sterling, R. Underground building design: commercial and institutional structures. (VanNostrand Reinhold, N.Y., 1983).

(37) Taschek, J., Thomas, L. & Lusk, P. Lens-guide viewing systems: an innovative use of daylighting at the University of New Mexico, School of Architecture & Planning. Passive Solar Conference Proceedings, Santa Fe, New Mexico, 1983.

(38) Rohles, F.H. & Wells, W. The role of environmental antecedents on subsequent thermal comfort. ASHRAE Transactions (83) 1977, 21-29.

(39) Rohles, R.H. & Krohn, R.J. Thermal comfort as affected by chair style and covering. Proceedings of the Human Factors Society 26th Annual Meeting, Seattle, Washington, 1982.

(40) Janowski, W. "light" and "airy" underground. Lighting Design and Application (Jan., 1980), 14-18.

(41) Laviana, J.E., Mattson, R.H. & Rohles, F.H. Plants as enhancers of the indoor environment, in Pope, A.T. & Hough, L.D. (Eds), Proceedings of the Human Factors Society Vol II, Norfolk, Va., 1983.

(42) Stanwood, L. Specifying color: is it really a science? Construction Specifier (June, 1983) 50-55.

(43) Kleeman, W. The challenge of interior design. (CBI Publishing Co., Inc., Boston, 1981).

Human Factors in Organizational Design and Management
H.W. Hendrick and O. Brown, Jr. (Editors)

THE EFFECTS OF VDT WORKSTATION DESIGN ON OPERATIONAL PRODUCTIVITY, WORK BEHAVIOR, AND COMFORT. HAWTHORNE REVISITED?

Louis Tijerina
OCLC, Online Computer Library Center, Inc.
Dublin, Ohio 43017 USA

Dr. Thomas H. Rockwell
Department of Industrial and Systems Engineering
The Ohio State University
Columbus, Ohio 43210 USA

A study of VDT workstation design was carried out in an operational environment to determine the effects of workstation layout, seat design, and glare control filters on operator productivity, work behavior, and comfort. Results revealed no significant differences in productivity as a function of changes in workstation design variables. Workstation design changes did affect time spent at the workstation and comfort assessments. The pattern of results suggests that job design and social/organizational factors may dominate any influence of ergonomic changes in the workplace.

INTRODUCTION

Video display terminal (VDT) workstations are being implemented in an ever-increasing number of settings, often with little attention given to well-established ergonomic principles (National Academy of Sciences, 1983). In response to this, numerous articles have appeared advocating improved work-station design features, and bills have been brought before various state legislatures demanding ergonomically sound workstation design. However, there has been little published research which empirically evaluates the effects of VDT workstation design factors on operator performance at work. Like the famous Hawthorne studies (Roethlisberger and Dickson, 1939), social/organizational conditions may influence human performance more than workplace design in a VDT-based environment.

There is evidence to suggest that ergonomically designed VDT workstations affect operator productivity and comfort. For example, Grandjean, Nishiyama, Hunting, and Pidermann (1982) evaluated the effects of adjustable VDT workstation settings (e.g., seating, keyboard height, etc.), in a laboratory study. Subjects reported feeling more relaxed working at workstations they adjusted than working at workstations with fixed dimensions; however, the effects of workstation adjustability on productivity were not reported. Habinek, Jacobson, Miller, and Suther (1982) used a legibility task to evaluate the effectiveness of VDT antireflection treatments in the laboratory. Performance and preference data revealed little difference among the three types of treatments tested and all were effective in reducing screen reflections relative to an untreated VDT screen. Dainoff, Fraser and Taylor (1982) evaluated good and poor VDT workstation design by simulating a data entry task and reported a 24.5% performance increase under best-case ergonomic design conditions and a decrease in musculoskeletal complaints when compared to a worst-case control. Studies like these suggest that improved VDT workstation design could increase productivity and/or comfort.

The major research objectives of the present study were to evaluate the effects of workstation layout, improved chair design, and glare control filters on operator productivity, behavior at work, and comfort. Unlike previous laboratory experiments, this study was conducted in an operational environment under normal conditions of workload, and system performance. The study was conducted in an area consisting of 51 VDT workstations at which two shifts of operators each worked 7 hours per day. The operators were engaged in converting cataloging information from library cards into a database.

For each card, an operator would search the database for a match and if one was found, edit the record as appropriate. If a record for that card was not found, the operator would call up a workform and key in all the information on that card. The work was fast-paced, keyed to system response time, and visually demanding, requiring operators to spend extended periods of time sitting and keying in data at terminals.

METHOD

Subject. Sixteen VDT operators were selected to participate in the study. Their performance was monitored one week prior to the start of the experiment and these data were used to group the operators into four teams of four persons each. Two of the operators did not complete the entire study and their replacements were not included in the final analysis. The final sample thus consisted of 11 females and 3 males between the ages of 18 and 37 years. Work experience ranged from 3 weeks to 10 months.

Apparatus. Sixteen workstations were used in the study. All were equipped with the same type of work desk and terminals. These terminals had 15" monitors with etched-glass screens and detachable keyboards. Micromesh glare reduction filters were attached to eight of the terminals to assess their effects on visual comfort and performance.

Workstations were configured wither in a side-by-side layout of 4 workstations or in clusters of 4 workstations arranged in cartwheel fashion (workstations set at right angles to one another). The cartwheel layout put a greater physical distance between operators and required more floor space to accommodate all the workstations. The side-by-side layout put operators in greater proximity to one another and made more efficient use of floor space.

The two types of seating used at the experimental workstations were termed "old" and "new." The eight "old" seats were standard secretarial seating without armrests and with screw-type adjustment mechanisms which were functionally non-adjustable. The eight "new" seats were equipped with armrests, full contoured backrests, and pneumatic seat height adjustments. Adjustments were easily made with pushbutton or lever control mechanisms.

Procedure. The collection period was four weeks long. Teams of subjects worked one week each at four different configurations according to the experimental design given in Table 1. In an effort to equate system response time for all experimental workstations, steps were taken to insure that they were all serviced by the same communications processor and mainframe.

Table 1
Experimental Design: Workstation Configurations and Assignments

Experimental Workstation Configurations			Treatment Schedule			
Layout	Seating	Glare Control	Week 1	Week 2	Week 3	Week 4
Side-by-side	old	no filter	A			B
Cartwheel	new	no filter		A	B	
Cartwheel	old	filter		B	A	
Side-by-side	new	filter	B			A
Cartwheel	old	no filter	C			D
Side-by-side	new	no filter		C	D	
Side-by-side	old	filter		D	C	
Cartwheel	new	filter	D			C

NOTE: A, B, C, and D refer to 4-person teams.

Measures collected during the study fell into one of three categories. Measures of productivity were taken from daily reports completed by each operator. These report forms provided information on the type and number of inputs processed, estimates of worktime, system downtime, and time spent on activities other than data entry. Two videocassette recorders and cameras with 6-hour tapes were used to determine the proportion of time each operator spent at the workstation daily. Finally, subjective assessments of each workstation arrangement were collected by means of a short survey distributed weekly which included open-ended questions about impressions of the workplace followed by several 7-point comfort ratings and an item on seat adjustment behavior.

RESULTS

During the course of the study it became apparent that task specialization was occurring. Three of the 14 subjects were doing original inputs almost exclusively while the remaining 11 subjects were doing non-original transactions such as searching and editing. To deal with this task heterogeneity, all analysis of variance were performed on data from the 11 subjects performing the non-original inputs only.

Surprisingly, analysis failed to uncover any significant effects in productivity, either in terms of rate of output or amount of output, as a function of seating type, use of glare reduction filters, or workplace layout.

Truetime represents the total time spent at the workstation, as derived from videotape analysis. An analysis of truetime spent at the workstation per day indicated significant effects as a function of workstation design. An analysis of variance revealed significant main effects for layout ($p<.02$), seating ($p<.04$), and glare control ($p<.04$), as well as a significant seat x filter interaction ($p<.01$). Mean truetime at the workstation is 17 minutes per day, longer with the side-by-side layout than with the cartwheel arrangement (5 hours 58 minutes and 5 hours 41 minutes, respectively). Mean truetime per work day is approximately 14 minutes greater with new seats than with old (5 hours 56 minutes and 5 hours 42 minutes, respectively), and operators remained at the workstation approximately 13 minutes per day longer when no micromesh filter was used than when one was present (5 hours

56 minutes and 5 hours 43 minutes, respectively). The seat x filter interaction indicated that with no add-on filter present, operators remained at the workstation about 33 minutes per day longer with new seats than with old seats (6 hours 12 minutes and 5 hours 39 minutes, respectively). However, differences in the effects of seating diminish with the presence of filters (5 hours 46 minutes with old seating and 5 hours 41 minutes with new seating).

Seat comfort ratings were analyzed to assess the impact of seating on operator comfort. The new seating (average rating = 6.3) was judged significantly more comfortable ($p<.005$) than the old seating (average rating = 2.9). The effects of the micromesh glare reduction filter on visual comfort were assessed by analysis of visual comfort ratings. Visual comfort with filters (average rating = 5.6) was judged significantly more satisfactory ($p<.05$) than without filters (average rating = 4.8).

DISCUSSION

Intuitively, one might expect that greater comfort would lead to greater time spent at the workstation. The results of the study indicate that, on the average, an operator spends about 14 minutes per day longer at a workstation equipped with new seating than with old seating. Even with this small difference a cost/benefit analysis based on hourly wages indicated that new seating would pay for itself in reduced unscheduled rest pauses (additional personnel hours) within a year. Coupled with the unanimous preference among subjects for new seats, the high comfort ratings given during use of these seats, and the positive impact they would have on employee morale, a recommendation to provide new seating seems justified.

At first glance, the experimental results with the glare control filter seem counter-intuitive. That is, subjects rated their visual comfort higher with add-on filters than without, yet spent an average of 13 minutes per day less at workstations equipped with such devices. However, operators noted in post-study interviews that when the filters got dusty or the monitor brightness was insufficient, the filters were often worse than nothing at all. Perhaps the reduction in reflected glare could not offset the negative effects of image degradation caused by dust and reduced character brightness, effects which would force the operator away from the terminal. A similar line of reasoning may be applied to the glare control x seat interaction.

Subjects spent an average of 17 minutes/work day longer at workstations arranged side-by-side than in a cartwheel. The comments made during post-study interviews indicate that while the subjects have different preferences for the layouts studied, they said they like the side-by-side layout because it made communication easier.

The pattern of results in this study indicate that the workstation design factors examined had no effect on productivity but did affect work behavior and comfort. The causes of this pattern of results cannot be ascertained at this point. However, the data suggest a hypothesis in which job design and social/organizational factors will dominate the influence of workstation design on productivity. This hypothesis merits further investigation and may prove useful when evaluating the effects of VDT workstation design changes in actual work settings.

Insignificant differences in productivity across workstation configurations indicate that performance stayed within a fairly narrow band, not going below or above certain limits. It is hypothesized that this stability in performance can be attributed to worker motivation, operational constraints, and work organization. The social/organizational conditions involved consisted of published productivity standards which could have set a floor to output. Job advancement depended on meeting these standards and the weekly productivity of each operator was posted on a public display, providing further incentive not to fall below the minimum standards. Similarly, the worker motivation could have set a ceiling on productivity. The hypothesis of par (Smith, 1978) predicts that, unless motivated to do otherwise, people will do what is required of them even though they could do more. In the present case, the motivation to perform at a peak may have been low because the operators in the study were primarily entry-level workers at the bottom of the pay scale performing a fairly repetitive task. System response time could also have limited the degree to which productivity could vary with changes in workstation design. Variation in system response time and performance are outside the operator's focus of control and may serve, de facto, to pace the worker.

Given that social/organizational factors and system constraints appear to dominate workstation design effects on productivity, Becker (1981) argues that it may be necessary to rethink productivity and consider the secondary consequences of good workstation design. Among these secondary consequences are: increased comfort and decreased complaints; fewer or shorter unscheduled rest pauses; decreased turnover, absenteeism, and error rate. These secondary consequences may be harder to assess than daily production, but the benefits may well prove to be non-trivial.

Job design and organizational factors have been implicated as a principle source of stress among VDT operators (National Academy of Sciences, 1983; Kirk, 1983; Arndt, 1982). Many complaints are filed by workers in jobs in which a single task (e.g., data entry) dominates the workday, the pay is relatively low, and a worker's responsibility is limited primarily to working constantly and avoiding errors. These factors may also serve to veil the effects of workstation design improvements on productivity. If, as the Hawthorne studies suggested, motivation can mask the effects of poor workplace design, then perhaps it can also mask the effects of good workstation ergonomics.

REFERENCES

(1) Arndt, R. Working postures and musculoskeletal problems of video display terminal operators: A review and reappraisal. American Industrial Hygiene Journal, 1983, 44(6), 437-446.

(2) Becker, F. B. Workspace: Creating Environments in Organizations. New York: Praeger, 1981.

(3) Dainoff, M., Fraser, L., and Taylor, B. Visual, musculoskeletal, and performance differences between good and poor VDT workstations: Preliminary findings. Paper presented at the 26th Annual Meeting of the Human Factors Society, 1982. 10p.

(4) Grandjean, E., Nishiyama, K., Hunting, W., and Pidermann, M. A laboratory study on preferred and imposed settings of a VDT workstation. Behavior and Information Technology, 1982, 1(3), 289-304.

(5) Habinek, J., Jacobson, P., Miller, W., and Suther, T. A comparison of VDT antireflection treatments. Proceedings of the 26th Annual Meeting of the Human Factors Society, 1982, 285-289.

(6) Kirk, N. S. Organizational Aspects. In Ergonomic Principles of Office Automation. Stockholm: Ericsson Information Systems, 1983.

(7) National Academy of Sciences. Video Displays, Work and Vision. Washington, DC: National Academy Press, 1983.

(8) Roethlisberger, J., and Dickson, W. Management and the Worker. Cambridge, MA: Harvard University Press, 1939.

(9) Smith, G. L. Work Measurement: A Systems Approach. Columbus, OH: Grid Publishing, 1978.

Human Factors in Organizational Design and Management
H.W. Hendrick and O. Brown, Jr. (Editors)

OPTIMIZING HUMAN-MACHINE INTERFACE IN AUTOMATED OFFICES

Marcia L. Johnston

Lieutenant
United States Navy

The human-machine interface of the management information system used in U. S. Navy Pay/Personnel Administrative Support System offices was evaluated because low user performance and nonacceptance of the system was evident. A checklist of optimum ranges of workspace conditions was compiled and applied to the hardware, software, and workspace arrangement in one personnel office to identify problem areas. Strengths and weaknesses with possible alternatives and positive solutions are discussed.

The automated workspace, a result of applying the latest data processing technology to the office environment, is the focus of increasing attention. Management has invested large sums of money in hardware and software which promise to increase productivity and effectiveness. Meanwhile, the workforce finds itself under increasing pressure to perform on expensive, complex systems. The result can be positive but many offices have realized less than expected progress. Dr. Frederick Herzberg describes the factors which motivate the avoidance of work or the attraction to work in his two-factor theory [1]. Intrinsic factors, associated with the job itself, lead to job satisfaction and the degree of attraction to work, while extrinsic factors, such as working conditions and company policy, lead to job dissatisfaction. The dissatisfiers have no positive effect, their absence means only the absence of job dissatisfaction. Attention to the interface between humans and the machines they must utilize to perform their jobs can eliminate factors which lead to job dissatisfaction. This element is critical to the success of otherwise correctly designed automated systems.

The primary focus of this study is to examine one office which introduced computers into the workspace and how the human-machine interface was optimized. The most current research indicates an optimum range of conditions for this interface and a checklist (Figure 1) was compiled to compare with the existing conditions. The Navy Pay/Personnel Administrative Support System was implemented in newly consolidated offices worldwide, using a management information system to provide automated support for a personnel data base. This data base is the only local automated source of specific personnel and pay information available to Navy organizations supported by the personnel office servicing Navy members in the area. Assigned to manage the system four months after implementation and training was complete, this observer found the system under-utilized and not accepted by the operators. The system was being utilized successfully by other sites and the application programs had been accepted by the requirements group. Interviews conducted with management and users revealed the need

FIGURE 1 OPTIMUM CONDITION RANGES
FOR THE 5TH - 95TH % OPERATOR IN THE AUTOMATED OFFICE

I. ENVIRONMENT	
A. Lighting	Indirect, not on screen or in eyes
B. Temperature, Humidity	74-81°F, 80-0% relative humidity [3]
C. Noise, degree of annoyance	Acoustically treated, isolated
II. HARDWARE	
A. Furniture	
1. chair	5 legged, armless, contoured
a. adjustable height	Seat 15-18" tall, 15" max. depth [4]
b. lower back support	6" min. backrest, adjust ht. 8-10"
B. Keyboard Video Display	
1. keyboard location	26-30" desk, 16" min. knee depth
a. height	Parallel to elbow, detachable
b. angle	0-15° from desk to top back row
2. displays	Viewing distance 18-30" to read
a. screen characters	0-30° below horizontal sight line
b. width, height, spacing	Width 70-80% ht., space 20-50% ht.
c. color	Black characters on white [2]
d. contrast	Similar letters distinctly different
e. keyboard characters	Contrasting, legible, logical layout
f. controls	Distinct, isolated, critical:2 hands
g. status indicators	Visible, illuminated, well labeled
h. highlighting	Distinct cursor, critical functions
i. cluttering	No extra labels, distractions
3. screen adjustability	Contrast, angle 0-20° back on top
4. built-in glare control	Overhanging casing, non-glare glass
5. ease of use of controls	Textured knobs, relative action
C. Data Storage Materials	
1. vulnerability	No accidental disturbance, handling
2. within reach	Accessible from operator position
3. status monitoring	Labels, on/off indicators visible
D. Printers	
1. location	
a. status monitoring	On/off, paper, print status visible
III. SOFTWARE	
A. Relates to keyboard design	Uses keys as intended
B. Easy access with security	Accessible without vulnerability
C. Special functions	Help:tutorials on-line [5] Backspace:correct without reentry Scan:forward, back, screen, document
D. Response time	15 seconds maximum holds attention

IV. WORKSPACE ARRANGEMENT

A. Master Operator Position All controls, status indicators and materials for system operation must be easily visible and within reach of master operator position, secure, phone nearby.

B. Data Entry Position Located relative to in/out box but out of traffic flow, input materials in comfortable position (accessories), other necessary components visible. Provide privacy.

C. Function/Paper Flow Terminals separate from noisy high speed printers, paper separaters, distribution workspace and report distribution points, if possible. Analyze workflow and arrange components, in/out boxes, supplies, distribution centers, control positions logically.

for a human factors study. Personnel Support Activity Detachment Pearl Harbor, the largest and most varied configuration with the U. S. Navy Personnel Support Activity Hawaii organization, was selected for study.

The system consists of a CADO brand, model 40/IV minicomputer with a central processing unit, three floppy disk drives, three keyboard video display terminals, a letter quality printer and a high speed line printer. The high speed line printer is free standing. The terminals and letter quality printer are placed on existing furniture. The system is located in a centralized area of the office with three operators assigned to maintain the data base, produce reports and forms, and to perform word processing functions.

Lighting is fluorescent overhead and direct from draped windows on two sides of the room. Venetian blinds were resurrected and used to replace the existing draperies which provided inadequate blocking of afternoon sun. A small window air conditioner keeps temperatures within a comfortable range and decreases humidity in the room. The high speed line printer generates a high level of noise even though noise reducing material was used to line the printer casing. The letter quality printer had been installed in a noise reducing hood which significantly reduced the annoying noise level. The existing furniture was the most easily solved problem. Furniture throughout the office was gathered during the consolidation phase from local organization personnel offices as they were dissolved--a huge collection of used government furniture. The tables were 35" tall, chairs were armed and not adjustable or on swivels. One workstation put the keyboard near chin level. This contributed significantly to the physical discomfort of the operators. The letter quality printer table was so unstable that it swayed while printing. Workstation tables of the optimum dimensions replaced the incorrect ones. Chairs were obtained which provided adjustable lower back support, adjustable height, contour and the appropriate dimensions. A mobile printer stand of adequate stability, designed to accommodate boxes of paper and a paper catcher in the rear, was introduced.

The hardware components of the computer system reveal evidence of design with attention to the human-machine interface. The proper keyboard angle is built into the terminal, the screen position and angle are adjustable beyond what is necessary. The screen typography strokewidth to height is 1:9 using a 7x9 dot pattern with 9 scan lines/mm which is within established optimum standards of 8 to 10 or more. Character width to height is 2:3, with green characters on a black background. Although shifting from the white background of input sheets to the black background on the screen requires continuous eye adjustment, green letters score highest by users for resolution on a black background [2]. Glare control features include a layer of non-glare glass on the face of the screen and a protruding casing around the screen which blocks direct overhead light. The screen background is fixed with a moving cursor. The cursor is distinct: a solid, illuminated rectangle. Characters appear as typed in, added to the existing text. Symbols and brightness of light are used for highlighting the commencement of a critical function with intensifying brightness at normal and one twice as bright level: *CRITICAL FUNCTION--DO YOU WISH TO CONTINUE?*. A bell rings when an unacceptable entry is made.

Keyboard labels are cluttered with three symbols on most keys. Over half of the keys are function keys for word processing, labeled on the facing edge of the key. In addition to the letter label, another, unused code is labeled across the top of the key. This absolutely bewilders trainees. Because the equipment is leased, altering the keyboard is not recommended but masking the unused codes would reduce cluttering. Letters follow the typical typewriter configuration with additional keys on each side. The system controls appear in a line across the top of the keyboard, separate from the other keys, with labels on each key stating its function clearly and legibly. The on/off switch, an easily operated toggle switch, is located on the back panel of the terminal and rarely viewed. The location reduces the opportunity for accidental deactivation. Controls for adjusting contrast and angle of the screen are not distracting. The knobs are textured and use action relative to the amount of adjustment required.

The disk drives are located in the lower right front section of one system workstation. The controls for opening the drives are 18 inches from the floor, flush with the front edge of the table. Very vulnerable to accidental activation, this presents a danger to the integrity of the disks since the spring switch ejects the disk when pushed. A red light is positioned in the center of the switch to indicate when the disk is in use but is not easily monitored by the operator. In this configuration, placement of the three floppy disks is critical. Disk labels, however, are located on the wrong side of the disk for viewing while inserting, creating a problem for the operator. Marketing brochures revealed that the drives had been installed on the wrong side of the system workstation. If installed correctly, labels could be easily monitored from the seated position. This problem was eliminated with a subsequent upgrade but placing labels on the back side of the disks to be viewed during use is a solution. Positioning of the disk drives is critical to an application in which frequent switching is required. Proper handling of the disks is displayed on the disk jacket very effectively in five different languages, aided by diagrams which illustrate the caption.

The controls on the letter quality printer are not easily visible from the seated position because of the noise reducing hood. The existing arrangement did not allow for monitoring the status of paper, whether the printer was on or off, or printing correctly, from the operator position at the terminal. As a result, a critical security report was frequently misprinted. Controls with settings not to be changed are located just inside the printer under a detachable plate removed for servicing the printer. This plate is often removed by the operators to interrupt print because no pause feature was designed in the software. Consequently, the settings are often accidentally changed, causing an incorrect interface with the rest of the system and jumbled print. A label was attached near the switches illustrating the proper settings for easy monitoring and proper reset. This printer was moved to a more easily monitored position. The high speed line printer on/off toggle switch is located on the rear panel. Unless printing, the printer makes little noise and the status is hard to determine. So prior to using it, the operator must get up and check the printer. The paper feeds from the front and is easily monitored. Generally, the printer is left on during the day to ensure that reports print out as expected.

The software interface, controlled by the contractor, is the main source of problems and the most difficult to adjust from the front end. The screen interface is controlled by the application software, not hardware engineering. More care in designing human-machine interface is needed because users are impatient and critical when handicapped by a clumsy design. Software use of the keyboard does not relate to the hardware design. Only one of three cursor direction control keys are utilized. The other three direction controls are letter keys with dual functions. The select function key, used during data entry to move from one field to another, is not the return key for word processing but is used to change functions when followed by another letter key with a dual function. Volumes of text have been lost because of this arrangement. The application software is menu driven and easily accessible with critical functions limited to master operator access. Functions are easily controlled and even terminals can be restricted to perform only authorized functions. No tutorials are available on-line but a separate word processing tutorial was obtained from the contractor two years after implementation. During data entry, a backspace key, which erases as it moves, is utilized only within the field being entered. If mistakes are noticed after the field has been exited, the operator has the option of cancelling the entire record or edit, or completing the entry, recalling the record and correcting the mistake. This safeguards the data and it provides security for the operator against accidentally destroying data. A scan function is used both in data entry and word processing. The key action is difficult to control in word processing as the stimuli is collected faster than it is used. Response time varies depending on the function performed. When operators perform text storing or report sorting during data entry, response time slows considerably, falling outside the maximum optimum range. Scheduling can be used to reduce this problem if it becomes critical.

The workspace arrangement was studied and redesigned to provide a logical flow of processes and paper based on the operations of the office. Careful placement of components on furniture of proper dimension ensured that the master operator could perform necessary system functions without leaving the position or straining to monitor the status of critical elements. Data Entry positions were provided a clipboard attached to a flexible arm which adjusts to place input sheets within comfortable view. These workstations were positioned away from the door, in/out baskets and the traffic flow. Space for input sheets reserved after data entry for a set time was created separate from new information collection points with a cycling system established. A word processing workstation was set up to enable monitoring of the letter quality printer status without strain. A report distribution area, with a decollater placed next to the high speed line printer and a worktable in line after it, was created near the entry.

User dissatisfaction decreased significantly and interest in system capabilities and possible uses increased after removing possible sources of annoyance. Training, aids and guides lessened the effect of unsolvable problems. However, management interest in working conditions alone was a significant factor in decreasing user dissatisfaction.

[1] Herzberg, F., Mausner, B., & Snyderman, B., The motivation to work (2nd ed.) (Wiley, New York, 1959).

[2] Jones, K., Examining the human aspects of technology, Mini-Micro Systems 14 (1981) 49-50.

[3] Fanger, P. O., Thermal comfort, analysis and applications in environmental engineering, in: McCormick, E., Human Factors in Engineering and Design (McGraw-Hill, New York, 1976).

[4] Rafferty, C. and Keener, J., Providing terminal comfort, Mini-Micro Systems 14 (1981) 119-124.

[5] Chafin, R., Software ergonomics, Human Factors in Applications Programming, Infosystems 30 (1983) 102.

Human Factors in Organizational Design and Management
H.W. Hendrick and O. Brown, Jr. (Editors)

PARTICIPATION IN TECHNOLOGICAL CHANGE : ARE THERE ONLY UNEXPECTED CONSEQUENCES ? THE CASE OF OFFICE AUTOMATION

Marco Diani
Centre National de la Recherche Scientifique
Centre d'Etudes Sociologiques
82 rue Cardinet 75017 Paris
France
Sebastiano Bagnara - Raffaello Misiti
Consiglio Nazionale delle Ricerche
Istituto di Psicologia
Via dei Monti Tiburtini 509, 00151 Rome
Italy

Participative methods of technological change reveal new and more complex problems of Office Automation leading to an increase in organizational mental load, conflictual cooperation and stress. A more realistic relation between the technological limitation of computer-based information systems and its reflects on cognitive-mediation of work can provide more adequate indicators of organizational performances.

I. INTRODUCTION

During the last 20 years, the study of the consequences of office automation has been mainly oriented to :
- The physical characteristics of new technologies, in particular all the issues arising from long terme use of video display terminals (1).
- The effects of office automation on the structure of total employment and the distribution of skills and qualifications (2, 3).
- The elaboration of systematic analysis allowing to redesign office functions in an automated setting (4, 5).

The new systems contributed very often in the past to routinize job content, providing no financial incentive, isolating workers, and threatening the elimination of jobs. The consequence of behaviorally unsophisticated office automation is now clear. Rational human resistance to change will prevent any economic advantage (6). Thus, the study of the "unexpected consequences" of the diffusion of office automation appeared as a very important issue for management, worker's organizations, and also to a certain extent, for producers. Year after year, the generalization of the new technology provoked an incresaing series of phenomena difficult to control, forecast and manage, whose common characteristics are :

- the resistance to adopt and use the new technology
- the underutilization of the machines and reduced economic performance
- the rise of social conflicts and recurrent organizational breakdowns.

A wider worker's participation in technological change appeared as a realistic solution (7).

2. WHAT IS MEANT BY PARTICIPATION ?

In order to prevent and to reduce the importance of these phenomena, participative methods of technological change have been developed, varying mainly according to the different strenghts of unions in Europe, USA and Japan (8). The solutions found to prevent the "unexpected consequences" of office automation can be divided into two categories : procedural and substantive. The first ones are more like a normative set of regulations based on legislation, standards and rules, mostly promoted by unions and concerned with the methods of introducing the new technologies. Substantive issues relate more to operational conditions once office automation has been implemented. The fundamental aim of participative methods is to establish a strong consensus around the new technology and the new organizational setting of the office, by promising a democratic decision of implementation of office automation, providing (9) :

- informations to the unions at an early stage of the process, long before the final decision is to be taken and it is still possible to influence the choice ;

- joint union-management bodies to discuss, negociate and to supervise the changes, with the possibility of consulting independent experts ;

- user's involvment in the design of future organization.

Substantive issues are oriented to the protection of existing status, salary and qualifications providing :

- the same level of employment existing before the introduction of new technologies will be maintained ;

- in-house displacement is allowed only if associated with important retraining programs ;

- an increase of quality of working life ;

- a limitation of the time to be spent working on VDTs.

3. AN ANALYSIS OF A PROTOTYPICAL CASE OF PARTICIPATION.

In spite of important results, in particular a reduction in the number and intensity of visible conflicts, it appears that the participative methods of office automation reveal some new problems, more difficult to solve. Though much field work remains to be accomplished, some observations from a recently conducted study can be relevant to the discussion of these problems (10). We analyzed a prototypical case of participation in a large nation-wide bureaucracy, highly unionized at every level of the hierarchy. The degree of cooperation between unions and management is very high. The tasks accomplished are of two types :

- structured, formalized and foreseable : invoices, uptaking and processing files, payments ;

- unstructured, difficult to formalize : communications, coordinations with others institutions or units, special cases.

Before the generalization of office automation, the structure relied heavily on a centralized EDP system, whose speed, accuracy and performances declined very fast, both for technological reasons and for the expanded quantity and complexity of tasks to be performed. The participative introduction of the new system was considered by all the institutional actors involved as the best mean in order to attain :

- an increase in productivity and a better quality of services,

- an improvement of the quality of working life, through the elimination of repetitive and low-skilled tasks,

- more autonomy in decision-making process at all levels of the structure.

More than in other cases, all the standard procedures of participative methods were followed with a strong emphasis on two points :
- the preservation of employment, salary and previous qualifications, despite the important modifications created by the new technologies,

- the limitation to four hours of dayly work on VDTs, and the creation of new organization of work, based on binomes, i.e. employees alternating on the same display during the day.

4. PROBLEMS BEYOND THE SCOPE OF PARTICIPATION.

In this prototypical case, we did observe phenomena particularly pertinent to the study of future trends in office automation, that cannot be solved only with the application of participative methods (11).

First, the division of work created by the binome, based on external contraints and measurements, is more concerned with health and safety regulations than with the content and the nature of office work. In the automated workplace the quantity and the flexibility of the informations available is almost beyond the physical limits of the machines : then, one of the most radical changes is that, paradoxically, operators do not know what to do during the second half of the working day, the most important operations and informations being available only through the VDTs. Moreover, particularly complex cases or special operations require impredictable quantities of time, and are almost impossible to plan in advance, leading to a very rigid and unbalanced task-allocution between members of the binome. The solution proposed in this prototypical case of participation probably reduced the fears of impersonal control, but introduced the pressure of the "peer-group" control of individual performances. In the second place, all the organizational structure of the office was changed by the new technology, but the hierarchical structure of the office was unchanged and we observed some micro-conflicts based, for instance, on :

- the differences among the types of technological culture of employees,
- the difficult of integrating the traditional office functions within the new knowledge required to insure the maintenance of the machines.

5. THE ORGANIZATIONAL MENTAL LOAD AND THE COGNITIVE MEDIATION OF WORK.

Above all, we observed the increase of mental work-load related to the cognitive mediation of organizational factors. This organizational mental-load is very difficult to manage, measure and reduce:the incompatibility between organizational models and computerized procedures can lead to a paradoxical and forced development of forms of social competition, cognitive conflicts and "conflictual cooperation." (12)

With computer based technologies, a component of mental work-load that has been insufficently studied becomes visible : it is represented by all the variables that define the social organization of aims and the division of work. For instance, given the same conditions, the number of errors varies as a function of the control, directly exerced both by the hierarchy or the peers-group : an operator devotes more attention to the "controller", is less concentrated on the task and the likelihood of errors increases. Furthermore, the importance of the error itself is immediately expanded by the organizational dimension created by the social comparison of the two members of the binome.

This organizational work-load is relevant for two main reasons (13). First of all, the market of office automation provides hardware and software that present very little flexibility, based on analysis of work procedures according importance almost uniquely to structured activities. This technological rigidity of office automation devices makes difficult for operators to respond to uncertainty, to perform non-structured tasks and to consolidate every type of informal and learning-by-tasks expertise. Formal negociations and informal arrangements with the hierarchy and the managers became much more difficult : a number of mental activities shift from accomplishing the tasks of the office to the standardized rules, trying to adapt behavior to the goals and tasks of the computer-based system.

Once goals and rules are standardized, less space is left to bargaining process : the integration of procedures does not increase the richness of tasks and job satisfaction, but leads instead to an increase of attention, concentrated on control, perceived as direct and non negotiable. Errors are perceived as organizational failures (14, 12).

6. THE LIMITS OF COOPERATION.

The organizational mental load has very important consequences on the reliability and performances of automated office systems, leading to a diminution in quality and quantity of expected work. It has been observed (15) that an increase in social cooperation can mediate and reduce the pressure and the task monotony or uncertainty can be offset by other kinds of responsability such as employee participation in decisions regarding work flow, standards hours, office layout. Social cooperation is then viewed as new organizational perspective and a consequence of a successful participation in technological change. Judging from our experience, the forms of social cooperation observed in automated offices are paradoxical creative responses provided by informal segments of the office in order to increase their bargaining power and to share informations concerning functions not recognized by the new organizational and technological systems.

More than the result of a "shared culture", this paradoxical cooperation is a temporary adaptation to a stressing working condition : it constitutes an indicator that parts of the traditional office organization, not integrated in the computer-mediated system and not negotiable through collective bargaining, are cognitive-mediated by individuals, leading to an increase of their mental workload.

CONCLUSIONS.

With office automation, tasks and goals of the traditional office organization too difficult to be standardized, become increasing cognitive-mediated, with little possibility of bargaining and negotiation concerning :
- the division of work and organization of goals,
- the organizational ambiguity of tasks and boundaries.

This tendency of office automation to exasperate the non-negotiable part of the work reduces the possibility of human performance and adaptation in a very impredictable work environment, difficult to standardize.

Participative methods of introduction of office automation , and the democratic division of work, at least in the type of organization that we have observed, instead of establishing a mutual confidence among operators seem to reinforce these elements of rigidity. It is likely that the effectiveness of participation in technological change methods will be improved by paying less attention to formal aspects of work redesign, while establishing a more realistic relation between the technological nature of computer-based information systems, its reflects on cognitive-mediation of work and goals, so to provide more adequate indicators and measurements of individual and organizational performances.

REFERENCES

(1) Grandjean, E., and Vigliani, E., (eds.), Ergonomics aspects of visual display terminals (Taylor & Francis, London, 1980).
(2) ETUI, The impact of microelectronics on employment in Western Europe in the 1980 (Euroinst, Brussels, 1979).
(3) Sirbu, M., Understanding the social and economic impacts of office automation (The Japan-Usa Office automation Forum, Cambridge, Mass, 1982).
(4) Otway, H.J., and Peltu., (eds.), New office technology. Human and organizational aspects, (Frances Pinter, London, 1983).
(5) Bjørn-Andersen, N., (ed.), The human tide of information processing, (North-Holland, Amsterdam, 1980).
(6) Driscoll, J.W., How to humanise office automation, Office : Technology and people,1(1982) 167-176.
(7) Mumford, E., and Sackman, H., (eds.), Human choice and computers, (North-Holland, Amsterdam, 1975).
(8) Mowshowitz, A., (ed.), Human choice and computers,2 (North-Holland, Amsterdam, 1980).
(9) ETUI, Bargaining and new technologies, (Euroinst, Brussels, 1982).
(10) Diani, M., and Bagnara, S., Unexpected consequences of participative methods in the development of information systems : The case of office automation, in : Grandjean, E., (ed.) Modern office jobs, (Taylor & Francis , London, in press).
(11) Shackel,B., (ed.), Man-computer interaction : Human factors aspects of computer and people, (Sijkoff and Noordhoff, Rockville, MA., 1981).
(12) Bagnara, S., and Diani, M., L'organizzazione nel carico di lavoro mentale, Studi Organizzativi (1), 1984, 156-167.
(13) Bagnara, S., Il lavoro umano nell' ufficio automatizzato, in Occhini, G.,(ed.), L'automazione nell' ufficio, (Franco Angeli, Milano, 1984).
(14) Crozier, M., Implications for the organization, in Otway, H.J., and Peltu, M., (eds.), New office technology. Human and organizational aspects, (Frances Pinter, London, 1983).
(15) Zuboff, S., Computer-mediated work : The emerging managerial challenge, Office : Technology and people, 1 (1982), 237-243.

Human Factors in Organizational Design and Management
H.W. Hendrick and O. Brown, Jr. (Editors)

ORGANIZATIONAL ANALYSES: IDIOSYNCRATIC METHODOLOGIES SEARCHING FOR HOLISTIC SOLUTIONS, OR WHAT'S WRONG WITH THE WAY THEY ARE DONE

Alan L. Carsrud

Department of Management
Graduate School of Business
The University of Texas at Austin
Austin, Texas 78712
U.S.A.

The problematic outcomes of organizational analyses as typically done by consultants, social scientists, managers, and engineers are reviewed in this paper. The focus is on organizational analysis based solutions where changes or equipment are implemented without adequate consideration of the interrelationship of jobs and informational needs of the people in the organization. The social and psychological impacts of typical organizational solutions are reviewed as they relate to the development of User Systems Analyses.

This paper is the initial presentation in the panel session, "User Systems Analyses: Network Analyses of Organizational Structure, Tasks, Personnel, Equipment, and Procedures".

INTRODUCTION

"What you recommend doesn't work." This is a phrase frequently spoken and/or heard by a variety of individuals who are concerned with the design and management of the work place. Organizational consultants, academics in the social and behavioral sciences, human factors engineers, information systems designers, managers, and marketing representatives are often accused of not understanding or appreciating the complexity and inter-relatedness of events and people within office settings. This is even the case when they attempt to perform consulting, conduct research, design equipment systems, manage a business, or sell office products.

The typical results of the failure to attend to detail and to understand are: structural or procedural recommendations that are inappropriate, or not implemented; research studies that are at best esoteric, or useless; management solutions that may be debilitating; and/or equipment/software solutions that are ineffective, frustrating, and costly. There is a proclivity on the part of all those involved to blame either the narrow-mindedness of the organizations, and/or the stupidity of the consultant, academic, engineer, salesperson, manager, or office staff.

The real culprits are not the people or organizations involved, but rather the popular, yet limited, conceptualizations of the organization (or the work-place), and the narrowly focused, or empirically weak, methodologies for studying the organizations and people. Recent advancements, in the form of User Systems Analyses (USA), may offer some relief from the problems of previous conceptualizations of organizations and the work people who work within them. Certainly the area of organizational development, as currently applied, needs a methodology that takes it beyond its current clinical/idiographic approach to organizational change.

The purpose of this and following papers on User Systems Analyses (Glick & Beekun, 1984; Krus, Carsrud, & Glick, 1984; Brady, 1984; and El Sherif, 1984) is to present a unique conceptual, interdisciplinary, multi-level methodology for studying organizations and the people within them. User Systems Analyses (USA) attempts a systematic, analytic, and empirical integration of a wide variety of organizational factors such as information-work flow, job tasks, individual personnel characteristics, equipment utilization, administrative procedures, and managerial structure. USA attempts to address the concerns of Hackman (1978) about the general weaknesses of previous conceptualizations of socio-technical theory in terms of its incompleteness, lack of causal specificity, and inattention to measurement issues.

PREVIOUS THEORETICAL CONCEPTUALIZATIONS

In order to better understand the rationale behind USA it is important to consider briefly some of the approaches currently used in the study of organizations (or the office environment). Each of these methodologies is strongly influenced by the specific traditions, or academic disciplines that developed them. They are analogous to the ancient proverb of blind men describing an elephant. Each has his own unique, and valued perspective, yet none describes the whole.

Despite the desirability of more holistic approaches in the study of organizations, Miner (1980) has argued they have been less fruitful in producing convincing empirical research support. This may be the result of better methodologies and data collection systems for more micro-oriented theories. It is also possible that because of rigid disciplinary boundaries, most theories and methodologies related to the study of different organizational behaviors do not cross various levels of analyses (Miner, 1980). Only a few conceptual attempts, such as the concept of integration in goal congruence theory (Argyris, 1973; Argyris & Schon, 1974) have been made to bridge various levels of analyses.

Approaches, such as Organizational Control Theory (Tannenbaum, 1968), Psychological Open Systems Theory (Katz & Kahn, 1978), Decision Theory (Simon, 1977), and Sociological Control Theory (Thompson, 1967), are examples of academic approaches to the study of various organizational behaviors that take very specific views of the "organizational elephant". Although each of these conceptual approaches has produced a wealth of empirical, or case data, each is strongly colored by the particular academic discipline of the theorist and researcher. What is needed is a method of analyzing organizational behavior that is either atheoretical, or multi-theoretical in basis.

Glick and Beekun (1984) provide a review of sociotechnical approaches to the study of organizational behavior as they relate to the development of USA. Despite the potential problems with sociotechnical theory as typically applied (Hackman, 1978), the concept of blending levels of analyses may be the only way to obtain a fully integrated view of the organization and the people in it. The panelists on USA represent a variety of academic and applied disciplines, including: personality and social psychology, sociology and administrative sciences, human factors engineering, electrical engineering, and computer systems.

Using these various academic disciplines as a conceptual background, USA addresses Hackman's (1978) concerns that the typical sociotechnical approach: 1) does not specify the attributes of group and individual tasks that are required for effective work groups; 2) does not take into account specific individual differences of the workers; and 3) does not address the internal dynamics of the work groups being studied. Brady (1984) reviews the development of the USA as it addresses the needs for a multi-level methodology for organizational analyses. Krus, Carsrud, and Glick (1984) discuss the development of a computer assisted USA program and the application of USA in an existing organization to address personnel issues.

ORGANIZATIONAL ANALYSES: PROBLEMS IN APPLICATION

Organizational analyses have typically been used for a variety of applied purposes, as well as for purely academic outcomes. These applied purposes have included: organizational development; strategic planning, job analyses; organizational designing; and managerial and procedural changes. Additional uses have included: job design and enrichment; equipment purchases; employee participation in management; improving organizational productivity; studying staff satisfaction; and designing office layouts and equipment solutions (Bowers & Hauser, 1977; Cummings & Molloy, 1977; French, Ross, Kirby, Nelson, and Smyth, 1958; Goodman, 1979; Porras & Berg, 1978; and Seashore & Bowers, 1970).

The following are brief descriptions of typical problems resulting from the application of solutions based on inappropriate and/or limited organizational analyses. These problems are not the result of bad recommendations per se. Instead, they are the result of not being aware that any one behavior may be contingent upon several behaviors. Likewise, any one behavior may affect many other behaviors exhibited within the organization.

Management, consultants, and equipment salesmen have frequently recommended the use of computer systems and word-processors in the various organizations. Often these installations have not resulted in the time savings promised, and often have been cited as contributing to increased dissatisfaction of employees. The reason seems to be that when the equipment was placed into the office environment and procedures, little concern was taken about the social and work relationships the equipment was disrupting. Often these relations were between the secretary and the principal he/she served. A related issue is when one job is automated, while support positions, or subsequent positions in the workflow are not automated to deal with the increased output, thus overloading or underutilizing others in the organization. This failure to note the interrelatedness of jobs in the work place is a critical problem that occurs more often than it should.

Another frequently noted problem with organizational redesign occurs when changes are made in how, or what, a person does in the office without looking at the other activities in which that person is involved. Often these changes may be viewed as job enrichment by management, but as a stress producing information overload by the employee. A related issue is the tendency for organizational changes to be made that do not take into account the personalities or skills of the participants. Such changes are often "doomed from the start" because of such "personnel problems".

Often managers and consultants alike assume that adding more people will improve the productivity of an organization. Even with improved automation, etc., such additions ignore the reality of developing relationships between the new workers and the existing organization and personnel. The addition of a single new worker adds the potential for multiple new interaction and information exchanges. Failure to properly integrate the new worker into the organizational system often negates the potential value of the additional manpower.

RESEARCH ON USER SYSTEMS ANALYSIS

The joint research project between the University of Texas at Austin (UT) and International Business Machines (IBM) has attempted to overcome the problems with traditional organizational analysis solutions that have been discussed above. The focus of USA, as Glick and Beekun (1984), Brady (1984), and Krus, et al. (1984) note, is to make the workflow diagram a more useful tool in a broad range of organizational problem solving. During the course of the joint UT-IBM project, a number of benefits for USA have become obvious. Several of these benefits have actually been implemented in various organizations (Krus, et al, 1984). These include the marketing and purchase of new equipment, improved training packages for new employees on organizational work procedures, and job redesign. El Sherif (1984) reviews the applications of USA and the potential of its computer assisted analysis to improve the job satisfaction and performance of individuals, while increasing productivity and profitability of the organization. Although USA is still a developing methodology, its potential in both basic and applied research settings is encouraging. The whole picture of the organizational "elephant" may yet develop.

REFERENCES

[1] Argyris, C., Personality and organization theory revisited, Administrative Science Quarterly, 18 (1973), 141-167.

[2] Argyris, C. & Schon, D. A., Theory in practice: Increasing professional effectiveness. (Jossey-Bass, San Francisco, 1974).

[3] Bowers, D. G & Hausser, D. L., Work group types and intervention effects in organizational change, Administrative Science Quarterly, 22 (1977), 76-94.

[4] Brady, L., Review of the systems methodology of User Systems Analyses, in: Hendrick, H. W. and Brown, O. (eds.), First international symposium on organizational design and management, (North Holland, Amsterdam, 1984).

[5] Cummings, T. G. & Molloy, E. S., Strategies for improving productivity and the quality of work life (Praeger, New York, 1977).

[6] El Sherif, H., Implications and applications of User Systems Analyses, in: Hendrick, H. W. and Brown, O. (eds.) First international symposium on organizational design and management, (North Holladn, Amsterdam, 1984).

[7] French, J. R. P., Ross, I. C., Kirby, J. R., Nelson, J. R. & Smyth, P. Employee participation in a program of industrial change, Personnel, 35 (1958), 16-29.

[8] Glick, W. H. & Beekun, R. I., A theoretical overview of User Systems Analysis, in: Hendrick, H. W. and Brown, O. (eds.) First international symposium on organizational design and management, (North Holland, Amsterdam, 1984).

[9] Goodman, P.S., Assessing organizational change: The Rushton quality of work experiment (Wiley, New York, 1979).

[10] Hackman, J. R., The design of self-managing work groups, in: King, B., Steutfert, S. and Fiedler, F. W. (eds.), Managerial control and organizational democracy (Wiley, New York, 1978), 61-91.

[11] Katz, D. & Kahn, R. L., The social psychology of organizations (Wiley, New York, 1978).

[12] Krus, E. J., Carsrud, A. L., & Glick, W. H., Interactive organizational workflow analyses: An approach to problem solving in organizations in: Hendrick, H. W. and Brown, O. (eds.) First international symposium in organizational design and management (North Holland, Amsterdam, 1984).

[13] Miner, J. B., Theories of organizational behavior (Dryden Press, Hinsdale, Ill., 1980).

[14] Porras, J. L. & Berg, P. O., Evaluation methodology in organizational development: An analysis and critique, Journal of Applied Behavioral Science, 14 (1978), 151-173.

[15] Seashore, S. E. & Bowers, D. G., Durability of organizational change, American Psychologist, 25 (1970), 227-233.

[16] Simon, H. A., The new science of management decision (Prentice-Hall, Englewood Cliffs, N.J., 1977).

[17] Tannenbaum, A. S., Control in organizations (McGraw-Hill, New York, 1968).

[18] Thompson, J. D., Organizations in action (McGraw-Hill, New York, 1967).

Human Factors in Organizational Design and Management
H.W. Hendrick and O. Brown, Jr. (Editors)

A THEORETICAL OVERVIEW OF USER SYSTEMS ANALYSIS

William H. Glick and Rafik I. Beekun

Department of Management
University of Texas
Austin, Texas 78712

User System Analysis is an evolving approach for assessing organizational needs for information processing equipment and evaluating task and organizational design. This paper presents User Systems Analysis as an integration of the sociotechnical systems approach and social network analysis. Substantial productivity gains and improvements in organizational effectiveness may be captured by reorganizing work procedures and investing in information handling equipment.

In the previous paper, Alan Carsrud identified a series of limitations with traditional approaches to analyzing and changing organizations. Most importantly, he illustrated the problems encountered when new information processing equipment and procedures are introduced following limited analyses that: focus on a single unit of analysis such as individuals or departments; neglect the interdependence of individuals in organizations; and either adopt a highly specific atheoretical approach that results in numerous ill-supported recommendations for change or an overly abstract theoretical approach.

Although the sociotechnical systems approach (Susman, 1976; Trist, 1981; Trist & Bamforth, 1951) was not developed specifically to evaluate the need for information processing equipment, it overcomes many of the limitations of previous analysis strategies. This paper reviews the contributions and limitations of the sociotechnical systems approach to organizational analysis and proposes social network analytic techniques for assessing patterns of social and technological interdependence and evaluating sociotechnical system fit. The evolving theory and methodology, User System Analysis, suggests several propositions for improving task and organizational design by introducing new information processing equipment and/or reorganizing work procedures.

THEORETICAL DEVELOPMENTS LEADING TO USER SYSTEM ANALYSIS

SOCIOTECHNICAL SYSTEMS

Sociotechnical systems theory developed as a reaction to two separate approaches to task and organizational design: the scientific management approach that emphasized technological determinism in the structuring of organizations and the social psychological and human relations approaches that only considered social and organizational factors. The basic tenet of sociotechnical systems is that both social and technological factors must be considered in the design of tasks and organizations. The social and technological subsystems must be jointly optimized to achieve effective overall system performance.

Although it is difficult to find a parsimonious set of specific system design factors in the sociotechnical literature, several general trends may be observed. First, the adoption of an open systems perspective leads to the assumption that system characteristics at one level affect the functioning of systems at other levels (DeGreene, 1973). Thus, multiple levels of analysis are used to analyze task and organizational designs. Sociotechnical analysis frequently proceeds on four levels; individual, workgroup, organizational, and macrosocial (Kelly, 1978; Trist, 1981). Each level is important to an understanding of the whole.

Second, the systems perspective has encouraged efforts to model tasks and organizations (DeGreene, 1973). The models have varied from rich verbal descriptions of operations (Rice, 1958; Trist & Bamforth, 1951) to rudimentary flow diagrams of work processes (Kolodny & Kiggundu, 1980) to analytical models (DeGreene, 1973; Rousseau, 1977). The accuracy and level of sophistication of these models has varied.

A third common element in sociotechnical research is the introduction of autonomous workgroups (Kelly, 1978; Sussman, 1976; Trist, 1981). Autonomous workgroups are designed to produce a meaningful whole unit of output with minimal supervision and minimal dependence on the rest of the organization. Group members are typically allowed to allocate tasks, to set their own pace for task completion, and to become multi-skilled individuals to eliminate bottlenecks in group production.

SOCIOTECHNICAL SYSTEMS ANALYSIS OF INFORMATION PROCESSING NEEDS

The sociotechnical systems approach offers several contributions to the assessment of organizational needs for new information processing equipment and procedures. First, it highlights the need to consider both technological and social factors. The recent book on computer system acquisition and implementation by Lynne Markus (1984) repeatedly underscores this point. Second, sociotechnical systems research has shown the necessity of multilevel analyses for designing tasks and organizations. Finally, sociotechnical systems research has demonstrated the utility of rudimentary workflow diagrams and multiple techniques for modeling organizational systems.

Although joint optimization of social and technical factors is clearly desirable, sociotechnical systems theory has not provided a sufficiently explicit definition of joint optimization (Kelly, 1978; Trist, 1981). The critical dimensions of social and technical subsystems are inadequately specified. Some authors suggest that the pattern of workflow interdependence is the important characteristic of the technical subsystem (Rice, 1958; Rousseau, 1977), however, existing descriptions of these patterns of interdependence tend to be highly specific to the individual case study (Rice, 1958; Trist & Bamforth, 1951) or too abstract for developing appropriate solutions to information processing needs (Rousseau, 1977). The social subsystems have been described in terms of roles, communication patterns (Taylor, 1975), and friendship and kinship ties (Trist & Bamforth, 1951). Unfortunately, no precise definitions of social or technical subsystems have received widespread acceptance.

Sociotechnical systems research is also limited by its advocacy for autonomous workgroups in all situations. Kelly (1978) noted that almost all sociotechnical research has been conducted in organizations with high process uncertainty, a condition that does not exist in many clerical and information processing operations. Thus, autonomous workgroup design may be inappropriate in many settings that are concerned with investing in new information processing equipment.

TECHNOLOGICAL INTERDEPENDENCE

A more explicit definition of the social/organizational and technical subsystems is provided by J.D. Thompson's (1967) open systems theory of technological interdependence. It is argued that effective system performance requires a match between the pattern of technological interdependence (pooled, sequential, and reciprocal) and the method of coordination (standardization, planning, and mutual adjustment). The similarities between this macro-organizational theory and the sociotechnical approach should be readily apparent (Rousseau, 1977). Both argue for a match between two subsystems for effective overall system performance. Unfortunately, Thompson's (1967) form of analysis is limited by the possibility of multiple patterns of interdependence in a single organization and the existence of many hybrid patterns of interdependence that do not fall neatly into any of the defined categories.

SOCIAL NETWORK ANALYSIS

Social network analysis was developed out of the early work on sociometry by mathematical sociologists (Burt, 1982), who have revived interest in this methodology for organizational analysis (Lincoln, 1982; Tichy, Tushman, & Fombrun, 1979). In this context, "network analysis" refers to a set of descriptive techniques, it should not be confused with the prescriptive optimization routines found in management science. Network analysis is a very flexible, yet precise analytical method for describing patterns of relations among actors in a social system. The content of the relations among actors may vary from verbal communication to authority relations to workflow interdependence. Thus, organizational needs for information handling equipment and/or organizational restructuring can be assessed by analyzing work and information flows between individuals, work stations, and equipment.

Following the sociotechnical systems perspective, we can hypothesize that the greater the overlap between the patterns of work (ie., technical subsystem) and information flows (ie., social subsystem), the higher the system performance. Social network analysis provides a precise mathematical definition of the degree of this overlap by calculating the average multiplexity between two or more networks (see Burt, 1982 for a more thorough discussion). It may also be used to assess the relative importance of different contents of the social and technical subsystems such as the communication, friendship, and authority networks. In contrast with Trist and Bamforth's (1951) emphasis on friendship and kinship ties, a network analysis approach may show that overall communication patterns are more important determinants of system effectiveness.

Additional propositions are suggested by system and positional characteristics such as workflow centrality and criticality (for lists of additional network characteristics, see Burt, 1982; Lincoln, 1982; Tichy et al., 1979). For example, highly central individuals in the work and information flows are likely to become very overloaded and exhibit symptoms of stress and underperformance. Further, bottlenecks and organizational problems are likely to occur when the organization becomes very dependent on a few individuals or work units who are centrally located in critical junctures of the workflow. Network analysis can be used to identify these highly critical positions so that system designers can restructure the patterns of interdependence and increase overall system effectiveness. Unfortunately, individuals in these critical, central locations are also likely to be the most resistant to proposed changes in the work and information flows. Network analysis can at least identify these potential problem areas.

This quantitative form of descriptive network analysis is frequently supplemented by graphical displays of simple networks. The network diagrams are easily interpreted, while the quantitative analyses provide a precise, accurate method for summarizing the information in complex networks. This precision should facilitate the development and empirical verification of a stronger theory for organizational analyses. The combined use of graphical workflow diagrams and social network analysis can help organizational analysts to go beyond anecdotal descriptions of the work flow. This should lead to stronger recommendations for improving organizational effectiveness by changing work and information flows and/or investing in new information processing equipment.

CURRENT STATE OF USER SYSTEM ANALYSIS

User System Analysis attempts to integrate the strengths of these different approaches to the assessment of organizational needs for information processing equipment and/or restructuring of work and information flows. The following diagram outlines the steps involved in a comprehensive User System Analysis:

FLOW DIAGRAM OF USER SYSTEMS ANALYSIS

Collect Data From:	Summarize Data Into:	Evaluate Organization by Comparing Summarized Data With:	Recommend System Changes to the:
*Interviews *Direct observation *Job analyses *Production and personnel records *Design specifications for equipment	*Workflow diagrams *Network statistics *System profile statistics *System effectiveness	*Common sense *Past experience *Organizational theories *Similar data from other organizations	*Equipment *Workflow *Staffing *Coordination methods *Office layout

User System Analysis is an improvement over previous methods of assessing organizational work and information flows for several reasons. First, data are collected from a number of sources, not just one or two informants or equipment design specifications. The following paper by Brady elaborates on this critical step. Second, the methods for summarizing and presenting the data are vastly improved. Work and information flow diagrams provide an extremely useful graphical summary of operations. Krus, Carsrud, and Glick, in the fourth paper of this panel session, discuss alternative formats for presenting this information and a partially developed computer program that draws workflow diagrams from raw data. Third, past organizational change efforts have frequently relied on idiosyncratic evaluation techniques, while User System Analysis is a systematic empirical and theoretical approach to evaluating organizations. As suggested above, sociotechnical systems theory, Thompson's (1967) theory of technological interdependence, and social network analysis should contribute greatly to improving the evaluation of organizational needs for information processing equipment and/or restructuring of work and information flows. Future work in this area should strengthen the contribution of these theories to User System Analysis and test the empirical validity of more specific propositions about the relationship between system effectiveness and the patterns of work and information flows. Fourth, User System Analysis provides suggestions for system changes in several areas. Some organizations may need new word processors and computer equipment, while others only need to reorganize the patterns of work and information flows to use their current equipment more effectively. User System Analysis should help America work smarter.

REFERENCES

[1] Burt, R.S., Towards A Structural Theory of Action (New York, Academic Press, 1982).

[2] DeGreene, K.B., Sociotechnical Systems: factors in analysis, design and management (Englewood, New Jersey, Prentice Hall, 1973).

[3] Kelly, J., A reappraisal of sociotechnical systems theory, Human Relations 31 (1978) 1069-1099.

[4] Kolodny, H. and Kiggundu, M., Towards the development of a sociotechnical systems model in woodlands mechanical harvesting, Human Relations 33 (1980) 623-645.

[5] Lincoln, J.R., Intra- (and inter-) organizational networks, in: Bacharach, S. (ed.), Research in the Sociology of Organizations 1 (Greenwich, Connecticut, JAI Press, 1982).

[6] Markus, L.M., Systems in Organizations (Boston, Pitman, 1984).

[7] Rice, A., Productivity and Social Organization: the Ahmedabad experiment (London, Tavistock Publications, 1958).

[8] Rousseau, D., Technological differences in job characteristics, employee satisfaction and motivation; a synthesis of job design research and sociotechnical systems theory, Organizational Behavior and Human Performance 19 (1977) 18-42.

[9] Susman, G.I., Autonomy at Work (New York, Praeger, 1976).

[10] Taylor, J., The human side of work: the sociotechnical approach to work system design, Personnel Review 4 (1975) 17-22.

[11] Thompson, J.D., Organizations in Action (New York, McGraw-Hill, 1967).

[12] Tichy, N.M., Tushman, M. and Fombrun, C., Social network analysis for organizations, Academy of Management Review 4 (1979) 507-519.

[13] Trist, E., The evolution of sociotechnical systems as a conceptual framework and as an action research program, in: Van de Ven, A. and Joyce, W. (eds.), Perspectives on Organization Design and Behavior (New York, Wiley, 1981).

[14] Trist, E. and Bamforth, K., Some social and psychological consequences of the longwall method of coal getting, Human Relations 4 (1951) 3-38.

Human Factors in Organizational Design and Management
H.W. Hendrick and O. Brown, Jr. (Editors)

INTERACTIVE ORGANIZATIONAL WORKFLOW ANALYSES: AN APPROACH TO PROBLEM SOLVING IN ORGANIZATIONS

E. J. Krus, Jr., A. L. Carsrud, and W. H. Glick

Department of Management
The University of Texas
Austin, Texas 78712
U.S.A.

Previous papers described User Systems Analysis (USA) as a new approach for evaluating task and organizational design and assessing organizational needs for information processing equipment. This paper describes an early application of USA in a personnel office. The USA workflow diagrams contributed significantly to improving the effectiveness of that personnel office. A computer aided design program is introduced that greatly reduces the costs of drawing workflow diagrams for User System Analysis.

Organizational researchers, consultants, managers, and computer sales people have tried various approaches to assessing organizational needs for information processing equipment and evaluating work and information flows, organizational structure, and personnel utilization. The three preceding papers in this panel session have discussed: several strengths and weaknesses of previous assessment techniques (Carsrud, 1984); a new, more comprehensive approach, User System Analysis, that adopts material from sociotechnical systems theory and social network analyses (Glick & Beekun, 1984); and some of the trials, tribulations, and triumphs connected with developing data collection strategies for User System Analysis (Brady, 1984).

This paper extends the discussion of "User System Analysis: Network Analyses of Organizational Structure, Tasks, Personnel, Equipment, and Procedures", by presenting an early application of this approach in a personnel office of a large school district. A major focus of this study was a workflow diagram that was laboriously compiled and extensively used in recommending changes in office operations. Although this project was very successful, the costs of compiling the workflow diagram instigated a number of innovations in User System Analysis. The second half of this paper details some of these developments, particularly the computer aided design program for drawing the workflow diagrams, Interactive Organizational Workflow Analyses (IOWA). This program can also be used to simulate hypothetical patterns of work and information flows to allow managers to explore "what-if" scenarios before investing in actual restructuring of workflows or new information processing equipment. Thus, IOWA can be used as a decision aid for managerial problem solving.

The major focus of this paper is on the development and use of workflow diagrams in User System Analysis. A workflow diagram shows how individuals, activities, workstations, equipment, information, inputs, and final products are connected in an organization. A good diagram can indicate who performs different activities, what inputs they need to perform those activities, what outputs are created, where the inputs come from, where do the outputs go, how long does each activity require, how frequently the activity is performed, and who initiates the activity. Additional information, such as friendships, lines of authority, and who goes to whom for work-related advice, may be represented in the diagram depending on the purpose of the analysis.

The workflow diagram can be used independently or in combination with the social network analyses described by Glick and Beekun (1984). Both analyses summarize the same set of relational data on the networks of tasks, personnel, equipment, and procedures. Workflow diagrams provide graphical presentations of these networks, while the social network analyses furnish a quantitative summary of the same information. A good workflow diagram is a picture that is worth a thousand words and a supplementary social network analysis supplies the analytical precision to satisfy those who are less artistically inclined.

AN EARLY APPLICATION OF USER SYSTEM ANALYSIS

One of the authors, Alan Carsrud, was initially contacted by the Director of Personnel of a large southwestern school district to evaluate and improve the personnel operations. The office was under fire to recruit and select more highly qualified teachers to meet teacher demand in the expanding school district. The school district agreed to conduct an extensive self-study to improve their operations and allow us to test an initial version of the work and information flow analyses that are now called User System Analysis.

The analyses involved direct observations of activities, examination of job descriptions for the personnel office, questionnaire administration, extensive interviews with all members of the department and several individuals in related departments, collection of all forms and reports used in the office, and consultation with other experts from personnel offices in the private sector. Most of this data collection, particularly the interviews, focused on compiling a detailed description of the work and information flows. More than 60 hours of the researchers' time was devoted to collecting this information. An additional 80 hours was involved in translating this information into a fairly comprehensive workflow diagram that traced their activities from recruiting to selection to placement of teachers. Additional supporting activities such as training, teacher counseling, and evaluation of teachers' administrative potential were also included in the diagram. The resulting diagram covered approximately six by twenty feet of paper with small print.

This diagram was extremely useful in subsequent analyses. Based on the formal job descriptions, we had expected to find a high degree of consistency across the five Assistant Directors of Personnel and their support staff. We thought that the analysis would be very simple. The diagram revealed, however, that the office was extremely decentralized with each Assistant Director performing very different activities. The Assistant Director who worked most closely and effectively with her secretary, was one of only two Assistant Directors who shared a secretary. The workflow diagram indicated that this secretary was extremely overworked and under severe stress. This observation was confirmed by questionnaire responses, direct observation, and reports that this secretary was exhibiting obvious physical health symptoms of stress.

The workflow diagram also helped to solve the mystery of temporarily inaccessible teacher applications. Very few, if any, files were being misplaced, but the office was entirely dependent on paper files that several people were trying to access during the same week. The inaccessible files were not lost, but merely in transit.

The final recommendations from this study led to a major redesign of the office policies and procedures and the design of a new data and word-processing system (Carsrud & Glick, 1982). Although the administration expected a request for several new personnel staff, only two new employees were needed following the analyses. The new procedures eliminated a number of high stress, high conflict bottlenecks from the office, equalized the workload, improved employee morale, and reduced absenteeism and turnover. Current reports suggest that overall system effectiveness was greatly improved with positive effects on teacher recruiting and selection.

A modified version of User System Analysis was developed in a subsequent study at the school district. The goal of this study was to design and create a classified personnel office for the school district. Similar data collection procedures were employed with greater emphasis on questionnaires rather than interviews for assessing the workflows. Unfortunately, several of the respondents felt that the questionnaires were too complex and time consuming. Fairly useful data, however, was collected at a great savings in researcher time. The final report from this study (Carsrud, Gronlie, Glick, & Krus, 1983) was favorably received and resulted in the creation of a highly effective classified personnel office with minimal disruption of activities in the autonomous departments that had previously performed their own personnel functions.

INTERACTIVE ORGANIZATIONAL WORKFLOW ANALYSES

The two school district studies demonstrated the utility of User System Analysis, particularly the workflow diagrams. They also revealed many of the difficulties of this approach. Both researchers and organizational personnel devoted numerous hours to constructing the workflow diagrams and the resulting diagrams became very unwieldy for some of the analyses. For some purposes, the details of the whole diagram were important, but frequently the recommendations were based only on sub-portions of the diagram or summaries of the whole picture. Further, the diagrams were organized by individuals and by the temporal sequencing of activities. It would have been useful to have additional diagrams with the same information organized by the physical layout of the work stations and/or by area of the school district that was being served. Given the high labor costs of drawing each diagram, however, a single multi-purpose diagram was drawn for each study.

Interactive Organizational Workflow Analyses (IOWA) is a partially developed computer aided design program that overcomes many of these difficulties. The primary inputs to this program are relational data on the networks of tasks, personnel, equipment, and procedures. The output is a set of diagrams of varying focus and detail. Users can request diagrams that organize the workflows by individuals and temporal sequencing of activities as in the school district studies, or they may request a diagram that uses the physical layout of the organization as a blueprint that contains the location of individuals, workstations, and equipment in the organization. Similarly, users can get a general picture of the organization by drawing a diagram showing only the strongest (most frequent or most important) relations or they may prefer to get full detail on the activities of a single individual. The IOWA computer program greatly reduces the costs of producing diagrams so that multiple, special purpose diagrams can be generated from the same data set with a few swift keystrokes.

The workflow diagrams generated by IOWA can be used for many purposes. First, they are an integral part of the User System Analysis. As shown in the USA flow diagram by Glick and Beekun (1984), the workflow diagrams are used in combination with social network analysis to summarize the relational data. System profile characteristics and measures of system effectiveness are also analyzed to provide a comprehensive assessment of organizational needs for information processing equipment and evaluating the networks of tasks, personnel, equipment, and procedures. Review of the User System Analyses by managerial decision makers should suggest ways of improving the mix and organization of these components.

Second, the flexibility and reduced costs of IOWA make it ideally suited for simulations. Managerial decision makers can view the current workflows then try out alternatives for improving organizational effectiveness by changing the relational data inputs. This will create revised workflow diagrams that can be compared and evaluated before actual changes are ever implemented. This allows the decision maker to play "what-if" games with the workflow diagrams rather than with actual operations.

Third, the workflow diagrams are an excellent complement to detailed job descriptions. They can be used to help determine staffing needs, skill requirements, and salary levels for various jobs. They can also be used for training new employees. Orientation sessions can start with a general overview of the organization's workflow, looking only at relations between and within various departments. Then, the individual's new job and its connections with other workers can be examined to show how that individual is a critical component of the system. Attention can be directed to the sources of inputs and output destinations so that the new employee can notify the right people to go to immediately if anything goes wrong.

CONCLUSION

Interactive Organizational Workflow Analysis reduces costs and improves accuracy of workflow analyses in User System Analysis. As a comprehensive approach to organizational analysis, USA holds considerable promise. As a tool for organizational development, USA has potential utility to the organizational consultant. As a method of determining the effects of proposed equipment purchases, USA through IOWA could be invaluable to both equipment providers and the managers making the purchases. For the academic researcher, USA and IOWA could provide accelerated study of various organizational characteristics. Finally, for the individual and the organization USA should help them "work smarter."

REFERENCES

[1] Brady, L., Review of the systems methodology of User Systems Analyses, in: Hendrick, H. W. and Brown, O. (eds.) First international symposium on organizational design and management, (North Holland, Amsterdam, 1984).

[2] Carsrud, A. L., Organizational analyses: Idiosyncratic methodologies searching for holistic solutions, or what's wrong with the way they are done, in: Hendrick, H. W. and Brown, O. (eds.) First international symposium on organizational design and management, (North Holland, Amsterdam, 1984).

[3] Carsrud, A. L. and Glick, W. H., Final report on the professional personnel office of the Austin Independent School District.(Carsrud, Glick and Associates, Austin, 1982).

[4] Carsrud, A. L., Gronlie, D. A., Glick, W. H., and Krus, E. J., Final report on the evaluation of classified personnel operations of the Austin Independent School District and recommended organization and procedures for a unified classified personnel section. (Carsrud, Glick and Associates, Austin, 1983).

[5] Glick, W. H. and Beekun, R. I., A theoretical overview of User Systems Analysis, in: Hendrick, H. W. and Brown, O. (eds.) First international symposium on organizational design and management, (North Holland, Amsterdam, 1984).

Human Factors in Organizational Design and Management
H.W. Hendrick and O. Brown, Jr. (Editors)

USER SYSTEM ANALYSIS

Leo Brady

Human Factors Engineering
IBM Corporation
Endicott, New York 13760
U.S.A.

INTRODUCTION

"USER System Analysis" provides the methodology and technique for systematic evaluation of new and existing customer markets and for adapting the right technology to meet user needs with excellent probability of customer acceptance.

There are four major stages (Figure 1) to the analysis which produce detailed requirements for system design and integration. These stages are:

1) development of a "USER" environment overview scenario,

2) detailed analysis of this scenario to identify key "USER" function/tasks which might be applicable for automation,

3) development of "USER" requirements to perform these function/tasks, and

4) establishment of key system requirements, based on "USER" needs, integrated into the environmental scenario, with development usability criteria identified.

A. The "USER" and Their Environment (1, Figure 1)

The best way to understand customer requirements for future needs and to develop criteria for effective solutions is to learn from the customer how the business is conducted today. To do this, a general system overview diagram (scenario) of the customer environment is created. The overview diagram will provide a vehicle for the developer to understand in customer terms what environment the system will operate in.

This is specially critical for the small, first time "USER" as they do not have the expertise to evaluate the impact a complex system can have upon their work. The creation of the overview diagram is also a training exercise for all participants and establishes a common goal for any system implementation.

The first step in preparing such an overview consists of interviewing customer employees and principals to obtain high level flow diagram information describing the business as it operates today.

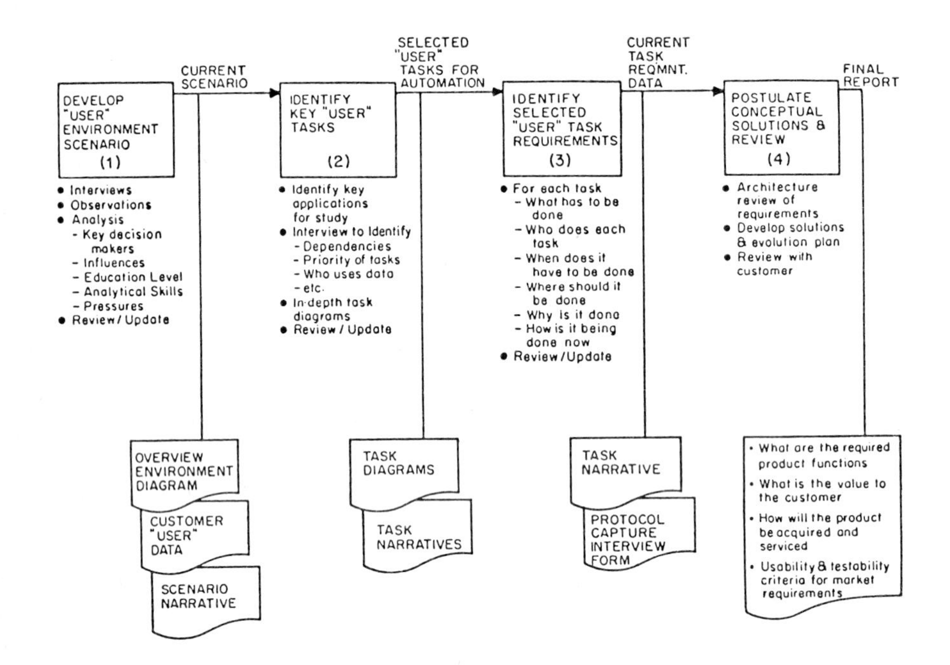

Figure 1
User System Analysis Overview

In support of the overview, certain key customer data must also be gathered to aid in definition and selection of major function/tasks for system implementation. The following items are documented in narrative summary form by each interviewer:

Identification of -

- The key decision maker(s),
- Influences upon the key decision maker(s) (people, events),
- Education level of users,
- Analytical skills of users, and
- Environmental pressures on key decision makers, and those who influence decision makers.

Once the diagram and data collection task is completed, it must be reviewed with the principal contributors for their approvals. A diagram and narrative update is then performed to provide an agreed upon baseline for further analysis of the customer requirements.

B. Key "USER" Task Identification (2, Figure 1)

After the overview scenario has been completed and reviewed, those functions involved in a specific application are identified. Additional flow diagrams detailing these selected functions for their supporting tasks are then created. These diagrams are prepared for today's scenario to identify major points of entry (user interfaces) for tomorrow's system. This will aid the customer, who may not be a data processing specialist, to relate the product to their needs.

Analysis of these diagrams will identify the following: specific tasks for automation, interdependencies of tasks, key contributors to the tasks, information being processed, sequence of tasks, and other tasks or functions which could become candidates for automation in the future.

Once again the customer must be involved. To support the analysis, interviews are conducted with the "USER" population to gather detailed task information such as:

- Individual employee tasks and dependencies,
- What information exists today to support tasks,
- How is it gathered today,
- Who uses the data,
- Size and organization of data,
- Priority of tasks, and
- Identified interfaces.

The customer review of these diagrams will make any potential new areas for involvement very clear and help structure an evolutionary plan for modular implementation of future improvements. This will allow a systematic capability growth, for new systems, to be planned and even scheduled at initial design time. It also permits a customer to grow in automation capabilities with minimum impact on their existing system.

C. "USER" Task Requirements (3, Figure 1)

Once the key tasks for automation are identified in Step B, many data items must be gathered to support a system design implementation. This data must be captured in an organized manner and in a format/content which is readily transferrable to usability criteria. For each identified task the following information must be known:

(1) What has to be done?

- Volume of transaction,
- Media requirements,
- Frequency of transactions,
- Interdependency on other tasks,
- Priority (now vs. later),
- Accuracy and quality of information, and
- Number of file cabinets, desks, and data in each.

(2) <u>Who does each?</u>

- Skill level required,
- Physical requirements,
- Value of time,
- Turnover rate, and
- Education level.

(3) <u>When does it have to be done?</u>

- Priority,
- Frequency,
- Interdependency on other tasks, and
- Time sequence.

(4) <u>Where should it be done?</u>

- Primary and secondary location,
- Supporting requirements, and
- Interdependencies.

(5) <u>Why is it done?</u>

- Any formal government regulation requirement,
- Required by other industries, and
- Required by customer only.

(6) <u>How is it being done now?</u>

- Equipment currently used,
- Forms and documentation currently used,
- Media requirements, and
- Time requirements.

Once again upon completion of this step in the analysis, customer review is required. It is necessary that the customer be satisfied that this data being developed is accurate, and the system design (if available), is in agreement with the user's perceived needs.

D. <u>The Solution</u> (4, Figure 1)

Detailed analysis of the customer scenario, interviews, diagrams, and data collection sheets by key system development people, will establish responsive customer market requirements which reflect a particular customer set. Once a "USER" study has been completed for a particular set, such as a law firm, the study results may be used as a "selling tool" for additional firms within a customer set.

The "USER System Analysis" final report will contain answers to the following questions:

(1) <u>What are the required product functions?</u>

- What are the conceptual product functions needed,
- Who will use the product,
- What training would be needed by the user,

- Usability criteria,
- Which costs will be displaced,
- User documentation requirements,
- How important will the product become, and
- What product support and enhancements are needed?

(2) What is the value of the product to the customer?

- What are displaceable costs,
- Is productivity increased, and
- Is improved quality of service provided by the customer?

(3) How will the product be acquired and serviced?

- What communication method is used,
- What selling channel will be employed,
- What service and maintenance channel will be used,
- What support (programming, education, etc.) is to be provided, and
- What evolution plan will be implemented?

The "USER System Analysis" technique performs a structured analysis for determining market requirements that are "USER" driven. It provides a vehicle for verifying design concepts and requirements for any integrated processing system, ensuring that the system will gain customer acceptance through friendly and optimized "usability" characteristics.

Human Factors in Organizational Design and Management
H.W. Hendrick and O. Brown, Jr. (Editors)

INCREASING THE EMPLOYMENT OF DISABLED PEOPLE

Shinichi Iwasaki, Koji Mihara, Akinori Komatsubara and Yoshimi Yokomizo
School of Science and Engineering
Waseda University
Tokyo Japan

This paper deals with the expansion of employment opportunities for disabled people. The employment conditions of disabled people working at companies in Tokyo were researched case by case in detail, using questionnaires and interviews. Based on the results, we have developed recommendation for employing disabled people in companies.

INTRODUCTION

Formerly in Japan, disabled people have often been treated as peculiar beings. Accordingly, disabled people have been isolated from society and their life zone has been limited to their home, or welfare facilities (an industrial home for disabled people etc).

However recently the concept called Normalization, which may be very popular in European countries and the U.S. has gradually penetrated into Japan. In Japan, Normalization is understood as follows: There are always handicapped as disabled people in the society. Therefore we should create a society where disabled people are able to live in a healthy condition, both mentally and physically.

To this end, the Japanese government adoped a law to promote the employment of disabled people in 1960. In this law, the rate of employment of disabled people is set at 1.5% of all the employee in a private company. Moreover this law was revised in 1976, to this effect, the private companies whish employ a smaller percentage of disabled must pay, but that the companies which employ more than 1.5% disabled are given compensations.

However, at the present time, in most companies disabled are less than 1.5% of the employees. Among the reasons for this unsatisfactory situation, we can point out the following important two factors.

(1) Because many companies do not understand the disabled people (their abilities, skills, etc), they hesitate to employ them. Therefore they can not understand how to employ and how to arrenge thier jobs.

(2) The job hunting method for disabled people is poor. Authorized agencies which introduce the newest employment informations in companies for them have not been established and disabled can not find jobs easily.

In order to promote the employment of disabled people, these two factors must be removed. The procedures or methods of the disposition from employment for disabled people must be developed. Based on the about point of view, we started our study to develop such procedures.

METHODS

We hsve investigated the employment conditions of the following kinds of disabled people working in Tokyo city case by case in detail for three years.
(1) amputees
(2) cerebral palsy victims
(3) the blind
(4) the deaf
(5) visceral handicapped
(6) mentally retarded

In this paper, we report on the amputees and cerebral palsy victims. We made our investigation in line with the procedures shown in Figure 1.
The each step of this procedure include the following study.

(1) Basic Research of the Disabled
We have been investigated the following items about the disabled for more than 20 years.
Measurement of disabled people's capacity and capability, such as motion study, workload assessment, etc.
Development of self-help tools and equipment for disabled people's, etc. Without these studies we could not start the study dealing in this paper.

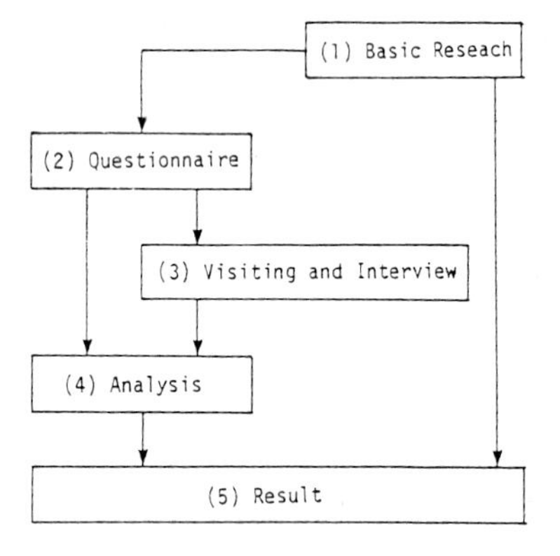

Fig.1 Investigation procedure

(2) Questionnaire
1788 private companies in Tokyo city in Japan which employ diabled people were investigated by using questionnaire. We obtained 426 effective samples from these companies.
In the questionnaire, we investigated the disabled employees as follows.

1) kind and degree of his disabilities
2) his character or personality
3) contents of his job
4) his employment condition (e.g. wages)
5) education and job training method and contents for him

We also asked the employer's opinion of the disabled employee in the questionnaire.

(3) Visiting and Interview
Among the effective samples, we chose companies which employ the seriously disabled. We visited the companies and investigated the cases in great detail.

(4) Analysis
The results were analyzed at the development of employment procedures.

RESULTS

(1) Results of Questionnaire

We analyzed the relations between the kind of disabilities and jobs. The results are shown in Figure 2.

We also analyzed the relations between the degree of disabilities and job, as shown in Table 1.

(2) Results of Visiting and interview

The detail of each of the seriously disabled employee are shown in Table 2.

Discussion and Conclusion

(1) Figire 2 and Table 2 show that there is little correlation between the kinds of disabilities and jobs.

(2) It is similarly shown in Table 1 that there is little correlation between the degree of disabilities and jobs.

(3) In the cases in which the disabled employee can do his job in spite of his disability, companies have little need to improve their facilities.

(4) No set procedure is used for the employment of disabled people.

When we examine the results, we find that the kind and degree of his disabilities should not limit the kind of occupation. It can be said that his character or personality of qualification are more important in his employment, the same as with healthy people.

Accordingly employers must take care of not only the kind and degree of his disabilities but also his total capacity for jobs. Besides they must do task analysis and work study which is

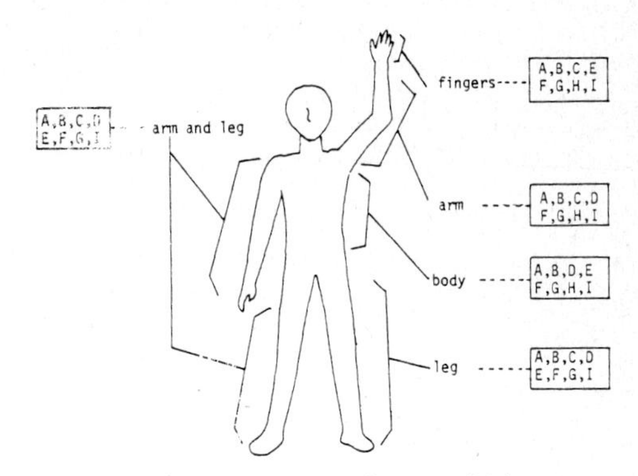

A:technician , B:manager , C:clerk , D:salesperson , E:driver
F:guard , G:recepion clerk , H:workingman , I:engineer,skilled operator

o This Figure shows that he takes the occupation show in the ☐, even if he has the disabled place of the body.

Fig.2 The relations between the kind disabilities and jobs

JOBS \ DEGREE*	1	2	3	4	5	6
technician	3	10	14	19	9	12
manager	2	3	5	15	12	5
clerk	6	31	56	68	65	45
salesperson		2	2	9	4	5
driver			2	9	8	8
guard		1	3	6	8	5
reception clerk			1	2	1	
workingman		6	12	20	18	15
engineer skill operator	1	12	40	60	55	53
TOTAL	12	65	135	208	180	148

*This is difined by the Law of welfare (Fukushi-hou) which is controlled by the ministry of Health and Welfare in Japan.

Table 1 The relations between degree of disabilities and jobs

necessary to relate the remaining capasity of disabled people.

Of course, disabled people must do their best to enhance thier capacity or skills which are necessary for their jobs.

Based on the above discussion, in conclusion, we propose the procedure of employment enlargement of disabled people, shown in Figure 3.

This procedure has been accepted by the metropolitan government, which has recommended its use to companies in Tokyo when employing disabled people.

question number	jobs	kind of lesion: amputation(arm)	paralysis(arm)	amputation(leg)	paralysis(leg)	functional disease (body)	paralysis (one side of the body)	degree	self-help tools: wheelchair	crutches	stick	artificial arm and leg	posture: sitting	standing	both sitting and standing	improvement in the workshop: desk and chair	toilet	tools and equipment	job rotation	others
1	clerk				○			2		○			○			○			○	○
2	clerk	○						2							○			○		
3	clerk					○		2	○				○				○			○
4	clerk	○						2				○	○					○		
5	clerk		○					2					○							
6	clerk						○	2					○					○		
7	clerk				○			2	○	○			○			○	○			○
8	clerk					○		2			○		○					○		
9	clerk			○				2				○			○					○
10	clerk		○					2					○					○		
11	clerk	○						1				○			○				○	
12	clerk				○			2			○		○							
13	clerk	○						3							○				○	
14	clerk						○	2			○	○	○							
15	clerk				○			1	○				○							○
16	clerk				○			4						○						○
17	technician	○						2					○				○			
18	technician			○				1	○	○		○	○				○			○
19	technician			○				4				○	○							
20	technician				○			2			○		○						○	
21	technician				○			3					○							
22	technician						○	1		○			○							○
23	printer	○						2						○						
24	printer		○		○			6					○				○	○		
25	printer	○						2						○						
26	engineer, skilled operator				○			4					○							
27	engineer, skilled operator		○					2					○					○		
28	engineer, skilled operator				○			1	○				○			○	○	○	○	
29	engineer, skilled operator				○			3						○						
30	engineer, skilled operator						○	3					○							
31	engineer, skilled operator				○			2				○	○					○	○	
32	engineer, skilled operator				○			2		○			○							
33	engineer, skilled operator		○					5							○				○	
34	engineer, skilled operator			○				6				○	○				○			
35	engineer, skilled operator				○			7					○				○			
36	engineer, skilled operator				○			4				○			○					
37	others						○	1					○					○		
38	others		○		○			3						○						
39	others				○			4						○						○
40	others					○		5						○						
41	others						○	2					○							

Table 2 Results of visiting and interview

References

(1) Increasing the Employment of Disabled Peopled , Report of Tokyo Metropolitan Government, No.1-189, 1982.

(2) Yokomizo Y., A case study of Employment of Disabled People, Report of Production System Group, No.6, The Japan Ergonomics Research Society, 1982.

(3) Examples of Employment of Disabled People (Shintaishogaisha Hogo shiryo), (Tokyo: ZENKORO), 1980.

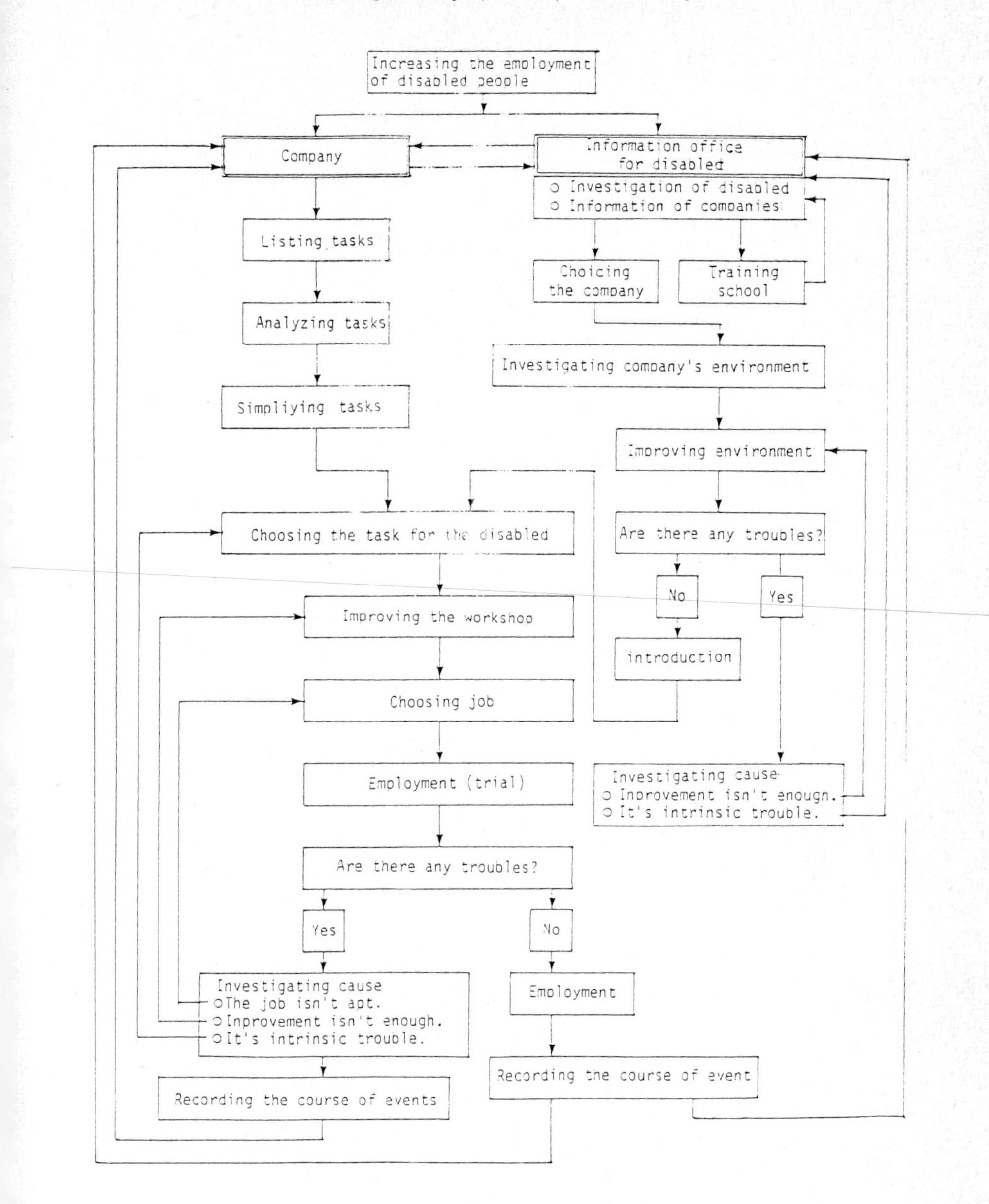

Fig.3 The procedure of increasing the employment of disabled people

HUMAN-COMPUTER INTERACTION

Human Factors in Organizational Design and Management
H.W. Hendrick and O. Brown, Jr. (Editors)

QUANTITATIVE ANALYSIS OF EYE MOVEMENT AT VDT TASK

Hiromu Matsumura, Yoshimi Yokomizo
School of Science and Engineering, Waseda University, Tokyo Japan

Takeshi Kurabayashi
Japan Institute of Synthetic Technology Ltd. Tokyo Japan

VDT tasks have been increasing recently. In these tasks, a miss of character sometimes causes a serious accident. Some reports say that this miss occurs when eye moves disorderly. Though these irregular eye movement can be observed with using EOG, there seems no effective method to measure and evaluate them quantitatively. Accordingly, in this research, EOG and misses of character during VDT task were analysed and the relation between them was studied. As a result, it comes clear that the four items of EOG can be the effective indexes to evaluate the work strain.

INTRODUCTION

VDT tasks, which have been increasing recentiy, are a new kind of visual task. In these tasks, just a trivial miss of character which are produced by fatigue sometimes causes not only accumulation but also a serious accident. With using EOG(electro oculogram) that could be measured during task, analysing the condition of eye movement when misses of character occur, evaluating the work stress of tasks, it will be possible to detect and prevent misses of characters.

Noro, Yamamoto, et ai.(1983) reported that eye movement began to be disarranged with the passage of time, and, misses of characters increased. Bahill and Stark(1975) reported that saccadic eye movements were affected by fatigue, and Schmidt et al.(1979) reported that peak velocity of saccade fell in proportion to mental fatigue.

As it is above, there are many reports that eye movements were affected by fatigue. But, many of these reports were from data were measured in laboratories, in which the head of the subject was supported by a head rest. With settling a part of body of subject, the subject can't work in this position. there will be a way to measure eye movement and analyse eye movement quantitatively while the subject works normaly.

The purpose of this short study was to measure EOG of subject with no support of parts of the body, and to analyse those data quantitativity,to investigate variability of eye movement when the subject misses characters, and to obtain the indices to evaluate the work strain.

As a result, it became clear that the variability of the following items would be effective indices to evaluate the work strain.

(1) fixation duration for watching
(2) peak velocity of eye movement
(3) angle of eye movement
(4) absolute amounts of eye movement

METHOD

Six normal subjects, females ranging in age from 26-46, were required to do 1 hour-VDT tasks twice at intervals of ten minutes. At the second task, their EOG were measured by PolyGraph. All subjects were subjected to measurement twice. The PolyGraph's electronics had a total system bandwidth of DC-30Hz. The CRT used had a 12-inch display, the color of characters was green(26-32cd/m²), the background was black(8-12cd/m²). nothing supported the head of the subject. To check up on the distance from eyes to display, picture were taken of the subject by Video Tape Recorder from the side. The task was basically a data-search task in which the subject searched for required information from imformation displayed on CRT after key-operating.

Method of quantitative analysis and items of investigations were shown as follows.

(1) fixation duration for watching

there were three key-operating that reqired in addition to normal key-operating for each operation (the one making operation was difined as from getting requiredimformation until finding out after key-operating). The variavility of these three fixation duration of eye movement for watching key were investigated for all operations.

(2)peak velocity of eye movement

Since the required information was different in each operation, eye movement for each operation was different. But, there were three parts that eye points of subject acted the same movement in each operation. The variability of peak velocity of eye movement (saccade) of these three parts that gained by differential calculus amplifier of PolyGraph were investigated for all operations.

(3) angle of eye movement

The variavility of angle of eye movement of those three parts which were explained in item(2) were investigated.

(4) Absolute amounts of eye movement

About 8 operations among about 70-80 operations that were carried out in 1 hour-task were operations in which the subject searched for the same imformation. Measuring EOG of these particular operations by time-constant 0.03sec,passinr through integrator amplifier of PolyGraph, like EMG, the variavility of absolute amounts of eye movement were measured.

Beside these, such factors as the condition of the subject and performance of task, the rate of misses of characters and operation time were recorded. And all subjects were required to answer a short questionaire about fatigue before and after the task.

RESULT

Responses to the questionaire showed that all subject were fatigued after the task.

Fig 1 shows the variability of mean rate of misses of characters and mean time that were needed to finish one operation. Perhaps, because of fatigue, misses of character and operation time tended to increase in the latter half of the task.

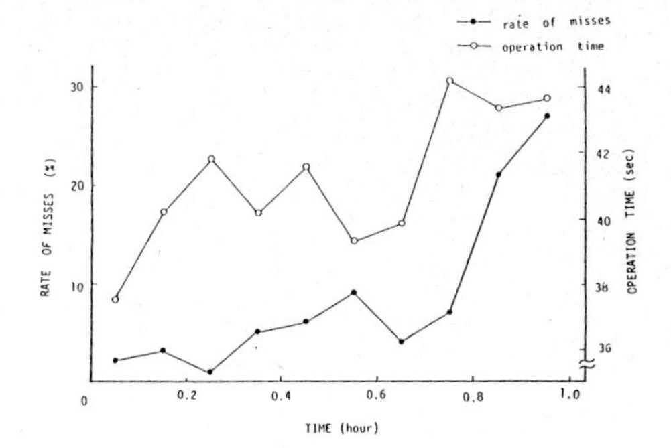

Fig.1 Variability of Rate of Misses and Operation Time

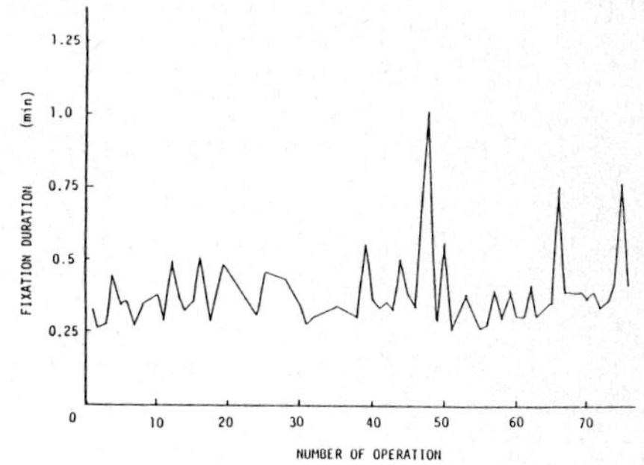

Fig.2 Variability of Fixation Duration (subject A)

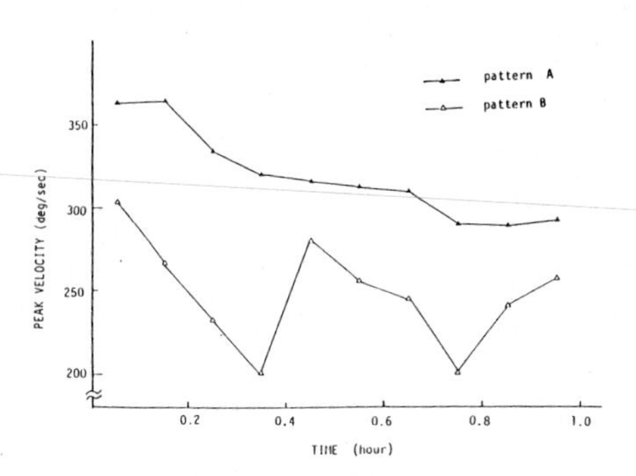

Fig.3 Variability of Peak Velocity (subject B)

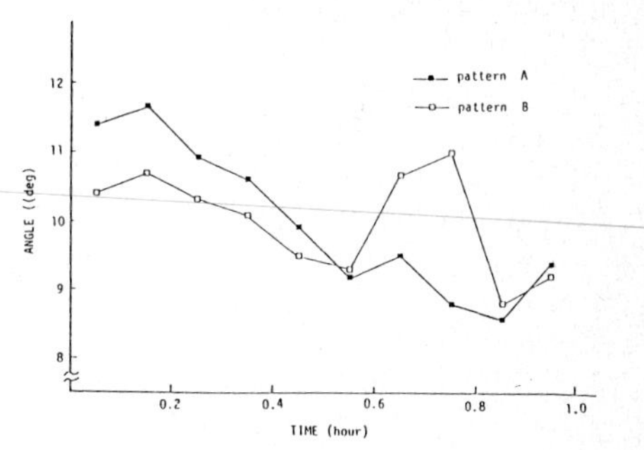

Fig.4 Variability of Angle of Eye Movement (subject B)

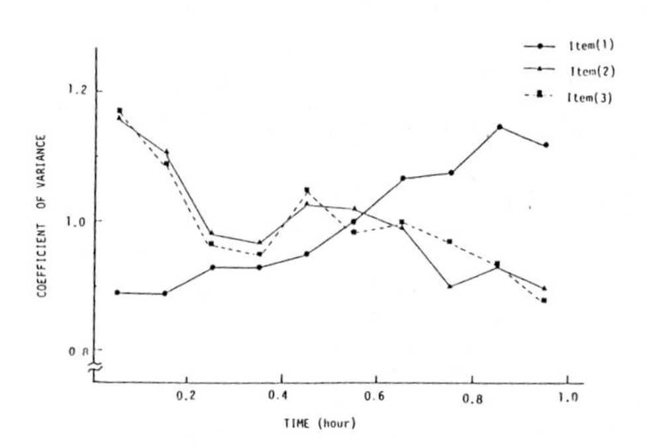

Fig.5 Variability of Mean Data of Item(1)-(3)

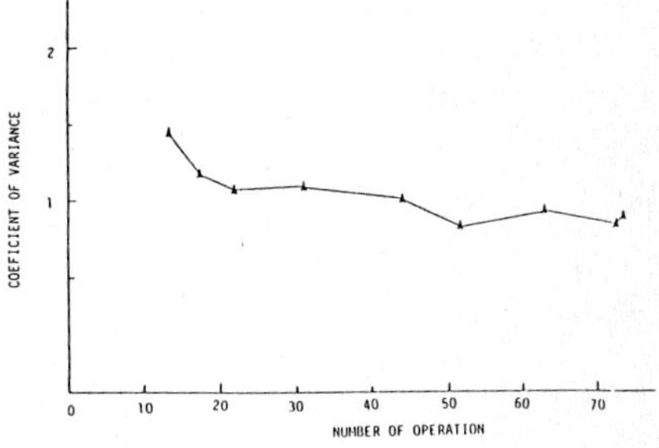

Fig.6 Variability of Absolute Amounts of Eye Movement (subject C)

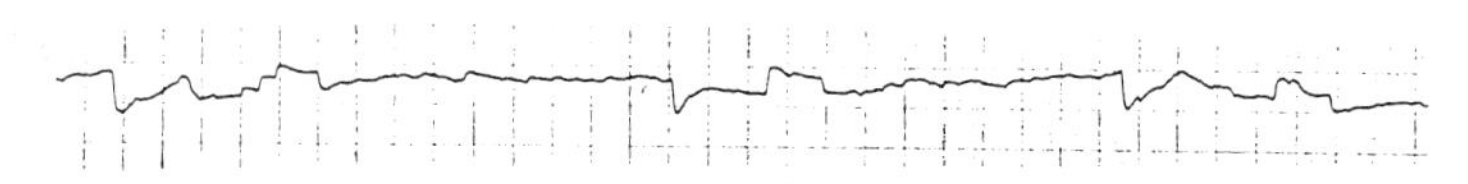

Fig.7 EOG of 11th-operation (subject C)

5deg
1.0sec

Fig.8 EOG of 72th-operation (subject C)

According to item(1) fixation duration for watching, for example, a result of subject A was shown in Fig.2. Shown in Fig.2, the fixation duration had atendency to indicate a long duration occasionally in the latter half of the task rather than to extend throughout. This tendency was observed for all subjects.

According to item(2) peak velocity, item(3) angle of eye movement, results of subject B were shown in Fig.3,4. Pattern A in Fig.3,4 was data of the part where subjects required the fastest eye movement in three measured parts of each operation. The peak velocity and angle of eye movement of this part had atendency to decrease as time goes on.
Pattern B in Fig.3,4 was data of the part where subject didn't require fast eye movement. Peak velocities and angles of eye movement of this part tended to decrease, but, there were not a definite tendency.

Variability of mean data of item(1), mean data of pattern A of item(2), (3) were shown in Fig.5. The coefficient of correlation between variability of the rate of misses of characters, operation time and variability of mean data of item(1)-(3) were 0.60-0.85.

Fig.7,8 were EOG of the subject C. These EOG were measured when the subject watched information displayed on the CRT, and Fig.7 was measured in the 11th operation, Fig.8 was measured in the 72th operation.
The required imformation of the 11th operation and the 72th operation were the same. But,EOG of the 11th operation was sharper than that of the 72th operation. For quantitative analysis of the variability of eye movement like these, item(4) absolute amounts of eye movement was used.
The result of item(4) of the subject C was shown in Fig.6. Fig.6 showed that eye movement became dull and lacked sharpness in the latter half of task. The coefficientof correlation between the variability of the rate of misses of characters was high as with the other items.

DISCUSSION

The theoretical background of the four items of investigation of this short study was that eye movement became dull because of eye fatigue and mental fatigue, and showed the following tendency.

(1) eye movement that was stopped once for fixation should demand time for next movement.
(2) peak velocity of eye movement should decrease.
(3) angle of eye movement should decrease, together with movement of the neck.

(4) eye movement should lack sharpness

As a result of the experiment, there were some differences between individuals, but, eye movements of all subjects indicated similar tendencies. Item(1) fixation duration of eye movement had atendency to increase with lapse of task, item(2) peak velocity of eye movement and item(3) angle of eye movement had a tendency to indicate uniform decrease where fast eye movement was inquired.

Item(4) absolute amounts of eye movement resulted in the same tendency as item(2),(3). And, there was a interrelation between increase of misses, operation time and these items.

The way of quantitative analysis of eye movement of this paper needed no support of a part of the body of the subject. Subjects can perform the task in the usual way. And, data of each item were minute that subject can't change cosciously.

But this short study was nothing but an investigation of a sample of VDT tasks.

Many more investigation of samples will be needed. Depending on the accumulation of data, it will be possible to define the state of eyemovement when the eye is fatigued. Then evalution of work strain, and prevention of misses will be possible.

REFERENCE

Noro,Yamamoto,et al.(1983) , characteristic of vision at VDT task,The Japanese Journal of Ergonomics 1983,Jun vol.19, Supplement pp206-207

Bahill and Stark,(1975) , Overlapping Saccades and Glissades are produced by fatigue in the Saccadic Eye Movement System Experimental Neurology 1975 vol 48,pp 95-106

Schmidt,et al. (1979) , Saccadic Velocity Characteristics:Intrinsic Variability and Fatigue , Aviation,Space,and Enviromental Medicine April, 1979

Human Factors in Organizational Design and Management
H.W. Hendrick and O. Brown, Jr. (Editors)

A New Objective Measurement Method of Visual Fatigue in VDT Work

Tsunehiro Takeda*, Yukio Fukui*, Takeo Iida*
Kazuo Karasuyama**, Toshiko Kigoshi**

*Industrial Products Research Institute, M. I. T. I., 1-1-4 Yatabe-machi Higashi, Tsukuba Science City, Ibaragi-ken, Japan
**Electronic Business Machines Development Center, Canon Inc., 2-1, Nakane 2-Chome, Meguro-ku, Tokyo, Japan

A new objective measurement system for visual fatigue is developed by improving the Campbell type optometer. Then a measure which is defined by accommodation area(integration of response curve to sudden change of target) is proposed to use to represent visual fatigue. The measure is examined by five subjects' simple VDT/paper works, and found to be proper in agreeing with subjective report of subjects' fatigue. Each measurement takes only several minutes.

INTRODUCTION

As the use of computers has become increasingly common, the number of people who work with VDTs has risen dramatically. This increased usage of VDTs has generated a greater concern for the safety and comfort of people using these VDTs. Many studies have been done to investigate whether or not VDT users are at greater risk in terms of health than nonusers(Matura, 1981). However, no clear conclusion has been reached and the issue is still in controversy(Starr, 1982). Most studies have adopted questionnaire research or subjective experiments such as visual acuity, near point and flicker fusion frequency. Some researchers have tried to determine objective criteria for visual fatigue employing measures of eye fixation and regression, pupil diameter, blink rate and so on. Östberg has reviewed these studies and concluded that the most promising lines of research was to employ measures of the refractive power of the eye and has done this with a laser optometer (Östberg, 1980). The study was very attractive and suggestive, but it was also another subjective measurement which depended upon responses of subjects and hence required longer measurement time and yielded lower precision. This report presents a new objective accommodation measurement apparatus and a method for its utilization.

MEASUREMENT SYSTEM

Figure 1 shows the configuration of the measurement system. Inner target is driven by a micro-computer which also controls the entire measurement timing. A monitor TV is attached to the optometer to make setting easier and the monitored iris picture is used to measure its diameter. Acquired data are stored at measuring time in the micro-computer after A/D conversion, monitored with X-Y recorder and at the same time recorded by data recorder and VTR for safety and later analysis. Figure 2 illustrates the optical system of the dynamic refractometer.

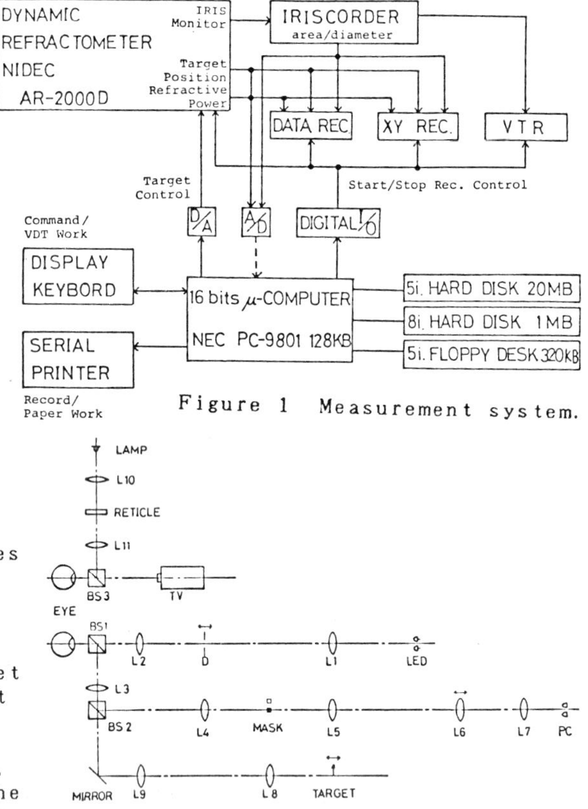

Figure 1 Measurement system.

Figure 2 Optical system.

The light of alternatively driven infrared LEDs projects images of diaphragm(D) on the retina through lenses L1 and L2. The reflected light of the images is gathered through lenses L3-L7 and produces second images on the photo cell(PC). Utilizing these images, D and L6 are servo-controlled to match the images on the retina. Refractive power is calculated from displacement of D and L6 from origin. Lenses L3, L8 and L9 compose the Badal lens system which is modified to hold the image magnitude of the target on the retina constant in spite of the eye's accommodation (patent pending). Lenses L10, L11 and BS3 make up the monitor of the eye for setting. The apparatus has the following specifications: measurement range ±12.5D. (Diopter), resolution ±0.05D., linearity ±0.1D., and cutoff frequency 4.7Hz.

MEASURE OF FATIGUE

If the inner target is moved slowly as shown in Figure 3, it is difficult to find out apparent change between responses of before and after works. On the other hand, the step change of the target position causes more apparent differences in responses. Figure 4 shows one typical step response of subject F.T. (see next section). The differences of responses can be analyzed to six major factors as indicated in the figure: (A) increase of dead time, (B) increase of rise time, (C) occurrence times to show difficulty of maintenance of near accommodation, (D) increase of accommodative lag, (E) increase of blink, (F) shift of far point toward subject. Though the factors were statistically analyzed, variances were too large to justify firm conclusions. It was later found that the areas surrounded by the mean of five far points at 0 hour(1), target locus of its change from near to far(2) and accommodation responses(3) as indicated in the figure were rather stable, had smaller variances and could be supposed to represent unified effect of the six factors. Therefore the decrease of the area is defined as a measure of fatigue. The step stimulations were fixed to be four times at each measurement and the duration of each stimulus was decided to be 15 seconds to extract information as much as possible in a short measurement period. Near and far point of the target were determined from the subject's accommodative power measured from ramp responses.

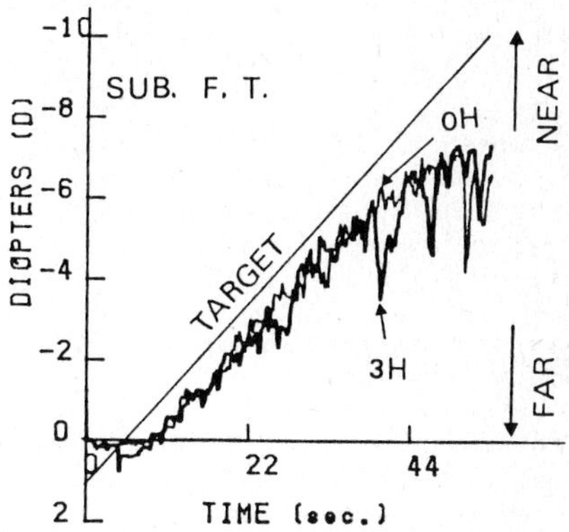

Figure 3 Typical ramp resposes before(1H)/after(3H) the work.

The area was normalized by dividing with stimulus height and time duration to make the comparison meaningful, i.e. a perfect response becomes unity.

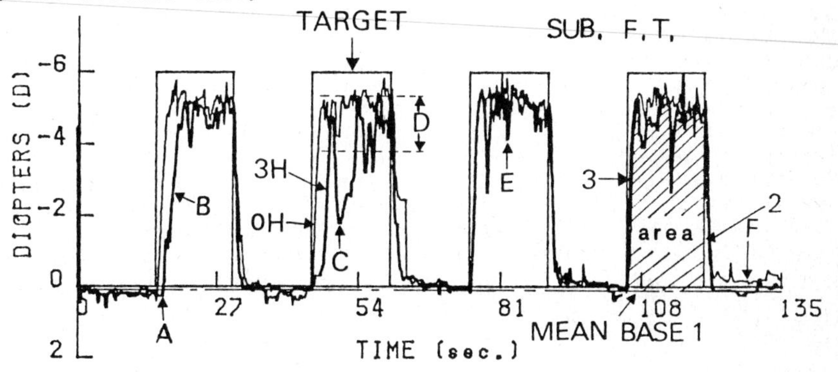

Figure 4 Six major changes of step responses before(0H) and after(3H) the work and the accommodation area(shaded portion).

EXPERIMENT

To confirm the reliability of the method, the accommodation areas were measured and subjective feeling of eye fatigue on 25 items were examined before and after three hours' VDT and paper work. The works were to add up to the total of a specified two digit number out of the displayed 266(19x14) numbers. Special apparatus were designed for paper work to simulate VDT work as closely as possible except that the figures are displayed on papers(Picture 1).

SUBJECTS: Five emmetropic subjects were selected from some 200 new employees of Canon Inc. who had visual acuity 1.0 and more. they were all 18 years old and three of them were female.

PROCEDURE: The paper work was carried out on the first day and the VDT work was done the next day. Exercises for measurement and the work were done before the work in the morning. Ophthalmological tests such as degree of astigmatism, near and far point was done in the first morning. Schedule was planned to offer subjects sufficient rest time to eliminate the influence of the day before. From one o'clock, each subject underwent measurement and began the works. It took about 15 minutes to gather all five subjects' step responses. The measurements were carried out every one hour for three hours.

Picture 1 Used display for paper work.

RESULTS AND DISCUSSION

Figure 5 shows mean numbers of items checked on the questionnaire. Change is evident before and after the work, but the difference between the works was not statistically significant. Figure 6 shows mean numbers of correct/error answers for every 30 minutes. The correct answers decreased and the errors increased. Again there was no significant difference between the both tasks. Figure 7 shows the change of mean areas of four accommodative step responses at each hour. In the figure, the curves of two subjects were omitted to avoid complication, hence the mean of the subjects are also the mean of the three remaining subjects. The datum of subject F.T. at 0H is apparently abnormal because it must approximate the datum of VDT at 0H. F.T. seemed not to be accommodate with the measurement for the first two measures. The change of the mean area of the three subjects and the mean area of the five with respect to time were rather stable in spite of those factors. Three dimensional analysis of variance

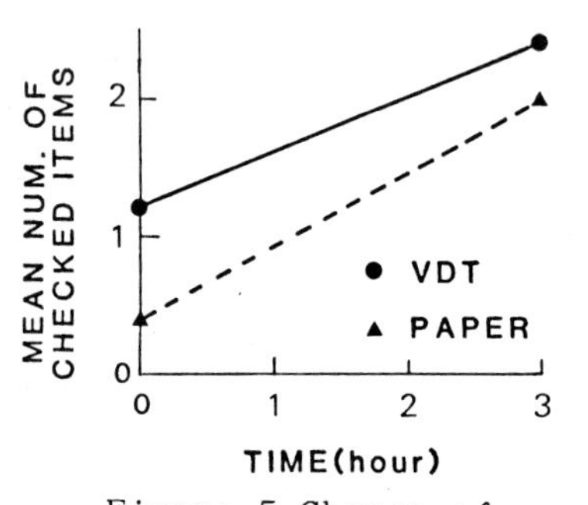

Figure 5 Change of subject reports.

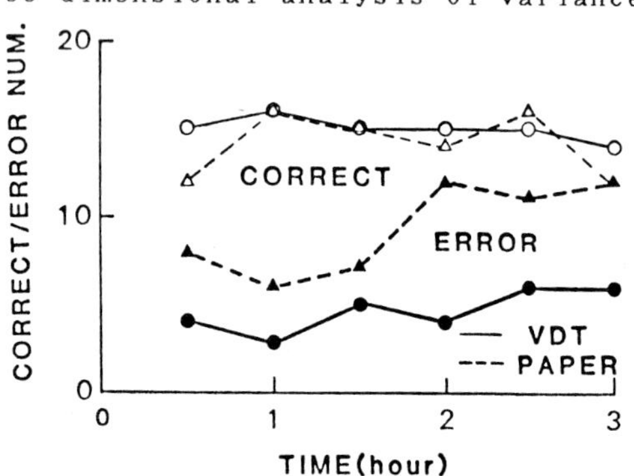

Figure 6 Change of scores.

Table 1 analysis of variance of accommodation area

	A	B	C	AxB	AxC	BxC	AxBxC
No. 1-3	.292	**51.3	**10.2	**13.[illegible]	2.85	*2.26	.905
No. 1-5	.0382	**187.	**16.0	**12.9	2.00	1.07	**2.50

* $p<0.05$, ** $p<0.01$; A -- the difference of work; B -- the difference of subject; C -- the effect of time

was carried out for both cases as shown in Table 1. The difference concerning time was highly significant($p<0.01$; C) and agreed with the result of the questionnaire. Hence it was concluded that the method was applicable to measure visual fatigue. The difference between individuals was also highly significant(B). But unexpectedly, there was no statistically significant difference between the two forms of work though it agree with the subjective report(A, AxC). The reasons might be as follows: (1) the height of the paper work apparatus was higher by 10 cm and it was somewhat difficult to establish comfortable eye distance from the display because of dimensional limitation, (2) the used CRTs(CanoWord 35) had less reflective glare compared with popular low cost CRTs. The effects of reflection and colors of CRTs and so forth are expected to be investigated in the near future.

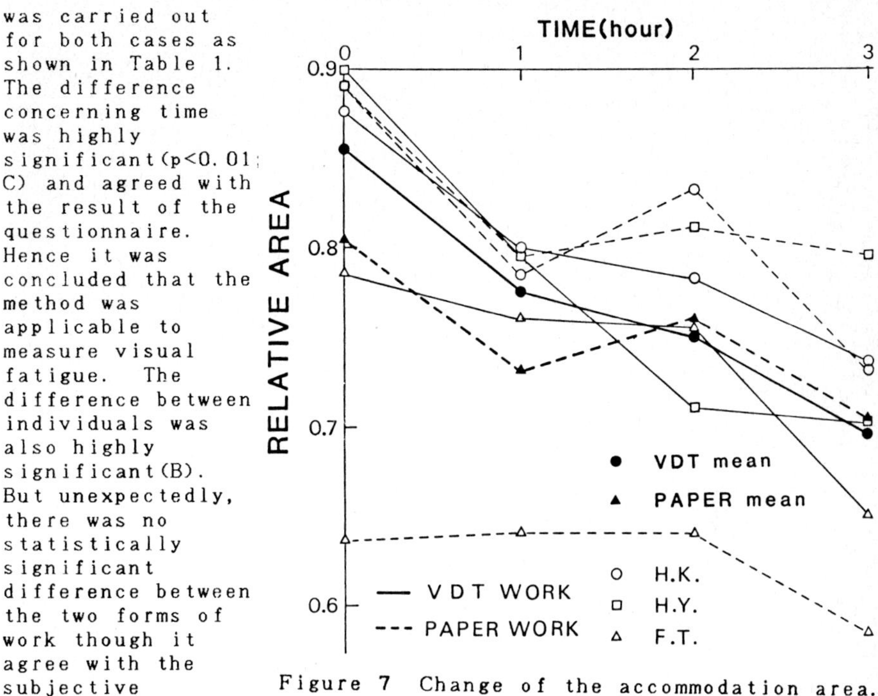

Figure 7 Change of the accommodation area.

CONCLUSIONS

A new apparatus to measure accommodative response was developed and the normalized area of response to step changes of target was found to be satisfactory to measure visual fatigue in VDT work objectively. More intensive and precise experiments are needed to clarify the influences of VDTs on operators.

REFERENCES

(1) Campbell, F. W. and Robson, J.G., High-Speed Infrared Optometer. J. Opt. Soc. Am., 49 (1959) 268-272
(2) Matura, R.A., Effect of visual display unit on the eyes: A bibliography(1972-1980). Human Factors, 23 (1981) 581-586
(3) Östberg, O., Accommodation and visual fatigue in display work. Proceedings of the International Workshop, Milan, March, 1980. (Taylor and Francis, London, 1980, 41-52)
(4) Starr, S.J., Effect of video display terminals on telephone operators. Human Factors, 24 (1982) 699-911

Human Factors in Organizational Design and Management
H.W. Hendrick and O. Brown, Jr. (Editors)

Human Factors in Voice-Input Applications

Rinzou Ebukuro

C&C Sensor Systems Department
The NEC Corporation
Minatoku Shiba, Tokyo
Japan

Word set problems are very common in the voice input application field. This paper discusses voice-input learning skills as related to word set properties and focuses on the word lengths of Japanese voice-input word sets. According to our studies acquisition of voice-input skills for words of appropriate lengths, three or more moras, was faster than that for shorter words. Experimental and application examples appear in the section on the practical application of a voice-input system.

INTRODUCTION

A voice-input system is dependent upon people and is thus greatly affected by these people. The sensing of objects by the five senses, translating phenomena into suitable words, and saying these words to the machine as input are the three major steps for human beings in the voice-input operation.

There are many factors in the relationship between voice-input performance and human behavior; a small part of which, verbal errors, will be discussed in this paper. This paper also focuses on word length and the acquisition of voice-input skills by human beings. This relationship is classified by our experimental results and by field operation examples. Further, the importance of well-timed instructions, i.e., instructions given at the appropriate times, and the allowance of sufficient training time will also be demonstrated.

WORD LENGTH AND ACQUISITION OF VOICE-INPUT SKILLS

Japanese words are characterized by vowel sounds. The word length is determined by the number of moras strings in which vowel sounds are found. These are some exceptions to this rule.

If we use voice inputs of one mora, poor machine performances can be expected. Further, audiences hearing these one mora words will experience listening difficulties. Some typical two mora Japanese utterences are /ICHI/ meaning one, /ROKU/ for six, etc. A continuous word input will take more training to learn than isolated spoken input. Three mora utterences are much easier to learn than shorter words. However, similar sounding words must not be put into the word set. A lower probability of similar word occurrences can be expected if the word set is composed of longer words. Hence, a more relaxed training method is used during long word set operations containing four or five mora.

THE VOICE-INPUT LEARNING PROCESS

Experimental results concerning the relationship between word length and the learning process are presented here.

Table 1 Confusion Matrices for Mora Dictation

(A) Dectation for the human ear

	U \ D	A	I	U	E	O
A	ka	→ na				
	ta	← ka·ha				
	na	→ ra				
	ya	→ na·wa				
	na	← ta·ra·wa				
I	i				→ te	
	ki		→ ti			
	si		→ ni			
	ti		→ ki·si			
	ni		→ i·hi·mi·ri			
	hi		→ i·ki·si			
	mi		→ i·ni·i			
	ri		→ i		→ te	
U	u			→ ku hu ru		
	su			→ tu		
	nu			→ u mu nu	se	
	hu			→ u·ku		
	mu			→ u		
	ru		→ ri		→ te	
E	e				→ ne	
	ke		→ ki		→ ne	
	ne				→ se·he·me	
	me				→ ne re	
O	ko					→ no
	so					→ to
	no					→ ro
	ho					→ o
	mo					→ o
	ro					→ o·to

(1) Media; Telephone (4) Listeners; male 7 people
(2) Utterance; Discrete
(3) Speakers; Male 5 people

(B) Recognition by voice input device

	U \ D	A	I	U	E	O
A	ta	→ na·ha				
I	ki		→ ti·mi			
	ti		→ si			
	ni		→ mi·ri			
U	ku			→ hu		
	nu			→ mu		
E	te		→ ri			
	ne				→ me re	
	he				→ re	
	me				→ re	
O	ko					→ ko
	so					→ ro
	yo					→ o
	ro					→ yo

(1) Media; Microphone (4) Listener; Voice input device
(2) Utterance; Discrete
(3) Speaker; Mr. A (1 person)

Mora [1]: Confusion matrices for mora dictation for the human ear and a voice input device are shown in Table 1 as (a) and (b) respectively. The same errors are shown to occur in the same vowel group in both cases. These are unavoidable hearing errors that occur when people have no recourse to the context that a word is used in. This is why the mora recognition area is insufficient for modern voice input applications.

Numerals [2]: Japanese numerals are generally composed of one or two moras: 1/ICHI/, 2/NI/, 3/SAN/, 4/YON/, 5/GŌ/, 6/ROKU/, 7/NANA/, 8/HACHI/, 9/KYŪ/, 0/ZERO/. (/SHI/ for four and /SHICHI/ for seven are deleted.) Even small changes in an utterance or missing phonemes will affect the recognition of a word. Therefore, careful enunciation should be requested of a voice-input operator throughout an operation. If the operation is a discrete utterance numeral input, fewer difficulties in the acquisition of speaking skills will be found. Conversely, for continuous speech inputs, operators will need to cooperate with the machine more.

Fig. 1 shows the learning curves for two unexperienced males designated as A and B, for three digit continuous numeral inputs. The two curves are similar and clearly reflect the operators' ability and efforts to acquire the voice-input skill. This data was collected at the parcel sorting site.

The basic instructions given to the operators were as follows:

(a) Avoid /SHICHI/ for seven and /SHI/ for four. Use /NANA/ for seven and /YON/ for four.

(b) Each mora must be enunciated clearly.

(c) Each word should be uttered in three time, i.e., its rythm should be three time.

(d) Talk to the machine making your utterances even in tempo and clear.

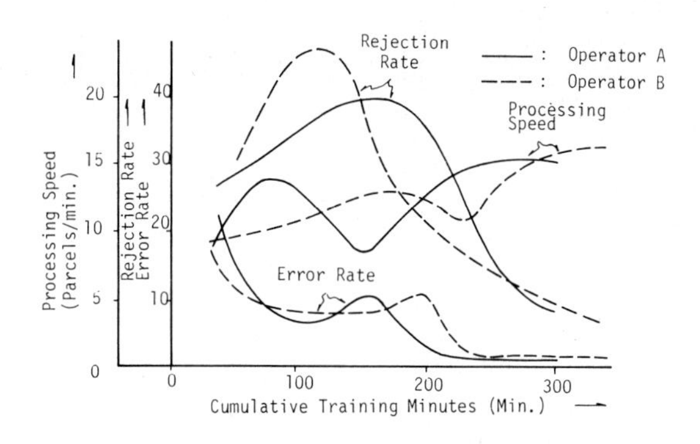

Fig. 1 Learning Curve of Numeral Input

Clear and carefully said utterances were made during the first training stage. However, after the operators acquired the basic skills, they tried to increase their input speed which led to an increase in rejections and thus to a decrease in processing speed. Note that the operators were especially trying to decrease the rejection rate during the 100 to 120 minute span but no improvement in the error rate could be obtained. On the seventh day of training, an instruction was given to the operators to maintain the basic rules for utterances for numeral input even at a loss of voice-input speed. Figure 1 shows that at the 183 minute mark for operator A and the 194 minute mark for operator B the above instruction was given and also that their performances rapidly improved. This result indicates the importance of giving practical instructions to the operators at the appropriate time and for the allowance of sufficient time for training.

Place name (3-mora words): Table 2 shows the result of discreet place name input of three mora word sets which include 20-place names and which contain similar words. Twenty place names were uttered by inexperienced people a total of ten times each. These utterances were recorded on magnetic tape. The third utterance was used as the standard (reference) pattern for voice recognition. Place name reading was input from fourth utterance. No differences can be observed between the tests except that operators E and F had higher rejection rates. This is probably due to the difference of inherent abilities between different people.

Table 2 Voice Input Test Results
(3 mora place names)

Op	P	1st	2nd	3rd	4th	5th	6th	7th	Total
A	E	0	0	0	0	0	0	1	1
	R	0	0	1	1	1	0	0	3
B	E	0	0	0	0	0	1	0	1
	R	0	1	0	0	0	1	1	3
C	E	0	0	0	0	0	0	0	0
	R	0	1	0	1	0	1	0	3
D	E	2	1	0	0	0	0	0	3
	R	0	0	1	0	1	1	0	3
E	E	2	0	1	0	0	0	1	4
	R	1	0	1	1	2	2	0	7
F	E	0	0	0	0	1	1	1	3
	R	1	1	0	2	3	1	1	9
TOTAL E	NE	4	1	1	0	1	2	3	12
TOTAL E	%	3.33	0.08	0.08	0.00	0.08	1.67	2.50	1.43
TOTAL R	NR	2	3	3	5	7	6	2	28
TOTAL R	%	1.67	2.50	2.50	4.17	5.83	5.00	1.67	3.33

Note

E : Error
R : Reject
NE: Number of errors
Nr: Number of rejects
P : Performance
Op: Operator
Number of utterances:
20 place names each

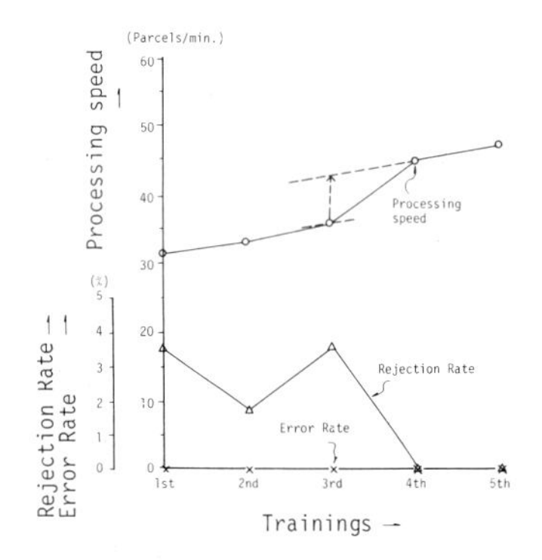

Fig. 2 Learning Progress for 56-Place Name Input

Place names [3]: Figure 2 shows the learning progress of inexperienced male adults with Japanese prefectural name word sets which include some control words. The error rate was zero percent from the beginning and the rejection rate became zero after the training was performed three times. Next, the input speed was increased from about thirty one utterances per minute for the first to the third training, to nearly fifty utterances per minute during the fourth training. The main reason for the rejections was the hesitation in place name enunciations.

FIELD OPERATION EXAMPLES [1]

Table 3 gives field operation examples created by the engineers who actually carried out the program from the very first. The following tendencies are observed in the table:

(a) Discreet place name input training requires a very short learning period.

Table 3 Voice input training properties

Case No.	Site	Word set		Treatment of word set	Training		Performance
		Control word	Data word		Trainees	Training period	
1	Parcel Sorting	Reset Cancel Same etc.	Two digit numeral 0 ∿ 9 Continuous speech	0 → Zero or "Maru" Cancel → "Dame"	About 40 people	2 ∿ 3 months	Both rejection and error rates are within the specified value.
2	Bag Sorting	Cancel Same etc.	Only plate names	Nothing	Several people	About 4.5H for each person	Ditto. However the rejection rate is 1/2.7 and the error rate is 1/4 as compared to case No.1
3	Data input for auction	Back Local Remote	The words for grading (Best, Medium etc.) and Numerals 0 ∿ 9	Nothing	A few people	About 1 month	Ditto. Error rate is at most 1/4 as compared to No.1.
4	Department store goods sorting	Cancel	Place names	Shortening of word length and replacement	Part timers are handling with voice input with no key board backup system		The data satisfy the specifications.

(b) Continuous recognition numeral input training requires a month. (learning period)

(c) A smaller number of operators will require a shorter training period and be more efficient than a larger group.

As previously mentioned, site management positively affects operation skill acquisition during voice input training by the operators. Note the number of trainees in Table 3.

CONCLUSION

The relationship between word length and the acquisition of voice-input operation skills has been discussed in this paper. Further, the importance of appropriate word set preparations and the method of management has also been emphasized. These principles will remain effective even after voice-input technology has been throughly explored. Finally, I would like to express my heartfelt gratitude to Mr. Tomoyuki Isono and his staff for their untiring help in the performance of these experiments and in the collection of the data.

[1] R. EBUKURO, Utilization of voice input from the viewpoint of word set property, Transactions of the symposium on human interface in computer system (in Japanese). Information Processing Society of Japan. (July 1983) 39-52.

[2] R. EBUKURO et.al., Investigation on voice input learning curve for numeral input (in Japanese). The 22nd SICE Academic conference (July 1983) No. 3603, 691-692.

[3] R. EBUKURO et.al., Introduction of voice-input system to the industrial field (in Japanese) Automation, Vol. 24 No. 6 (June 1979) 35-40.

Human Factors in Organizational Design and Management
H.W. Hendrick and O. Brown, Jr. (Editors)

VOICE OPERATION IN CAD SYSTEM

Tomoki Shutoh, Shichiro Tsuruta,
Ryuichi Kawai and Masamichi Shutoh

NEC Corporation
C & C Systems Research Laboratories
4-1-1 Miyazaki, Miyamae-ku, Kawasaki, Kanagawa 213, Japan

Experimental evaluation results for voice operation effectiveness in a CAD system are described. For this purpose, speech recognizer acceptability, operation capability in real work and subjective preference of voice operation efficiency were examined by both beginner and trained personnel.

These examinations showed that the voice operation method had higher performance in accomplishing input and editing work than other methods did.

1. INTRODUCTION

In a conventional CAD system , such as for IC mask design, manually operated devices, such as keyboard, data tablet and digitizer are generally used for data and command input in man-machine interaction. However, these devices are not convenient for designers, because it is necessary to remember a string of command and its menu arrangement for smooth operation. Therefore, long term training might be needed.

For improving this traditional man-machine environment and realizing more natural operation, voice operation effectiveness has been examined in regard to the CAD system[1],[2].

2. EXPERIMENTAL SYSTEM

This CAD system, CGDS, covers artwork generation in IC manufacturing. In this part, designers interactively input IC mask patterns one by one, by manipulating input devices at a workstation, while they check and correct patterns, as needed. The external appearance of Intelligent Color Graphic System (ICGS) , used in this experiment for a workstation, is shown in Fig.1. This ICGS operates as one of interactive editing stations in CGDS. It can be operated as a stand-alone system by its own microprocessors and highly specialized hardware and software.

Fig.1 ICGS external appearance.

In the workstation, like ICGS, a menu function is generally supported to eliminate troublesome command input. By this menu function, designers only indicate a point within a desired menu table, which is positioned on a part of data tablet face or displayed on a CRT, instead of entering a full command string through the keyboard.

DP-200 speech recognizer ,manufactured by NEC, is connected to ICGS by RS-232C interface The DP-200 is designed for speaker dependent recognizer, so designers have to register their utterance of each command as reference patterns before they use it. In this experimental system, the DP-200 was implemented so as to operate in the same way as pointing out locations on the menu. That is, one utterance corresponds to one position indicating operation. For examinee's advantage, Japanese commands, such as 拡大 (KAKUDAI:enlargement), 矩形追加 (KUKEI-TSUIKA:addition of rectangle) etc., were used for utterance in voice operation and explanation on the menu table.

3. EXPERIMENTAL METHOD AND ITS RESULT

To evaluate voice operation effectiveness , the following factors have been examined.

(1) speech recognizer acceptability
(2) operation capability in comparison with other devices in real work
(3) subjective preference of voice operation efficiency

3.1 SPEECH RECOGNIZER ACCEPTABILITY

To clarify the ease factors pertinent to speech recognizer utilization, a recognition experiment was carried out by 10 beginners and 4 trained personnel. First, each examinee registered 50 command utterances, 1 training sample for each command word, after one training session. Then, the examinee tried the recognition experiment 4 times, 50 test utterances in each trial, for a total of 200 test utterances.

Table 1 shows recognition experiment results. In the beginner's case, the average reject count is 9.0, and error count is 2.7. On the other hand, in trained personnel case, average reject and error values are 7.25 and 1.5 , respectively. This shows that experience is not an important factor in speech recognizer utilization. In addition, according to reject and error occurrence distribution for each command, most of these occurrence concentrated on about 5 commands. It has been confirmed that the correct response rate is greatly improved, by making registration of these command words, over again.

Table 1 Recognition experiment results.

Beginners

	A	B	C	D	E	F	G	H	I	J	Average
Reject Count	14	8	11	12	11	4	5	3	13	9	9.0
Error Count	7	1	4	1	2	2	0	6	0	4	2.7

Trained Personnel

	K	L	M	N	Average
Reject Count	7	4	8	10	7.25
Error Count	2	1	2	1	1.5

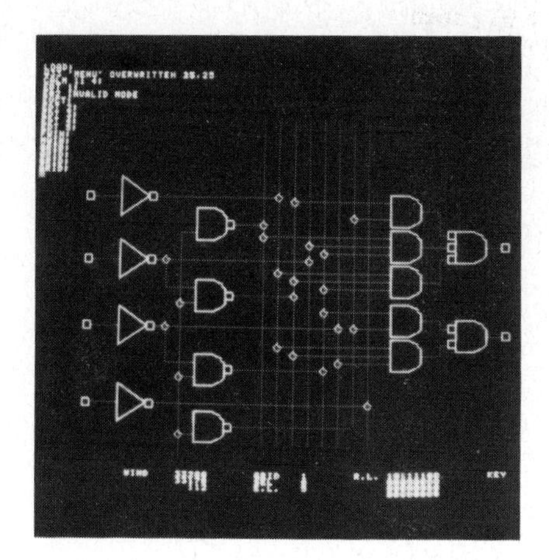

Fig.2 Logic circuit diagram used for input work evaluation.

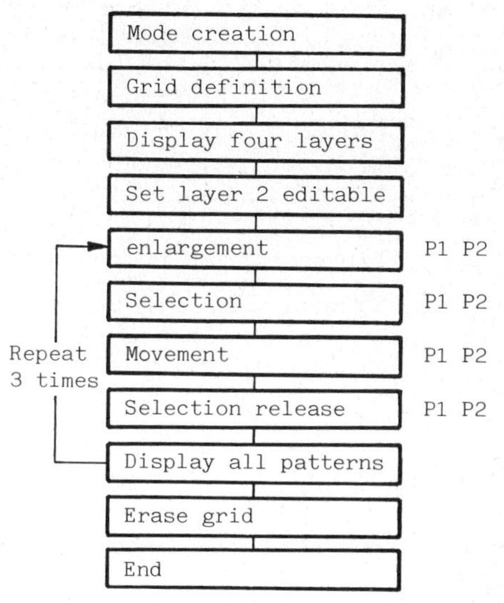

Fig.3 Operational flow for editing work. In this figure, P1 and P2 mean positioning by data tablet.

3.2 OPERATION CAPABILITY COMPARISON

To compare the operation capability for voice operation with that for generally used devices, two operations, input work and editing, were examined by three methods using keyboard, data tablet and voice. These operations are typically used for design work.

In input work, an attempt to feed in a circuit diagram, shown in Fig. 2, was tried 5 times by two examinees. In accomplishing this work, the examinee positioned about 400 points and entered about 120 commands, for each trial.

In editing work, a pattern editing sequence, shown in Fig.3, was examined using IC mask pattern data by 5 examinees, 2 inexperienced persons and 3 persons trained in CAD system use. Each examinee corrected three wiring patterns one by one using enlargement operation. These were only a part of 2000 patterns. In this case, 24 points positioning and about 35 commands entrance are needed in each trial. Each examinee tried this correction flow 5 times in each operation method. In each trial, an elapsed time was measured for accomplishment and an error operation was counted.

Table 2 and Table 3 show the measured elapsed time and error count in each work. In editing case, the recorded time is an average of 5 trials and beginner's average and trained personnel average means an average for each groups, respectively. For example, beginner's average is 18'30" using key operation, 9'54" using menu operation and 7'00" using voice operation. Beginner's speed in voice operation is higher than trained personnel's speed in menu operation.

Figure 4(a) and (b) show a training effect trend for beginners in each situation , respectively. When using voice, the training effect is somewhat small, in comparison with using the menu. It shows that memorizing voice commands is easier than memorizing command positions in the menu.

Table 2 Examinee's measured time of 5 trials for two groups in input work.

	Menu Operation					
	1	2	3	4	5	Average
Beginner	19'44"	16'46"	15'42"	14'05"	13'17"	15'55"
Trained man	14'17"	13'54"	13'05"	-	-	13'45"

	Voice Operation					
	1	2	3	4	5	Average
Beginner	13'48"	12'31"	11'46"	10'16"	9'59"	11'40"
Trained man	11'31"	10'38"	9'50"	-	-	10'40"

Table 3 Examinee's measured total time in 5 trials and average time for beginners (A,B) and trained personnel (C,D, E) in three input methods in editing work.

	Key Operation		Menu Operation		Voice Operation			
	Elapsed Time	Input Error	Elapsed Time	Input Error	Elapsed Time	Input Error	Error Count	Reject Count
Beginner A	23'40"	10	10'20"	4	7'18"	7	2	5
Beginner B	13'19"	4	9'27"	2	6'42"	1	0	1
Beginner's Average	18'30"	7	9'54"	3	7'00"	4	1	3
Trained man C	12'05"	6	8'35"	5	8'17"	5	2	3
Trained man D	12'16"	9	8'10"	2	7'23"	9	2	7
Trained man E	8'57"	2	6'50"	1	6'06"	4	0	4
Trained man's Average	11'06"	5.7	7'52"	2.7	7'15"	6	1.3	4.7

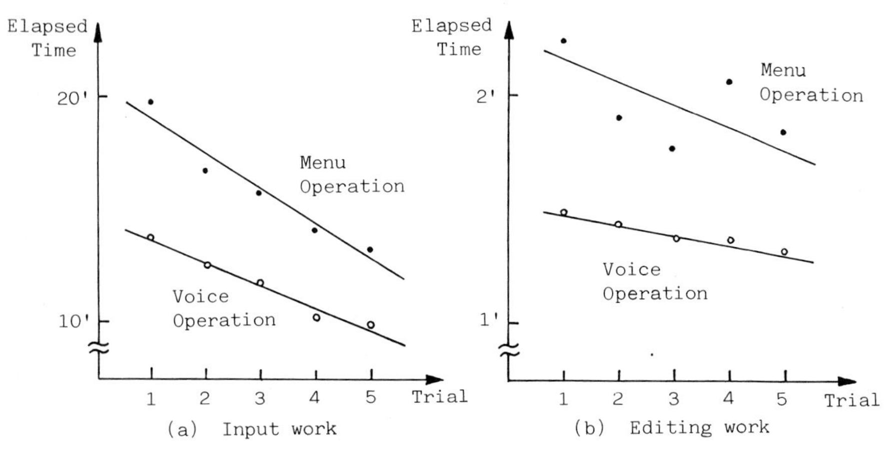

Fig.4 Beginner's training effects for input work (a) and editing work (b) in menu and voice operation, respectively.

There are no differences between the two groups in regard to difficulty in utilizing the speech recognizer. These evaluations indicate that voice operation makes it possible to achieve this work faster than for other cases, without any distinction between the two groups. The reason is that an operator can carry out two processes simultaneously. Namely, they can move the data tablet stylus to the next desired position, while speaking the command.

3.3 SUBJECTIVE PREFERENCE

Table 4 shows subjective preference data for voice operation efficiency. According to this result, good operation capability under any attitude is mostly prefered. The parallel operation capability, the second prefered choice, causes higher efficiency while executing the work, as previously mentioned. These capabilities are more useful when designers use the digitizer, because the designer must manipulate a cursor with both hands to position it accurately at a required point in the bending attitude.

A free definition capability is also important for Japanese designers, because they would not be familiar with English based command.

Table 4 Subjective evaluation of voice operation efficiency. A smaller value indicates a higher preference.

Voice Operation Efficiency	Evaluated Value
Operation capability under any attitude	1.8
Parallel operation capability	2.2
Easy operation	2.9
Free definition of utterance word	3.9
Ease in remembering the commands	4.2

4. CONCLUSION

Voice operation effectiveness in using the CAD systems was confirmed. However, from the human factors view point, there are still many problems to be evaluated. These experiments would not only improve man-machine interface, but also might be useful in designing a new speech recognizer.

5. ACKNOWLEDGEMENT

The authors would like to thank Mr. Y. Kato and Mr. S. Hanaki for their constant encouragement.

6. REFERENCES

[1]. Y.KATO, et al.,"Voice Input for CAD systems"
IFIP, Working Conference, Oct.1980, pp.150-159

[2]. T.SHUTOH, et al.,"CAD system for IC mask design"
Proc. CAM-I 10th annual meeting, Oct.1981, pp.93-106

INFORMATION SYSTEMS

Human Factors in Organizational Design and Management
H.W. Hendrick and O. Brown, Jr. (Editors)

Sources of Problem Solving for Computing End-Users: Human Computer Interface and Social Networks in Organizations

Elliot Cole
College of Information Studies
Drexel University
Philadelphia, PA 19104

ABSTRACT

How do end-users get the information they need in order to use today's relatively user friendly computing systems? This information may be obtained through user-friendly software -- a primary concern of human factors research-- as well as by social means -- social networks which facilitate sharing knowledge among users. This field study examines information sources end-users select. Organizational processes are found to be active in facilitating end-user computing. Most people used both personal sources and documentation (online and hard copy) in solving a computing problem. Three-quarters of the respondents found a personal source to be the most important in solving that problem. End users rate personal sources significantly more useful than documentation. These results suggest that organizational processes are an effective means of performing the human computer interface function, and can complement the cognitive processes currently being applied by interface designers.

INTRODUCTION

Interactive computing is viewed as an important means of making knowledge workers more productive in a competitive world. A critical problem in designing interactive computing systems is how to make the systems easy to learn and easy to use, objectives which are not particularly compatible. Making the system easy to learn involves conveying information to the end-user. This paper identifies some of the research areas involved in the delivery of knowledge, and reports results of a field study examining information sources selected by end-users when they needed information to solve a computing problem.

RESEARCH AREAS ADDRESSING INFORMATION FOR END-USER COMPUTING

There have been several streams of research addressing the problem of delivering technical information to people working in organizations. Most of the work on delivering this computing information has applied cognitive science cognitive psychological models and methodology. Objectives of this research have been to reduce learning time, to increase knowledge retained, to reduce working time, to enhance work state transitions (Card, Moran and Newell, 1983; Badre and Schneiderman 1982; Sondheimer and Relles 1982; and Reiner 1981). That approach has focused on the computer and the user in an asocial context without taking advantage of organizational processes which can also support the end-user.

Another stream has focused on organizational communication, including technology transfer mechanisms. (Allen, 1977; Gerstenberg and Berger, 1980; (O'Reiley and Pondy, 1976; Galbraith, 1977; Mintzberg, 1979; Rogers and Kincaid, 1981; Tushman and Scanlon, 1981; and Culnan, 1983; Culnan and Bair, 1983). Work in this area has addressed the problem of how information about technical topics can be conveyed to individuals and work groups in organizations. Problems have involved strategies to provide the right information at the right time to those who need it. Investigators have looked at the roles of different kinds of sources, e.g., printed and consequently static with respect to user and situational variables, as well as personal and thus potentially responsive to user and situational variables. This stream has looked at both the communication function and the social networks which support it.

A third stream has focused on diffusion of administrative innovations. Diffusion researchers have been interested in why some innovations become part of one organization's standard operating procedures while failing to thrive at another site (Zaltman et al., 1973; Rogers et al 1976; Perry and Kraemer, 1979; Teece 1980; Zmud 1982; Cole 1983; Rice, 1984). Computing practices have been setting for this research recently. A trend has been to examine the manner by which an organization adapts an innovation, i.e., end-user computing, to fit its organizational climate and culture.

Our research links these streams. Human-computer interaction is conceptualized as a problem in technology transfer. Individuals (both non-technical and technical personnel) acquire technical information by selecting from a set of information sources. Diffusion processes are involved in acquiring this information.

Information distribution may be through user-friendly software -- a primary concern of human factors research-- as well as by interpersonal means -- social networks which facilitate sharing knowledge among users.

METHOD

This is a field study applying survey research and the critical incident method (Allen, 1977), which focuses attention on situational variables. Fifty-five respondents in face to face interviews were asked to identify a recent computing "problem which completely stopped your progress in your work", excluding programming class assignments. The interview schedule examines information sources used, their usefulness, characteristics of the source, characteristics of the situation, problem complexity, computing background, time (elapsed, total working, and working with source), and relative time efficiency comparing use of manuals and online documentation to most important source. The sample consists of students and faculty using a timesharing mainframe and personal computers.

Two classes of information sources are examined in this study. One is prepared in advance by the designer, and is what Schneider (1982) calls static user assistance: online help, command summaries, syntax guidance, tutorials, etc.; the hard copy equivalents are included. This class is generally considered authoritative. A second class consists of personal sources such as a specific colleague, a specific manager (instructor in an academic setting), HELP desk consultant, HELP desk manager, and vendor's customer service staff.

FINDINGS

1. Respondents tended to use both personal and static sources: 67% used both types of sources, 20% used only social network sources, 6% used only manuals, and 7% used trial and error.

2. There are differences in the distributions of sources by most important, first used, and total use. For most important source, social network sources were nominated 74%, and static user assistance nominated 17%; 9% used trial and error. For the first used source, respondents turned to social network sources in 46% of the incidents, 38% for static user assistance, and 16% for trial and error. Of the 196 sources used by the respondents, 54% were social network sources, 34% static user, and 9% trial and error.

3. For most respondents (72%) it was important that the source knew about the substance of the work which had to be done, i.e., had knowledge of the task to be performed.

4. Finding a solution to one's problem using online documentation would take "somewhat longer" or "much longer" for 62% of the respondents.

5. Respondents found personal sources more useful than static user assistance. For the first used source, the mean for static sources was 2.47 (on a 1 to 5 scale with 5 high) while the mean for social network sources was 3.80 ($t=3.07$; $df = 44$ $p<.005$). For all sources, static user assistance mean usefulness was 2.38 and social network source mean usefulness was 3.30 ($t=3.90$; $df=175$; $p< .001$).

DISCUSSION

These findings have a bearing on mechanisms to deliver the knowledge base to end users, and on the nature of that knowledge base.

First, social network processes are at work in the dissemination of computing knowledge. Some of these social network relationships appear to develop spontaneously from within the end-user community, while other social network components, such as HELP desks, have been planned. The prospect of user networks forming independently of information systems management is likely to have technical and policy implications in both the short and long term.

Second, the substantial number of respondents using both social network sources and static user assistance indicates that both are involved together in delivering knowledge to end users. One should not look for one type of source or the other as a means of conveying information to users, but rather both together. This approach is supported by both technology transfer and organizational communication research. These areas would suggest the unit of analysis be the user clique and the network as a whole in addition to the single user.

Third, there is the suggestion that the knowledge base required by end-users may consist of more than one dimension. The finding that it is important to users that information sources have knowledge of the work at hand indicates the presence of a task dimension in addition to the command language dimension typically the objective of user assistance system designs. This finding may be related to end user computing, an approach which suggests that systems development personnel have expertise in one type of knowledge base while end users have expertise in a second type of knowledge base (See Rockart and Flannery 1983; Benson, forthcoming; and Rivard and Huff, forthcoming).

It is clear that when organizational processes and infrastructural social networks are at work, it is not desirable to rely only on individual-level human computer interfaces in designing information systems. Interpersonal processes also become appropriate potential mechanisms for the delivery of knowledge. One suspects that the effectiveness and appropriateness of interpersonal (social network) processes will be contingent on both the type of knowledge, characteristics of the individual, and characteristics of the situation. If this is the case, there is a need to better understand the structure of these social processes so that an effective interface strategy and design may evolve linking the individual-level tasks with interpersonal-level processes. This interface across levels of analysis becomes an important intersection of human factors research and organization design research.

REFERENCES

Allen, Thomas J. (1977). Managing the Flow of Technology. Cambridge: MIT Press.

Badre, Albert and Schneiderman, Ben. (1982). Directions in Human Computer Interaction. Norwood, NJ: Ablex Pub. Corp.

Benson, David H. "A Field Study of End User Computing: Findings and Issues", forthcoming in MIS Quarterly

Card, Stuart K.; Moran, Thomas P.; and Newell, Allen. (1983) The Psychology of Human-Computer Interaction. Hillside, NJ: Lawrence Erlbaum Associates, Pub.

Cole, Elliot. (1983). "New and Old Models for Office Automation," Journal of the American Society for Information Science. 34(3): 234 - 239.

Culnan, Mary J. (1983). "Environmental Scanning: The Effects of Task Complexity and Source Accessibility on Information Gathering Behavior", Decision Sciences 14(2): 194 - 206.

Culnan, Mary J.; and Bair, James H. "Human Communication Needs and Organizational Productivity: The Potential Impact of Office Automation", Journal of the American Society for Information Science 34(3):215 - 221 (1983).

Galbraith, Jay. (1977). Organization Design. Reading, Mass.: Addison-Wesley.

Gerstenberger, Arthur; and Berger, Paul. (1980). "An Analysis of Utilization Differences for Scientific and Technical Information," Management Science 26(2):165 - 179.

Mintzberg, Henry. (1979). The Structuring of Organizations. Englewood Cliffs, NJ: Prentice-Hall.

O'Reilly, Charles; and Pondy, L. R. (1979). "Organizational Communication," in S. Kerr, ed. Organizational Behavior. Columbus, OH: Grid.

Perry, James and Kraemer, Kenneth L. (1979). Technological Innovation in American Local Government: The Case of Computing. New York: Pergamon.

Reisner, Phyllis. (1981). "Human Factors Studies of Database Query Languages,", Computing Surveys 13(1):13 - 32.

Rice, Ronald (1984). "Reinvention in the Adoption Process: The Case of Word Processing," in Ronald, Rice, ed., The New Media, Beverly Hills, CA: Sage.

Rivard, Suzanne; and Huff, Sid L. "User Developed Applications", forthcoming in MIS Quarterly.

Rockart, John F.; and Flannery, Lauren S. "The Management of End User Computing", Communications of the ACM 26(10): 776 - 784 (1983).
Rogers, Everett; and Kincaid, (1981). Communication Newtorks.

Schneider, Michael L. "Models for the Design of Static Software User Assistance," Chapter 7, Pp. 137 - 148, in Badre and Schneiderman op cit.
Smith, Reid. (1981). A Framework for Distributed Problem Solving. Ann Arbor: Univ. of Michigan Press.

Sondheimer, Norman K; and Relles, Nathan. "Human Factors and User assistance in Interactive Computing Systems" IEEE Transactions on Systems, Man, and Cybernetics, Vol. SMC-12(2): 102 - 107 (1982).

Teece, David H. (1980). "The Diffusion of an Administrative Innovation," Management Science. 26(5):464 - 470.

Tushman, Michael L.; and Scanlon, T. (1981). "Characteristics and External Orientations of Boundary Spanning Individuals," Academy of Management Journal, 24: 83 - 98.

Zachary, Wayne; Goodson, John. (1982). "Human Factors in Distributed Intelligent System Design: Information Processing Constrainst on Human-Computer Cooperation." Technical Report 140026A. Willow Grove, PA: Analytics, inc.

Zaltman, Gerald; Duncan, R.; and Holbek, J. (1973) Innovations and Organizations. New York: Wiley.

Zmud, Robert W. "Diffusion of Modern Software Practices", Management Science 28(12):1421 - 1431 (1982).

Human Factors in Organizational Design and Management
H.W. Hendrick and O. Brown, Jr. (Editors)

AIDING AND ABBETING "MIS MANAGEMENT" APPLICATIONS OF ARTIFICIAL INTELLIGENCE TECHNIQUES TO MANAGEMENT DECISION SUPPORT SYSTEMS

Paul S. Reed

Department of Psychology
University of California, Santa Barbara
Santa Barbara, California 93106
U.S.A.

Artificial Intelligence research has successfully simulated the information processing operations involved in human problem solving and has the potential to contribute to the effectiveness of DSS through the development of "knowledge-based" systems. These systems have the capability of logically analyzing and interpreting patterns of data and can help to reduce the impact of information overload on managers. In addition, the application of AI to the user interface promises to contribute to the effectiveness of human-computer interaction by making DSS easier to use by non-experts.

One of the most important problems faced by managers in organizations today is the explosion of information available for facilitating decision-making. Although immense quantities of data are becoming available, it is also becoming more difficult to extract relevant information - knowledge, in other words - for aiding managers in making key decisions. This results from the ironic situation in companies which acquire more information and find that it is more difficult and costly to draw meaningful conclusions from huge, sprawling databases. Because one of the primary purposes of Decision Support Systems (DSS) is to facilitate effective decision-making, it is necessary to develop techniques for "distilling" relevant information (i.e. knowledge) from available data. A number of developments in artificial intelligence can assist in the development of tools that will allow managers to master the information explosion by performing "intelligent" and appropriate analyses on database information in response to the needs of the user.

The employment of artificial intelligence techniques to improve the quality of human computer interaction can greatly increase the likelihood of achieving some of the goals of decision support system identified by Alter (1980), namely 1) aiding decision making, 2) facilitating planning, 3) conducting on-line searches of databases, 4) training and simulation, and 5) interactive problem-solving. Advances in the field of cognitive simulation now make it possible to develop programs which carry out "thinking processes" involving the manipulation and analysis of data using specific models that allow managers to obtain decision-relevant information. This approach extracts key knowledge from massive databases to assist rational decision-making. It is now possible for a manager to state a concept or model and direct the decision-aiding system to examine the total information available in the databases, select the relevant information, and provide optional methods for verifying or implementing the original concept.

Developments in the area of "knowledge-based systems" have produced computers that are able to reason, or at least mimick the way humans employ information and past experience to solve new problems (Mueller, 1983). Such systems collect information in one or more storage areas (called knowledge sources), and facilitate problem-solving in a single, well-defined domain. For example, the set of logistical relationships between production, inventories, orders, and shipments allow the development of a knowledge source representing these relationships which can then be processed with artificial intelligence techniques to draw inferences and conclusions from this knowledge (Bonczek, Halsapple, and Whinston, 1981). Because artificial intelligence is primarily concerned with constructing machines which can behave "intelligently", it also offers considerable potential for improving the quality of the user interface by making a DSS easier to use for non-experts.

The human-computer interface in MIS is a critical factor in determining the overall success of a Decision Support System (DSS) by managerial personnel. The well-known phenomenon of "managerial resistance" underscores the need for interfaces with a high degree of "friendliness" to the non-expert user. The developments in user interface design may have a crucial impact on the acceptance of DSS by higher level management.

The hardware interface is a key element of the human-computer system because it has a major impact on the way a user interacts with the system. The type of device employed for operating the system is usually a keyboard, but the relatively recent development of systems utilizing peripheral devices such as the increasingly popular keypad (e.g. the Apple Lisa's "mouse") can substantially improve the ease of use for individuals with weak typing skills. The development of software with mouse-driven menus promises to simplify and quicken the selection of options. This type of device may facilitate the acceptance of DSS by managers by providing a simple but effective means of controlling the computer.

The software interface in MIS is also a crucial element in human-computer interaction, and may thus be part of the solution to managerial resistance. The design of displays, program logic, prompting, menu structure, and program structure must take into account the human factors issues in the managerial environment. The combination of the above elements determines to a large extent how a user will understand and interact with the system. A well-designed interface will facilitate information management by improving the efficiency and effectiveness of the user in obtaining critical information. The "distillation" of information from databases can be a useful and effective technique for coping with the "information overload" syndrome that is so common among managerial personnel. In addition, an intelligent interface can provide on-line help and feedback to users to make their task simpler and more satisfying.

The user's "mental model" of the computer system is another important component of the human-computer interface which artificial intelligence may help elucidate. Simulations of cognitive processes have increased our understanding of how users view the machines with which they interact. An ideal DSS would be designed so that the operation of the system is completely "transparent" to the user. Very few such systems exist and it is clear that DSS developers must design software that is compatible with the user's conception of their task. Research on the mental models employed by computer users is currently advancing our knowledge of the role played by "cognitive ergonomics" (or the mechanics of human information processing) in human-computer interaction and promises to contribute to the design of effective interfaces.

We have discussed how application of AI techniques to DSS may enhance the overall functioning of the system. Knowledge-based systems can improve the manager's utilization of databases and decreases the impact of information overload. The development of "smart" interfaces also promises to facilitate managerial acceptance by making DSS easier to use by non-experts.

References

(1) Alter, S. 1980, Decision Support Systems: Current Practice and Continuing Challenges, Addison-Wesley, Reading, Massachusetts.

(2) Bonczek, R. H., Holsapple, D. W., and Whinston, A. B., 1981, Foundations of Decision Support Systems, Academic Press, New York.

(3) Mueller, G. E., The potential of knowledge-based systems. In E. Boehm and M. Buckland (Eds.) Education for Information Management, Santa Barbara: The Information Institute, 1983.

Human Factors in Organizational Design and Management
H.W. Hendrick and O. Brown, Jr. (Editors)

INFORMATION SYSTEMS ETHNOGRAPHY: INTEGRATING ANTHROPOLOGICAL METHODS INTO SYSTEMS DESIGN TO INSURE ORGANIZATIONAL ACCEPTANCE

Wayne W. Zachary and Gary W. Strong
College of Information Studies
Drexel University
Philadelphia, Pennsylvania
U.S.A.

Allen Zaklad
Analytics, Inc.
Willow Grove, Pennsylvania
U.S.A.

Information Systems Analysis (ISA) seeks an information system's logical structure so information technology can be applied to increase productivity. Failure to employ relevant behavioral models is often cited for ISA failure. Since an information system is analogous to a culture, ethnographic techniques of cultural anthropology are highly relevant to ISA.

INTRODUCTION

An Information System (IS) may be broadly defined as a set of people and machines which has the collective task of transforming and disseminating information (as opposed to goods and services). As a result of the technological revolution of the last three decades, there has been a broad and accelerating shift of ISs from a manual basis to a strongly automated basis. The computer portion of an organizational information system (CIS) is frequently referred to as "the information system," as though it performed alone all the functions of the manual system into which it was introduced. Of course this is not true -- the overall IS is still (and probably will always be) a primarily human entity. But many of those involved in the design, implementation, and evaluation of the computer components of information systems have failed to realize this fact, with grave consequences.

WHY CIS's FAIL

"After more than two and a half decades of experience in implementing computer application systems a surprisingly large number of them still end in failure" [8]. In those failures where the computer information system is of little or no use upon completion, the reasons can usually be traced to behavioral issues, rather than technological issues [8]. In his seminal paper, Turner identified six such behavioral issues in CIS success: (1) Each organization requires a unique CIS, not an off-the-shelf product; (2) The CIS's capacity to provide information must match the actual needs of the organization; (3) The characteristics of the CIS must fit the critical organizational variables; (4) Implementation must be a dynamic process which converges upon the desired design; (5) Individual differences among users influence their responses to and acceptances of the system; and (6)

The politics of the organization are critical to the outcome. Several of these points have also been discussed by Keen [3]. These problems may seem insurmountable, when viewed from the perspective of a structured systems analyst untrained in behavioral issues. There are pressing needs for an IS analysis framework that considers the human side as primary and for analysis tools which are designed to deal with the "messiness" of human behavior rather than the "neatness" of the computer.

This paper introduces such a framework and accompanying set of tools. We call the entire package Information Systems Ethnography [7]. Its conceptual basis comes from cultural anthropology and its tools from the set of anthropological field methods known collectively as ethnography. Turner [8] has clearly set the mandate for IS methodologies that rely on "behavioral models," but he does not actually suggest one. We feel that IS Ethnography addresses the how-to side of incorporating behavioral models into existing system analysis and design methodologies.

THE CONCEPTUAL BASIS: AN ANALOGY

The first step in providing a behavioral basis for CIS design is to draw an analogy between information systems and human cultures as studied by anthropologists: both are adaptive systems which mediate between a group of people and an environment. In the case of a culture, the environment is the ecological setting. In the case of information systems, however, the environment is the business setting (Figure 1). Anthropologists divide a culture into realms of technology, subsistence economy and social organization [4]. Using our analogy we can make a similar decomposition of an information system into its tools, its information flows, and its organizational structure.

It is natural then to consider the tools of the anthropologist as appropriate for the information systems analyst when trying to understand an information system. Information systems analysis has many of the characteristics of cultural analysis or ethnography: (a) Both occur in an

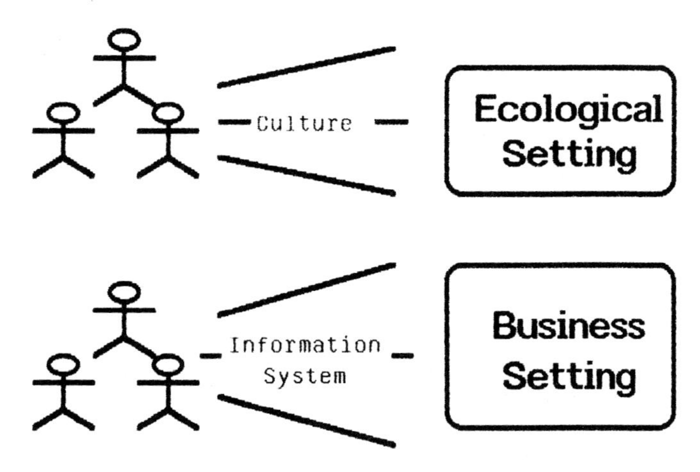

FIG. 1: CULTURE/INFORMATION SYSTEM ANALOGY

organizational setting; (b) Both are visible and threatening to the natives; (c) Both must consider "undiscussable" issues as well as the organizational infrastructure; (d) Both must consider information which has owners, has politics, and which flows along organizational lines; (e) And, most importantly, the information systems analyst as well as the cultural ethnographer is an outsider to the system being analyzed. A powerful set of tools has evolved in anthropology for aiding an ethnographer in understanding and analyzing the cultures he or she is studying, without threatening the validity of the study or the peace of mind of the natives.

ETHNOGRAPHIC TOOLS FOR IS ANALYSIS

The ethnographer is able to deal with the messiness of human behavioral issues because of the variety of his tools rather than any particularly powerful tool. We believe that a number of the tools in the ethnographer's toolkit can be expropriated for information systems analysis, including participant observation, use of key informants, the structured interview, the questionnaire, as well as non-verbal and other unobtrusive techniques. Each of these tools compensates for drawbacks of others. For example, the data derived from participant observation is limited in that the observer must interpret what is going on by himself. A few key informants can elucidate the meaning behind certain events. Yet, this is no way to acquire statistical validity such as can be acquired with the structured interview or the questionnaire. The structured interview provides the opportunity to follow up on interesting leads offered by certain questions; the questionnaire, on the other hand, insures anonymity. Non-verbal and other unobtrusive techniques of non-participative observation can provide an enormous amount of data because everything is considered relevant, but for particularly crucial events it may insure that no details are left out. These and other general-purpose ethnographic techniques are discussed in detail in Pelto and Pelto [6].

With the exception of questionnaire techniques, the general-purpose ethnographic tools are of an informal nature. To supplement the techniques listed above, a large number of more formal ethnographic techniques have been developed. Of these, we feel that two are particularly well-suited to information systems analysis: action-oriented social network methods, and semantically-oriented Multidimensional Scaling. The formal organizational structure of any IS accounts for only a small portion of its actual functioning. People always supplement the formal structure of organizations with less-visible infrastructure, a set of interrelationships which they develop and utilize to accomplish their work and meet their social needs. While the infrastructure is particularly difficult to analyze,it is crucial to actual IS operation, and thus to CIS design. Infrastructure can be examined effectively through social network analysis. Social network analysis originated as an ethnographic technique with the development of the action-oriented approach [1]. A set of canonical events is predefined by the systems ethnographer as triggers for chunks of information processing activity (e.g., receipt of a customer inquiry, or arrival of an order). The ethnographer records all the individual interactions occurring as a result of the trigger. He/she notes the nature of the relationship invoked by each interaction (this is the action set). He/she also records all the tasks undertaken and additional informational items encountered in the segment of information processing activity that follows the trigger (this is the action sequence). Formal network models of infrastructure can be built from these action set data, and formal models of the flow of action can be built from the action sequence data. It should be obvious that the trigger events must be carefully chosen. Preliminary participant observation provides insight into the system to define and select appropriate triggers. The action-oriented network view differs from that presented in manuals, by management, and/or in accounts given by participants.

It is also important to gain access to end-user perceptions or semantics. In database development the semantics of the data/information environment must be explicitly incorporated into the database design. Semantically-based Multidimensional Scaling (MDS) is a powerful technique for uncovering the semantic organization of information among end-users. As a statistical technique, MDS uses a set of inter-object dissimilarities to construct a multidimensional metric space and locate the objects in that space so their metric distances are a monotonic transform of their initial dissimilarities. MDS results are interpreted as defining a set of underlying dimensions of meaning that interrelate items in an information domain. The choice of items to be used in an MDS study must be based on some intuitive understanding of the IS environment gained from more general ethnographic methods. Semantic MDS, like network analysis, is a structured, formal follow-up to general participant observation. The data needed for a semantic MDS can be collected by any of a number of standard psychometric tasks, including paired comparisons, triad comparisons, constrained or unconstrained sorting, or criterion rankings (for more details on MDS see Kruskal and Wish [5].

AN INTEGRATED SYSTEMS ETHNOGRAPHY

The entire sequence for an integrated systems analysis methodology incorporating systems ethnography is shown in Figure 2. The methodology should be viewed as incorporating ample provision for reversion to an earlier step. To some degree, every methodology advocates "rethinking," but we believe it is of special concern here because <u>ETHNOGRAPHY FACILITATES RETHINKING</u>. Many methods, especially the structured ones, make rethinking difficult by placing too much order in the analysis, creating a false sense of validity. Ethnographic methods are oriented toward the fuzziness of human behavior. Specifically, ethnography facilitates rethinking by: (1)

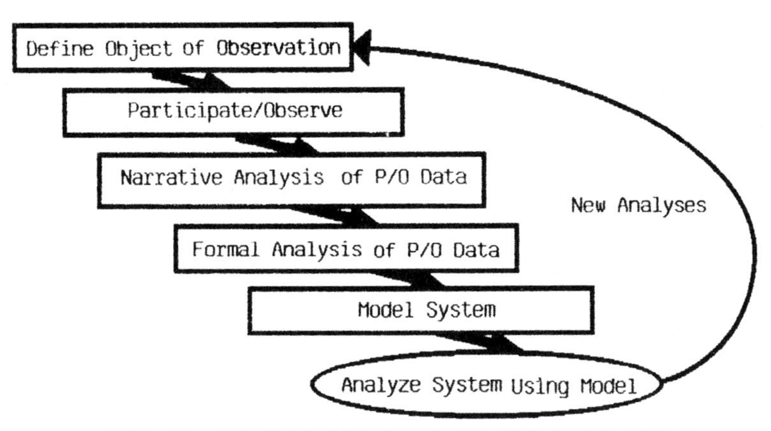

FIG. 2: INFORMATION SYSTEMS ANALYSIS CYCLE

allowing the analyst access to the infrastructure, and forcing him/her to deal with the complex nature of day-to-day work realities; (2) forcing the analyst to deal with how the system really works versus how the participants think it works and the frequent divergences between the two; (3) providing the analyst with a varied toolkit of formal and informal means of examining the same phenomena, thus permitting the customization of analysis; (4) forcing the analyst into direct, meaningful, and constant contact with the people who make up the IS and will become the ultimate users of the CIS being designed. This is the real point of systems ethnography -- to facilitate the rethinking and tailoring of general tools that must be part of any successful analysis process.

CONCLUSION

We believe IS ethnography enables structured methods to deal with relevant behavioral issues in information systems. Structured methods pay some measure of lip-service to the the analyst's need to interact with and understand end-users. For example, DeMarco places the "User Survey" as the first step in his structured analysis sequence [2]. He notes that this "survey" will take up roughly one third of the overall analysis effort without offering guidance on how to accomplish this. Since the IS analyst is by definition an outsider, it is necessary for her/him to quickly acquire at least the rudiments of an insider's understanding of the IS under study. While the general-purpose ethnograhic methods listed above are ideal for this, they are not enough by themselves -- their results are informal and prosey. The problems of such "Victorian novel specifications" which might result from ethnograhpic analysis alone are all too well known. The more formal ethnographic tools of action/sequences and semantic MDS introduce a much-needed rigor into the more narrative general ethnographic approaches. They also generate formalisms which are very useful in structured analysis: for example, action set/sequences can be used to build Data Flow Diagrams, and MDS results can be used to develop Data Dictionaries. The formal ethnographic analysis is thus a "natural" intermediary between unstructured ethnography and formal structured analysis.

REFERENCES

[1] Barnes, J.A., Networks and political processes, in: M. Swartz (ed.), Local Level Politics (Aldine, Chicago, 1968).
[2] DeMarco, Tom, Structured Analysis and System Specification (Yourdon, New York, 1979).
[3] Keen, Peter, Information systems and organizational change, Communications of the ACM 24 (1978) 24-33.
[4] Keesing, R., Theories of culture, in B. Seigel (ed.), Annual Reviews of Anthropology (Annual Reviews Press, Palo Alto, 1974).
[5] Kruskal, J. and M. Wish, Multidimensional Scaling, Sage University Paper Series on Quantitative Applications in the Social Sciences, No. 11, (Sage Pubs., Beverly Hills, 1980).
[6] Pelto, Perti and Gretchen Pelto, Anthropological Research: the structure of inquiry (Cambridge University Press, Cambridge, 1978).
[7] Strong, Gary and Wayne Zachary, Information Systems Ethnography: applied anthropology in industry, paper presented at the Society for Applied Anthropoology annual meetings, Toronto, Canada (1984).
[8] Turner, Jon, Observations on the use of behavioral models in information systems research and practice, Information and Management 5 (1982) 207-213.

Human Factors in Organizational Design and Management
H.W. Hendrick and O. Brown, Jr. (Editors)

EXPERIMENT MANAGEMENT SYSTEM
A Software Tool for Human Factors Research

Charles Lloyd and Tzvi Raz

Department of Industrial and Management Engineering
The University of Iowa, Iowa City, Iowa

EMS (Experiment Management System), is a package written for personal computers, designed to assist the experimenter in conducting human information processing experiments involving discrete-trial reaction time tasks. EMS manages the assignment of experimental conditions, the presentation of stimuli, and the recording and compiling of response data for all subjects participating in an experiment.

THE COMPUTER IN THE HUMAN FACTORS/PSYCHOLOGY LABORATORY

The computer in the human factors/psychology lab can offer the researcher many advantages. Experiments often entail the presentation of hundreds stimuli, the recording of large amounts of response data, and subsequent calculations. In many experiments, stimuli and responses must be timed to a high degree of accuracy. Further, for experiments of any type, the above activities must be conducted in a highly controlled environment and with a high degree of consistency across subjects.

It is exactly these types of functions which computers can support. Computers are inherently consistent, are designed to record and manipulate large amounts of data, and can time events with great precision. Computers have been used in the human factors/psychology laboratory in the past to perform exactly these types of functions. In addition, computers are being increasingly used by researchers to perform many tasks besides experimental control. In a case study discussing the influence of computers on psychological research, Church (1983) lists nine research activities in which he uses computers:

1. Search of Literature

2. Control of Experiments

3. Recording of Results

4. Storage of Data

5. Analysis of Results

6. Development of Theory

7. Comparison of Data With Theory

8. Preparation of Figures

9. Preparation of Manuscript

Describing the computer as a "general purpose device", Church points out that the computer is quickly replacing many special purpose devices such as relay circuts, calculators, and typewriters. In fact, Church mentions that in his laboratory, five different computer systems and five separate languages were used to concuct all nine of the research activities listed.

One serious problem which is common to many laboratories in which a computer is used is that a great deal of training is often required of the experimenter in order to be able to set up and run experiments. In a "typical" mini-computer controlled laboratory, an experimenter must be able to: read and write code in assembler and a high level language such as FORTRAN or PASCAL, know enough editor commands to be able to create and modify files, and know enough about the computer's operating system and architecture to be able to compile, link, load, copy, and delete files. For a person with a minimal background in computers, obtaining the above mentioned skills is no trivial task. Such a person may have to take several courses in programming and computer organization or spend months gaining enough "hands on" experience to be able to use the laboratory equipment efficiently. In laboratories as sophisticated as that described by Church, training new personnel must be an onerous task indeed!

EMS DESIGN FEATURES

The primary goal in developing EMS was to increase the throughput of the laboratory by concentrating on the needs of the researcher who does not have an extensive computer background. Using EMS, such a researcher can very quickly become a productive member of a research group. This person will be capable of setting-up and running a wide range of reaction time tasks with only a minimum of time investment by the more senior members of the research group. As this person's familiarity with computers and programming increases, he/she will become able to modify the system to fit a wider range of experimental needs.

In order to fulfill this goal, the following features were incorporated into EMS:

1. EMS is menu-driven. The user does not need to remember any commands in order to operate the system.

2. All responses are checked for the correct data type and range, making it hard to enter invalid data.

3. Error messages are provided for invalid and out-of-range data.

4. EMS is well documented and highly modularized, making it easy to modify to fit varying needs.

5. EMS takes advantage of current trends toward "personal computers" which are generally much easier to operate than laboratory computers used in the past.

The researcher with considerable programming experience will also find EMS useful. EMS performs many of the functions which need to be done in any computer controlled experiment. EMS can save the more experienced researcher much programming and de-bugging time, thus allowing more time to be spent expanding the system to fit changing experimental needs.

FUNCTIONS OF EMS

EMS is divided into three major options as follows: Define New Experiment, Continue Old Experiment, and Obtain Hardcopy. (See figure 1) These three options will be discussed separately below.

EMS Menu

A Define New Experiment

B Run Subjects

C Obtain Hardcopy

D Exit Program

To make a choice, enter A, B, C, or D

Figure 1: EMS Menu

1. OPTION A, "Define New Experiment", will provide the experimenter with all of the instructions, prompts, warnings, and error messages necessary to make entry of the required data easy and accurate. This option prompts the experimenter for the following four types of information:

   a) Experimental Identification, which includes the title of the experiment, the name of the experimenter, the department with which the experiment is associated, and the date that the experiment was initalized.

   b) Experimental Dimensions, which includes the number of subjects who will be participating in the experiment, the number of experimental treatments which each subject will be presented, the number of practice and test trials per treatment, and the total number of treatments per experiment.

c) The sequence of stimuli which will define each experimental treatment.

d) The sequence of treatments which will be presented to each subject.

To perform the functions required in OPTION A, the experimenter is required to know how to turn on the computer and how to insert and copy diskettes. The experimenter does not need to know how to program. If the experimenter would like to create his/her own stimuli, then knowledge of the appropriate graphics commands is required.

2. OPTION B, "Continue Old Experiment", controls the presentation of stimuli, and the measurement and recording of responses for each subject. This option also prompts each subject for any personal information required by the experimenter such as: name, age, identification number, whether or not the subject wears glasses, etc.

   In a typical experimental situation, the experimenter will start OPTION B running and then let the subject be seated at the keyboard. The program will first prompt the subject for the above mentioned personal information. When this information has been correctly entered and verified, the program will present instructions to the subject as to the procedure to be followed during the experiment.

   The subject will then be presented practice trials in order to familiarize him/her with the task. Subsequently, test trials for that experimental condition will be presented. This sequence of presenting instructions, practice trials, and test trials will be repeated for each experimental condition which the subject will be presented.

   After all trials have been presented to the subject, the subject is informed that the experiment is over and a "thank you" message is presented.

   To participate in the experiment, subjects may be required to type their name and identification number. They are not required to know anything about programming or how the computer works.

3. OPTION C, "Obtain Hardcopy", provides hardcopy so that a printed record of the experiment can be obtained. (See figure 2) Five types of information may be chosen from the Hardcopy Menu:

   a) Experimental identification and dimensions.

   b) The treatments which will be presented to each subject.

   c) The sequences of stimuli which make up each treatment condition.

   d) The response data obtained from each subject.

   e) Summary statistics for the data collected.

Hardcopy Menu

A Experimental Identification and Dimensions
B Treatment Sequences for Each subject
C Stimulus Sequences for Each Treatment
D Response Data for a Single Subject
E Summary Statistics for a Single Subject
F All Data, Sequences, Dimensions, and ID
G Return to EMS Menu

To make a choice, enter A, B, C, D, E, F, or G

Figure 2: Hardcopy Menu

EMS Version 1 is capable of presenting graphic and textual stimuli via a CRT with no special hardware or modifications. With the addition of an I/O board, hardware clock, and assembler subroutines, EMS could be efficiently used to control a wide range of external devices such as buttons, joysticks, graphic display terminals, slide projectors, tone generators, etc.

EMS is currently being used to conduct an experiment which examines the effects of prior stimulation on a second stimulus in a double stimulation, choice reaction time task. In this application, EMS has proved very useful in that set-up time for this type of experiment has been reduced from weeks to hours.

REFERENCE:

(1) Church,R.M.,The influence of computers on psychological research: a case study, Behavior Research Methods & Instrumentation. 15 (1983) 117-126

Human Factors in Organizational Design and Management
H.W. Hendrick and O. Brown, Jr. (Editors)

PROTOTYPING - A METHOD TO EXPLORE HUMAN FACTOR ASPECTS IN APPLICATION SOFTWARE

E. Keppel, G. Rohr

IBM Germany Scientific Center Heidelberg
Tiergartenstrasse 15
6900 Heidelberg, FRG

This paper presents prototyping techniques as a methodology to setup rapidly human factors experiments and to investigate user behaviour during real-life working sessions. A prototype system is described which aims at investigating the usability of icons in software interfaces. It consists of a text editor model, embedded in a basic file handling system and driven by means of an interchangeable set of iconic or verbal commands. APL is used as modelling tool. The continuous adaptation of the system to the evolving experimental needs is achieved by adherence to prototyping techniques, made up of functional modularity, parameterization, and stepwise refinement.

INTRODUCTION

The rapid growth of computer-based systems in work-organizations involves a broad range of varying user types, with different skills, motivations, and attitudes to communicate with these systems. As a consequence, traditional software engineering methods and tools are re-assessed in order to promote those techniques which are closer to the humans' way of thinking and handling.

Prototyping is a development methodology of increasing popularity. It has been originally introduced in software design as a stopgap to the life-cycle discipline in order to bypass uncertain user requirements, incomplete specifications, or to investigate unpredictable systems' behaviour. It emerges nowadays as a valuable and efficient engineering methodology, competing successfully with the traditional ways of designing, i.e., subsequent steps of predictable development phases (phase-model).

There is neither an exact and general definition of prototyping nor exists an exhaustive range of application areas. Depending on the target product, different methods and tools will be applied. As a common characteristic, prototyping is seen as a complementary set of strategies and methods to achieve a full or partial functional simulation of a target system, focusing on the rapidity of development, the ease of extension and modification, and the completeness of the user interface [1].

Prototyping aims at observation, evaluation, learning, and iteration. Its appeal resides on this close similarity with fundamental principles of human cognition [2]. Adequate prototyping supposes communication and feed-back, involving the whole organisation: designers, customers, endusers, managers. Its goal is human modelling, i.e., to verify and predict behaviour of people.

Prototyping techniques have penetrated various fields of applied sciences. Three broad goals and areas for prototyping can be distinguished:

1. Designing information systems. The benefit is the exploration of design choices during evolutionary development [3] [4]. Various aspects of the information system impact on the organization may also be explored [5].

This includes psychological and social factors when simulating system functions and tasks during the development process.

2. Evaluating product usability. There is evidence that early endusers involvment in prototyped software interfaces will improve acceptance of the final product [6].

3. Experimenting human factors. The next section describes prototyping methods applied in experimental psychology. It illustrates the benefits of this methodology to run comprehensive and versatile experiments.

A SYSTEM TO EVALUATE ICONIC INTERFACES

A prototype software system has been developed as part of human factors research conducted at the IBM Scientific Center Heidelberg. The main project objective was to perform a systematic experimental evaluation of iconic commands in software interfaces: Are icons better understood than verbal codes? Is it possible to derive design principles for them? Results of this experiment are presented at this conference, see Reference [7]. As additionnal objectives, it was desirable to gain insight in the endusers ability to learn a new computer system, to understand and to manipulate icons, and to navigate inside a complex set of system states. Various learning, teaching, and guidance techniques should also be investigated, applying a set of complementary means like on-line helps, tutorials, message prompts facilities, and pictorial state indicators.

To meet the project objectives, the following experimental approach has been defined: A software system was necessary to model an editor, made up of six elementary commands (insert, delete, replace, move a string, and input a line). This editor was embedded in a system of basic file handling commands (print, show, send, rename, erase, save, and edit a text file). To complete the system, a set of control commands (return, scroll, etc...) will allow test persons performing predefined tasks, operating in a realistic terminal session environment.

Several sets of icons were developed depicting the system commands under investigation according to varying design principles. The system was supposed to be driven by means of any set of iconic or verbal commands.

An overview of the states and commands of the target system, configured in a specific experimental setup, is displayed in Figure 1. It consists of the following states: (1) Initial state (2) File selection state (3) File handling state (4) Renaming and classifying a text file (5) Editor state

Additionally, to teach the test persons in system usage and to measure as well as to evaluate their learning success during a session, helps are available at every system state.

The amount of initial requirements put upon that system suggested strongly that prototyping techniques only could fulfill the experimental objectives. APL has been selected as implementation tool. Its merits as a highly productive, interpretative, multi-purpose interactive language offers all necessary facilities for a robust and rapid implementation. The modelling capabilities of APL permitted to provide the full functionality achieving a real-life perception of the system: Test persons are able to perform efficiently editing tasks: create a text file, correct it, rename it, print and send it, etc... Reformatted text is immediately fed back to the terminal user.

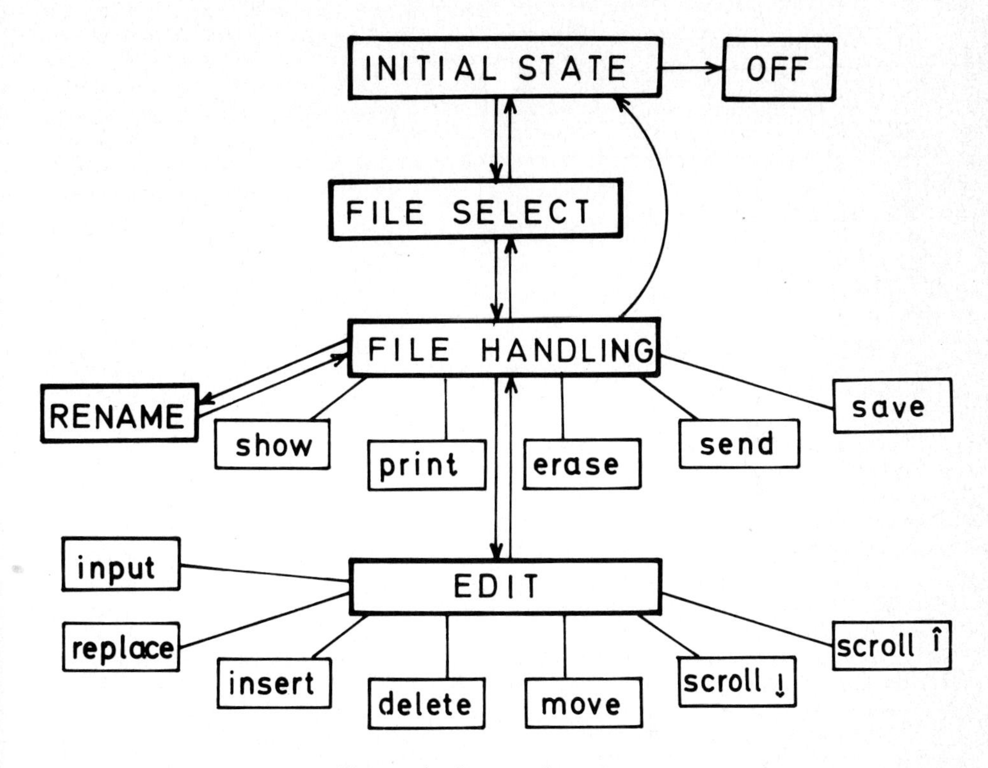

Figure 1: System Overview

The system is implemented with VS APL, under IBM VM-CMS Operating System [8]. To obtain high quality graphic representation of the icons, IBM 3279 Color Terminals were used. The underlaying file handling functions are executed via auxiliary processing by means of host CMS commands.

The terminal screen for two sample states, File Handling and Editing, is shown in Figures 2a and 2b. They display the basic structure of the interface:

1. The icons driving the system are disposed on both lateral borders of the screen, the 'Menu' area. To execute a system function the test persons must either point onto the icon with the light pen or position the cursor on it.

2. The 'State Area', located on the top portion of the screen, displays, by means of icons, the three most recently activated sequence of system states. This area is used to investigate the users' ability to navigate through system states.

3. The main central area displays either the text to edit or, in the file selection state, the menu of available text files

4. The bottom 'Control Area' is used to display verbal and iconic user guidance and system messages. This area serves also to prompt the user for text input.

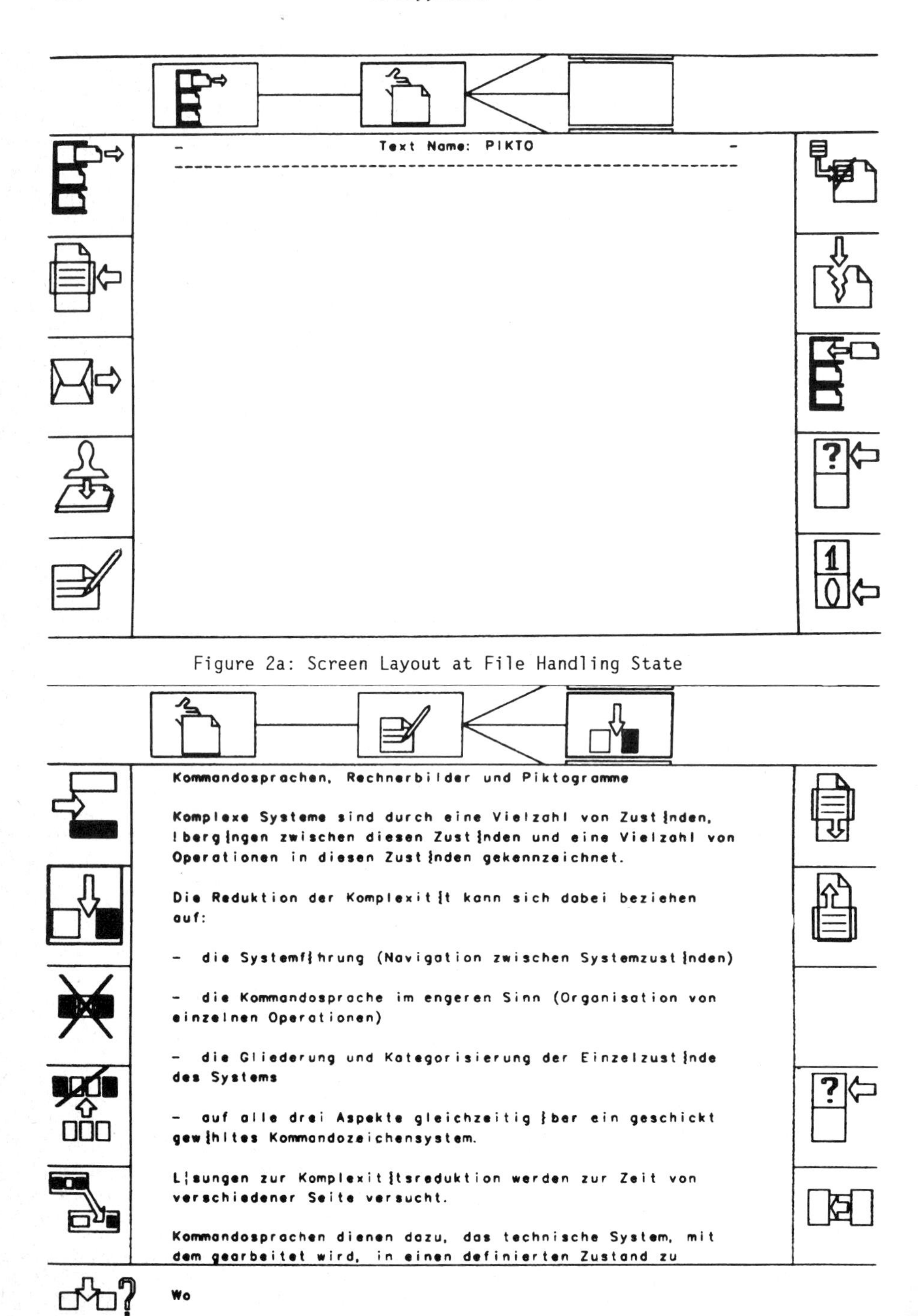

Figure 2a: Screen Layout at File Handling State

Figure 2b: Screen Layout at Editing State

Parameterization and modularity are extensively used to achieve a high degree of flexibility adapting the system to the experimental needs:

- Any new set of icons is easily integrated. Type, size, and disposition of the icons on the screen is controlled by program parameters.

- Modularity allows for an easy re-arrangement of the system state sequences. Alternative system structures are easily implemented to evaluate the impact of system control complexity on users

- Help and tutorial facilities can be modified and extended in various ways to investigate advanced help and teaching strategies.

To facilitate the evaluation of experiments, the system maintains automatically two data files:

1. A file recording all relevant test person interactions with the system: type of user detect, errors, idle time in helps, reaction time, etc...
2. A directory file of terminal sessions summaries overviewing the test persons results

The continuous adaptation of the system to the experimental needs is easily mastered by an experienced APL programmer.

CONCLUSIONS

Usage of system in first experiments has proven the various benefits of prototyping as a flexible and efficient methodology to conduct experiments in applied psychology. Its power resides mainly in the flexibility to include rapidly new ideas due to intermediate results arising in the course of the experimentation.

REFERENCES:

[1] Floyd, C., A Systematic Look at Prototyping,
Working Conference on Prototyping, Namur (Belgium), October 1983

[2] Jorgensen, A.H., On the Psychology of Prototyping, cf. Ref. [1]

[3] Dearnley, D.A., Mayhew P.J., In Favor of System Prototypes and their Integration into the Systems Development Cycle, The Computer Journal, Volume 26, No.1 (1983)

[4] Nosek, J.T., Organizational design choices to facilitate evolutionar development of prototype information systems, cf. Ref. [1]

[5] Naumann, J.D., Jenkins, A.M.: Prototyping: The New Paradigm for Systems Development, MIS Quarterly, September 1982

[6] Clark, I.A., Software Simulation as a Tool for Usable Product Design IBM Systems Journal, Vol. 20, 272-293 (1981)

[7] Rohr G., Keppel E., Iconic Interfaces, where to use and how to construct. Paper presented at this Conference.

[8] IBM Virtual Machine-System Product: General Information Manual, GC20-1838, and VS APL for CMS: Terminal User's Guide, SH20-9067

Human Factors in Organizational Design and Management
H.W. Hendrick and O. Brown, Jr. (Editors)

USABILITY TEST METHODOLOGY

Lynn C. Virta

Human Factors Assurance Department
International Business Machines Corporation, Inc.
Austin, Texas
U.S.A.

Usability means different ideas to different people. The complexity and subjectivity of testing systems for usability makes quantification of test results a challenge. Meaningful test results may be gleaned by a variety of analytical methods and graphic representations. Usability scores also aid in determination of problem areas. Management can be made aware of usability issues by benefitting from a combination of methodologies. Thus can human factors considerations be corrected in a product before it is released.

° What is the problem in usability testing?

Usability means different ideas to different people. Doing iterative testing of systems with contract operators using sets of scenarios immediately produces too much data, and subsequently too little time to analyze it. Inconsistent methods bear less credibility when reporting on test results or making a case to management to improve usability in a product.

° What is the solution?

The answer to the problem is to creatively apply criteria given by Planners and to standardize partitions of these criteria. Using a word processing system, charts giving meaning to test results may be produced promptly at test conclusion. The use of SAS (Statistical Analysis System from SAS Institute, Inc., Cary, North Carolina) can produce tables and graphs which indicate whether criteria were met by task and by test measurement.

° Why should we use it?

We should use this method for data reduction and graphic representation of the problems found in test. There is more time available for assessment of test results when a customized standard program exists for each test and machines do the work for us rather than individual test conductors duplicating efforts of others, or doing analysis by hand. A smooth analysis of test results is critical to augment the leverage that usability testing groups might have to improve the product and to make solid presentations to Planning and Development personnel who can implement the changes. The desired response when making a case for altering code or making a publications change in the final product phase requires rigorous proof. Management is more apt to respond to test results on a graph rather than to read a large number operator comments concerning a problem.

° How is data input?

Hundreds of individual pieces of information are generated in test. The same types of test measurements are taken for each user, who performs a certain number of scenarios as designed by the test conductor. This testing is reiterative in the sense that a set of users perform the test, test input generates corrections to the product, and changes are made which dictate further testing. Information collected from test on data collection sheets includes learning and doing times, percent task completion, assists, and unfound references. An attitude question is phrased above an unmarked line segment one decimeter in length, ten being the positive value. Attitude is measured linearly and rating is calculated by taking the mean of five to eight responses to a questionnaire particular to the task. The data collection sheets are collected by the test conductor, who then inputs the data easily into a SAS data set via column input done in the same order as the data collection sheets. Efficient input would be ordered thus:

Task Num	Op Num	LTime	DTime	% Success...
1	1	1:30	:14	100
2	1	:20	:13	100
3...				

° How do we sort the raw test data into partitions determined by criteria from Planning?

- Chart the criteria by task number and test measurement. Refer to "Sample Criteria From Usability Test", Figure 2 below.
- Calculate the partitions. Example: time criteria for a task is 60 minutes. Objective would be 1 hour ± 15 min. (15 min. = 25% times 1 hour) Anything better than 45 min. (25% better than criteria) becomes Exceeds. Now figure 50% more than the 60 min; 100% more; and 150% more. These figures become the center value of the fails criteria partitions. The boundaries of the partitions become 25% more or less than the center value. Worst case is the outer limit for objective + 150%.

SAMPLE CRITERIA FROM USABILITY TEST

TASK	LEARNING TIME	DOING TIME	% TASK SUCCESS	ASSISTS	UNFOUND REFERENCES	ATTITUDINAL RATING
1	1:00	:30	100	1	0	6
2	:45	:60	100	1	0	6
3	:20	:15	100	1	0	6
4	:30	:15	0	0	0	6
5	1:30	:45	100	0	0	6
6	:20	:15	100	0	0	6
7	1:00	:30	100	1	0	6
8	1:00	:30	0	1	0	6

Figure 2

- This partitioning can be done by hand, or it can be done by a SAS procedure already on the system.

- A SAS procedure could easily be customized to determine any time criteria, and then calculate the partitions. The same is true for the other test measurements.
- An example of partitioning is shown for typical time criteria in Figure 3 below, "Sample Partitioning of Time Criteria".
- Linear metric value is assigned to an unmarked line segment to determine the attitude ratings to the nearest tenth. Criteria is 6.0 on a scale of 0.0 to 10.0, 10.0 being the positive value. Hence:

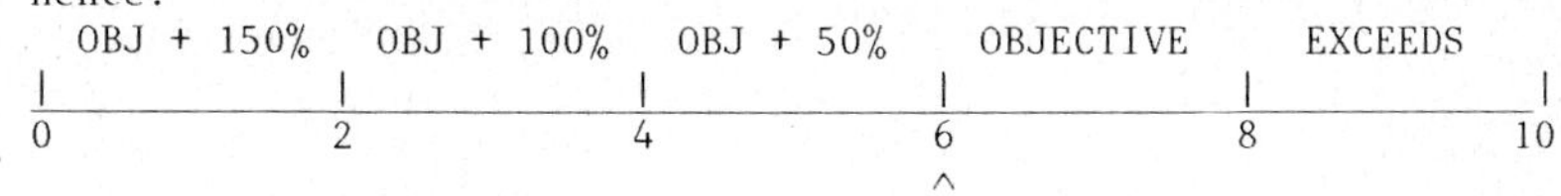

- The information in the data sets can then be used to produce charts. The charts are persuasive in convincing management that a problem exists that needs to be corrected.

SAMPLE PARTITIONING OF TIME CRITERIA

If time criteria is 15 minutes,	......< 11 min.....	is EXC
	......11 - 19......	is OBJ
	......20 - 25......	is OBJ + 50%
	......26 - 35......	is OBJ + 100%
	......> 35.........	is OBJ + 150%
If time criteria is 30 minutes,	......< 23 min.....	is EXC
	......23 - 37......	is OBJ
	......38 - 50......	is OBJ + 50%
	......51 - 67......	is OBJ + 100%
	......> 67.........	is OBJ + 150%
If time criteria is 45 minutes,	......< 36 min.....	is EXC
	......36 - 55......	is OBJ
	......56 - 74......	is OBJ + 50%
	......75 - 100.....	is OBJ + 100%
	......> 100........	is OBJ + 150%
If time criteria is 60 minutes,	......< 45 min.....	is EXC
	......45 - 75......	is OBJ
	......76 - 105.....	is OBJ + 50%
	......106 - 135....	is OBJ + 100%
	......> 135........	is OBJ + 150%

Figure 3

° How might a quick chart graphically represent test measurement results at the conclusion of test?
  - Use Tasks as horizontal headings. Sub-headings are the partitions: EXC, OBJ, +50%, +100%, and +150%. Test measurements are the

vertical headings: Time, Per Cent Task Success, Requests for Assistance, Attitude.
- Fill the chart with the tallied totals of end users whose test data indicated placement in that partition. Lead operator assistance is especially useful in this task.
- An example is shown in Figure 4 below, "Usability Test Results". This chart is drawn on a word processing system. It is useful as a tool for tracking usability test results, not for reducing test data.
- For example, if this chart were to immediately pinpoint problem areas in "time to learn" and "time to do" criteria being met, then the test conductor would likely turn first to the documentation as a problem source. Quick input and suggestions for appropriate fixes may aid in the improving the usability of the product, particularly if the manual is due at the printers.
- If circumstances of the test make this suggestion not workable (too many tasks or test measurements) then another graph idea may work better, as illustrated in Figure 5. This chart nets out results by depicting task descriptions on the ordinate and the test measurements on the abscissa.

o How might two sets of attitudinal ratings be overlayed to represent meets/does not meet criteria?

- "Attitudinal Ratings" example is Figure 6 below. The original was created by SAS and can be customized.
- The horizontal headings along the bottom of the chart are the tasks. The vertical headings are the partition names. A horizontal line separates the meets/does not meet criteria areas of the chart. One symbol represents one set of attitudinal ratings (for example attitude at the first time a task is done), and the second sample represents the second set.

o How might the frequency of occurrences in each criteria partition be graphically represented by test measurement?

- A frequency bar graph may be used to provide a graphic representation of the number of occurrences within each partition. Read the horizontal scale; the number represents one user doing one task. A numerical table shows frequency, cumulative frequency, per cent and cumulative per cent.
- "Frequency Bar Chart", Figure 7 below, is an example result of the SAS procedure graphically representing two test measurements (References to other documentation, Unfound References) and how test measurement compared to criteria. The test conductor may immediately determine which test measurements are meeting criteria. Similar horizontal bar charts were produced for all the test measurements.

o How may the tasks be aligned to compare attitudinal ratings, to see changes in attitude before and after experience with the product?

A SAS procedure created the chart below in Figure 8, which illustrates that 5 out of 8 tasks met or exceeded criteria in attitudinal rating in a sample

USABILITY TEST RESULTS

Test Measurement:	TASK1 Exc	TASK1 Obj	TASK1 +50	TASK1 +100	TASK1 +150	TASK2 Exc	TASK2 Obj	TASK2 +50	TASK2 +100	TASK2 +150	TASK3 Exc	TASK3 Obj	TASK3 +50	TASK3 +100	TASK3 +150	TASK4 Exc	TASK4 Obj	TASK4 +50	TASK4 +100	TASK4 +150
Time	8					8					5	1	1							
Per Cent Task Success		8					8													
# Requests	7	1				5	3				5	2	1				8			
# References	7	1				7		1			7		1			8				
Unfound References		7	1				7					8					8			
Attitudinal Ratings: TASK	4		1				4	1				4	1				5			
EXIT	5					5					5					5				

Data entered in the EXC or OBJ columns meet criteria; the others do not. The numbers represent the number of operators whose task performance scored in that criteria partition. Exception is attitudinal ratings which are measured by centimeter on a scale of 0 to 5, 5 being positive. Partitions: EXC means exceeds criteria. OBJ is objective, same as criteria. Objective + 50% is half again more than criteria; OBJ + 100%, twice criteria; OBJ + 150% is the most negative partition.

Figure 4

TASKS THAT DID NOT MEET CRITERIA

(▲ means at least 60% of the operators task performance did not meet criteria)

TEST MEASUREMENT : TASK:	Time To Learn	Time To Do	% Task Success	Assistance	Unfound Reference	Exit Attitude	Overall Satisfaction
Learning DOS	▲	▲			▲		
Backup System	▲						
Backup Work		▲					
Erase File							
Using Messages		▲					
Copy File							
Cancel Jobs		▲					
Print File							
Print Directory	▲	▲					▲
Recover	▲	▲		▲	▲	▲	▲
TEST MEASUREMENTS :	Time To Learn	Time To Do	% Task Success	Assistance	Unfound Reference	Exit Attitude	Overall Satisfaction

Figure 5

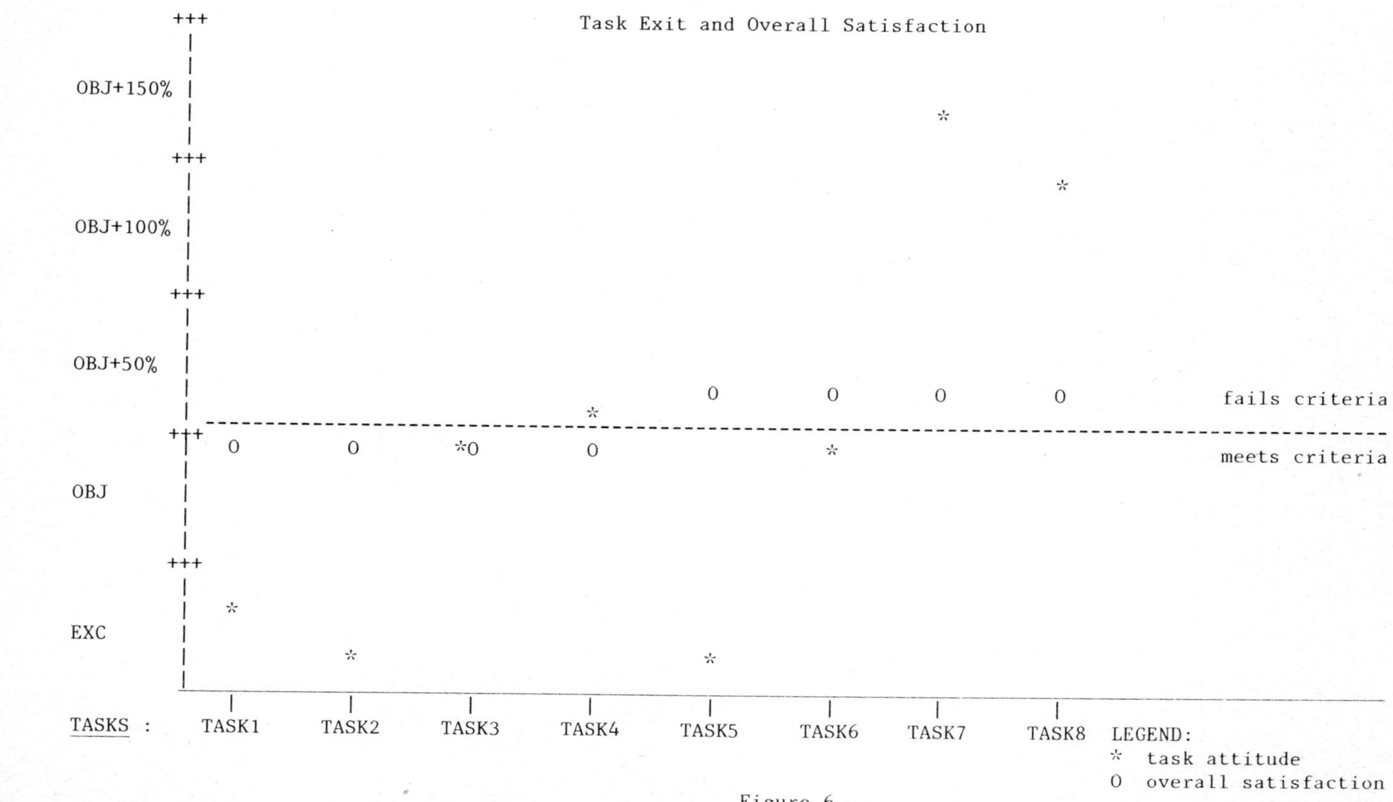

Figure 6

Note: Tasks 1 - 4 comprise Test Case 1; tasks 5 - 8 form Test Case 2.

FREQUENCY BAR CHART

CUMULATIVE FREQUENCIES FOR ALL TASKS AND ALL OPERATORS FOR REFERENCES TO OTHER DOCUMENTATION

FREQUENCY BAR CHART

REFCRIT		FREQ	CUM. FREQ.	PERCENT	CUM. PERCENT
EXC	XXXXXXXXXXXXXXXXXXXXXXXXXXXXXXXXXXXXXXXXXXXXXXXXXXXXXXXXXXXXXXXXXXXX	68	68	85.00	85.00
OBJ	XXXX	4	72	5.00	90.00
OBJ + 50%	XXXX	4	76	5.00	95.00
OBJ + 100%	XX	2	78	2.50	97.50
OBJ + 150%	XX	2	80	2.50	100.00

5 10 15 20 25 30 35 40 45 50 55 60

FREQUENCY

CUMULATIVE FREQUENCIES FOR ALL TASKS AND ALL OPERATORS FOR UNFOUND REFERENCES

FREQUENCY BAR CHART

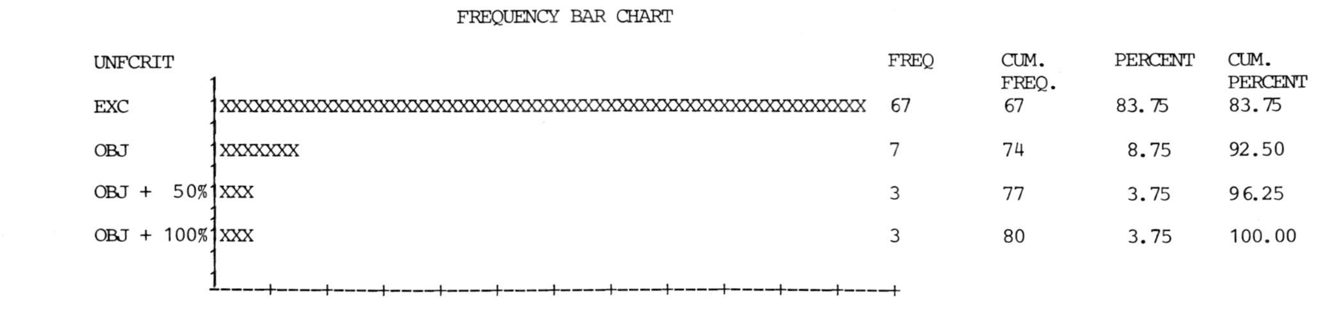

UNFCRIT		FREQ	CUM. FREQ.	PERCENT	CUM. PERCENT
EXC	XXXXXXXXXXXXXXXXXXXXXXXXXXXXXXXXXXXXXXXXXXXXXXXXXXXXXXXXXXXXXXXXXXX	67	67	83.75	83.75
OBJ	XXXXXXX	7	74	8.75	92.50
OBJ + 50%	XXX	3	77	3.75	96.25
OBJ + 100%	XXX	3	80	3.75	100.00

FIGURE 7

size of eight users. Generally, repeated use tended to raise a user's opinion of the product if it changed at all. Tasks 4, 6, and 7 did not meet criteria in attitudinal ratings. Rating1 represents first trial; Rating2, the last trial.

ATTITUDE IN CRITERIA PARTITIONS

TASK	SCENARIO	RATING1	RATING2
1	Learning	EXC	EXC
2	Backup Sys	OBJ	EXC
3	Backup Work	OBJ	EXC
4	Erase File	OBJ + 100%	OBJ + 50%
5	Copy File	EXC	EXC
6	Print File	OBJ + 50%	OBJ + 150%
7	Print Dir	OBJ + 50%	OBJ + 50%
8	Recover	OBJ	OBJ

Figure 8

° How may a subjective idea such as usability be quantified?

Calculate the weighted average of the goodness of the procedure. The test measurements have been weighted by Planning to give special importance to the attitude measurement, which is usually 40% of the total usability score. Task success usually is weighted at 20%, while time to learn and to do is generally 10% to 20% apiece. The weights supplied by Planning have been used as a factor to multiply by the percentage computed by the number of operators who met criteria divided by the total operators who attempted each task. These products are added together to form a usability score for comparison purposes. If all test criteria have been met 1.00 is a possible score; 0.0 is the worst case score. Goodness is anything better than .60. The score is best used to judge troubled areas. Figure 9 exemplifies usability score calculation.

USABILITY SCORES

TEST MEASUREMENT:

TASK:	Attitude		Time: to Learn		To Do		% Task		Assists		Score
Task1	.4(1.00)	+	.1(1.00)	+	.2(.83)	+	.2(1.00)	+	.1(.83)	=	.95
Task2	.3(.20)	+	.1(.40)	+	.2(.40)	+	.3(.80)	+	.1(.80)	=	.54
Task3	.3(1.00)	+	.2(.00)	+	.2(.50)	+	.2(1.00)	+	.1(1.00)	=	.70

Figure 9

Calculate the score by dividing the total number of operators who attempted the task into the number of operators who completed it. Multiply that by the weight given the test measurement. Add all the products together to get the usability score, which is between 0.00 and 1.00.

° How may criteria for the usability score be determined?

A criteria for the score may be assigned by Planning to be .60. The logic follows: each criteria partitioning gets 20%. Thus, Exceeds is 20%, Objective is 20%, Slightly Failed Criteria is 20%, Failed Criteria Severely is 20%, and Worst Case is 20%. Sixty percent to 100% of the Usability Score indicates a positive response, which also fits in with the metric scaling of attitude measurement and criteria partitioning of the Usability Test Results. Any score that fails criteria deserves further scrutiny as to the reasons for failure.

° How can the usability score be used?

The usability equations may be analyzed to determine problem areas. Numbers near zero in "time to learn" and "time to do" might indicate problems in either the documentation or the design on the user interface (which the publications cannot overcome). Task3 "time to learn" indicates none of the six operators' task performance passed criteria. Attitude on Task2 failed, as did both time measurements; the score indicates the whole task failed. Making a presentation to management to get a problem fixed is effective if test data justify attention to problem areas. Reallocation of people and time to get more usability into a product requires as strong a case as can be presented.

° How can a usability test group improve effectiveness?

The problem of producing too much data in too little time can be handled with more consistent methods. Usability groups across sites can apply the criteria partitioning and SAS procedures to make graphic representations of usability problems from test. Data reduction may be accomplished with less time involved and less duplication of effort. Test data can be used to prove a problem exists and to get management attention to recommendations.

Human factors considerations may be corrected in a product before it is released. Since product schedules are normally tight, a solid case for usability improvements must be made. One way to do this is by quantification of test measurements appropriate to the product. Suggestions from test may contribute to making making the product easier to learn, easier to use. Such nebulous but vital considerations are advanced by these various analytical methods and presentation tools.

Human Factors in Organizational Design and Management
H.W. Hendrick and O. Brown, Jr. (Editors)

THE CRAWFORD SLIP METHOD AND PERFORMANCE IMPROVEMENT

Richard A. Rusk and Robert M. Krone

Institute of Safety and Systems Management
University of Southern California
Los Angeles, California
U.S.A.

THE SETTING

Why an old method is in a renaissance mode?

Most managers are underutilizing their most readily available source of performance improvement--their own workers. People working the problems on a day-to-day basis have know-how needed by decision makers. Organizational structures, internal politics and personal characteristics often stop needed communications from bottom to top of the company or agency.

The University of Southern California is the focal point for the re-emergence of the Crawford Slip Method. It is a low-cost, high-speed and high quality knowledge recording and processing system. We believe it to be the finest qualitative systems analysis tool on the market. In making that statement we include for comparison purposes: quality circles, Delphi, Nominal Group Technique, Brainstorming, survey questionnaires, creativity enhancing techniques, suggestion boxes and many other idea generating methods being used.

Dr. C.C. Crawford, of USC, invented the method in 1925. Since he resumed his professional activities in 1979, he and a team of analysts in Los Angeles and Dayton have been achieving increased productivity in business, government, military, health sciences and educational organizations. Prior to 1979 Dr. Crawford taught and consulted with his method through a wide spectrum of applications. He authored or coauthored 19 books and hundreds of articles, manuals and reports on his method or about research done with the Crawford Slip Method. The bibliographic listing provides publications since 1979.

Our goal with this article is to provide the Human Factors community of scholars an introduction to the Crawford Slip Method by presenting the underlying theory followed by brief descriptions of the method's four major steps: (1) The planning process; (2) Getting know-how on slips through a workshop; (3) Organizing the data obtained; and (4) Producing the professional product that affects an organizational improvement.

Those wishing to apply the Crawford Slip Method for their organization's performance improvement, after this limited exposure, should begin with Crawford and Demidovich (1983) and Crawford, Demidovich and Krone (1984).

THEORETICAL BASES OF THE CRAWFORD SLIP METHOD

Workers everywhere have much knowledge about problems, remedies for those problems and procedures to effect the remedies that have not been recorded.

That know-how is both internal and external to the organization due to the individual's experience. As technological complexity increases the importance of that know-how to the success of the enterprise rises. But time limitations, lack of good methods for extracting that know-how and political constraints often prevent access of leadership to that critical source of expertise.

Given an opportunity under circumstances of anonymity and controlled mental focusing (called "targeting") much of this needed know-how, opinion, judgement and recommendations can be recorded in a short time period. The Crawford Slip Method consistently gets valuable needed input from people on a specific problem in fifteen minutes. A longer time is usually preferable as input for several targeted subjects is the norm.

Workshop participants write simultaneously, before any discussion of the subject, to avoid group-think or biases for their mental searches. They write their ideas on small slips of paper, 2-3/4 x 4-1/4 inches. There are important substantive reasons for that size as well as practical, logistics and classification reasons. Writers average one idea per minute. After ten minutes it is normal for the mind to slow down its generation of ideas on that one targeted subject. By using the Crawford Slip Method to open your next staff meeting you could generate 80 ideas in ten minutes from 8 people. How many ideas did you get in your last staff meeting? That is a mini application of the method. We often get thousands of ideas in a workshop.

Anonymity removes the fear of reprisal or political backlash by participants. Independent and simultaneous writing insure the least amount of biased results and the highest possible penetration of the subject. Having the ideas in written form guarantees that the important nuances of each individual's thinking will be preserved. Once slips are written they become a permanent data base for any number of future analytical goals.

Participants need to learn only a few fundamental procedures related to writing their ideas concisely on slips. However, this form of writing is counter to previous experience and may feel unnatural at first. Without the right motivational introduction workshop participants could perceive this as a trivial exercise similar to itemizing their grocery shopping needs. This method is neither trivial nor entertainment. It is hard work for the slip writers and very difficult for the analyst to administer correctly. We always recommend experimentation with this method in a non-career critical situation until experience is gained. It is the most efficient know-how gathering and processing system we have used. But, like any other systems analysis tool, it must be done right or the results may be disappointing.

PLANNING FOR A CRAWFORD SLIP METHOD WORKSHOP
How to get your best brains to produce needed ideas?

People selected for a Crawford Slip Method workshop must have familiarity with the organizational mission and product(s). They need not be experts in the Delphi sense. You will want to have your experts in your workshop. But if you and your organizational leadership knew all your current troubles and how to fix them you would not need performance improvement. We find no such public or private entities today.

There are no standard Crawford Slip Method workshop compositions. Each will be different and sensitive to a large set of variables surrounding your specific type of organization, its goals, its people, its products, its location, its clients, its size and its history. There will be times when a homogenious group of people are best and others when a mix of organizational levels, skills and responsibilities is advised. Although groups in the hundreds have been involved in workshops most needs can be met through twenty to thirty participants. There is always a trade-off between work time lost by your participants and the data you get from them in that time. A dozen very knowledgeable people will produce a surprising number of ideas in a fifteen minute slip-writing session on one subject. Six writing simultaneously will produce more than thirty gathered to provide oral inputs. The minimum number that can usefully use the Crawford Slip Method is one. In fact, doing it yourself is the best way to prepare for a workshop.

People invited to your workshop should be told that "boning up" or writing position papers is neither required nor desirable. They have the information stored in their personal memories. The mechanics of writing slips can be provided a workshop group in five minutes. From the participant's viewpoint no special training is needed. From the analyst's viewpoint this method gets very complex.

One of the critical parts of preparation for a Crawford Slip Method workshop is the design of the "targets" which focus the minds of slip-writers. A correctly prepared targeting sheet is the key to a successful workshop. See Crawford and Demidovich (1983).

GATHERING DATA ON SLIPS
How to conduct a workshop?

A typical first Crawford Slip Method workshop is a "Diagnostic." We use the term in the same way as Human Factors consultants or medical specialists Target #1 is designed to produce slips on current performance problems and Target #2 asks for remedies. Doing a diagnostic workshop first avoids too narrow a concentration on a preconceived set of problems and failing to identify the whole set which can only be assembled from the minds of individuals working in different parts of the enterprise. Once that data base is available additional targets addressing the specific problems can be designed. Slip writers can penetrate each target, general or specific, with an average of ten ideas before "mental slow down" occurs.

Workshops can be held almost anywhere but preferably in a place where participants can concentrate without disruptions on each target and where each person has a writing surface.

DATA ORGANIZATION
How to classify 1000 ideas?

The most difficult part of the Crawford Slip Method is classification of the hundreds or thousands of ideas collected one per slip of paper during a workshop. Space constraints here preclude any sort of adequate explanation. In principle, the analyst allows the ideas of each slip to suggest a category (inductive logic going from the data to sets of like ideas). We often get over 100 categories at the first rough classification. They are positioned alphabetically on a large table. The next step is to see which categories might be merged into a stable set of product sub-sections (e.g. chapters of a manual or sections of a report). This involves deductive logic as your concept for the final product comes into play. The final step is to prioritize the sub-sections thus created into a logical order. The sub-sections must meet the criteria of sufficiency (i.e. a systems approach to the problem) and mutual-exclusiveness (e.g. don't talk about training in every sub-section). The analyst continually reevaluates the ideas on the slips with the categories and sections created to insure that his or her data reduction does not interject biases into the final product. This is a professional activity that takes experience.

THE PERFORMANCE IMPROVEMENT PRODUCT

How the Crawford Slip Method produces closure?

The Crawford Slip Method is an entire system of idea generation, organization and professionally written product that can lead to improvements. Until you work through its labyrinth of passage-ways several times you cannot grasp how the pieces of this system, created by Dr. Crawford over six decades, fit together. It is an experiential method. Your first reaction after asking a group of your people to write slips will be one of amazement at the quantity, quality and diversity of useful ideas that are generated in a ten or fifteen minute writing session. That euphoria may fade as you face the much more difficult task of data reduction, editing and product preparation. As with any systems analysis method, no two applications will be just alike. Experience with one situation helps with the next but may not avoid a new unforeseen pitfall.

If you have done a good job of target design, workshop leading and classification the actual writing of the product will go rapidly. You will, of course, write it directly from slips. You work with your product prioritized sections one at a time and convert the stacks of related ideas into paragraphs and sentences. You write new slips to create the transitioning necessary between ideas. You can manipulate any idea, paragraph or subsection at any time by simply moving a slip or a set of slips. Classification trays made to accommodate the slips are very helpful here and are described in Crawford and Demidovich (1983). You will find that this method beats your word processor. We organize an entire product document completely on slips before inserting in the word processor. The Crawford, Demidovich and Krone book (1984) went directly from every sentence on a slip to the typesetting word processor. The Crawford, Demidovich, Krone and Rusk bibliography entries for this article were written using the Crawford Slip Method. Once you see the advantages we predict you will change your own writing technique to include the Crawford Slip Method.

THE FUTURE OF THE CRAWFORD SLIP METHOD

Applications of this method are sprouting in public and private organizations nationally. A few countries outside the United States have had limited exposure. The Crawford Slip Method is being used for long range planning, needs assessment, corporate new product identification, educational system goals, professional authorship, marketing, productivity improvement, theory testing, curriculum planning, business expansion and a wide spectrum of personal, family and social uses. See Chapter 17 of Crawford, Demidovich and Krone (1984) for our plans to institutionalize the method in a think tank environment.

A number of converging trends in post-industrial society make the Crawford Slip Method more relevant as time goes by. The merging of this method with electronic networking has exciting potential which is being pursued by members of Dr. Crawford's team. Future organizational needs will center about too little time for analysis, training or performance measurement and a requirement to penetrate more deeply into problems before they produce catastrophic outcomes. The Crawford Slip Method will find expanding applications in such a future.

The Crawford Slip Method easily relates to the purpose of this Conference in many ways. In this section we will attempt to lay out some of them.

Organizational Design and Management (ODAM) first characterizes its thrust as improving organizational effectiveness, productivity and Quality of Work Life (QWL). It does this through a sociotechnical systems approach aimed at making the best use of the psychosocial, technical and human-machine performance interface factors particularly in organizational design efforts. ODAM is likewise interested in related models for managerial decision making, in managerial roles and in the use of human factors specialists as consultants to management. Each of these areas will be briefly described in Crawford Slip Method terms, showing the system's versatility and comprehensive nature.

Using the workshop techniques outlined above, much rich data is generated. When classified and fedback to management, this intelligence produces guidance as to the problems discovered as well as for the solutions suggested. Inevitably the feedback leads directly to performance improvements and to productivity gains. Individuals who feel valued by the process are also more committed to implementing the solutions. Both are important to productivity. The worldwide interest in the efficacy of the Quality Circle Movement is predicated upon such seeminly simple values and dynamics!

Closely related is the Quality of Work Life (QWL) issue. Applying the Crawford Slip Method permits individuals to contribute significantly to the organizational effort heightening their sense of "being needed," and their output under the Crawford Slip Method can be directed towards design of jobs or work stations. The workshop simply has to ask the right questions (targets).

From the data developed in this instance, job enrichment or enlargement ideas can be implemented which then provide more interesting and challenging work and thus individual motiviation toward higher plateaus of productivity! Of course, letting people help in the design of their jobs just makes good sense, too. They know best what they can do and they most probably know best what needs doing as well. Why not ask and use what they say? The results are <u>all</u> positive for the organization.

Perhaps the most powerful use of the Crawford Slip Method is as a model for managerial decision making.

One of the more important aspects of management is the basing of decisions and actions on what Chris Argyris (1973) calls "Valid information." Yes, you and I have both made decisions on less than ideal data and many times. But how effective were our actions or decisions, particularly when the information upon which they were founded was frail or foul! The management literatute is replete with advice and examples calling for data to be given to the levels where the decisions are to be made (Koontz, et. al., 1984) but this doesn't always happen. Perhaps more significant is the issue of the validity of the information. Where did it come from? Is it good? Is it reality tested? Is it believable? Is it useful?

"How good is my unaided judgement?" managers often ask. Crawford's approach helps to mobilize the best of the ideas of the members of a system and to get those ideas into action; to do it in a way that provides a lessened fear of exposure for the individual as well as a greatly reduced risk of censure from peers. It is rapid, safe and independent. With today's technical complexity, turbulent environs, multiple interpersonal interactions, time urgency and the rising cost of a wrong decision, managers are looking for a more systematic way. This is one of them!

The managerial role, while changing, is also remaining the same in many ways. The Crawford Slip Method can serve to significantly aid and improve the classical managerial functions and processes. Think how rich employee inputs could positively impact for example, planning and goal setting, organizating and staffing or coordination and controlling.

The Crawford Slip Method as a tool for the human factors specialist serving as a consultant to management likewise makes good sense. They will always need efficient, yet flexible, diagnostic tools they can trust; they will naturally identify with the psychosocial dimensions of the system (participation, anonymity, commitment, etc.); the process is rational and logical leading to better managerial identity and it has a 60 year proven track record of success in a great variety of contexts and subject matter!

The Crawford Slip Method is a true systems method. It highlights the interdependencies and interrelationships of components, it considers questions wholistically by bringing in various levels of people, skills and ideas and fitting it all together. By proper targeting it can address problems of economics, culture, technology, administration, ethics, legalities or politics.

Wherever one respects the ideas of the individual and has need and a desire to use those ideas, the Crawford Slip Method has a role. In this application we have tried to demonstrate its utility to the arena of organizational design and management. We have found it to be a powerful method.

BIBLIOGRAPHY

1. Argyris, Chris, "Intervention Theory and Method: A Behavioral Science View." Reading, MA: Addison-Wesley (1973).
2. Crawford, C.C., "How You Can Gather and Organize Ideas Quickly," CHEMICAL ENGINEERING (July 25, 1983). Reprinted in IEEE TRANSACTION ON PROFESSIONAL COMMUNICATIONS, Vol. PC-26, #4 (Dec. 1983).
3. Crawford, C.C. and Demidovich, John W., "Think Tank Technology for Systems Management," JOURNAL OF SYSTEMS MANAGEMENT (Nov. 1981).
4. Crawford, C.C. and Demidovich, John W., CRAWFORD SLIP METHOD: HOW TO MOBILIZE BRAINPOWER BY THINK TANK TECHNOLOGY. Los Angeles, CA: University of Southern California School of Public Administration (Jan. 1983). Available through the USC Bookstore, Los Angeles, CA 90089-0892.
5. Crawford, C.C., Demidovich, John W. and Krone, Robert M., "Instructing and Communicating: How to Recycle and Improve Expertise by the Crawford Slip Method," EDUCATION AND PROFESSIONAL COMMUNICATION TRANSACTIONS, Joint Issues (3rd Quarter 1984) forthcoming.
6. Crawford, C.C., Demidovich, John W. and Krone, Robert M., PRODUCTIVITY IMPROVEMENT BY THE CRAWFORD SLIP METHOD: HOW TO WRITE, PUBLISH, INSTRUCT, SUPERVISE AND MANAGE FOR BETTER JOB PERFORMANCE. Los Angeles, CA: University of Southern California School of Public Administration (1984). Will be available July 1984 through the USC Bookstore, Los Angeles, CA 90089-0892.
7. Crawford, C.C. and Think Tank Team, "Complexity Crisis: How to Close the Gap Between High Complexity and Low Productivity," LOGISTICS SPECTRUM (Winter 1983).
8. Demidovich, John W. and Crawford, C.C., "Linkages in Logistics: How To Improve Coordination by the Crawford Slip Method," LOGISTICS SPECTRUM (Summer 1981).
9. Demidovich, John W. and Crawford, C.C., "Precision in Data Management," DATA MANAGEMENT, forthcoming.
10. Koontz, Harold, O'Donnell, Cyril and Weihrich, Heinz, MANAGEMENT (8th Ed). New York: McGraw-Hill (1984).
11. Krone, Robert M., "A Systems Improvement Method for Managers," SYSTEMS SCIENCE AND SCIENCE (Proceedings of the 26th Annual Meeting of the Society for General Systems Research, Washington, D.C., January 5-9, 1982, Len Troncale, Ed.).
12. Krone, Robert M., "A Pacific Nuclear Information Group: Prospects and Guidelines," JOURNAL OF EAST ASIAN AFFAIRS, VOL. III, NO. 2, (Fall/Winter 1983).
13. Krone, Robert M. and Rusk, Richard A., "The Crawford Slip Method: How to Increase Productivity Through Participation," EXECUTIVE MAGAZINE, Orange County edition (May 1984), forthcoming.
14. Krone, Robert M. and Rusk, Richard A., "Applying the Crawford Slip Method in Industry or Government," EXECUTIVE MAGAZINE, Orange County edition (June 1984), forthcoming.
15. Zachary, William B. and Krone, Robert M., "Managing Creative Individuals in High Technology Research Projects," IEEE TRANSACTIONS ON ENGINEERING MANAGEMENT, Special Issue on Managing Technical Professionals (February 1984).
16. Zemke, Ron, "CEO Concerns for the 80's Have HRD Implications," TRAINING/HRD, Vol. 17, No. 10 (October 1980).

NOTE: Between 1926 and 1979 Dr. Crawford authored or coauthored 19 books and hundred of articles, manuals and reports on the Crawford Slip Method or about research done with his method.

Human Factors in Organizational Design and Management
H.W. Hendrick and O. Brown, Jr. (Editors)

USER-ORIENTED STRUCTURED EVALUATION AND DESIGN OF DATA PROCESSING APPLICATIONS

James A. Carter Jr.

Department of Computational Science
University of Saskatchewan
Saskatoon, Saskatchewan
Canada

The evaluation and design of data processing applications can be structured using a hierarchy of user-oriented data processing functions in order to benefit the end user.

THE USER

The accelerated and widespread use of computers in recent years has put an increasing emphasis on designing systems that are oriented towards the needs of the end user rather than those of the computer professional. User Friendly, Easy to Use and a variety of other expressions have been added to more conventional ones such as Human Engineered and Ergonomically Designed to describe this new orientation. They are appearing more and more frequently in both sales literature and scholarly journals. To understand what they should or do mean, one must first consider the attributes of the end user of computing.

End users can be defined as persons whose main responsibility is to accomplish some task or tasks other than to make computers work. This definition does not preclude an end user's constant reliance on a computer as a tool to accomplish these other tasks. Similarily, computer professionals can be defined as persons whose main responsibility is to make computers work. Likewise, this does not preclude persons not constantly using computers from being computer professionals. The main distinction is whether the computer is a tool or a means to another end or whether it is an end in itself.

The range of potential end users of computers in an organization is as broad as the range of the individuals in an organization. Thus, end users may range from senior management to clerical workers and even unskilled laborers. This range of responsibilities and presumably of skills poses a major problem in designing systems that must meet the needs of a number of individuals in an organization. It is further complicated by the various levels of computing experience that end users in today's society may possess. This variation in experience results from the difference in the amounts and the kinds of use each individual makes of computers both within and outside of the organization.

The increase in computer price/performance resulting from the development of microprocessors and microcomputers has caused a proliferation of computing power throughout society. This has resulted in a much greater reliance on computers as tools in a greater variety of areas within most organizations. While the cost of computing power is falling, the salaries of individuals are growing. This cost relationship increasingly emphasizes the need to design systems to maximize the efficiency of the user rather than that of the computer.

A potential for substantial economies of scale in the availability of common software packages has come with this increase in utilization of microcomputers. With previous generations of large main frame computers, popular packages could expect sales to number in the hundreds of copies only. This led to high development costs being shared among relatively few purchasers. In order to keep package prices down many user oriented features were ignored and packages were relatively inflexible. With the potential sales of microcomputer packages in the tens or hundreds of thousands, package prices can be considerably lower while still supporting the high development costs associated with user friendly software. Furthermore, developers need to include high quality user oriented features if they hope to make the large number of sales at competitive prices that are necessary to support the development costs.

The proliferation of inexpensive software has greatly expanded the role of the computer as a tool for many end users. The tools provided to users, in most application areas, are constantly increasing in their sophistication and usefulness due to the increased competition among various popular packages. The increased variety of software packages also leads to the growing number of different areas in which a typical end user might make use of a computer as a tool.

The range and complexity of interactions between a typical user and a computer system has grown quite large. This results in the requirements for a different type of relationship than exists between an individual and other types of machines. With other machines, interactions are typically limited in number and complexity,and the user is merely an operator helping the machine to perform its assigned task. The user of a computer is considerably more than the operator of a machine and needs to be treated accordingly. Card et al (3) state that rather than operating a computer, the user communicates with it so as to perform the task together. Therefore, design emphasis needs to be placed upon the communication as well as the task.

AREAS OF HUMAN FACTORS IN COMPUTER SYSTEMS

Organizational, hardware, and software oriented human factors should be considered in the design and use of computer systems. Each of these areas has a major impact on the successful use of computers as tools within an organization.

Cooke and Drury (7) identified three organizational areas where human factors influence the success of use of computers. The introduction of computers can change political, sociological, and psychological components of an organization. Political changes include shifts in power between departments or other power blocks within the organization. Sociological changes include disruptions of work patterns and interactions between members of the organization. Psychological changes include the results of computerization on the attitudes and motivation of individuals within the organization. These changes are not unique to computerization but may occur with any change in organizational procedures. Therefore, despite the potential magnitude of these changes, most organizations should be well equipped to deal with them and they will be left for further consideration to other sources.

An end user may interact with a variety of hardware devices including input devices that vary between keyboards, touch panels, light pens, and mice; output devices that vary between video displays, printers, and plotters, and processing devices that vary in reliability and response time. Each device has a number of features that require ergonomic considerations in design. Many of them have been the subject of considerable study by both researchers and manufacturers. Recently some manufacturers have made efforts to improve the ergonomics of microcomputers. Other manufacturers have copied them for a variety of reasons including ease of imitation and a recognition of the desirability of some standardization. Most users are generally forced to choose between the few different models that the manufacturers choose to produce rather than being able to configure an ideal system based on ergonomic research. Detailed consideration of hardware related human factors, therefore, will be left to other sources.

Major areas of human factors concerns with software design include the sytle of dialogue, the syntax of commands, and the naming convention used within a system according to Card et al (3). A number of complex factors are involved within each of these areas and more are involved between areas. Various authors (1,3,11,13,16,21,23) have discussed general human factors guidelines for designing software interfaces between computers and end users. These guidelines represent a variety of direct and indirect experiences in using different systems combined with surveys of supporting research, where available. None of these sets of guidelines provide a complete framework of human factors standards for developing user-oriented software. Various of these sets of guidelines conflict in specifics, reflecting the diversity of opinions and styles possible regarding interactive software design.

Within the software industry, these conflicting opinions and styles have led to the feeling that standardization may be neither feasible nor even desirable. This provides the user with a greater variety of packages to choose from with little guidance to help in making the selection. This is further complicated when a user requires a number of different packages. The resulting confusion and difficulties for the user have led some some software producers to attempt to integrate a number of their products to provide the user with an all in one solution.

The current concepts of structured analysis and design do not go far enough in ensuring the development of high quality user-oriented interaction. While the analysis phase generally is sufficiently user-oriented, the design phase is not. Structural analysis concentrates on producing a logical description of the needs of the user as identified by the anlayst, according to Gange and Sarson (8). Structured design concentrates on producing a physical implementation of these logically defined needs, optimizing in terms of computer capability and structured programming concepts, according to Yordon and Constantine (26), rather than in terms of human factors concerns. The user is seldom involved in the details of the interaction beyond his initial input to the analysis phase. Thus, the quality of the interaction between the computer and the user almost always left to the judgement and experience of the designer and those who implement the design. The result can be a logically correct but physically difficult to use system. This can be especially troublesome to users who interact with a variety of systems, each with its own style of physical implementation.

Concerns with interaction between systems and users have partially contributed to the development of various integrated packages such as: Lotus 1-2-3, Context MBA, Knowledge Man, and others. Despite claims to be the only package that a user needs, an examination of such packages finds that each package generally has one or more features that some of the others are missing. Also, a number of basic functions, needed by the user, are often left to the operating system rather than being integrated within the package. These differences lead to the need for a set of criteria for users to apply in evaluating the completeness of such packages, as well as for designers of such packages to aim towards meeting.

There are two needs, therefore, one for guidelines in designing user-oriented interaction for systems in general, and the other for guidelines for design and selection of general purpose integrated systems in particular.

NAMING USER-ORIENTED FUNCTIONS

The basis of interaction in any system is dependent upon the selection of functions to be performed by the system. Most current systems are based on the system performing these functions in response to the user entering specific commands. These commands are generally named to aid user recognition and understanding, regardless of whether or not the actual command name is required to be entered into the system. One of the most important design concerns, therefore, should be the selection and naming of the various functions that the user requires for interaction with the system. This is also one of the areas where human factors research and existing design guidelines tend to be least specific. Existing guidelines generally include the following observations on the selection of an appropriate vocabulary:

USE OF NATURAL ENGLISH TERMS

Experiments have been conducted demonstrating the superiority of using English-like commands in user applications. Landauer et al (10) found that user-suggested English terms were superior to traditional computer-oriented commands in the areas of user completion, accuracy, and efficiency for a group of text editing tasks. Ledgard et al (12) found similar results in comparing English-like commands with notational commands. Robey and Taggert (18) suggested that the appropriate types of processing varies with the task. This leads to the question of how English-like commands should be used. Miller (15) stated that, "while we have no doubts at all concerning the capacity of natural English for specifying highly precise and complete procedures, we are concerned about people's abilities to use the language in this way, particularly for difficult problems". Regardless of the format that commands take, they should make use of a user-oriented English-like vocabulary for maximum efficiency and user friendliness.

STANDARDIZATION FOR ALL SYSTEMS

The current state of the vocabulary used in most information systems has been compared to the problem of discovering the appropriate commands in an adventure game by Carroll (6). This is due to the large number of different terms that are used for naming the same function in various systems as recognized by Schneider and Thomas (20). The "use of a number of different words to label the same computer function" negatively influences the overall performance of users according to Scapin (19). Treu (25) states that although, "the average user surely would prefer to speak one interaction language to accomplish essentially the same tasks in each of a variety of different computer systems" standardization is far from a reality since appropriate standards have not been developed and agreed upon yet.

Both the concepts and the terms that describe and name them should be standardized for all systems used by the same person. Use of various equivalent words or phrases to denote variations of a single concept should be avoided according to Sisson (23). The large number of different systems that a person may want makes the development and implementation of a set of standards both highly desirable and very difficult according to Treu (25). Morland (16) states that this standardization must include parts of the operating system that the user interacts with directly, as well as application programs.

STANDARDS BASED ON USERS

Benbasalt (2) states that any standard must incorporate the characteristics and needs of the user. The development and use of such standards depends on the design of user tasks that are based on a conceptual model of how the user will use the system to accomplish tasks. According to Carey (4), "this knowledge begins with the system's representation for information in the task domain (files, records, etc.) and the operations on functions that apply to them (storing, copying, etc.)" This suggests the use of object-oriented systems such as the Smalltalk system described by Goldberg and Robson (9).

STANDARDS THAT ARE NOT CONFUSING

Bailey (1) states that the various terms used should be semantically unambiguous to aid user learning and to reduce user errors. He also states that the use of these terms should not conflict with normal conventions.

Card et all (3) in discussing a user's working memory point out that the number of immediate actions possible should be limited to avoid user confusion. Ledgard et al (11) and Miller and Thomas (14) have found that where the number of options are large users will tend to only deal with a subset at a time. Shneiderman (22) suggested that a syntactic/semantic model of command formulation proceeds from a user's perception of the hierarchical organization of the task to be performed. Thomas and Carroll (24) found in research into factors involved in selecting names that, "names that people create are not arbitrary with respect to other names they create" but rather they are selected to be hierarchically consistent and to provide minimal distinction from other names.

DEVELOPING THE HIERARCHY

The identification of functions to be standardized can be based upon a consolidated set of the various commands and functions found in typical user-oriented systems, as was done in a study conducted by the author.

Selection of software systems of each type was done on the basis of selecting enough prominent examples of the type of software system so that no new functions were encountered in the last example investigated. In all, 42 systems were investigated in detail, consisting of 20 application packages, 17 operating systems and 5 query languages. Of the application systems, 3 were systems of accounting packages, 5 were spreadsheets, 7 were information utilities, and 5 were integrated application systems. Successive sets of consolidations and organizations of the identified functions proceeded until a hierarchy was reached that provides a logically organized description of the various user-oriented functions without redundancies. The hierarchy attempted to classify the functions based on what is done in general by the user rather than what is specifically done by the computer. Further refinements have been done to the hierarchy proposed in Carter and Everett (5) to make it compatible with the object oriented approach to data systems. These refinements have been done without compromising the hierarchy's user-orientation. The hierarchy consists of three major categories of generally required functions, two major categories of additional special functions and four major categories of application dependent functions. The application dependent functions were excluded from further investigation due to their application dependence.

THE HIERARCHY

The following is the hierarchy of user-oriented functions:

A. GENERALLY REQUIRED FUNCTIONS

1. DATA FUNCTIONS
    - 1.1 ADD OR ADD TO AN OBJECT FUNCTIONS
        - 1.1.1 Create an Object
            - 1.1.1.1 Create a Field
            - 1.1.1.2 Create a Record
            - 1.1.1.3 Create a File
            - 1.1.1.4 Create a Directory
            - 1.1.1.5 Create a Form
            - 1.1.1.6 Open a Window
        - 1.1.2 Add to an Object
            - 1.1.2.1 Add Data to a Record
            - 1.1.2.2 Add to File
            - 1.1.2.3 Link File to a Directory
            - 1.1.2.4 Join (Append) Directories
            - 1.1.2.5 Add Protection
        - 1.1.3 Copy an Object
            - 1.1.3.1 Copy a Field
            - 1.1.3.2 Copy a Record
            - 1.1.3.3 Copy a File
            - 1.1.3.4 Copy a Window
    - 1.2 CHANGE AN OBJECT FUNCTIONS
        - 1.2.1 Change a Field
        - 1.2.2 Change a Record
        - 1.2.3 Change a File
            - 1.2.3.2 Split File
            - 1.2.3.3 Sort File
            - 1.2.3.4 Translate Characters
            - 1.2.3.5 Encrypt a File
            - 1.2.3.6 Unencrypt a File
            - 1.2.3.7 Update a File with Another

- 1.2.4 Change Object Name
  - 1.2.4.1 Change Field Name
  - 1.2.4.2 Change File Name
  - 1.2.4.3 Change Form Name
- 1.2.5 Change Object Format
  - 1.2.5.1 Change Field Format
  - 1.2.5.2 Change Record Format
  - 1.2.5.3 Change File Format
  - 1.2.5.4 Change Form Format
- 1.2.6 Change Window
  - 1.2.6.1 Modify Window
  - 1.2.6.2 Link Windows
- 1.2.7 Change Protection
  - 1.2.7.1 Change a Password
  - 1.2.7.2 Change Access

1.3 DELETE OR DELTE FROM AN OBJECT FUNCTIONS

- 1.3.1 Delete an Object
  - 1.3.1.1 Delete a Field
  - 1.3.1.2 Delete a Record
  - 1.3.1.3 Delete a File
  - 1.3.1.4 Delete a Directory
  - 1.3.1.5 Delete a Form
  - 1.3.1.6 Close a Window
  - 1.3.1.7 Delete an Index
- 1.3.2 Remove From an Object
  - 1.3.2.1 Empty a Field
  - 1.3.2.2 Empty a Record
  - 1.3.2.3 Empty a File
  - 1.3.2.4 Compact a File
  - 1.3.2.5 Remove Trailing Blanks
  - 1.3.2.6 Remove Link to Directory
  - 1.3.2.7 Remove Protection

1.4 GET AN OBJECT FUNCTIONS

- 1.4.1 Find an Object
  - 1.4.1.1 Find a Field
  - 1.4.1.2 Find a Record
  - 1.4.1.3 Find a File
  - 1.4.1.4 Find Protection
  - 1.4.1.5 Obtain Data Item Statistics
  - 1.4.1.6 Count Words
  - 1.4.1.7 Compute Two or More Fields
  - 1.4.1.8 Find Object Type
- 1.4.2 List all Objects
  - 1.4.2.1 List all Field Names
  - 1.4.2.2 List all File Names
  - 1.4.2.3 List all Directories
  - 1.4.2.4 List all Access Permissions
- 1.4.3 Output an Object
  - 1.4.3.1 Output a Field
  - 1.4.3.2 Output a Record
  - 1.4.3.3 Output a File
  - 1.4.3.4 Output a Directory
- 1.4.4 Specify an Object
  - 1.4.4.1 Select a File
  - 1.4.4.2 Activate a Directory
  - 1.4.4.3 Designate an Index Key Field

1.5 COMPARE AN OBJECT FUNCTIONS
- 1.5.1 Compare Fields
- 1.5.2 Compare Records
- 1.5.3 Compare Files
- 1.5.4 Compare Directories

2. TASK EXECUTION FUNCTIONS

2.1 SIGNING ON

2.2 SIGNING OFF

2.3 CHANGE SIGNON PASSWORD

2.4 STARTING EXECUTION FUNCTIONS
- 2.4.1 Execute Now
- 2.4.2 Execute in the Future
- 2.4.3 Execute Again
- 2.4.4 Execute Batch

2.5 STOPPING EXECUTION FUNCTIONS
- 2.5.1 Leave Interactive Program Normally
- 2.5.2 Terminate Abnormally
- 2.5.3 Interrupt Processing
- 2.5.4 Continue Processing
- 2.5.5 Stop Batch Requests Before they Start

2.6 EXECUTION STATUS FUNCTIONS
- 2.6.1 Obtain Status of Current Processes
- 2.6.2 Obtain Status of Batch Requests

2.7 PRIORITY FOR EXECUTION FUNCTIONS
- 2.7.1 Change Priority
- 2.7.2 Obtain Queue Information

3. INFORMATION FUNCTIONS

3.1 USER ASSISTANCE FUNCTIONS
- 3.1.1 Assistance with Commands
  - 3.1.1.1 List Commands
  - 3.1.1.2 Explain Command Description
- 3.1.2 Assistance with Errors
  - 3.1.2.1 List Errors
  - 3.1.2.2 Explain Error Type
- 3.1.3 Provide Tutorial
- 3.1.4 Provide Help Screens
  - 3.1.4.1 Display Next Help Screen
  - 3.1.4.2 Display Previous Help Screen
  - 3.1.4.3 Return From Help
- 3.1.5 Provide Keyboard Assistance

3.2 OTHER INFORMATION FUNCTIONS
- 3.2.1 Provide Information on System Users
- 3.2.2 Provide Information on Last Signon of Users
- 3.2.3 Provide Information on Where User Is
- 3.2.4 Provide Information on Last Commands Executed

B. ADDITIONAL SPECIAL FUNCTIONS
(only high level of hierarchy described here)

4. PERIPHERAL FUNCTIONS

4.1 TERMINAL FUNCTIONS

4.2 DISK FUNCTIONS

4.3 OTHER PERIPHERAL CONTROL FUNCTIONS

5. SPECIAL FUNCTIONS

5.1 USER TO USER COMMUNICATIONS FUNCTIONS
5.2 DATE AND TIME FUNCTIONS
5.3 WRITING AIDS
5.4 ARITHMETIC OPERATIONS
5.5 SYSTEM UTILIZATION ACCOUNTING

C. APPLICATION DEPENDENT FUNCTIONS
(not further described here)

6. BIBLIOGRAPHIC FUNCTIONS

7. SPREAD SHEET FUNCTIONS

8. GRAPHICS FUNCTIONS

9. TEXT EDITING FUNCTIONS

HOW THE HIERARCHY CAN BE USED

The hierarchy can be used in a variety of ways. It can be used to analyze individual systems for completeness or to analyze groups of systems for compatibility and consistency. In systems analysis the hierarchy can provide a check list to remind the analyst to look for all of the logical needs that might correspond to the various functions. In systems design it can be used to standardize the design of the physical implementations of various logical systems and to provide checks for compatibility and consistency of functions and function names between various systems. It can also be used as a basis for further research into the most efficient names for each of the functions described within it.

REFERENCES

[1] Bailey, R.W., Human Performance Engineering: A Guide For System Designers. Englewood Cliffs, N.J. (Prentice-Hall, 1982).
[2] Benbasalt, I., The Impact of Cognitive Styles on Information System Design, MIS Quarterly, 2(2), (1978), 43-54.
[3] Card, S.K., Moran, T.P., and Newell, A. The Psychology of Human Computer Interaction, Hillsdale, N.J., (Lawrence Erlbaum Associates Publishers, 1983).
[4] Carey, T., User Differences in Interactive Design, Computer, (November 1982), pp. 14-20.
[5] Carter, J.A. and Everett, H.J.M., A Hierarchy of User-Oriented Functions, Proceedings of the Canadian Information Processing Society Session '84, (1984).
[6] Carroll, J.M., The Adventure of Getting to Know a Computer. Computer, (November 1982), pp. 49-58.
[7] Cooke, J.E. and Drury, D.H., Management Planning and Control of Information Systems, Hamilton Ontario, (Society of Management Accountants of Canada, 1980).
[8] Gane, C. and Sarson, T., Structured Systems Analysis: Tools and Techniques, (Prentice-Hall, 1979).
[9] Goldberg, A. and Robson, D., Smalltalk-80: The Language and Its Implementation, Reading, MA., (Addison-Wesley, 1983).
[10] Landauer, T.K., Galotti, K.M. and Hartwell, S., Natural Command Names and Initial Learning: a study of text-editing terms, Communications of the ACM, 26 (7), (1983), 495-503.

[11] Ledgard, H., Singer, A., & Whiteside, J., Directions in Human Factors for Interactive Systems, Berlin, (Springer-Verlag, 1979).
[12] Ledgard, H., Whiteside, J., Singer, A., & Seymour, W., The Natural Language of Interactive Systems, Communications of the ACM, 23 (10), (1980), 556-563.
[13] Martin, J., Design of Man-Computer Dialogues, Englewood Cliffs, N.J., (Prentice-Hall, 1973).
[14] Miller, L.A. and Thomas, J.C., Behavioral Issues in the Uses of Interactive Systems, Goos, G. and Hartmanis, J., (ed) Interactive Systems, Berlin, (Springer-Verlag, 1977).
[15] Miller, L.A., Natural Language Programming: Styles, Strategies, and Contents, IBM Systems Journal, 20 (2), (1981), 184-215.
[16] Morland, D.V., Human Factors Guidelines for Terminal Interface Design, Communications of the ACM, 26 (7), (1983), 484-494.
[17] Osgood, C.F., Suci, G.J., and Tannenbaum, P.H., The Measurement of Meaning, (Urbana, IL., University of Illinois Press, 1957).
[18] Robey, D. and Taggert, W., Human Information Processing in Information and Decision Support Systems, MIS Quarterly, 6, (2), (1982), 61-73.
[19] Scapin, D.L., Computer Commands in Restricted Natural Language: Some Aspects of Memory and Experience, Human Factors, 23 (3), (1981), 365-375.
[20] Schneider, M.L., & Thomas, J.C., The Humanization of Computer Interfaces, Communications of the ACM, 26 (4), (1983), 252-253.
[21] Schneiderman, B., Software Psychology: Human Factors in Computer and Information Systems, Cambridge, MA., (Winthrop Publishers, 1980).
[22] Schneiderman, B., Direct Manipulation: A Step Beyond Programming Languages, Computer, (August, 1983), 57-69.
[23] Sisson, C., Criteria for Evaluating Ease-of-Use, Proceedings of the Thirteenth Annual Conference of the Society for Management Information Systems, 13, (1981), 121-127.
[24] Thomas, J.C., & Carroll , J.M., Human Factors in Communication, IBM Systems Journal, 20 (2), (1981), 237-263.
[25] Treu, S., Uniformity in User-Computer Interaction Languages: A Compromise Solution, International Journal of Man-Machine Studies, 16, (1982), 183-210.
[26] Yordon, E. & Constantine, L., Structured Design, Prentice-Hall, (1979).

Human Factors in Organizational Design and Management
H.W. Hendrick and O. Brown, Jr. (Editors)

ICONIC INTERFACES: WHERE TO USE AND HOW TO CONSTRUCT?

Gabriele Rohr, Eric Keppel

IBM Germany Scientific Center Heidelberg
Heidelberg, W. Germany

It is assumed that iconic interfaces are a useful tool in displaying system states and operations in complex working environments of a high technological level. Relevant features of icons representing such states and operations are investigated. According to these results icons are constructed and investigated against verbal commands in a real system and task environment. First results suggests that iconic interfaces are better to impart an adequate mental model of the system structure.

INTRODUCTION

Working environments of a high technological level are characterized by operating with machines having a multiplicity of states, possible operations in and connections between these states, and objects to be worked on. These machines represent complex systems in which complex tasks have to be organized by the user in a certain way to perform them successfully. Consequently the user needs to be always aware of the structure of the system, its different states and operations, and furthermore has to reorganize and adapt the task structure to them. Hence, mental workload grows in the same way the the system and the task complexity increases, especially because of the greater demands on mental memory. Reducing mental workload means reducing complexity in some way.

How can we reduce complexity? First, complexity can only be regarded with respect to the user of a system, and consequently it can only be defined for a given representation of a system and not for the system itself /1/. So reduction of complexity must take place at the level of symbolic representation of the system, at the human-machine interface, i.e. the organization of the single displays and their reference/relation to each other. A hierarchical organization has been proposed in /1/. Furthermore symbols should represent concepts which are related to the causal structure of the system, i.e. should help to acquire a mental model of the system structure. They should be equally related to the structure of the task to be performed by means of this system. Scapin /2/ e.g. showed that learning command names together with structuring rules by means of a machine model, led to higher retention then learning without.

The use of iconic interfaces in place of verbal command sets has been discussed recently. The basic idea why iconic interfaces is: icons can be constructed in such a way that the user has a better chance to acquire implicitly a model of the system structure when looking at the whole set of icons than when he would have dealing with a verbal command set. This assumption is based on the fact that there exist special areas of information presentation where complex information could be presented more condensed and in a wholistic way by graphical symbols (e.g. navigation: maps, manipulation: operating instructions /3/). Hence, for a special composition of graphical symbols it seems to be possible to impart a generalized model of the system structure more efficiently.

One example for this special nature of icons condensing complex information and imparting a mental model of the system structure is the iconic solution for the "erase file" command in the office system of APPLE LISA*. Here the user depicts an icon representing a file (a paper sheet) and puts it into a picture of a paper basket. The "paper basket" is a strong symbol because it represents simultaneously the "erase" function (throw away) and the "undo" function (take it out again) which is possible in this context.

There is some empirical evidence /4/ /5/ that people are able to represent more or less abstract concepts by means of pictorial signs. It is interesting that people introducing new abstract concepts , very often use pictorial allegories defining them. Paivio /6/ during his research on imagery mnemonics suggested that the mental representation of the referent of an abstract concept is executed by a process of "concretization" whereby a link is made to a concrete situation or object by means of associative chaining. So one pictorial symbol can represent a class of events as a prototype, e.g. the picture of a church can represent the concept of religion.

Howell an Fuchs /4/ found that the more abstract a concept became the more abstract the visual signs and symbols chosen tended to be (e.g. slash, asterisk, arrow etc.). So a diagonal slash was used universally for negation, in combination with a concrete object as a prohibitive signal, and an arrow for direction. Composed terms consisting of a mixture of concrete and abstract concepts tended to be represented by at least one concrete symbol accompanied by abstract symbols, mostly in common usage. On the other side results of Jones /6/ showed that subjects preferred visual representations for more abstract concepts composed of several concrete symbols, e.g. the concept "interview" by the picture of two heads and a microphone in the middle. Less stereotypy occured the more abstract the concepts became.

In general only little is known about the elements representing abstract concepts visually. It was our concern to investigate iconic interfaces in view of the advantages assumed by us. Two main questions had to be answered. The first of them is how must visual symbols look like , i.e. what features they must be composed of, to impart a special meaning. Second : can a special set of icons composed of relevant features with respect to the meaning lead to a better understanding of the system structure, i.e. helping to acquire an adequate mental model of the system, than a comparable set of verbal commands?

FIRST EMPIRICAL STUDY: RELEVANT FEATURES OF ICONS

Because very little is known about human imagery of certain processes, neither about the elements nor their combination with respect to a special meaning, it was necessary to construct a large set of pictorial symbols (icons) varying with respect to the kind of elements (features) and their combination, and furthermore to create a specified system and task environment of which the functions could be related to pictorial symbols. Since the main concern was obtaining the functional meaning of certain features and their combinations rated by a group of subjects - it was not possible to combine each feature element with each other systematically because of the resulting large amount of pictorial symbols - a special method had to be developed. This method was constructing a feature vector containing as elements all features used for the composition of any of the pictorial symbols, assigning a "1" to the element if the particular icon contained this feature, and a "0" if not. So each icon was described by a feature vector containing a special distribution of ones and zeros.

METHOD 1

An icon set of 72 icons was constructed on the basis of 16 elementary features and their combinations. Four classes of features were used (following the classes proposed by Howell and Fuchs /4/):

1. rectangles, varying in size, number, and spatial combination;
2. abstract symbols, i.e. arrows in different direction, slash, cross, points, question-mark, arrow-top, and rupture;
3. numbers and letters differing in combination
4. pictorial symbols in a stronger sense, i.e. sketchlike pictures of a stamp, pencil, envelope, paper sheet, hand, and a box.

A questionary was developed introducing the set of icons as belonging to a special system and task environment, a text processing system, representing functions of this system. This kind of a system was chosen because the task of text processing is known generally.

A list of questions with regard to the functions occuring in the system described was allocated to each icon. Each question contained a rating scale where the subjects had to rate with "YES" if the function requested was represented by the icon or with "NO" if not. In addition they had to indicate their decision on a five point rating scale (questionary 1).

Furthermore a questionary was constructed where each icon was asked for the amount of functionality, manipulative information, and pictorial quality it contained, again on a five point rating scale (questionary 2).

10 subjects (19 - 23 years old, 6 women, 4 men) without any computer experience or any experience with text processing systems participated at the test. Each subject had to answer all the questions with respect to each icon given in questionary 1 and questionary 2. Test duration was about 2 - 3 hours.

To compute the results each feature vector representing a special icon was weighted by the mean value of the rating it yielded for a special question, i.e. function in the described system and task environment. All weighted feature vectors were than added for a special function respectively. So each requested function yielded a weighted feature sum vector with a profile of values for each function. Those features having the significant highest values were then regarded as the relevant features of a special function.

RESULTS 1

The evaluation of the data of questionary 2, i.e. the ratings of the icons with respect to their functionality, manipulative information and pictorial quality showed only one effect. There was only one feature with significant high values for functionality and manipulative information: the arrow.

The evaluation of the data of questionary 1, i.e. the ratings of the icons according to their represented function in a text processing system, showed more differentiated results. Generally speaking there was a tendency to represent functions grouped as "file handling" operations by means of more pictorial symbols and rectangles in a special combination. In contrary functions grouped as "editor" operations had a tendency to be represented by means of more abstract symbols and letters and their combinations. Nearly each command yielded distinct features by which it was represented, e.g. for "send file" it was an envelope, for "erase file" it was a paper sheet with a rupture, for "save" several little rectangles with a box and an horizontal arrow, for "move a string" a diagonal arrow with two rectangles, for "delete a string" a cross on letters etc.

Summarizing, the results allowed to construct a specified set of icons varying with respect to functionality which can be tested in a system and task environment.

SECOND EMPIRICAL STUDY: EXPERIMENTAL APPROACH

In the second empirical study we wanted to test specified sets of icons varying with respect to certain features only, the functional or manipulative ones, against a verbal command set in a real system and task environment. Our main concern was to gain information: (1) how people handle a set of pictorial symbols in comparison to a verbal command set (i.e. do they perform a given task faster under different conditions), (2) if it is easier to acquire an adequate mental model of the system structure by an iconic interface.

METHOD 2

Three command sets were constructed, two icon sets and a verbal command set, with respect to a text processing environment. The icons were constructed according to the relevant features given by the results of the first study. The two icon sets differed only in the functional or manipulative aspect respectively, i.e. with file handling commands they differed only in having an arrow as additional element (functional) or having it not (non functional). With editing commands those elements were marked (manipulative) which represented cursor actions and were not marked or missing (non manipulative) in the other set. Figure 1 shows an example of the icon sets.

Figure 1: An example of the different command sets (functional, non-functional, verbal) for file handling and editing commands

For the experiment a text processing prototype was developed by which it was possible to implement different sets of icons or verbal commands. This prototype system was structured into three main states: an initial state, a file handling state, and an editing state. Additionally there were two substates "rename" and "select", subordinated under the file handling state for evaluation. The system structure is described more detailed by Keppel and Rohr (in this volume).

For each state or substate a help state exists allowing the user to inform himself about the meaning of the symbols (iconic or verbal) occuring in this state with respect to their operating function. Only to the initial state a special help state was subordinated containing general information about the structure of the whole system. This partition of helps was implemented to observe if persons do not understand the special command symbols (helps of the file handling, editing etc. states) or do not understand the structure of the system, i.e. cannot acquire an adequate mental model.

All the operations the subject are doing on the system will be registrated on-line, so that at the end of each experiment a complete protocol of all operations the subject did and the time between them will result.

15 subjects (20 -24 years old, 8 women, 7 men) without any computer experience or experience with text editing systems participated at the experiment. They were randomly distributed to the three groups of command sets, five subjects per group, balanced with respect to sex.

Each subject got five tasks to perform. These tasks contained: creating a new text, changing an old one, printing a text, send a text etc. The number of necessary operations for file handling and editing tasks was equalized. The subjects received only a brief instruction at the beginning of the experiment, and were told to seek for help in the system's help states if they would not know how to proceed.

For the purpose of data evaluation the time each subject spent in file handling and editing state was computed separately. The same was done for the number of operations. These values were related to the least number of necessary operations to perform the tasks. By this method values resulted representing the time and the number of faults needed for each necessary operation. Furthermore the number of help requests was computed for each help state separately (i.e. general help, help file handling, help editing).

RESULTS 2

An analysis of variance was computed for time, number of faults per operation (commands x task states: 2x3 factorial, having repeated measurements on the last factor) and number of help requests (commands x help states: 3x3 factorial with repeated measurements on the last factor) separately.

There was no statistical significant difference between the icon set and verbal command groups concerning time and faults per operation in task performance. There was only a significant difference (1% level) between file handling and editing operations, editing operations requiring more time and producing mor faults. This seems to indicate that the editing operations produce more mental workload than file handling operations whereby the kind of symbols have no influence.

Different results came from the analysis of variance for help requests. There was a significant effect for the number of help requests between the three help states (1% level) on the one side, but more interesting is the significant interaction (5% level) between commands and help state. According to our hypotheses the verbal command group showed a greater number of requests for general help information, i.e. the system structure, than the icon groups. On the other side the icon groups needed more help requests for file handling symbols (see figure 2). This would mean that the verbal command group had more difficulties to acquire an adequate mental model of the system structure than the icon groups, but has an easier understanding of the single commands in the file handling state. Verbal nouns can better represent concepts of processes like print, rename, etc. For editing functions this advantage of verbal commands is lost. Furthermore it is an interesting fact that in the editing state the group with the functional icon set needed less help requests than the nonfunctional group according to our hypotheses, but that the verbal group needed still less requests than the functional group. More detailed analyses have to be done to explain these effects, e.g. to compare the first tasks performed with this ones at the end of the test, resulting in better understanding of learning effects.

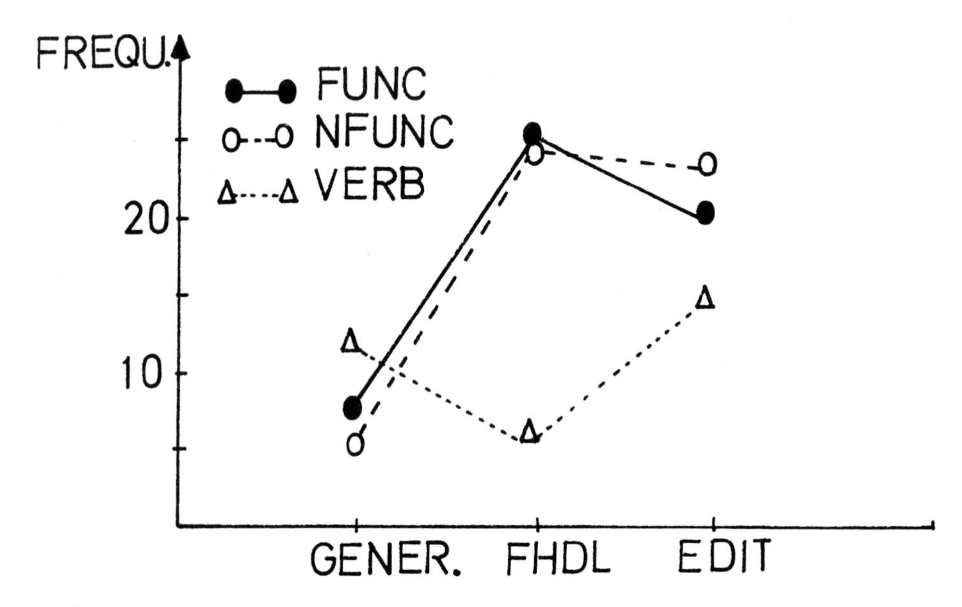

Figure 2: Mean frequencies of help requests in the different help states (general help, help file handling, edit)

CONCLUDING REMARKS

We could show that in general it is possible to represent operations and states of a system by means of pictorial representations. The icons representing single commands were as good understood as verbal commands though not better in our case.There was only one advantage they showed in comparison to verbal commands: They impart better a mental model of the system structure to be worked on. This could mean that they reduce complexity and consequently mental workload. In the future more detailed research is needed to confirm our findings, and to find out causal explanations. It is very important also, to investigate learning effects in more detail.

REFERENCES:

/1/ Rasmussen, J. and Lind, M., Coping with complexity, in: Stassen,H.G. and Thijis, W. (Eds.), Proceedings of the First European Conference on Human Decision Making and Manual Control (Delft University of Technology, Delft, 1981)

/2/ Scapin, D.L., Computer Commands labelled by users versus imposed commands and the effect of structuring rules on recall, in: Nichols, J.A. (Ed.), Human Factors in Computer Systems - Proceedings (Gaithersburg, Maryland, 1982)

/3/ Kalkofen, H., Probleme und Ergebnisse der Lehrfilmforschung - I. Bemerkungen zur Methodologie, Research Film 7 (1972) 561 - 572

/4/ Howell, W.C. and Fuchs, A.F., Population stereotype in code design, Organizational Behaviour and Human Performance 3 (1968) 310 - 339

/5/ Jones, Sh., Stereotypy in pictograms of abstract concepts, Ergonomics 26 (1983) 605 - 611

/6/ Paivio, A., Imagery and Verbal Processes (Holt, Reinhart and Winston, Washington D.C., 1971)

ORGANIZATION AND MANAGEMENT OF ERGONOMICS

Human Factors in Organizational Design and Management
H.W. Hendrick and O. Brown, Jr. (Editors)

ORGANIZATION AND MANAGEMENT OF ERGONOMICS IN DIVERSE ORGANIZATIONS (PANEL)

Ogden Brown, Jr., Chair
University of Southern California
Los Angeles, California, USA

The Technical Group on Organizational Design and Management is interested in examining the organizational characteristics of successful companies which have ergonomics groups or departments within their formal structure. Of particular interest is the organizational design utilized by these firms and how the ergonomics function is incorporated. Also of interest is the degree of differentiation and the extent to which integrating functions are present. One method of inquiry is to examine organizational programs, noting initial design, subsequent changes, management practices, and problems encountered in initiating ergonomics efforts. Another method of inquiry to learn about organizational design, alternative management systems, and problems encountered is to consult those who have the responsibility for the management of ergonomics programs in large and diverse organizations which have incorporated the ergonomics function into their formal organizational structure.

ORGANIZATION AND MANAGEMENT OF ERGONOMICS AT AT&T BELL LABORATORIES

Brent Coy
AT&T Bell Laboratories
Whippany, NJ, USA

The organization and management of human factors activities in AT&T Bell Laboratories will be discussed.

ORGANIZATION AND MANAGEMENT OF ERGONOMICS AT HONEYWELL, INCORPORATED

Susan M. Dray
Honeywell, Inc.
Minneapolis, MN, USA

The organization and management of ergonomics at Honeywell, Inc. will be discussed, as will the long term human factors plan. An examination will be made of selected innovative programs, and resource allocations for human factors activities will be addressed.

THE ORGANIZATION AND OPERATION OF HUMAN FACTORS IN IBM

William H. Emmons and Alan S. Neal
IBM Human Factors Center
San Jose, California, USA

There are human factors departments in every IBM product development laboratory throughout the world. These groups vary in size from just a few human factors specialists to groups having a staff as large as twenty individuals. Several of the larger human factors groups, in addition to serving the human factors needs of the local laboratory, have corporate-wide responsibilities in certain functional areas.

In addition to the human factors departments in the development laboratories and the human factors group in the Research Division, there are several human factors specialists who hold staff positions in Corporate or Group headquarters. These people facilitate coordination and communications among the groups.

Since IBM is a rather de-centralized corporation, the human factors departments vary somewhat in how they are organized and operate. In this talk, I will concentrate on the operation of the Human Factors Center in San Jose and note those areas where it may differ significantly from other IBM human factors organizations.

The San Jose Human Factors Center (HFC) reports into a divisional headquarters organization because we support more than one laboratory in San Jose, the San Jose Development Lab and the Santa Teresa Program Development Lab. Most other groups in IBM report into various technical services areas within a specific laboratory. A few groups report into Design Centers, some are located in a Product Assurance function, and a few individuals work in manufacturing organizations.

The HFC is line funded (fully budgeted). We do not charge product groups within our division for our services. However, when we do work for other IBM divisions, we receive reimbursement for time and materials which provides a cost relief to our parent division. Some of the other IBM groups receive all or part of their funding from the product groups that they are supporting.

Most of the human factors specialists in IBM are psychologists, but there are a number with engineering degrees. The psychologists usually have advanced degrees, often PhD's, in experimental, cognitive, or perceptual psychology. They frequently have academic training or experience in other disciplines as well, such as engineering, programming, statistics, etc. The human factors specialist at many of the IBM locations has the title of Human Factors Engineer/Psychologist. In addition to the human factors specialists, some of the groups have some support personnel. For instance, the HFC has within the group systems and applications programmers, electronic engineers and technicians, and laboratory assistants.

The mission of the human factors groups is to provide the human factors services needed by the product development groups within the laboratory. In addition, they may provide some ergonomic assistance to various manufacturing organizations. Some of the human factors groups also conduct applied research on general human factors issues, particularly issues related to the general product mission of the laboratory.

The HFC currently has these major missions: (1) product support for the Santa Teresa Programming Lab, (2) applied research on software human factors, (3) research concerning the effect of VDTs on human vision, (4) development of and research supporting internal, national and international standards for the design and use of VDT workstations, (5) evaluation of data entry devices, especially keyboards, and (6) development of methodologies for evaluation and research on the human-computer interface.

To carry out the above diverse missions, the HFC is organized into three sub-departments: (1) the Operator Performance Evaluation Laboratory, (2) the Software Human Factors Department, and (3) the Physiological Optics Department. The Performance Evaluation Lab provides the other HFC departments with testing facilities, and programming, engineering, and computer services. In addition, this group conducts the data entry device evaluations and research. The Physiological Optics group is, of course, responsible for the VDT work, and the Software Human Factors group for the support of the Programming Lab and the research on the software side of the human-computer interface.

The HFC obtains work in a number of ways. Human factors groups in other divisions may request us to conduct studies for them when their facilities or staff are inadequate for the task. In this case, we are not responsible for providing all of the product support, but just design and conduct the requested research or evaluations. We get involved in projects in the local division primarily at the request of the project management or staff. There are human factors requirements in the IBM Product Development Guide which encourages them to look us up, but we try to promote ourselves as helpmates in the team-effort of building the product, not just another bureaucratic requirement. We like to get involved with a product very early in the design cycle, when the specifications are just being formulated. We provide advice to the developers based on our experience and the technical literature. We sometimes conduct experiments to compare alternate designs. As soon as possible, we build prototypes of the user interface as both a design aid and as a vehicle for human factors testing.

Almost all of the testing done in our laboratory is computer controlled with on-line data collection. We use test subjects who are typical of the people who will ultimately use the product. They learn the systems being tested from drafts of the product documentation and are asked to perform relevant tasks. Data are collected on speed and accuracy, and problems are identified. Subjects' preferences and suggestions are also recorded. This information is communicated to the product groups along with suggested design changes. As soon as the changes are agreed upon, we modify our prototypes and retest. The process is continued until the point where that product is developed sufficiently for us to scrap the prototype and start testing early versions of the product itself.

Communications is an important part of the job for all IBM human factors personnel. In addition to the meetings, presentations, and memos to product personnel, the staff prepares formal technical reports on major studies. These are circulated to a variety of groups within the company and to all other human factors departments. We also encourage publication in professional journals and participation in professional society activities and conferences. We are most active in the Human Factors Society, but also participate in the International Ergonomics Association, ACM Computer-Human-Interface group, Society for Information Display and a number of others. In addition, members of the IBM human factors community gather together for a technical conference once a year. We interchange information in a newsletter, and have informal communications via electronic mail and by phone.

Human factors has been an important function in IBM for many years (the HFC was founded almost 30 years ago). We have seen a dramatic growth in staffing and quality of facilities of the human factors groups in recent years. The future looks bright, and we hope to continue to help provide our customers with products that are easy to learn and to use.

HUMAN FACTORS AT HEWLETT PACKARD

Anna M. Wichansky
Hewlett Packard Company
Cupertino, California, USA

The Human Factors Engineering group was formed at Hewlett Packard in 1982 to serve the collective needs of its Computer Groups. As the company's interests shifted from instruments to computers, changes in customers, users and regulatory requirements demanded an increased awareness of human factors apects of product design. The Human Factors group was chartered to direct the integration of human factors into the design and marketing programs of the computer product development divisions.

The Human Factors group performs three basic types of activities at HP. First, we provide technical support and guidance on user interface issues common across computer products under development in the divisions. Second, we disseminate human factors information with the company to foster an internal awareness of the importance of ergonomics in product design. Third, we contribute to human factors activities outside HP, such as the ANSI VDT standards committee and sponsorship of university research, to establish HP as a leader in ergonomic product design.

The process by which we do our work is strongly influenced by our organizational position within the company. Our group is currently part of a staff organization responsible for Computer Groups Quality Control and Productivity, reporting directly to the Executive Vice President of Computer Groups. As such, we generally coordinate human factors activities for the product development divisions below us on the organizational chart, rather than providing direct support on specific projects. HP operations are decentralized, and individual divisions are highly autonomous. We work through an established network of people in the divisions who serve as human factors focal points. In addition, we support a variety of corporate groups, including Government Affairs, Public Relations, Quality Assurance, Standards, Manufacturing, and Engineering. We have also provided guidance to other lines of business within HP, including Instruments and Medical Groups.

The success of the Human Factors group at HP may be measured by the increased level of awareness within the company of human factors; the incorporation of ergonomic principles in design of computer products (including standardization of an ergonomically designed keyboard and display tilt and swivel as product features); and hiring of human factors experts within the divisions. Future plans include starting a laboratory for user evaluation of products and prototypes and joint human factors-market research studies of customer needs for ergonomically designed product features.

ORGANIZATION OF HUMAN FACTORS ACTIVITIES WITHIN INFORMATION PROCESSING COMPANIES

Paul T. Cornell
Burroughs Corporation
Plymouth, MI, USA

The increased importance of human factors within the electronics industry has occurred suddenly in most companies. Rather than having departments emerge in an evolutionary fashion, human factors groups have often had to start without a history or established record. This scenario presents both opportunities and obstacles. The degree to which the opportunities are exploited and the obstacles are overcome will often depend upon the individual manager as he will have much input on the modus operandi between HF and the various organizational groups.

A problem can emerge in that senior management may not understand ergonomics and human factors beyond a "buzzword" or "common sense" level. Unfortunately, HF has been sold by some as an application of common sense. The understanding of the discipline will depend not only upon the perspective of management, but the conditions under which human factors was introduced. For example, the climate may be more hospitable if the reason for HF was to improve product quality rather than combat legislative activities restricting product use. The onus is again on the HF manager, as he must convey to senior management the breadth and depth of the field.

Organizing the HF staff can also be troublesome. The emergence of many subdisciplines, each with its own literature base and theories, makes it difficult for one individual to "cover all the bases". The problem is one of assigning responsibility for products, lines-of-business, and/or corporate divisions. For example, a new CRT project would require input from specialists in several areas: vision, ergonomics, cognition and task analysis. This suggests a matrix approach, but problems with project responsibility and coordination may be encountered.

A final trouble spot concerns available information and research. Good literature is only beginning to become available in areas such as software, computer documentation, CRT use in offices, and training of naive computer users. This leaves HF personnel with the chore of integrating what is available and making "best guesses" about what is not. When laboratories are available to do original research, the time-frame is often not conducive to prolonged studies. Quick and dirty research may become the norm rather than the exception. There is not any easy solution to this problem, but creating a staff with a multidisciplinary background is one way of minimizing potential misjudgments.

ORGANIZATIONAL DESIGN AND ORGANIZATIONAL DEVELOPMENT

Human Factors in Organizational Design and Management
H.W. Hendrick and O. Brown, Jr. (Editors)

HOW TO REALLY MAKE THINGS WORK

Henry M. Taylor

R.E.C., Incorporated
Chester, Virginia
U.S.A.

Normal human cover-up of personal mistakes at work causes frustration among employees and can even lead to company failure. The number of personal blunders is strongly influenced by the nature of the organizational design and management. A four-year case study from a small manufacturing facility in Virginia is presented to highlight what they are doing to successfully bypass this frustration. Employees are minimizing the risk for process and personal errors while making gains in product quality. Finally, the control mechanism from a Rubik's cube demonstrates just how they manage it.

INTRODUCTION:

There seems to be a wide sense of frustration within this country that things don't work. Things cost too much for what we get! Businesses cry that other countries are underselling us. It's inferred that in the U.S.A. the quality of employee goods, productivity and services are not what they should be. Magazines and newspapers report that many Americans, while at work, suffer from a trend of despair. Almost everyone complains about the other person's mistakes.

In fact, defective pacemakers not already placed in people's chests were recalled. Many foreign cars do perform better, longer and for less cost. Teachers, railroad engineers and football players strike despite depressed economic times.

What's wrong? Most of our production elements, machines and institutions with their employees, seem at least to have the traditional potential for better productivity - yet as a whole we are not making it. All too often for a country whose history is one of industrial strength, our business organizations are failing. Paradoxically, we are told that part of the problem for employees within business is that failure isn't allowed to exist in his or her personal value system. The conscientious employee who honestly errs in one of our totalitarian manufacturing organizations works under a threat. The job will not be long secure once the organization discovers the blunder. Our businesses are failing, yet we fail to recognize that a major cause for this failure is the cover-up of mistakes by individual employees. The word employee means everyone hired for wages or salary. More about this later.

WHY THINGS DON'T WORK-PROBLEM RECOGNITION:

Human factors psychologists tell us that most employees, while at work, will not allow themselves to be dimished by their own personal errors. Whether personal error involves physical, biological or behavioral factors; no one wants to look bad. An employee will do almost anything within his or her power to hide error. Many individuals in our biggest companies as well as our government cannot handle personal mistakes. They unjustly cover up errors, frequently retaliating when questioned. Too often this personal injustice clouds business decisions and overrides the pride of others. Eventually, management is perceived by its employees as unwise, lacking both in basic courage and self-discipline. Then employee participation and productivity start to wane. In the long run the organization falls short of its potential.

To summarize problem recognition: *most employees - while at work - almost always - survive - personal errors - by any means.*

The place where the particular site case study presented here fits into a successful corporate management system - to achieve product qualtiy, productivity, and competitive position - is seen in Figure 1.

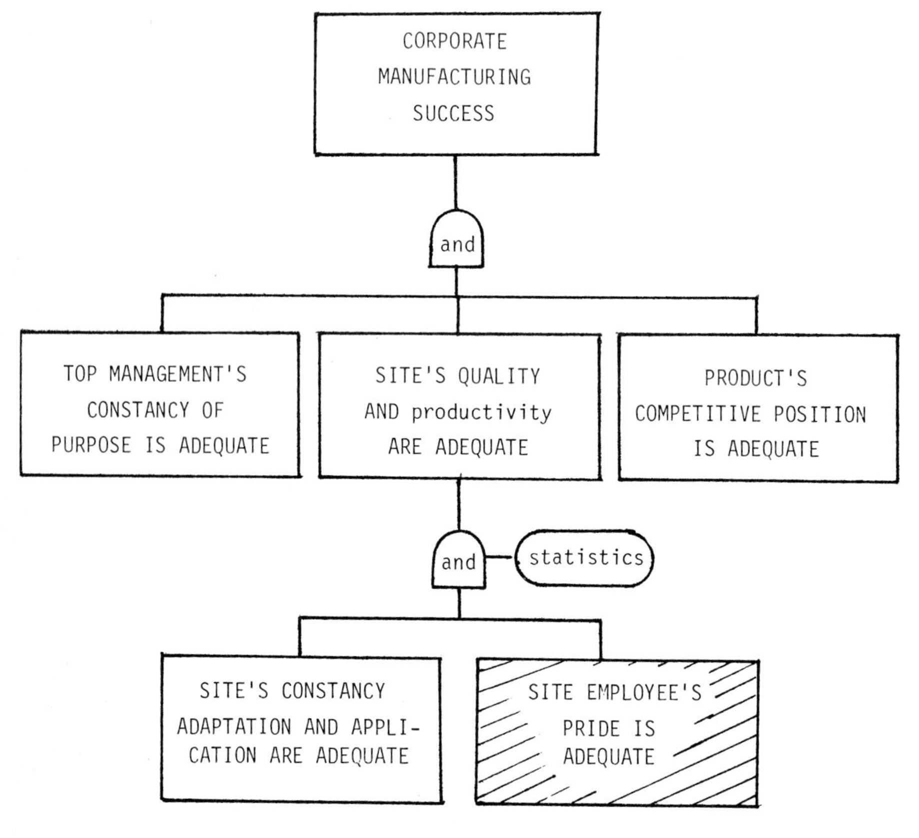

Figure 1

This Logic Tree Suggests The Overall Relationships That Cause Manufacturing Success (Deming 1982).

The case study herein deals with human factors that affect employee pride; ▨ in Figure 1. Reviewed here is one small organization's development of a site system whose management mechanism acts to reverse the trend for frustration and despair resulting from cover-up. The following review summarizes their organization from two perspectives. First, a historical evaluation summarizes *what* things have occurred at the site. Second, the control mechanism of a mechanical puzzle is functionally subdivided to model *how* things are made to work at their location.

WHAT HAS HAPPENED-CASE STUDY:

Infracorp, Ltd. and Lee Laboratories, Inc. located in Petersburg, Virginia manufacture pharmaceutical products to be distributed primarily under the A.H. Robins label. Approximately fifty-five employees are involved with making Robitussin and other fine drugs there. The plant is jointly owned by A.H. Robins Company and Boehringer Ingelheim of West Germany.

At its onset, top management started with a concept that safe work design and practices are profitable. Plant design demonstrates their attitude that safety in today's business climate means reliability which in turn makes for a more productive organization. They designed a very reliable batch process. In addition, the engineering design at the Petersburg location to control chemical emissions in the work place is very reliable. It provides a powered fresh air supply and also exhausts work air from many different key points within the plant. This system is engineered to be tamperproof and all components operate continuously. The design to control fire represents state of the art engineering reliability and insight that only develops after years of applied experience. A 400,000 gallon reservoir and twin pump system can deliver 2,500 gallons of sprinkler water deluge per minute into the manufacturing area.

As a new facility five to six years ago, their top management - with an honest desire to create a nice place to work - began to hire personnel. They established farsighted policies and practices. For example, all employees at the site are paid on a salary rather than an hourly base. There is little organizational structure to create an adversary condition between employees-technicians, supervisors, and managers. They provide very understandable instructions called batch sheets to detail major tasks stepwise for employee guidance. A word processor is used to ensure that the same process needs (add acid, caustic, water, etc.) for different major tasks/batch sheets are covered by the same instruction. The batch sheets also include special safety instructions for employees. This is a near perfect way to facilitate the conscientious effort from inexperienced technicians toward becoming skilled chemical operators. The end result of these human factors and procedural endeavors is to minimize the potential for personal error, thus minimizing the potential damage to personal pride.

Their toxicology, safety, industrial hygiene, and medical subsystems are also developing along the lines to reduce the risk for personal error. For, example, they developed an innovative three-tiered approach to acquire just the right amount of toxicological information which they share. Employees know when, why, and how to protect themselves. Employees conduct electrostatic tests as a precaution when transferring potentially explosive dusts. Supervisors educate other employees in muscle memory so that conscientious employees self-regulate safe lifts themselves. All site employees can also measure environmental stresses with expert guidance, conduct job safety reviews, and discuss the findings freely. Employees use meters to determine their pulse rates after physical work tasks to help educate eager people for a safe work pace. They have implemented an appropriate medical examination to help assure employees that they are not troubled by insidious work stresses - chemical or physical. Epidemiological comparison is made of the chemical exposure dose and biological responses of exposed employees with those not exposed - to search for any unexplained differences. No significant differences have been found except among smokers.

However, they do something about it when at times a risk seems borderline acceptable. For example, special shock absorbing seats have been installed in a fork truck to lessen unwanted employee stress. This was accomplished with a team approach which involved the entire materials handling department. Employees are expected to, and do initiate corrective work orders before catastrophic events occur. Supervisors have conducted select chemical analysis upon employees' exhaled breath at the end of a shift - both to educate everyone and to check that the unexpected was not happening. It wasn't. Recently, they have instituted a series of programs to enhance everyone's understanding about the process. This knowledge encourages optimum decision making on everyone's part. The commonality among these programs is a high level of employee participation and pride. The end result is reduced risk for personal error.

Everyone also has learned to accept the fact - and to be reasonably tolerant - that employees are only human and make mistakes no matter how well the work situation is positioned. Continual effort is made to develop systems to minimize the risk from personal error. System tree analysis techniques are used by a number of employees to remedy, but mostly to prevent system failure before it can cause personal mistakes. Recently, everyone in site management took part in a Decision Making Course to enhance voluntary participation by all employees.

Their managerial philosophy includes an honest effort to set an efficient mixture of the elements for risk control into place. They want the following six elements for management control of risk - (1) executive commitment, (2) documentation of all loss experience, (3) accountability for results and responsibility (Taylor 1980), (4) measurement of all stresses (Taylor 1979), (5) audits of what is going on and (6) communications to keep everyone aware of hazards and what is happening - in a proper organizational position to control loss within acceptable limits. However, they go further and also want everything oriented in a way so that most loss is prevented. To accomplish this they work to make everyone "actually look and feel good." They are systematically orienting the foregoing elements for risk control management toward a system to upgrade employee skills and to enhance intra-site personal relationships and employee pride. The total comprehensive and systematized effort to prevent loss at the site is captioned under the term system safety. The end result is to reduce the probability for human error. An overview of this system is presented in the Table on the following page.

The site has already benefited greatly from the foregoing program. For example, (1) the modest rates for the frequency of personal injury and property damage at the end of the first year of operation are steadily declining, (2) injury compensation cases are nil, (3) certain insurance rates have also been reduced recently, (4) the amount of product rework has decreased, (5) pride in workmanship is increasing as management constancy increases, and (6) productivity has increased significantly.

To summarize this case study evaluation: *most employees - while at work - almost always - prevent - (if it's "managed" to be the easiest way to look good) - damage - by positioning risk management in accordance with human factors and orienting for system safety.*

1. SAFETY POLICY: A comprehensive statement to establish (1) the site goal for loss prevention in terms that relate to the site hazards and (2) to broadly outline supervisory as well as the other employees' responsibility to recognize, evaluate, and control loss.
2. SUPERVISORY EMPLOYEES SAFETY RESPONSIBILITY: To provide commitment and enforcement to recognize, evaluate, and control loss involving:
   A. administrative policy implementation and cost control
   B. property safety
   C. raw materials; their availability, cost, quality, and safety (purchasing)
   D. product quality and safety
   E. environmental safety
3. MANUFACTURING, MAINTENANCE AND DESIGN EMPLOYEES SAFETY RESPONSIBILITY: To conduct/design jobs and report risks to recognize, evaluate, and control loss. (Maintenance work requests and Pin-Point HARM, Taylor 1981).
4. SAFETY OBJECTIVES: Measurable performance tasks for each employee to recognize, evaluate, and control loss.
5. SAFETY EDUCATION: To provide appropriate expert insight and training for all employees to recognize, evaluate* and control loss through:
   A. on site and company expertise and job safety reviews
   B. outside expertise (audits, reviews, evaluations, and tox information)
6. SAFE WORKING CONDITIONS: To provide appropriate internal work practices such as batch sheet information and when appropriate, internal and external reviews to recognize, evaluate* and control imminent loss. (To include job safety reviews, problem tree analysis and emergency planning as appropriate).
7. SAFETY SURVEILLANCE: To recognize, evaluate* and control loss resulting from long term exposure to insidious hazards by:
   A. monitoring personal exposures to physical and chemical agents
   B. preplacement physicals and appropriate periodic health exams
   C. comparison of exposure doses with biological responses
8. SAFETY RECORD KEEPING: To document maintenance reliability, injury trends, and the foregoing loss prevention program elements for employee pride and an improved insurance rate; to detect loss trends early and for compliance use.

*Loss potential is evaluated as negligible, marginal, critical or catastrophic

Table

Employee Loss Prevention Elements To Achieve An Industrial Orientation For System Safety

While the foregoing chronicle is useful evidence about an effective organizational system - it nevertheless doesn't explain how safe and productive behavior was, and is crystalized in the work place. A description of how this is accomplished follows.

HOW THINGS HAPPENED-SOLUTION CONTROL:

There are three major related factors at work within the developing organizations at Infracorp, Ltd./Lee Laboratories, Inc. The three factors interrelate with each other to reduce the probability for human error and consequently the "need" to cover-up. The first is to establish or position the preceding risk management elements within the organization. The second is to systematically orient these program elements to maximize the personal skills and effective intra-personal relationships (human factors and personal pride) as in the Table. The third is a mechanism to crystalize employee pride and participation from the re-oriented risk management elements. The following discussion compares the similarities between solving the participatory management problem and solving a Rubik's cube. It is important to first position and to then orient risk management elements organizationally. In both situations it is the best approach to solving the problem. The solution for the Rubik's cube is to change the uneven distribution of its six colors to a more uniform whole with a single different color display upon each of its sides. Similarily, organizationally the solution to prevent a jumble of antagonistic employees, production problems, and quality loss is to bring the organization to a uniform whole with management constancy and employee pride in participation.

The cube is most readily solved by using its mechanism to repeat a two-step process again and again; first, by a series of steps to bring a smaller cube into position, without regard for its color match. Second, after the cube is in position, to orient its colors to fit with the color of its adjacent center cube. The puzzle cube, whose smaller cubes are only positioned properly, is only partially solved. It represents a frustrating "misfit" condition for goal oriented employees.

In a similar manner, Japanese business critics infer that many companies are frustrated because employees are not turned or oriented organizationally to make everyone "actually look and feel good." Such companies seem to be frustrated from their full potential for self-esteem, quality, and profit. The same dismal inference applies to companies whose conduct does not act with justice about the probabilities for personal error and employee pride.

The foregoing is all well and good. But what is the mechanism for how Infracorp, Ltd./Lee Laboratories, Inc. is making it work? What is the magic third catalyst-type mechanism to orient organizational position into productive success? Let's examine the mechanism of the puzzle cube in Figure 2 on the following page. Each of the puzzle's smaller cubes appear to move about. In fact, only the cubes on the corners and edges actually move. The center side cubes are fixed and can only rotate in place. This is the key to understanding the control mechanism of the cube. Moreover, its mechanical mechanism is a model for the organizational mechanism to crystalize employee pride, productivity, and safety with constancy of management purpose as the catalyst. The right side of Figure 2 shows that the six center puzzle cubes are secured to six independent inner arms. Each arm can be infinitely twisted, yet it stems from a center union point in a locked relationship.

Infracorp, Ltd./Lee Laboratories, Inc. has the capacity to twist risk management from the incompleteness afforded by organizational positioning alone because its center union point - its top management - is perceived by all employees to be completely honest. With constancy (Figure 1) and systematically (Loss Prevention Table), the mechanism of this honesty reaches all employees in a fair and just manner. This conduct removes barriers to employee pride. Consequently, most employees make an honest effort to not cover up personal errors but to orient themselves by making a just contribution to their organization. This, in turn, expands employee pride.

To summarize solution control: *most employees - while at work - almost always - honestly - participate - by positioning risk management and orienting for system safety.*

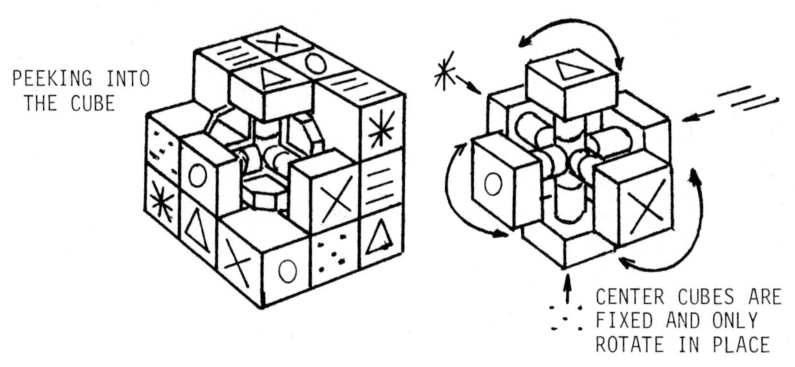

Figure 2

Mechansim Of The Cube

CONCLUSION:

In closing, it is important to present a few words about the timing for when "what things" were initiated at the site. It was and continues to be most effective if things are done one at a time. And done only when it is apparent to employees with expert advice that the extra effort necessary to accomplish the "what's" is justified. In essence, employees gradually recognize that their extra effort to control their destiny makes everyone look good. They are learning that loss prevention is overall the easiest way.

Infracorp, Ltd./Lee Laboratories, Inc. are hard at work applying human factors knowledge and using a system approach with honesty and justice to orient the elements of risk management. They continue to form loss prevention systems to avoid human process and design errors - for optimum employee pride, quality, competitive position, and everyone's peace of mind! The site's organizational design and management is better adjusting to maintain employee pride in their work. They are also constantly at work to make the work place surroundings technically capable of preventing normal human oversight and neglect. This is accomplished evermore as the work place is adjusting to the employees. Such design and adjustment minimize employee blunders and the need for cover-up. For the greater part, they are bypassing "cover-up" with its frustrations about why things don't work and showing how to make things really work effectively.

REFERENCES:

1. Deming, W.E., Quality, Productivity, and Competitive Position (Cambridge, MA, 1982)

2. Taylor, H.M., Occupational Health Management - by Objectives. Personnel 57, (1980) 58-64.

3. Taylor, H.M., Chemicals in the Air; How to Determine Exposure Dose, Professional Safety 24-8, (1979) 26-28.

4. Taylor, H.M., Pinpoint Harm, Hazard Prevention 17-4, (1981) 12-11,32.

BIOGRAPHICAL SUMMARY:

Henry M. Taylor is president of R.E.C., Inc., a management consulting firm located in Chester, Virginia, that specializes in a system approach to evaluate risks and control safety, health, and environmental hazards - particularly in the work place. Mr. Taylor has been associated with innovative tasks at Allied Chemical Corp., Stanford Research Institute, The Ethyl Corporation, and TVA. He received a B.A. in chemistry and zoology at North Central College and an M.P.H. in environmental science from the University of North Carolina. In the past, he has also served as an adjunct faculty member at Virginia Commonwealth University where he established their advanced course, "Human Factors and Environmental Stresses."

Human Factors in Organizational Design and Management
H.W. Hendrick and O. Brown, Jr. (Editors)

IMPLEMENTING WORK SYSTEM CHANGE[1]

H. W. Gustafson

American Telephone and Telegraph Company
New York, New York
U.S.A.

In the literature, the sole model for implementing work system change is diffusion, i.e., waiting and hoping an innovation will endure and spread. Reliance on diffusion seems a major cause of frustration and disappointment in work system change. Project management is proposed as an alternative, and a case history of success at AT&T is presented. Human factors consultants would appear uniquely qualified to market and support project management.

"What good is nutritious dog food if the dogs won't eat it!" -- AT&T saying

INTRODUCTION

'Work System' Defined

To avoid circumlocution throughout this paper, a name was needed to designate the total complex of factors -- organizational, informational, physical -- that contribute to producing output in a work setting oriented toward rational goals. The name 'work system' was adopted, nothwithstanding that a search of behavioral science and trade journal abstracts turned up more than a dozen other current usages of this expression, including, mainly from the space and ocean sciences, 'unmanned' work system. The present usage appears to subsume most of these others, the bulk of which address components, or subsystems, of a work system, such as organizational structure, content of jobs, production technology, administrative methods, and incentive arrangements.

The term 'sociotechnical system,' which has gained wide currency in the past decade or two [see, e.g., 1], is similar in meaning and may be regarded by many as synonymous. Semantically, however, it is difficult to accept that every element of a work system, as here defined, is classifiable as social or technical in character.

This paper is concerned with implementing change in work systems. Axiomatically, therefore, it is concerned with changes in elements of work systems and will refer to all of these generically as work system changes. This practice is justified the more by the fact of nature that work system

[1]The views expressed herein are the author's and not necessarily those of his employer.

elements are interdependent, so that a change in one is bound to affect the rest. As has been recognized since the 'Hawthorne' studies of the 1920's on telephone equipment manufacturing [2], these interdependencies go typically deeper, and are more abiding, than meets the eye. An illustration from the present-day telephone business is given by Williams [3], in which a mechanization project related to the reporting and repair of telephone troubles unleashed an unanticipated and unwanted chain of consequences.

Diffusion of Planned Change

All work systems change perpetually, and always have. Whereas in times past, however, change tended to be sporadic and regarded as aberrant, today change tends to be continual and accepted as natural. Acknowledgment in the technical literature of the continuous, cyclical nature of work system change is, nonetheless, rare, the work of Mackenzie, Martel, and Price [4] comprising a recent exception. Other writers who discuss cycles of change [e.g., 5] stress the stages within a cycle rather than the periodic recurrence of those stages.

Changes in work systems derive, to borrow the language of Goodman and Kurke [6], either from adaptation or from planning. Adaptation and planning are not, though, a dichotomy, but end points on a continuum. Change ordinarily embodies a measure of each process.

At the adaptive extreme of the continuum, change simply happens, hence cannot be managed deliberately. At the opposite pole of planned change, on the other hand, opportunity for deliberate management of implementation is available. To judge from the literature, however, seemingly never is advantage taken of this opportunity. Instead, implementation of planned change normally goes undirected, same as adaptive change.

In the dominant paradigm, a work system innovation is designed and tested experimentally, then handed to operating management in the hope the design somehow will 'diffuse' [7, 8, 9, 10, 11, 12], diffusion here referring both to maintenance of the experimental trial over time and to spreading of the innovation elsewhere. To facilitate diffusion, features to encourage faster and broader adoption may be incorporated in the design of the innovation, but implementation all the same is left to happenstance.

Diffusing work system change bears analogy to thrusting a product on the market hoping it will sell itself. Just as such strategy seldom succeeds in marketing, so is it unavailing generally with work systems, the dearth of evidence for the effectiveness of diffusion being notorious [7, 11, 13, 14, 15, 16].

Further, the idea of diffusion on the heels of successful experimentation appears paradoxical in any work setting where rational goal seeking is encouraged. For if a work system change is demonstrably advantageous, one would think the people in authority, rather than await diffusion, would strive vigorously to implement the innovation everywhere as immediately as possible.

Were the purpose, say, to construct new facilities or introduce new accounting practices, it is unlikely the executives in charge would contemplate implementing the design specifications by indeterminate, uncontrolled means. More probably, they would appoint project directors to steer the designs into realization and to be held accountable for doing so efficaciously -- whereupon one would hear talk not of experimentation and diffusion, but of milestones, critical paths, delivery dates, get-ready activities, cut-over schedules, slippages, overruns, and related managerial concepts derived from established engineering practice.

Yet in the large and burgeoning literature on work system innovation, the reader will seek in vain for explicit mention of how to oversee and manage the change process. In explorations into the causes of success and failure, nowhere is suspicion pointed at how the parties responsible for a work system innovation directed and controlled its implementation or how they organized themselves to accomplish that end. Instead, the causes of success and failure invariably are said to lie in the nature of the innovation, the adequacy of its design, the goodwill and receptivity of the personnel affected, the organizational climate and culture, or miscellaneous environmental influences.

Vagueness in Accountability and Guidance

Although the literature refers frequently to executives in authority, i.e., 'sponsors,' and to consultants, i.e., 'change agents,' as well as alluding here and there to 'managers,' 'practitioners,' and 'analysts,' it is rarely evident which of these individuals, if any, is accountable for results; and never is there discussion of a project manager assigned specifically to plan and supervise the work of implementation [see, e.g., the extensive literature reviewed in 5]. In occasional instances where titles like 'project manager' or 'project leader' appear [e.g., 17, 18, 19], the context makes it clear that the role intended is either nominal or pertains to design and trial activities only.

It is tempting to conjecture, consequently, that either the executive in power or the consultant is supposed to manage implementation personally. Such conjecture could be hazardous, however, in light of assertions such as the following: "The analysis, preparation, and implementation of a sociotechnical design is...the property of no individual or set of individuals; it belongs to the members of the organization whose working lives are being designed" [20, p. 791]. If these words can be taken at face value, it would seem that under the prevailing doctrine of experimentation and diffusion, everybody -- or what amounts to the same, nobody -- is supposed to be in charge.

Additionally, if someone were designated to manage a work system change and were to solicit advice from the literature on how to proceed, it is not unlikely the guidance obtained would read more or less this way: "Exactly how the change process should be undertaken is not very clear. At this stage, it is probably best to study the models presented by experts, to read reports of actual interventions, and to develop an internalized model that is acceptable" [21, pp. 5-6]. If such advice epitomizes the state of the art in implementation, it seems small wonder that the history of planned work system change is a saga largely of frustration and disappointment.

PROJECT MANAGEMENT

The Project Manager Role

If not already obvious, the thesis of this paper is that to afford planned work system change a reasonable chance of success, the change process, just as every purposeful endeavor, needs to be pursued in systematic fashion. Further, if a change is of significant scope or complexity, it needs to be directed by a project manager dedicated to the assignment on a first priority basis.

By project manager is meant an individual formally appointed to assemble, coordinate, and supervise application of whatever resources are necessary to carry out an executive decision. The charge to a project manager is not to perform particular duties, but to bring about a desired result. Accordingly, anything that may bear on the result falls within the project manager's purview. If, for instance, resource shortages or prior constraints threaten the outcome, it is up to the project manager to find alternate resources or loosen said constraints.

With negligible exceptions, an example of which is noted below, a project manager's authority arises from the priority accorded his or her project, not from the individual's position in the power hierarchy. Hence, the best project managers seldom are rigid authoritarians but, rather, felicitous blends of planner, coordinator, socializer, technician, and scrounger.

As a rule, neither sponsors nor consultants can be effective as project managers. The sponsor, or executive in charge, already has a full-time job running current operations and, for that very reason, suffers conflict of interest between day-to-day exigencies and the long-run demands of the change project. Because their priorities conflict in this way, sponsors who try to serve as project managers, apart from lacking time to do so, all too often end up sacrificing the future to the present.

Consultants likewise are handicapped as project managers in that, being alien to the organization, they do not participate in its politics and usually do not know the administrative ropes. This may be less true of internal than external consultants, but both types are prone to conflict of interest. Consultants qua consultants aim to work themselves out of a job as quickly as possible, whereas project managers are tied to their projects as first priority responsibilities for the duration.

Possible Exceptions

The much cited Harwood project [e.g., 13, 22] may illustrate an exception to the rule that a sponsor cannot manage a change project effectively. Rensis Likert, in his Foreword to Marrow, Bowers, and Seashore [23], attributes the success of this effort to application of a total system development approach: "Changes in all major factors such as the management system, the technology and engineering, leadership training, communications, decision-making, compensation, earnings, and union-management relations were undertaken in a planned, coordinated manner and in an orderly sequence. The result was an integrated whole consisting of compatible parts" (p. ix). What Likert omits to explain, however, is how this extraordinary coordination was achieved. The answer, it seems reasonable to hypothesize, is that A. J. Marrow himself served as project manager and, in his capacity as chairman of the board of the Harwood company, either found time or made time to plan and oversee the implementation. So long as he or she is not at once the chief operating official of the firm, give me the chairman of the board as project manager every time!

Bardach [24] reports a related example, from the public arena but similar in principle to the Harwood case, where a legislator sponsoring a bill exhibited evidently unique behavior in pursuing and bringing about implementation of the bill subsequent to enactment. The role of the legislator is summed up thusly: "We have...said of him...that he 'oversaw,' 'intervened,' 'followed up,' 'corrected,' 'expanded,' 'overhauled,' 'protected,' and so forth" (p. 31). While Bardach summarizes these activities as 'fixing' a game people play, what they add up to more precisely is de facto project management.

At least one instance also is on record where internal consultants performed productively as project managers to effect work system change [25]. In this case, the consultants belonged functionally to an organizational unit that made them available to the client on a long-term basis, thereby enabling them to participate directly in, perhaps supervise, implementation. To carry out such office, however, goes beyond the conventional consulting role. The situation depicted is more akin to the organizational arrangements businesses commonly use to bring new products to market, identified by Lorsch and Lawrence [26] as 'coordinating departments,' 'liaison individuals,' 'short-term project teams,' and 'permanent cross-functional coordinating committees.' Where such arrangements -- which comprise, of course, the very stuff of project management -- are routinely operative, the coordinators and liaison people are thought of as mainstream participants, not consultants.

To summarize, it seems that in singular circumstances, a sponsor or consultant may serve with effect as a project manager, but the odds against such an eventuality appear long.

Additional Points

In discussing experimentation and diffusion, Cole [7] quotes the former labor leader, Irving Bluestone, as saying: "I don't see the need for all this complex measurement and evaluation. If the local union leader and the plant manager like the new work arrangements, then it (sic) works" (p. 197). Taking Bluestone at his word, it may be conjectured that such agreements between union and management tend to lead as a matter of course to appointment of implementers -- in effect, project managers -- to ensure that the new arrangements become securely established. Inasmuch as project management is employed widely for marketing and other purposes -- indeed, books on the subject fill whole shelves of libraries -- it is not unlikely that ad hoc project management is fairly popular in work system change, if unheralded in the literature.

Cummings [8] likens work system change to agricultural change, observing that after an innovation in agriculture has proved beneficial, "...the results are disseminated to interested farmers by county extension agents who serve as change agents for farming practices" (p. 170). In fact, however, no such analogy pertains. County agents, like project managers, have the assigned duty of causing new practices to be implemented, whereas no counterpart role exists in models of work system change -- although some writers [27, 28] have proposed such a function for university extension agencies.

The absence of reference to project management in writings on work system change is not a little ironic in that project organization is among the work system innovations that consultants make available to clients [e.g., 29, 30], drawing to mind the parable of the shoemaker's children.

It is not, by the way, that the literature lacks advice on the activities to perform and human relations techniques to follow in attempting work system change. To the contrary, the seeker of such information runs severe risk of early satiety [for starters, see 8, 11, 13, 20, 31, 32, 33, 34, 35, 36, 37, 38, 39, 40]. The problem is that no one explains for whom all this good advice is intended or how it is expected to be carried out.

WORK SYSTEM CHANGE AT AT&T

The Support Framework

During the 1970's, some 150 planned work system change activities were undertaken at AT&T (American Telephone and Telegraph Company) on behalf of the operating telephone companies of the Bell System (the quondam U.S.A. telephone giant headed by AT&T, sundered since into multiple unaffiliated parts). The purpose of the changes was either to introduce new technologies in the workplace or to inaugurate new concepts of marketing and service. All changes entailed creation of new work organizations or major modification of old ones. Each change involved design of a prototype organization, such as a billing center or a repair center, to be followed by replication -- as it were, cloning -- of the prototype anywhere from dozens to hundreds of times throughout the operating companies.

Half these undertakings were adjudged too small, too commonplace, or too limited in impact on people to require formal project management. The other 75 activities, which, incidentally, addressed work changes affecting over 450,000 employees, were assigned to the charge of 75 project managers, proven performers in field operations but inexperienced in project work. The following support framework was provided:

1. A master plan showing the project managers how all the 150 change activities interfaced;
2. A master schedule showing phases and current progress of the 75 project-managed efforts;
3. A guidebook suggesting how to organize a project team, what work steps to carry out, and what kinds of product to generate;
4. A formal budgeting process enabling teams to be assembled and work executed;
5. Suggested technical methods for conducting individual work steps (e.g., how to optimize span of control, how to estimate staffing requirements);
6. Elective training courses in project management (e.g., PERT, cost-benefit analysis);
7. Elective consultative assistance;
8. An advisory committee for every project consisting of the project manager's functional boss and all peers of that boss having subordinates involved in the project;
9. For most projects, a steering committee composed of the functional bosses of the advisory committee members;
10. For projects requiring unusual resources, an executive council composed of the bosses of the steering committee members;

11. For every project, one or more volunteer operating companies to be the first users of and testbeds for the products of the project team;

12. Finally, each project manager was urged to form one or more user committees of operating company personnel representing the project's customers.

This support framework by no means sprang into being overnight, but evolved in response to demand during a period of approximately five years. Neither did the projects all commence at once. In the beginning, the erstwhile Bell System having been highly functional in organization and mentality, considerable indifference to project management, if not to say resistance, was encountered. The idea caught on only after some carefully nurtured pilot projects had demonstrated the easy acceptance of project management by all parties concerned in addition to its power to effectuate results. Still, in less than two years from inception of the pilot efforts, project management was the accepted approach at AT&T; and by the end of the third year, newly assigned project managers blithely assumed that project management of work system change was a venerable company tradition.

The Success Orientation

Characteristic of Bell System culture, the project management ideology that emerged entertained no concept of experimentation or of success and failure. Success was a foregone conclusion; failure was not an option. Implementation became known as 'controlled introduction.'

Projects could be cancelled if they turned out cost disadvantageous, or could be merged with others in the event of significant overlap, or could be modified extensively to correct mistakes or grasp unforeseen opportunities. Moreover, individual operating telephone companies could elect not to implement a given change; and if enough such decisions were registered, the project would be terminated. Also, no project manager was expected to be infallible in judgment or to meet all objectives all the time. Errors, slippages, and overruns were considered normal though regrettable. Yet no thought was harbored that a project in any sense could fail. The raison d'être of project management was, after all, to eliminate obstacles to success.

By the time divestiture of the Bell operating companies was announced in January, 1982, to be effective January 1, 1984, the 75 project-managed work system change activities had been reconfigured and thereby reduced somewhat in number, but all originally planned changes remained in flower and were making progress acceptable to management. Although systematic evaluation was not done in most cases, and a number of projects were only verging on implementation at the time, no instance of implementation denial or implementation disaster had occurred as of divestiture. In four projects that were evaluated with some precision, both the economic advantages and service benefits to customers exceeded expectation, and the morale of employees was either sustained at former levels or improved.

A Case Illustration

One of the evaluated projects focused on correcting the negative conditions outlined by Williams [3], cited earlier, when a mechanized customer record system was interjected summarily into the telephone repair process. Among the adverse consequences of the new system was a strained relationship between the people who receive trouble reports from customers and those who do the repair. No fully satisfactory solution to this problem ever was found; the best that could be done was to help the opposing sides recognize their difficulty and accomodate it. However, every other deficiency described by Williams was corrected by ferreting and eliminating the root cause, after which the redesigned work system was implemented in 20 of 22 operating telephone companies. Before-and-after surveys using control groups showed significant improvements in measures of customer satisfaction, and likewise with measures of employee job satisfaction and quality of work life.

In this instance, conditions at the outset were sufficiently dire that any corrective effort might have diffused and improved matters noticeably without project management. The same is unlikely, however, of the remaining change projects. In general, the work system changes under discussion were introduced to take advantage of new technologies or business opportunities, not to remedy existing maladies.

GENERALIZABLE LESSONS

Essential Support Elements

A support framework for project management as elaborate as that above can be sustained only by an enterprise having a sizable administrative staff, hence lies beyond the means of many organizations. On the other hand, the framework was erected largely because the available project managers were short on experience. In all probability, more surefooted project managers would not need such comprehensive support, and even might reject it. For that matter, the project managers at AT&T did not take long to begin perceiving the support framework as confining and to start moving in directions of their own.

The support framework elements believed indispensable, even with veteran project managers, are master planning, formal budgeting, and coordinated oversight. Whether one or many work system innovations are undertaken, it seems imperative to inform everyone as to what the innovations are about and what benefits they are expected to bring in what time frame (the plan), as well as to predetermine how the costs will be estimated and approved and who will bear them (the budgetary process). Vagueness or indecision in either area can lead only to confusion and controversy.

In addition, hope of success appears faint unless the project work is overseen by interdisciplinary advisory and steering groups representing all affected constituencies. The utility of such groups in facilitating work system change has been noted more than once [11, 18, 41], and virtually all writers on change appeal for participatory management of the process. Project management affords an appropriate vehicle not merely for organizing and administering participation by consultative, oversight, and user groups, but also for ensuring that the contributions received are duly reflected in the project's end results.

Institutionalization

The notion of institutionalizing a work system innovation to make it endure and spread is emphasized by Goodman, Bazerman, and Conlon [42], and again by Goodman and Dean [14]. A contradiction inheres in this suggestion, however, in that institutionalization no sooner begins than a new change comes along to drive out the previous one. Thus, if an innovation becomes too strongly entrenched, it may create undesirable resistance when the time arrives to replace it. Even social and cultural aspects of work systems cannot survive forever. It would seem more logical, consequently, to stress the temporary nature of most changes so that receptive attitudes will exist for subsequent ones.

Accordingly, the counter-suggestion is offered that what is wanted is to institutionalize not innovations themselves, but the means of dealing with recurrent innovation; and it is believed this is exactly what the AT&T support framework for project management helped to do. The framework helped people to grow used to the idea of change, to perceive a rational approach to coping with change, and to develop favorable views of its probable success and value to the organization.

The Power Base

Cole [7] observes that to be an effective agent of change, a consultant needs a power base. So does a project manager. Universally in the literature, the requisite power base for innovation is said to be the sponsor of the change together with other senior executives. It was learned quickly at AT&T, to the contrary, that the essential base of power was the middle managers who direct operations in the field and are in a position tactically to make things happen or, conversely, stonewall them. Though true that senior management must support any innovation, it was discovered that the way to secure such endorsement was through the middle managers destined to end up in the vortex of the change. These are the people who, after all, earn the top executives their bonuses and enjoy their confidence, and a large part of whose expertise lies in the care and feeding of upper management. Therefore, once middle management was enlisted in the cause, it was found that no further attention had to be given to pursuing the backing of senior executives.

Besides, to rely on the patronage of one or more top managers would seem to particularize more than institutionalize. Among other considerations, the power base for change, if lodged in the sponsor, disappears whever the sponsor does. Moore and Ross [39] comment, e.g., on the fragility of Scanlon plans in the face of executive turnover, while Crockett [43] describes a sweeping work system change in a government bureau under his direction that was abandoned after his departure. The project management approach, on the other hand, when adopted as the institutional way of dealing with change, can survive turnover among project managers very readily.

Implementation Aids

Among already mentioned features of the AT&T support framework that helped project managers minimize stonewalling (better known in the whilom Bell System as 'gracious noncompliance') were the advisory, steering, and user committees, the volunteer companies to serve as first users, and a pervasive positive attitude of implementation inevitability. Some further devices that worked to ensure timely and appropriate implementation were as follows:

1. Establishment of a cut-over (implementation) team in each company committed to implementing a given innovation, the team directed

by a project manager from that company;

2. Early introduction of work system changes at one or more so-called model locations in each implementing company, to allow firsthand familiarization by field personnel and to set a pattern for the rest of the company;

3. A certification program in which, at the option of the implementing company, the AT&T project team would audit the company's model locations with a view to certifying them as in compliance with design;

4. An operations review program in which each operating company would regularly audit its own installations subsequent to AT&T certification of its model locations.

The point to this seemingly ponderous regimentation was neither to promote standardization for its own sake nor to nourish the headquarters bureaucracy. In the Bell System -- as, one surmises, in most organizations -- the operating units would accept bureaucratic uniformity and direction only to the extent they felt these were beneficial to their own self-interest. They were articulate, well informed customers who exploited the project management approach to change, not the other way around. Whenever AT&T tried to push project management to the point of overkill, the companies were quick to react with their time honored response of gracious noncompliance. The important lesson from standardizing and regimenting an approach to work system innovation is that it helped instutionalize the change process. For approximately a half decade prior to Bell System dissolution, that process prevailed and, to all appearances, proved both successful and advantageous.

CONCLUSIONS

The Engineering Approach

What has been dubbed project management above was known more commonly at AT&T as 'the engineering approach,' for most of its prime movers had backgrounds in engineering and found a disciplined, coordinated process vital to their mental health. Otherwise, project management was called the 'hard-nosed' approach, in contradistinction to the human relations methods of organization development, sometimes referred to irreverently as 'touchie-feelies.' There is, however, no inherent conflict between the two orientations. Organization development techniques can be employed liberally within a project management framework, and may be quite helpful in facilitating implementation. A number of project managers at AT&T favored such techniques and engaged consultants, mostly internal, to assist in applying them.

As a result, the present paper does not conclude against any extant tools or methods of change whatever. Project managers need all the tools and methods they can lay hands on. The one thing concluded _against_ is the idea of diffusion, herein depicted (perhaps not altogether fairly) as waiting and hoping for implementation to happen.

What is concluded _for_ is an engineering approach to implementing change, of which the principal ingredients are:

1. A specific person is in charge of and accountable for bringing implementation about; and if such person has other responsibilities, this one carries first priority;
2. A plan exists for all to see showing how the institution is expected to operate after the change has been implemented;
3. The zeitgeist recognizes that change is cyclical and accepts project management as the routine way of handling it;
4. A formal budgeting procedure exists to provide and account for resources to be expended on the project;
5. An attitude of unrelenting positivity exists toward the potential value and success of implementation.

Human Factors Implications

The writer may have misread the AT&T experience, or reading it correctly, may have misconstrued the causes. But if the foregoing account is substantially accurate, favorable opportunities in the field of work system change are indicated for human factors consultants, the majority of whom come equipped with three important prerequisites: a) An engineering orientation, b) Project management familiarity, and c) A repertoire of 'war stories' about implementation failures for use in marketing. Other breeds of professional consultant also have stories about failure but tend to lack the other two essentials. Hence, a human factors background should provide competitive edge in persuading sponsors of the necessity for project management in work system change and in advising them on how to get such an approach underway.

Human factors traditionally has been torn between doing hands-on work and advising others how to do it. In the arena of work system change, the latter posture is strongly recommended. First require sponsors to display enough commitment to appoint project managers and institute at least a rudimentary support framework. Then act as privy councillors -- if need be, alter egos -- to the project managers.

REFERENCES

[1] Pasmore, W. A., and Sherwood, J. J. (eds.), Sociotechnical Systems: A sourcebook (University Associates, San Diego, 1978).

[2] Roethlisberger, F. J., and Dickson, W., Management and the Worker (Harvard University Press, Cambridge, Mass., 1939).

[3] Williams, T. A., Technological innovation and futures of work organization: A choice of social design principles, Technological Forecasting and Social Change, 24 (1983) 79-90.

[4] Mackenzie, K. D., Martel, A., and Price, W. L., Human resource planning and organizational design, in: Mensch, G., and Niehaus, R. J. (eds.), Work, Organizations, and Technological Change (Plenum Press, New York, 1982).

[5] Kimberly, J. R., Managerial innovation, in: Nystrom, P. C., and Starbuck, W. H. (eds.), Handbook of Organizational Design, Vol. 1: Adapting organizations to their environments (Oxford University Press, Oxford, 1981).

[6] Goodman, P. S., and Kurke, L. B., Studies of change in organizations: A status report, in: Goodman, P. S., and Associates, Change in Organizations (Jossey-Bass, San Francisco, 1982).

[7] Cole, R. E., Diffusion of participatory work structures in Japan, Sweden, and the United States, in: Goodman, P. S., and Associates, Change in Organizations (Jossey-Bass, San Francisco, 1982).

[8] Cummings, T. G., Sociotechnical systems: An intervention strategy, in: Pasmore, W. A., and Sherwood, J. J. (eds.), Sociotechnical Systems: A sourcebook (University Associates, San Diego, 1978).

[9] Greiner, L. E., Patterns of organization change, Harvard Business Review, 45(3) (1967) 119-137. (Reprinted in Dalton, G. W., Lawrence, P. R., and Greiner, L. E. (eds.), Organizational Change and Development (R. D. Irwin, Inc., & The Dorsey Press, Homewood, Ill., 1970).)

[10] Rogers, E. M., and Shoemaker, F. F., Communication of Innovations: A cross-cultural approach (2nd ed.) (The Free Press, New York, 1971).

[11] Walton, R. E., The diffusion of work structures: Explaining why success didn't take, Organizational Dynamics, 3(3), 1975, 3-22. (Adapted in Mirvis, P. H., and Berg, D. N. (eds.), Failures in Organization Development and Change: Cases and essays for learning (Wiley, New York, 1977).)

[12] Walton, R. E., Innovative restructuring of work, in: Pasmore, W. A., and Sherwood, J. J. (eds.), Sociotechnical Systems: A sourcebook (University Associates, San Diego, 1978).

[13] Burke, W. W., Organization development, in: Michael, S. R., Luthans, F., Odiorne, G. S., Burke, W. W., and Hayden, S., Techniques of Organizational Change (McGraw-Hill, New York, 1981).

[14] Goodman, P. S., and Dean, J. W.., Jr., Creating long-term organizational change, in: Goodman, P. S., and Associates, Change in Organizations (Jossey-Bass, San Francisco, 1982).

[15] Kahn, R. L., Conclusion: Critical themes in the study of change, in: Goodman, P. S., and Associates, Change in Organizations (Jossey-Bass, San Francisco, 1982).

[16] Mirvis, P. H., and Berg, D. N. (eds.), Failures in Organizational Development and Change: Cases and essays for learning (Wiley, New York, 1977).

[17] Bowers, D. G., Scientific knowledge utilization as an organizational systems development problem, in: Kilmann, R. H., Thomas, K. W., Slevin, D. P., Nath, R., and Jerrell, S. L. (eds.), Producing Useful Knowledge for Organizations (Praeger, New York, 1983).

[18] N. V. Philips Europe, Participation: Various ways of involving people in their work and work organization, in: Pasmore, W. A., and Sherwood, J. J. (eds.), Sociotechnical Systems: A sourcebook (University Associates, San Diego, 1978).

[19] Ross, E. H., Human factors in system development: Project team approaches & recommendations, Proceedings of the Human Factors Society --27th Annual Meeting, Vol. II (1983) 974-978.

[20] Cherns, A., The principles of sociotechnical design, Human Relations, 29 (1976) 783-792. (Reprinted in Pasmore, W. A., and Sherwood, J. J. (eds.), Sociotechnical Systems: A sourcebook (University Associates, San Diego, 1978).)

[21] Pasmore, W. A., and Sherwood, J. J., Organizations as sociotechnical systems, in: Pasmore, W. A., and Sherwood, J. J. (eds.), Sociotechnical Systems: A sourcebook (University Associates, San Diego, 1978).

[22] Katz, D., Kahn, R. L., and Adams, J. S., Conclusion, in: Katz, D., Kahn, R. L., and Adams, J. S. (eds.), The Study of Organizations (Jossey-Bass, San Francisco, 1982).

[23] Marrow, A. J., Bowers, D. G., and Seashore, S. E., Management by Participation (Harper and Row, New York, 1967).

[24] Bardach, E., The Implementation Game: What happens after a bill becomes a law (MIT Press, Cambridge, Mass., 1977).

[25] Kobylak, R. F., Human factor engineering - Agent of change, Proceedings of the Human Factors Society--27th Annual Meeting, Vol. II (1983) 561-565.

[26] Lorsch, J. W., and Lawrence, P. R., Organizing for product innovation, Harvard Business Review, 43(1) (1965) 109-122. (Reprinted in Dalton, G. W., Lawrence, P. R., and Lorsch, J. W. (eds), Organizational Structure and Design (R. D. Irwin, Inc., & The Dorsey Press, Homewood, Ill., 1970).)

[27] Beyer, J. M., and Trice, H. M., Current and prospective roles for linking organizational researchers and users, in: Kilmann, R. H., Thomas, K. W., Slevin, D. P., Nath, R., and Jerrell, S. L. (eds.), Producing Useful Knowledge for Organizations (Praeger, New York, 1983).

[28] Petersen, J. C., and Farrell, D. Extending the extension model: New institutional forms for useful knowledge, in: Kilmann, R. H., Thomas, K. W., Slevin, D. P., Nath, R., and Jerrell, S. L. (eds.), Producing Useful Knowledge for Organizations (Praeger, New York, 1983).

[29] Davis, S. M., Lawrence, P. R., Kolodny, H., and Beer, M., Matrix (Addison-Wesley, 1977).

[30] Kingdon, D. R., Matrix Organization (Tavistock, London, 1973).

[31] Beckhard, R., Strategies for large system change, in: Pasmore, W. A., and Sherwood, J. J. (eds.), Sociotechnical Systems: A sourcebook (University Associates, San Diego, 1978).

[32] Beer, M., The technology of organization development, in: Dunnette, M. D. (ed.), Handbook of Industrial and Organizational Psychology (Rand McNally, Chicago, 1976).

[33] Cummings, T. G., and Srivastva, S., Management of Work: A socio-technical systems approach (Kent State University, Kent, Ohio, 1977).

[34] Dalton, G. W., Lawrence, P. R., and Greiner, L. E. (eds.), Organizational Change and Development (R. D. Irwin, Inc., & The Dorsey Press, Homewood, Ill., 1970).

[35] Dalton, G. W., Lawrence, P. R., and Lorsch, J. W. (eds.), Organizational Structure and Design (R. D. Irwin, Inc., & The Dorsey Press, Homewood, Ill, 1970).

[36] Emery, F. E., and Trist, E. L., Analytical model for sociotechnical systems, in: Pasmore, W. A., and Sherwood, J. J. (eds.), Sociotechnical Systems: A sourcebook (University Associates, San Diego, 1978).

[37] Gruber, W. S., and Niles, J. S., How to innovate in management, Organizational Dynamics, 3(2) (1974) 30-47.

[38] Huber, G. P., Ullman, J., and Leifer, R., Optimum organization design: An analytic-adoptive approach, Academy of Management Review, 4 (1979) 567-578.

[39] Moore, B. E., and Ross, T. L., The Scanlon Way to Improved Productivity: A practical guide (Wiley, New York, 1978).

[40] Nutt, P. C., Implementation approaches for project planning, Academy of Management Review, 8 (1983) 600-611.

[41] Nolan, R. L., Managing information systems by committee, Harvard Business Review, 60(4) (1982) 72-79.

[42] Goodman, P. S., Bazerman, M., and Conlon, E., Institutionalization of planned organizational change, in: Staw, B. M., and Cummings, L. L. (eds.), Research in Organizational Behavior, Vol 2. (JAI Press, Greenwich, Conn., 1980).

[43] Crockett, W. J., Introducing change to a government agency, in: Mirvis, P. H., and Berg, D. N. (eds.), Failures in Organization Development and Change: Cases and essays for learning (Wiley, New York, 1977).

Human Factors in Organizational Design and Management
H.W. Hendrick and O. Brown, Jr. (Editors)

WORK SYSTEMS AS A LARGE SCALE SYSTEMS DESIGN PROCESS

This paper summarizes the theories, research, and work in organization design and change developed by a group at AT&T headquarters during the period 1977-1983. This "Work Systems" effort was performed under the leadership and guidance of M. B. Gillette and H. W. Gustafson in support of various AT&T groups. The products which resulted were successfully implemented throughout the Bell Operating Companies. The group changed personnel occasionally, and included for various periods of time the following members (listed alphabetically): R. S. Didner, F. L. Ficks*, S. Gael, L. T. Martin, J. V. Pilitis, D. A. Ryba, M. Saunders, J. W. Suzansky*, and R. J. Voegele, working with a host of project managers and taskforce members who contributed to the success of the Work Systems approach.

Organization design has historically been performed from perspectives of hierarchical arrangement, communication flow, quality of worklife, political expediency, and so on. Most techniques have treated organization theory as a unique field that requires some special methods and approaches. This paper proposes that by viewing organizations as "work systems", their design and implementation can be successfully completed by the same types of analysis, documentation, and project management that have been so successful with other systems efforts.

The use of the term "work systems" is not new; a literature search uncovered the term's use in hundreds of applications ranging from undersea exploration to unmanned space projects. For consistency, the following definition of a Work System is offered:

> An interrelationship of workers, their tools, and processes intended to convert some input into an output, operating in a specified environment.

There is a clear similarity between this definition and classical definitions of "systems", such as those provided by Hall and Fagen (1956): "a set of objects together with relationships between the objects and between their attributes", Gagne (1962): "system, then, implies a goal or purpose, and it implies interaction and communication between components", and Morgan, Cook, Chapanis, and Lund (1963): "It is the concept of a group of components designed to serve a given set of purposes". This similarity is intentional, to enable bridging the conceptual gap between traditional approaches to organization design and change and the more engineering and project management oriented applications successfully applied to systems. Simply stated, organization design is systems design, and can be subject to the same development techniques and management structures. What follows is a description of how generic system design and development methodologies were applied to organizations through the "work systems" approach at AT&T.

* Coauthors and presenters of this paper.

PROJECT MANAGEMENT

Perhaps the most valuable contribution of systems methodologies to the work systems approach was the manner in which major system changes were carefully planned and organized. The work systems approach followed this example, closely tieing organization change to strategic business planning. Since the details of this structure are elaborated by Gustafson (1984) elsewhere in these proceedings, only a brief summary is provided here of the key ingredients:

a. All major work system changes were derived from and integrated with long-range business operation plans.

b. A project management framework was developed to support and track all major work activities.

c. A full-time project manager from the end-user department was assigned to each major work system to oversee its design, development, and implementation.

d. The process and structure were developed to require a high level of participation by end-users in all activities.

WORK SYSTEMS DESIGN PROCESS

Again borrowing liberally from earlier systems design efforts, an eight-step work systems design process was developed. Each step is completed by a project team of designers, end-users, consultants, and subject matter experts assembled and coordinated by the project manager. While described sequentially below, several steps are often completed in parallel and development is generally an iterative process, with the outcome of a later step frequently requiring revisions to earlier ones.

1. OBJECTIVES

Using as input a broad description of the functions to be performed by the organization, the project team begins its analysis to determine the business values of the organization, the work functions that will have to be performed to realize these business values, and the specific, measurable objectives that can be assigned to this organization. Project planning is accomplished at this time, resulting in projections of budgets, time schedules, and personnel requirements. The results of this effort are a clearly stated mission, an assessment of business indicators that will be impacted, an estimate of potential payback, a proposed location of the organization in the hierarchy, and an estimate of penalties to be paid if the organization change is not implemented. Although an organization change may have initially appeared to be worthwhile, this analysis may demonstrate that payoff is insufficient to justify the change.

2. INTERFACES

The interface analysis is done to prevent duplication of work with other organizations and to specify the outputs and inputs required.

Each of the interfacing organizations is analyzed to determine media requirements, expected delivery times, and priorities for inputs and outputs, and to establish a common work vocabulary. "Contracts" are frequently arranged which become part of the implementation plan.

3. JOB AND ORGANIZATION DESIGN

With information about the inputs to the organization and outputs required of the organization, specification of the tasks required to convert these inputs to outputs can begin. Information is gathered from support system designers and job incumbents performing similar jobs. Standard techniques of job and task analysis are used to clearly specify all of the tasks that are required. These tasks are grouped into work functions and then jobs for both management and nonmanagement employees. Consideration is given to such factors as accountability, potential performance measures, minimal duplication, clear interfaces with other jobs, and job satisfaction. Job descriptions, organization charts, lists of skill and knowledge requirements, work methods, force management procedures, flowcharts showing job relationships, and any other materials necessary for the day-to-day operation of the organization are prepared.

4. ORGANIZATION STAFFING AND SIZING

If the new organization is to be maximally successful, the staffing (the number of people in each job) and the sizing (the number of times the organization will have to be replicated to accomplish the work assigned to a given geographical area) must be optimized. However, if this organization is to be replicated a number of times under different conditions, or if the operating conditions frequently change, an absolute number will in some cases be inappropriate and in others quickly become obsolete. Therefore, the purpose of this activity is not to provide a specific number of people or organizations, but to provide a procedure that will make this determination when local conditions are considered. Two methods for developing staffing and sizing procedures are presently available, one manual and one mechanized. Both methods are based on estimates of the time to perform each function and how frequently each is done in a given period of time. Designed as aids to management judgment, the procedures can be modified to account for local differences in the time available to accomplish the work, the skill levels of the workers, and environmental conditions such as travel requirements or equipment conditions. The resulting procedures can be applied by local staffs or managers to estimate the actual staff required to perform the work in the local situation.

5. PHYSICAL LAYOUT

To assure that the new organization will make optimum use of available space, that jobs with high interaction requirements will be adjacently located, and that there will be minimal interference between functions and people, the physical layout of the new organization must be considered. Obviously, no single workspace layout will be appropriate under all conditions. A computer-based procedure was developed that considers the configuration of the available space, the space required for each job, and the interaction requirements between jobs and functions. The output is an optimized layout diagraming the spatial location of each person in the organization.

6. MEASUREMENTS

Earlier in the process, a clear set of measurable objectives was established. In this step, specific measurement systems are developed to assure accomplishment of all the organization's objectives. Each measure must include a diagnostic component to provide the organization's management the capability of making corrections. Work flows are analyzed to determine where best to measure; jobs are simulated to determine quantities that can be produced; and similar activities in other organizations are studied. In some cases, final agreement on the measures and standards to be implemented may not be reached until after the new organization has been trialed. Care must be exercised to insure that the results being measured are clearly under the control of the measured organization.

7. TRAINING

During the Job and Organization Design activity, lists of skill and knowledge requirements were prepared. These lists are compared with the skill and knowledge developed by existing training, and a list of the requirements for new training is developed. As part of this activity, a specification is provided to the training developers, including prerequisite skill and knowledge, the sequence for training, and the timing and constraints for its delivery.

8. IMPLEMENTATION

The transition to the new organization must be guided by a plan developed sufficiently in advance to enable coordination and timely integration of all components, including employee assignments, physical arrangements, training delivery, and hardware and software installation. Generally, a phased transition over three to six months enables orderly change and continued performance of day-to-day tasks. The plan should include one or more initial trial or "controlled introduction" sites. Used as initial test sites, these locations are studied throughout implementation to uncover unanticipated problems in the work system design or the implementation plan. End-user participation throughout helps to secure future acceptance of the change by developing "champions" within the changing organization, and thorough data collection and analysis help to identify the successes of implementation and areas requiring further improvement.

CONCLUSION

Applications of the Work Systems approach to organization change in AT&T and the Bell System have resulted in hundreds of millions of dollars of savings and many orderly transitions. Successful major changes following this approach have been implemented in diverse service, sales, engineering, and administrative organizations. In future applications, less formal structure to support the process is likely, whether in the former Bell System or new organizations applying the approach. Key ingredients that should be retained, however, are the presence of a dedicated project manager in charge of development and implementation, the commitment of resources to accomplish the work, and thorough implementation planning and tracking which includes a high level of end-user participation.

REFERENCES

[1] Gagne, R. M. (ed.), Psychological Principles in System Development (Holt, Rinehart and Winston, New York, 1962).

[2] Gustafson, H. W., Implementing work system change, Proceedings of the First International Symposium on Human Factors in Organizational Design and Management (North-Holland, Amsterdam, 1984).

[3] Hall, A. D. and Fagen, R. E., Definition of System, in von Bertalanffy, L. and Rapoport, A. (eds.), General Systems Yearbook of the Society for the Advancement of General Systems Theory (1956) 18-28.

[4] Morgan, C. T., Cook, J. S. III, Chapanis, A. and Lund, M. W. (eds.), Human Engineering Guide to Equipment Design (McGraw-Hill, New York, 1963).

Human Factors in Organizational Design and Management
H.W. Hendrick and O. Brown, Jr. (Editors)

PERFORMANCE APPRAISAL: An area for HF interventions*

Henry E. Bender**
Concept Applications, Ltd.
Laurel Hollow, New York 11791

Evelyn Eichel
Manager, Management Development
PepsiCo
Purchase, New York 10577

Judy Bender
Concept Applications, Ltd.
Laurel Hollow, New York 11791

Performance Appraisal is important to employees, managers and organizational functioning. As such, it becomes a management function of critical importance. Techniques for implementing performance appraisals vary widely from company to company. This research identified the various techniques currently employed in U.S. corporations. Uses and organizational characteristics were analyzed. The results of the study are presented with implications for areas where human factors personnel could assist the Human Resources Department.

INTRODUCTION

Performance appraisal is frequently identified as a crucial management process, one most management practitioners concede is difficult to perform. Due to its importance in the sensitive areas of salary increases, promotions, and retention/dismissal decisions, it is often criticized by employees. Complaints focus on the potential of favoritism/bias, inappropriate nature of appraisal items and categories/non job relatedness, and non-standardization between the appraisal process among incumbents in similar jobs. For these reasons, performance appraisal is often considered an "art."

Brinkerhoff & Kanter (1) have reviewed the changing purposes to which performance appraisals have been applied. Prior to 1960, this process was used to control employees via its influence on salary and retention, discharge, or promotion decisions. The 1960's proved to be a time for

---

*This research study was conducted by the Management Research group of the American Management Associations. The complete research study entitled "Performance Appraisal: A Study of Current Techniques" was distributed to members of the Human Resources Division. It was guided by Dr. John W. Enell, Vice-President for Research (Ret.) and Dr. Henry E. Bender, Director of Research.

**Dr. Bender is presently employed by Bell Communications Research Inc., 6 Corporate Place, Piscataway, New Jersey, 08854.

expanding the uses of performance appraisal data. In a time of lowering productivity, the appraisal proved useful in employee development, organizational planning, and quality of work life improvement. It was used to impact on employee productivity, effectiveness, efficiency and satisfaction.

Equal Employment Opportunity legal pressures further changed the purpose of performance appraisals in the seventies. Organizations were now required to document and justify their administrative actions concerning the sensitive areas of salary, promotion, and retention or discharge decisions. It was recognized that performance appraisals could provide supportive documentation to the organization when threatened with legal actions. When conducted in accordance with Federal regulations, it provided required documentation in an orderly fashion (2, 3).

Human Resource professionals and others in the management heirarchy recognize the importance of performance appraisals. When conducted in an appropriate manner appraisals have the potential of increasing employee productivity by increasing morale and worker efficiency. Organizational benefits include: leadership, organizational stability, profit and growth improvements. Meidan (4) has noted the multi-organizational purposes of managerial appraisal in the following manner:

> "Generally speaking, the objectives of managerial appraisal are (1) to evaluate performers or improve performance in order to make promotion/dismissal/transfer decisions, review salaries, and identify training needs; (2) to allocate financial, production, technical, and marketing resources; (3) to aid in business planning; and (4) to make possible changes in organization and control systems."

RESEARCH DESIGN

Data for this study were obtained using a multiple data gethering approach. These included an intensive review of the literature; an extensive survey to determine current performance appraisal practices; and in-depth interviews with management professionals.

The literature review covered the past two decades. Primary focus was directed to the past decade due to the emergance of legal implications to the performance appraisal process.

The literature review guided the construction of the survey questionnaire. It attempted to gain information deemed most critical in the literature. The questionnaire was mailed to a systematic sample of members of the American Management Associations Human Resources, Finance, Marketing and Information Systems Divisions. In all, 2400 questionnaires were mailed. Useable responses were obtained from 588 members, representing a 24.5% response rate. The majority of responses were obtained from members of the Human Resources Division. They represent the basis for this presentation.

The respondents appeared to represent the broad spectrum of business in the United States. Fifty-three percent represented companies grossing under $100 million dollars; 21.1 percent grossed between $100 to $499 million; 25.7 percent had revenues exceeding $500 million dollars. Company size varied greatly. Sixty percent employed fewer than 100 employees; 23.7 employed more than 100 but less than 5,000 persons. Finally, 16.3 percent had payrolls exceeding 5,000 employees.

Ancillary data was obtained by the over 50 performance appraisal forms submitted by respondents.

In-depth interviews were conducted with 12 respondents and performance appraisal specialists. These were performed in person at the person's place of employment. The information obtained provided clearer insights and understanding of the data obtained.

RESULTS

The Human Resources/Personnel Department is usually assigned the responsibility of overseeing the performance appraisal process. This appears logical due to the numerous functions capable of using information provided. As can be seen in Figure 1, Employee Relations, Affirmative Action Programs, Organization and Employee Development, Labor Relations and Safety and Health Services significantly benefit from this activity. Secondary benefits are derived by the employee benefits and planning and administration sub-units.

Figure 1

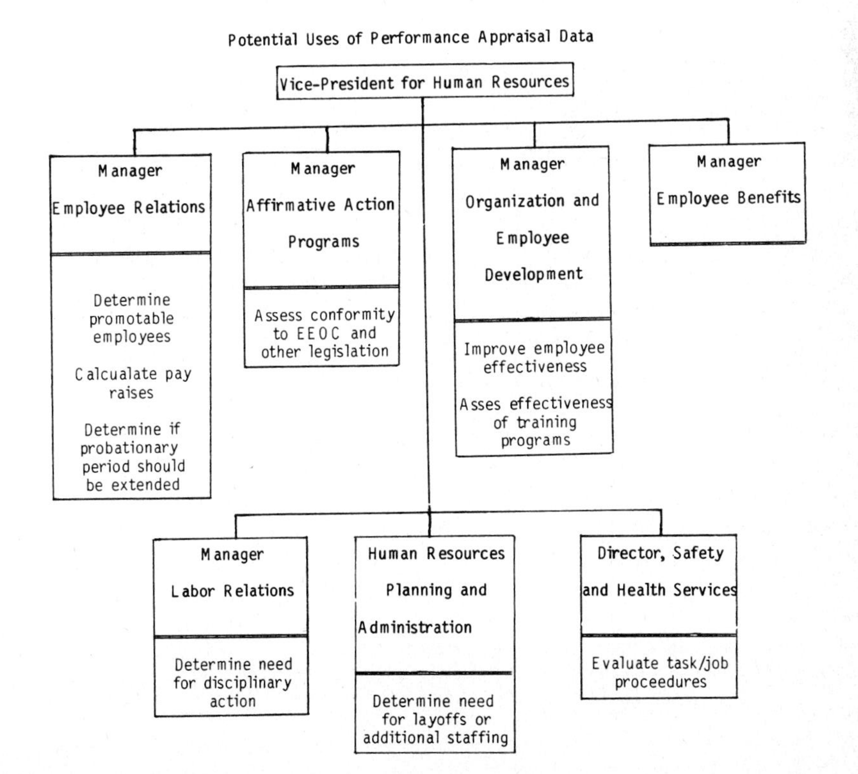

It was interesting to note that the purposes for which responding organizations use performance appraisal data has varied in accordance with shifting external influences. As noted in Table 1 and in the original AMA monograph (4), "performance appraisal currently exists in organizations for one or more of the following reasons identified in the survey: administrative, developmental, and/or legal purposes. More than 85.6% of the respondents in the present survey use performance appraisal for compensation purposes. Nearly two companies out of three use appraisal for counseling (65.1%) and training and development (64.3%). Other purposes are promotion (45.3%), manpower planning (43.1%), retention/discharge (30.3%), and validation of a selection technique (17.2%)." External validity exists for these findings. Comparable findings were reported by Hay Associates (6) and Locher and Teel (7).

Table 1

PURPOSES FOR WHICH RESPONDING COMPANIES USE PERFORMANCE EVALUATION

Purpose	Percentage of Those Responding
Compensation	85.6%
Counselling	65.1
Training & development	64.3
Promotion	45.3
Manpower planning	43.1
Retention/discharge	30.3
Validation of selection technique	17.2

Performance appraisal is best understood in relation to elements critical to the process. These include goal setting, job analysis, performance standards and performance feedback. They form the basis of an integrated performance appraisal process.

Goal setting occurs at all levels of the organization. It is the activity of "translating organizational goals into divisional, departmental, and finally, into specific job objectives, from a top down perspective. ... It focuses management's attention on needs; it helps to identify and define key areas of organizational performance and activity" (5). Goals specify performance, activities and results to be accomplished at all levels of the organization, within specified time frames.

Goal setting is an activity undertaken by over three-quarters of our respondents. Of these, 82% claimed that goal setting was part of their performance appraisal process. These responded stated that departmental goals were derived from corporate goals. Further, they believed that consideration was given to organizational/departmental interactions, and the effects these can have on specific jobs when developing their goals.

Job Analysis is concerned with the component tasks of jobs required to achieve departmental goals. These tasks are sub-elements of activities occuring in the organization. Job analysis focuses on the study and collection of information relating to task elements, operations, and responsibilities of identifiable jobs (8, 9). The job analysis forms the basis for job descriptions, a critical document to the Human Resources/ Personnel organization.

As seen in Table 2, only 27.1% of the respondents believed that the job analysis focused on employee behaviors. The majority (60.1%) stated that the job analysis performed in their organization was task based. Over 89% of the respondents noted that the job analysis resulted in a statement of job responsibilities, while 84% further noted that it states the objectives and purposes of the job. Since the majority of the respondents claimed that the focus of the job analysis was task based, it is not surprising to see that over three-quarters of the respondents state the resulting job descriptions specify job tasks. A slightly lesser percentage (72.1%) state that the resulting job description also states knowledge, skills and ability requirements of the job incumbent.

Table 2

FOCUS OF THE JOB ANALYSIS

	Not	Agree			Disagree	
	Applic.	Strongly	Moderately	Neutral	Moderately	Strongly
Focuses on						
Behaviors	22.8%	3.9%	23.2%	20.9%	21.3%	7.9%
Tasks	14.0	13.6	46.9	16.7	6.2	2.7
States						
Responsibilities	5.1	42.2	46.9	2.3	3.1	0.4
Objectives and purposes	4.7	39.5	44.5	5.1	5.5	0.8
Results in job description which states						
Tasks	5.1	30.6	44.7	9.4	8.6	1.6
Knowledge, skills abilities	7.8	29.0	43.1	11.4	7.1	1.6

Performance standards provide the basis for evaluating how well an incumbent's actual performance meets required/desired standards developed for the job. Standards can be developed as an outgrowth of the job analysis procedure. Over fifty percent of our respondents (55.5%) indicated that a series of standards or specific goals of performance is developed for each major segment of individual jobs. A major portion of these individuals further noted that the standards of performance were developed through a negotiation process that occurred between the employee and the supervisor. From this, it is concluded that in the majority of cases, while performance standards are developed, they are not based on an analytic technique. Rather, they are predicated on past performances and on the negotiation skills of the employee and the supervisor. Supervisorial influence is noted in that performance standards are linked to departmental objectives in organizations represented by nearly 70% of the respondents. This data is presented in Table 3.

Table 3

CHARACTERISTICS OF PERFORMANCE STANDARDS

	Not Applic.	Agree		Neutral	Disagree	
		Strongly	Moderately		Moderately	Strongly
Standards are						
Linked to departmental objectives	5.8%	22.2%	47.5%	15.6%	7.8%	1.2%
Developed for each major segment of individual's job	11.6	20.2	35.3	14.3	15.5	3.1
Negotiated between empl. & supv. and incl. in performance evaluation program	9.3	16.3	41.5	12.0	17.8	3.1

Performance appraisal instruments generally can be grouped into three categories including: comparative; absolute and outcome oriented. Each category contains several techniques. A discussion of these is beyond the scope of this paper. Such a discussion can be found in the original research study document (5). A summary listing of techniques and purposes for which they can be used is presented in Table 4. Our respondents indicated that the importance of each technique to their organization's performance evaluation process, in descending order of importance were:

- Essays
- Graphic Rating Scales
- Critical Incidents
- Weighted Checklists
- Ranking
- Behaviorally Anchored Rating Scales
- Forced Distributions
- Forced Choices
- Paired Comparisons.

Table 4

RELATIONSHIP BETWEEN APPRAISAL TECHNIQUE AND PURPOSE

Techniques	Purpose		
	Administrative	**Developmental**	**Justificatory**
Comparative			
Simple Ranking	X	—	—
Paired Comparison	X	—	—
Forced Distribution	X	—	—
Absolute			
Essay	X	X	—
Weighted Checklist	X	—	—
Forced Choice	X	—	—
Critical Incidents	X	X	⊗
Graphic Rating Scale	X	X	⊗
Behaviorally Anchored Ratings	X	X	⊗
Outcome			
Direct Index	X	—	—
Performance Standards	X	X	⊗
Management by Objectives	X	X	⊗

Legend:

X = the technique has the ability to be used for that specific purpose.

— = the nature of the technique limits its suitability for that specific purpose.

⊗ = the technique has the potential to be used for that specific purpose.

[Adapted from DeVries, Morrison and Shullman, 1980]

Performance appraisals are conducted at various intervals during the working year. Figure 2 indicates that 77.7% of the responding companies con-

Figure 2

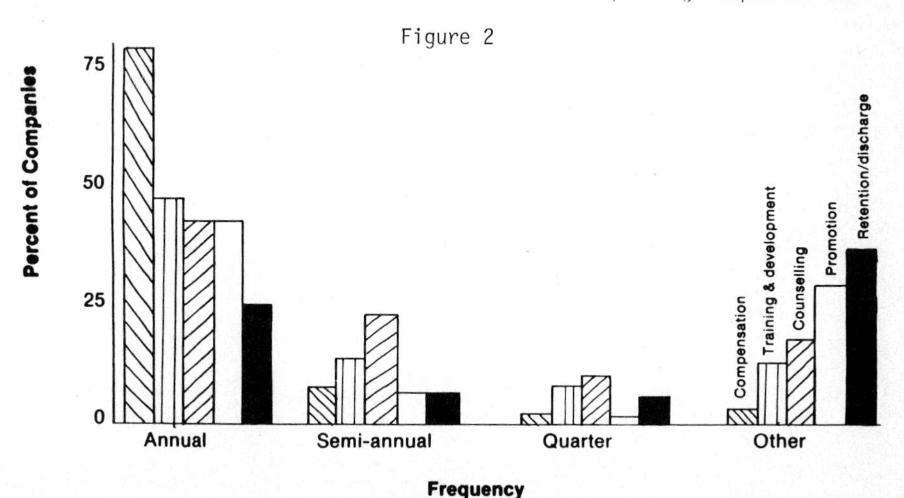

duct employee reviews yearly for compensation purposes. Fewer companies conduct such reviews yearly when they are oriented to counselling and training/development (39.2% and 45.3% respectively). For these purposes, appraisals occur more frequently, either semiannually, quarterly, or at some other interval.

DISCUSSION

A review of the study results indicated that behavioral analysis/task analysis provides the basic data upon which job descriptions and performance appraisal formats should be predicated. Our in-depth interview findings indicated that these data are frequently inaccurate or dated. Due to their background and experience in behavioral analysis techniques, this is the area in which the human factors professional can provide intervention services to Human Resources Management. Most are familiar with various approaches to job analysis including: task-oriented, worker-oriented, and abilities-oriented approaches. Further, they recognize the limitations of the various analysis techniques and realize the need to implement multiple techniques to achieve a complete overview of the jobs to be performed. This multiple analytic approach also provides the most complete information to the various sub-units of the Human Resources/Personnel group.

The value of the human factors specialists to the Human Resources/Personnel group is enhanced by their ability to evaluate productivity and determine the effects of environmental, organizational, and equipment factors. As such, their inputs to the organizational development process can be invaluable.

Finally, it is important to note that the majority of human factors personnel have extensive training in psychology. This background would be an asset in the development of performance appraisal recording forms, especially those that are critical incident, graphic rating scale, behaviorally anchored ratings, performance standards, or management by objective oriented. These techniques have been identified since they meet administrative, developmental and justificatory purposes.

SUMMARY

The findings of an American Management Associations study were presented. In the presentation of findings, it was noted that there are numerous points of intervention at which human factors professionals, with their background in behavioral analysis, form design/psychometrics can be extremely helpful to human resources/personnel professionals. Their experience in behavioral analysis would be an asset in the job analysis/job description development phase of the performance appraisal process. Their strong background in psychology would be a further asset in the development of appraisal forms, especially those that are based on the critical incident, graphic rating scale, behaviorally anchored, performance standards or management by objectives techniques.

FOOTNOTES

(1) Brinkerhoff, D.W. and Kanter, R.M., Appraising the performance of performance appraisal, Sloan Management Review, Spring, 1980, 3-16.

(2) Beacham, S., Managing Compensation and performance appraisal under the age act, Management Review, January, 1979, 51-54.

(3) Ford, R.C. and Jennings, K.M., How to make performance appraisal more effective, Personnel, March-April, 1977, 51-57.

(4) Meidan, A., The appraisal of managerial performance, AMA Management Briefing, New York: AMACOM, 1981.

(5) Eichel, E. and Bender, H.E., Performance appraisal: a study of current techniques, AMA Management Research Study, New York: AMACOM, 1984.

(6) Hay Associates, Survey of human resource practices, New York: Hay Associates, 1975.

(7) Locher, A.H. and Teel, K.S., Performance appraisal - a survey of current practices, Personnel Practices, 1977, 245-247+.

(8) Flippo, E.B., Personnel Management, New York: McGraw-Hill, 1980.

(9) Klinger, D.E., When the traditional job description is not enough, Personnel Journal, 1979, 243-248.

Human Factors in Organizational Design and Management
H.W. Hendrick and O. Brown, Jr. (Editors)

PARTICIPATIVE THEORY IN ACTION IN A FAMILY PRACTICE RESIDENCY PROGRAM: A CASE STUDY

Richard A. Rusk, Ph.D., David D. Schmidt, M.D. & Stephen J. Zyzanski, Ph.D.

Institute of Safety and Systems Management
University of Southern California
Los Angeles, California
U.S.A.

INTRODUCTION

The field of Organizational Behavior (OB) was born of a need for organizations to be more isomorphic with society. Perceptive people in the 1950's were beginning to see that a widening philosophical gap was rapidly becoming a chasm between systems and their populations. As large and ever-increasing segments of the populace became better advised of the system's dynamics, they were simultaneously becoming better equipped to challenge its vagaries. Education and communication were the obvious prime movers in the trend.

Industrial workers were less willing to accept the 1930's mentality of managers, perpetuated simplistically "because it worked during the depression." The uninformed tended to believe that workers only wanted money for their outside activities and security from job loss and would endure to get them. But as awareness increased attitudes changed; dogmas were rejected, methods often questioned; management paradigms of the era were severely strained. As put by one author, "if people have changed, then management techniques have to change. It can't be management by power, by status or fear anymore."[1]

Perhaps, before we look forward, it is proper to reflect the past. While we tend to believe our omniscience has created new managerial awareness, looking back brings doubt. For example, the "father of industrial psychology," sought ways to match people mentally and psychologically with their work context in a productivity sense, but was also highly attuned to a better individual "level of life."[2] This sounds strangely like today's "ergonomics" or "quality of work life" (QWL) efforts.

The Hawthorne Studies of the late 1920's[3] served to spawn interest on the part of the behavioral scientists, who then rapidly expanded our knowledge in such diverse areas of motivation, leadership, group dynamics and a host of others. What they really did is best said in "These scholars and others have shown how human beings bring to their task aspects of behavior which the effective manager should profitably understand.[4] The lead author undertook the study of OB with just such an intent.

Only a few short years ago, departments of OB in major educational institutions were sparse. In a 1978 survey of members of the International Association of Applied Social Scientists (IAASS)[5] eighty-nine sources were listed where quality professional or academic training could be obtained in the United States alone. Eighty-two of these offered graduate level degrees and forty-three offered terminal degrees. By 1982 it had grown to one hundred four sources.[6]

A similar expansion is occurring in industry as companies add speciality staff and organizations dedicated to monitoring and aiding effectiveness and efficiency from the people and process sides of business in the tone of Human Factors in Organizational Design Management (ODAM).

Thus, we have seen in our time a literal explosion of interest, research, empirical knowledge and practical application in a field which began its distillation from the social and behavioral sciences only in the 1950's.

BACKGROUND

In the summer of 1977, Dr. David Schmidt contacted the School of Management at Case Western Reserve University searching for aid in establishing the University Hospitals' Family Practice Residency Program. The main interest was in creating first and foremost a program which was financially viable. The OB department located a graduate student whose undergraduate preparation and experience was oriented to business and economics. Mr. (now Dr.) Dick Rusk worked with Dr. Schmidt and the two of them took the program from its infancy into full productivity. Without undue elaboration of details, this involved planning the entire program (a PERT network was developed and used), the creation of facilities, the acquisition of government grants, the establishment of university and hospital linkages, organizing (including complete and detailed system policies and procedures), staffing (including, of course, National Resident Match successes for residents), and finally, putting it all together. It was obviously quite a job and feat.

In those formative years, it was recognized that a top flight residency program is a organization composed of highly educated and motivated faculty (physicians and non-physicians), idealistic and spirited residents and talented support staff. Any attempt to manage such a group with an autocractic leadership style, would be like trying to control a spirited thoroughbred horse with brute force alone. A conscious effort was made to develop a work environment in which each member of the organization would become internally motivated to do his/her part to help the entire group realize its articulated goals.

THE IDEA

It soon became apparent, given the autonomy that was present (from the Chairman of the CWRU Department of Family Medicine) that the practice could innovate in many ways to test its ideas. The plan to utilize the theories and research findings of OB both as a means toward higher productivity and effectiveness and as a pragmatic test of their usefulness in the peculiar sitting.

The two prime theories utilized in this study and context were:

1. Participation Theory[7]--involving the creation of an organizational environment recognizing the potential contributions of all system members to enterprise objectives while achieving their own goals and the motivational importance of the process.

2. Organizational Climate[8]--"contributors" to its formulation, the primacy of the leader behavior, the presumed causal chain and the resultants.

The relationships between the use of these ideas and the managerial systems and processes in the ODAM sense will be demonstrated.

Participative theory suggests that an effective means of creating such an organizational environment is essentially management by consensus.

Attempts to operationalize this theory on a day-to-day basis are discussed below. In an educational or service enterprise, it is difficult to measure productivity. We utilized a standard instrument for the measurement of organizational climate as an indication of the health of the residency programs during a period of rapid growth. The data from this evaluation of a management style is reported and discussed.

GENERAL PARTICIPATIVE METHODS

Participative theory was implemented in our systems in various ways: through regular meetings, in decision making processes, in goal-setting techniques and in the day-to-day operational matters.

Meetings, though emphasizing task and orderliness (prior agendas, keeping on the subject, making arrangements and the rapid production of minutes) nonetheless contained ample opportunity for individual or subgroup input. Everyone knew the norms that existed creating a situation where: input for everyone was not only permitted but was desired; high clash of ideas was expected and utilized to further creativity and innovation; but where interpersonal differences were confronted, worked through and not permitted to detract from the group effort. There was also the sense that the structure had stretch or flexibility, so that where people needed room for expression, venting of frustrations or digressions into related (but not highly relevant) areas, these excursions would be permitted.

The net effect was that most people came to group meetings with positive expectations about the pertinence of them and the results. And when decisions were made, they were supported and followed through, because each member had had his/her own "say" (or the opportunity) and thus commitment was high.

The decision making model was one of decentralization; that is, authority for making decisions was vested in those individuals or positions where the best information was available with which to decide. Errors resulting were treated as problems to be solved or as lessons learned, rather than with scorn and punishment; as a result, errors became fewer. While policy decisions were made at the top or in consensus, growth demanded a continued emphasis on decisions at lower levels, appropriate to contexts and timing. Response times were continually shortened and respect for this appeared in several external quarters.

The goal-setting process called for special attention. Initially, the leadership decided the major courses and methods but as the number of people and sub-elements expanded, better definition and verifiability was needed. The response was off-work site workshop series (facilitated by the senior author, acting as a Human Factors consultant) where individuals first wrote their own impressions of organizations' prime mission and then contributed to a nominal group technique[9] in arriving at a collective definition. This mission was further developed by the various subgroups (residents, physicians, faculty, nurses, behavioral scientists, administrative staff) and translated into specific, measurable, verifiable goals.[10] Tracking devices and calendar target dates were then developed and used in monitoring and controlling progress. The team development aspects of this workshop also soon became apparent.

As for Organizational Climate (OC), the ideas of Litwin and Stringer[11] were used. They consider OC as an intervening viable. The givens are management systems (policies, procedures, structures developed to handle operational requirements), norms and values, and individual manager practices (specific managerial actions and behaviors). The dependent variables are employee health and retention, motivational arousal and organizational performance and development. The causal linkage is presumed to be management systems, managerial practices and norms and values combined to create OC. Climate creates expectations, expectations create arousal and reinforce certain kinds of desired motivations while inhibiting others which are undesired; motivation combined with skills (plus effectiveness of the team) results in productive behavior. The research of others tend to support these connections[12]

Litwin's "most important finding" is that "combinations of specific management practices are the most effective single prediction of climate in an organization or work-group." These practices include a wide range of behavioral factors: "the extent to which the manager rewards innovations, conducts team meetings and provides informal feedback, the honesty of communications during discussions of individual performance, the method(s) of decision making and many others."

As a function of intentional organizational design, the primary managers sought to continually increase their awareness and enlightened practices toward organizational members. Plans, policies and procedures were developed collectively with particular attention to the concerns of those most closely connected or involved. Structured arrangements were studied, reexamined and tested before implementation with special consideration to points of authority and skills of these crucial incumbents; those that seemed inappropriate after time were changed and improved. People were selected into the system based upon general congruence with existing system dynamics and values; evictions though rare were made in cases of inaccurate judgement. Norms were rapidly evident and uniformly supported.

The managers were basically open, honest, trusting and ethical people to begin with but were ready to learn improved techniques in dealing with interpersonally with team members. Praise and "strokes" were abundant in support of desired behavior;[13] undesired behavior, while not reinforced, was viewed more as opportunity for improvement never subject to raw punishment.

People on the team soon recognized the prized psychological environment and responded favorably. The unit was assessed the best of three in the city on many dimensions and reached national prominence in a very short span of time.

THE COST OF THIS TYPE MANAGEMENT STYLE

A conservative estimate of the cost of this management style is high. The salaries of the faculty, residents and staff for the time spent in group meetings and loss of practice income is also involved.

There are additional non-monetary costs. Management must develop great patience. Many issues could be settled by executive decree without the sometime laborious airing of the perspectives and opinions of each member of the group. Some members of the organization interpret such a leadership style as lack of leadership. They become angry with the leadership for not simply taking a position and using his/her authority to enforce this position. Some individuals would rather be told what to do and when to do it. No practice has universal appeal to all members of the social system.

There is at least one other strong plus for the use of management by participative theory and this is in implementation. While decision making itself may take longer, implementation is swift.[14] People are committed to the decision ahead of its effect. Contrast the more autocratic perspective of individual decision making where the decision is rapid but the implementation may be severely drawn out, even subverted. Here lies a major tradeoff to be considered by managers, in defining their roles.

RESEARCH METHOD

The "Profile of Organizational Characteristics," commonly known as the Likert Organization Climate Instrument,[15] was used as the prime data collection device. It was reproduced for research purposes and given to each member of the program usually during the July retreats. Respondents were asked to indicate both an actual state of current affairs and the condition which would represent the ideal state for them.

The instrument contained 48 items. Each item was rated on a 20 point scale with equal intervals and polar extremes. The items clustered into 7 major categories. Leadership, motivation, communication, interaction-influence, decision making, goal setting and control processes.

The instrument was administered to the population in July of 1978 (t_1), July 1980 (t_2) and in July 1982 (t_3). The majority of the sample had all three measures. There are some new personnel added and a few departures but this is not considered to be detrimental to the data.

RESULTS OBTAINED

Respondents routinely completed the questionnaire in 30 minutes or less. Thus, with a reasonably small time investment considerable information was gained. The results shown here are those selected to illustrate the usefulness of the information gained from this approach.

At the first administration of the instrument, the data showed that the highest 10% of the items (in terms of the mean scores) were in categories of interaction/influence and goal setting. The mean scores were fairly high as well indicating early strength in the organization on these parameters.

Conversely, the low 10% items were mainly in communication and decision making areas. This gave the unit its first indication of where some of its problem areas were and primed it for corrective interventions.

The changes which took place between t_1 and t_2 mostly showed overall improvement. Five of eight items were from communications or decision making clusters (the initial lowest areas). Four of these showed improvement that was statistically significant; level of the decision making, accuracy of the information in decision making, the accuracy of material superior-subordinate perceptions and the accuracy of upward communication. One variable, adequacy of upward communication, changed significantly but in a negative direction. It was also noted that there were no important changes in leadership or goal setting, areas where evaluations were already strongly positive. T_3 data showed basically a leveling off effect as the organization matured.

Thus, from the ODAM sense, this study considered the design of organizations (provision of a proper organizational climate) process aspects of the managerial role (how he behaved, ran meetings, etc.) and a sociotechnical model (medical/participative, goal setting, etc.) for managerial decision making. Productivity results were impressive over the period of the longitudinal study and the improvement in process measures were supported statistically. People felt better about their organization and its performance as well as exhibiting high job satisfaction themselves. A highly desirable combination.[16]

BIBLIOGRAPHY

1. Moreland, Pamela, "Work Integrated With Life Style Displaces Organization Man." Los Angeles Times, Part V, p. 1, December 7, 1980.
2. Munsterberg, Hugo, Psychology and Industrial Efficiency. Boston: Houghton Miffen Co., 1949.
3. Mayo, Elton, The Human Problems Of An Industrialized Civilization. New York: The Macmillian Co., 1933.
4. Koontz, Harold, O'Donnell, Cyril and Weihrich, Heinz, Management (8th Ed.). New York: McGraw-Hill Book Co., 1984.
5. Campbell, Susan, Graduate Programs in Applied Behavioral Science: A Directory. LaJolla, CA: University Associate, Inc., 1978. Handbook for Group Facilitators.
6. Pfeiffer, J. William and Goodstein, Leonard D., "The 1982 Annual For Facilitators, Trainers and Consultants," University Associates, Inc., 1982, p. 231.
7. From: Likert, Rensis, New Pattern of Management. New York: McGraw-Hill, 1961; McGregor, Douglas, The Human Side of Enterprise. New York: McGraw-Hill, 1970, primarily.
8. Mainly from: Litwin, George H. and Stringer, R.A., Motivation and Organization Climate. Boston: Harvard Graduate School of Business Administration, 1968.
9. See Annual Handbook for Group Facilitators, Jones and Pfeiffer, 1975, page 35.
10. Rubin, Irwin M., Plovnick, Mark S. and Fry, Ronald E., "A Program For Health Team Development." Cambridge, Mass.: Ballinger Publishing Co., 1975.
11. Litwin, et. al., Ibid.
12. See for example: Porter, L.W. and Lawler, E.E., "Managerial Attitudes and Performance." Homewood, Ill.: Richard D. Irwin, Inc., 1968 and an adaptation in Koontz, Harold, O'Donnell, Cyril and Weihrich, Heinz, Management (8th Ed.). New York: McGraw-Hill Book Co., 1984, p. 488.
13. See Blanchard, Kenneth and Johnson, Spencer, The One Minute Manager. New York: William Morrow and Co., Inc., 1982.
14. Koontz, O'Donnell and Weihrich, Ibid, p. 353.
15. Likert, Rensis, The Human Organization. New York: McGraw-Hill, 1967.
16. Szilagyi and Wallace, Organizational Behavior and Performance (3rd Ed.). Glenview, Ill.: Scott, Foresman and Company, 1983.

Human Factors in Organizational Design and Management
H.W. Hendrick and O. Brown, Jr. (Editors)

SAFETY ORIENTED ORGANIZATION AND HUMAN RELIABILITY

John Lindqvist — Swedish State Power Board
Bo Rydnert — LUTAB Consulting Inc.
Björn Stene — SYPRO Consulting Inc.

This paper briefly describes the theorectical background of an organizational model and its application to a checklist procedure. Special interest is shown safety oriented organizations such as nuclear power plants but the results are applicable to all organizations concerned with safety. Some short results from case studies are given.

BACKGROUND

The Nordic Council of Ministers has through its Liason Committee on Atomic Energy initiated and sponsored projects on human reliability. This paper describes some of the preliminary results of one of the subprojects to be concluded at the end of 1984. The aim of the project is to arrive at a better understanding of organizational factors affecting safety of process plants and to present a qualitative method of observing and evaluating organizational performance. Since long time both the plant itself and the humans (as an operating staff) have been under constant improvements. Because of lack of knowledge and no tradition the same has not been true when dealing with the operational organization. This can be illustrated in principle as three adaptation processes indicated in Figure 1.

ORGANIZATIONEL DEFICIENCIES AND WEAKNESSES

TMI as well as other incidents clearly demonstrated that deficiencies of plant organization contributed to the origin and aggravation of process failure. In addition, many reports from commissions evaluating system disturbances and crashes point out the human factor as a critical and contributing component. However, scrutinizing these reports together with more close investigations of organizational performance often reveals deficiencies in organizational resources and functions as the ultimate cause. The human being does not recieve the necessary support from the organization e.g. in terms of relevant information and explicit responsibility. Even worse, organizational functions can prevent him from using all his capabilities.

An organizational deficiency or weakness (ODW) is defined as a negatively influencing factor in achieving the intended safety goals. The actual process state and organizational environment determines whether the effect is grave or not. For a safety oriented organization it is of vital interest to establish in what ways the safety is affected by an ODW.

Two immediate needs must be fulfilled in order to identify and judge the impact of an ODW

a) A descriptive model of the organization and its resources and functions

b) A systematic way of evaluating the organization and its resources and functions

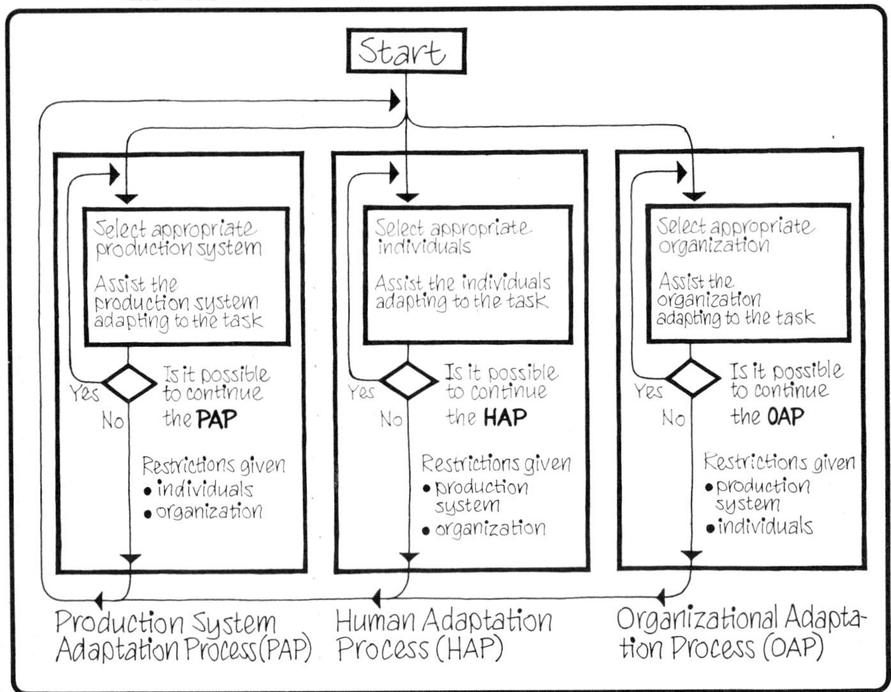

Figure 1: Principle iterative process for fulfilling a task.

MODEL DESCRIPTION

The aim of this paragraph is to define an organizational model and the interfaces between the individual, the organization and the plant. In this context safety is constricted not only to individual working enviroment but to the internal and external environment as well.

People working together establish common goals and tasks for their work. The basic reason for setting up an organization is that the common goal defines a task to big for one individual to handle. The breakdown of the tasks into subtasks is an always ongoing activity. By use of both formal and informal sets of rules, the tasks of the organization are governed. The rules are basicly the frame within the individuals work in order to reach their objectives. As a part of the organizational experiences obtained in fulfilling tasks the sets of rules are changed.

An organization consists of resources and functions. The resources are in this context systems for values, information and decision making and execution. The functions are grouped into management and administration, task realization and improvement and development. Since the organization exists in an internal and external environment the influencing factors are of essential interest. All of the above interacts and affects the safety achieved.

The resources and functions are related, so in order to fulfull the tasks set for the organization, both have to be established and in function. This is illustrated in Figure 2.

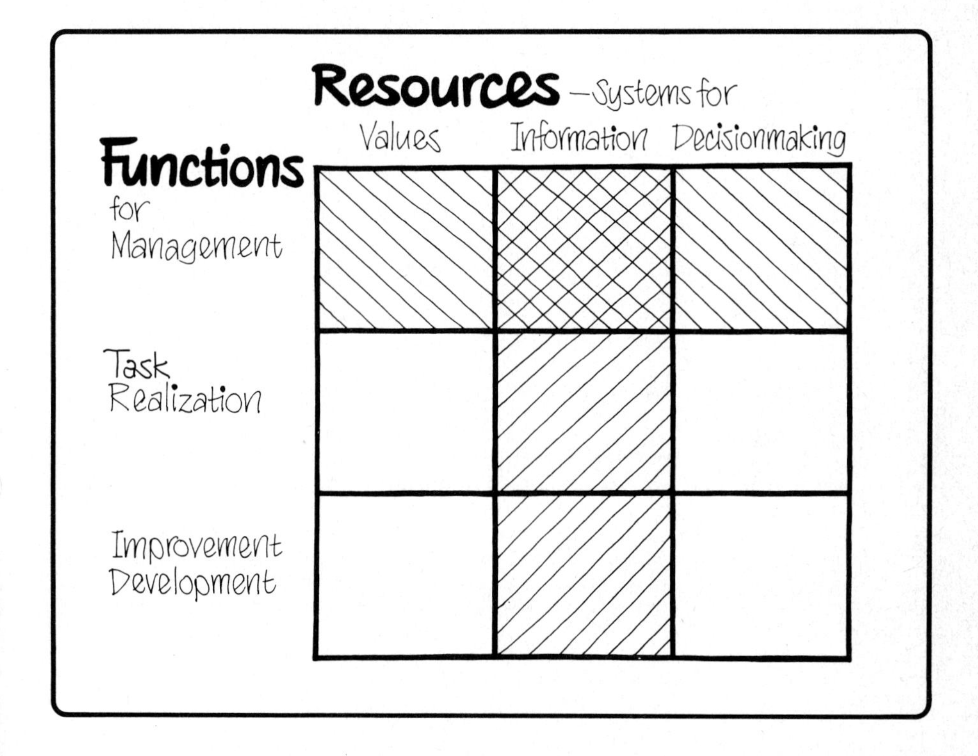

Figure 2: Relation between organizational resources and functions.

The members of an organization use the resources through their knowledge, experience, skill, motivation etc.

VALUE SYSTEM

The tasks of an organization are defined by a set of values. These values constitute the ground for all the activities achieved through the organization. The organizational goals express the content of the values.

The goals give rise to the activities of the organization. In this context, the safety related activities are of primary interest in conjunction with the activities defined as primary to the organization.

If the organization is of the size and quality needed for operation of a nuclear power plant, the value system will contain

- Technical and economical values (high production quality, acceptable safety etc)
- Social and humanistic values (stimulating tasks, generous reward system etc)
- Ethical and cultural values (promoting safety, regarding conservation of resources etc)
- Political variations of values (informing the public and politicians, accepting interest groups etc)

The value system reflects the basic views such as attitudes, traditions, habits, insight and adaptation of changes. Affecting how the systems for information and for decision making and execution are built. The usefulness of these systems is also determined by the quality of the value system.

It is in no way possible to disregard that two value systems are at hand, namely the value system of the individual and the value system of the organization. An effective organization is relying to a very high degree on the congruence between these value systems. This is certainly valid for the values concerning the product and its quality, the basic rules for interaction, and the factual content of issues and problems. When conflicts of interest become to large the activities can be interlocked. This results in a decreasing organizational capability to achieve the defined goals.

INFORMATION SYSTEM

One of the criterias for an effective organization is a total information system working properly. In this context the information system consists of all the various ways of collecting, transfering, distributing and validating information throughout the organization. It is not constricted to the computerized information systems. One sector of the information system is composed of the so-called informal system. The informal information system is involved in all the implicit interactions between the parts of an organization as well as between people. The importance of the informal system is usally not considered when reviewing organizational performance.

Information is defined as the merger between data and its context. Information will increase the knowledge of the reciever. (If not it is considered as desinformation). This knowledge can be used instantly or later to increase the possibilities for reaching a wiser decision. The optimal flow of information is derived from the physical and the psychological capability and need of the individual. As an example, a rather common situation in a control room is a flow of information beyond the human capability. The value of the information is dependent on the evaluation made by the organization as well as the individual as a result of their value systems.

Both the quality and the quantity of information will influence the effectiveness and the well-being of the individual. Therefore the organization must actively and constantly prepare information that is right in volume, time, place, price and directed to the proper user. Since cost is involved, it will not be possible to supply the optimal information at all instances. The individuals will need different sets of information due to their different knowledge and experiences. For an organization assigned to the operation of a plant the various operating states will also require different kinds and qualities of information. The information system must be able to meet this demands in a flexible way.

The information needed by any organization has different properties such as time dependent, spatially oriented and is in various stages of context. Figure 3 illustrates the relations and completeness of an information system.

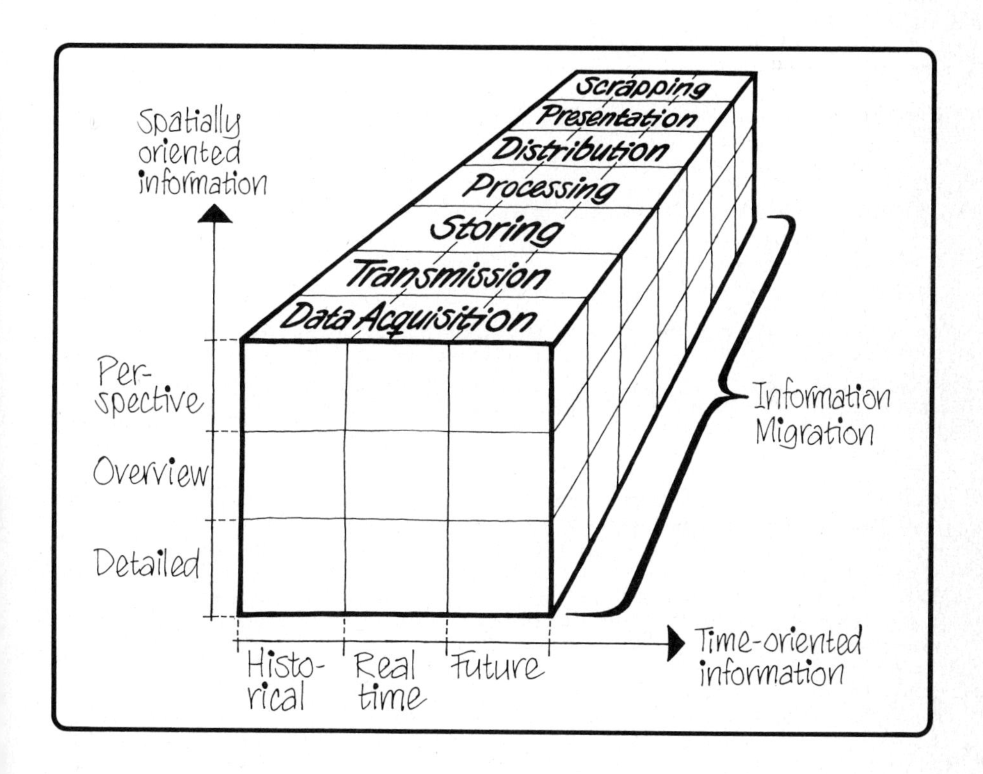

Figure 3: Matrix on the completeness of an information system.

The formal and informal information system shall fulfill the following requirements on quality

- Selection of information must be adequate for the actual situation.
- Comparison between actual information and earlier decisions must be made in such a way that discrepancies can and will be notified and treated.
- Distribution of information both to individuals within the organization and to surrounding organizations must be accurate and adequate.
- The presentation of information must be suited to the needs of the actual situation.

DECISION MAKING AND EXECUTIVE SYSTEM

The decision making and executive system are used by all supervisors within the organization. Realizing that decisions are made in different settings and needing different procedures is of fundamental importance in achieving the safety goals. For a plant operating organization dealing with safety the decision making and executive system must be flexible enough to cope with unexpected events. In order to facilitate this, the organization must constantly improve its system. The rational decision making consists of the following activities

- Identifying a problem. This implies observation of descrepancies
- Collecting and evaluating necessary information
- Defining the different decision alternatives
- Making the decision
- Executing the decision made
- Evaluating the result of the decision

Of course it will not be possible to follow these steps in a nice sequential order. Usually an iterative process is at hand breaking down the decision making in parts. This process seems to be a natural way for individuals to create problem size feasible to handle.

ORGANIZATIONAL FUNCTIONS

The organizational resources are used through organizational functions to achieve the intended objectives. If either the resources or the functions are ill conditioned, it will not be possible to reach proper results.

The set of organizational functions is divided into three basic groups. Each group is associated with activities needed for every organization, namely

- Management and administration

In this group the ways of organizing, co-ordinating, delegating, giving priority etc are defined. Moreover, there must exist methods for distributing responsibilities and resources of various kinds. In a safety oriented organization identification and evaluation of possible safety hazards and risks plays an important role. The application of safety regulations and definition of safety levels is also essential.

- Task realization

The working routines of an organization define the factual standards in respect to the settled goal for quality. These are also one of the means individuals use to evaluate their own achievements. This group also contains methods for detecting deviations from intended goals to be used by the other two groups of functions.

- Improvement and development

If the organization includes no functions for improving its own ways of organizing it will stagnate. Moreover, this applies also for the ways of increasing the knowledge of the staff. Even if the working routines and methods can be judged to be apropriate, a systematic way of reevaluating and improving is necessary. New methods of developing organization, plant and humans as a result of research must constantly be carried through. Normally, an organization dealing with operation of a plant is not in itself directly involved in research but it must have functions for adopting results into the organization.

FACTORS INFLUENCING THE ORGANIZATION

Every organization is under the influence of various factors both of internal and of external nature. These factors affect the organization through its members and vice versa. Since external factors such as society and authorities are involved, the effect is of great importance. For an organization operating a plant also corporate headquarter and various contractors will influence the plant organization in its activities. The individuals manning the organization will through their own value systems affect the content of the sets of rules defined by the organization. The production system such as the plant will put restrictions and requirements on the organization. Figure 4 gives a simplified view of how the influencing factors and the organization are related.

During the built up of an organization the resources and functions are established. How well these perform is determined during the operational state. In this state the organizational deficiencies and weaknesses will show up as various phenomena, such as practices, myths, spills of information etc.

A CHECKLIST PROCEDURE

As stated above a systematic way is needed for evaluating the organization and its functions. With respect to this, a lack of procedures seems to be at hand in analyzing organizational processes relating to safety. One solution to increase the understanding of organizational performance in terms of safety management and problem solving capacity would be to develop a checklist procedure. The purpose would be to methodically observe and analyze the organization based on a theoretical framework and experiences from plants.

In order to test how such information - and its quality - could be collected a preliminary checklist procedure was designed. The principal idea was that plant management should have an interest in understanding and making full use of the organizational resources and functions in reaching and developing the plant safety goals.

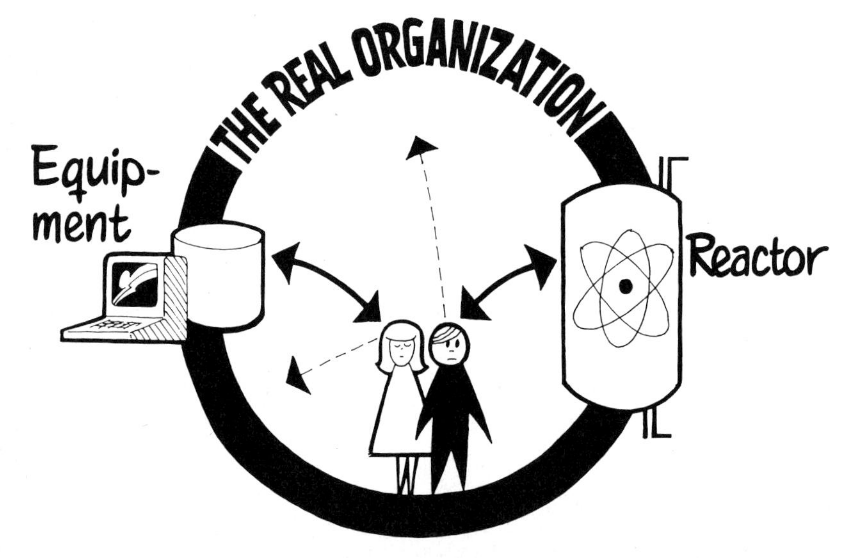

Figure 4: Factors influencing plant organization.

Every organization will rely on its members. It is important that the organization is able both to develop and to use its human resources. The interaction between the individuals and the organization could be seen as a complicated interplay between the organizational rules and the capabilities of the individuals.

The checklist was applied through interviews of plant personnel selected as representatives of organizational functions and, in case of disturbance evaluation, people responsible for decisions made.

Figure 5 illustrates the main topics of questioning set up in the checklist. Compared with conventional checklists, this one consists of questions related to fundamental resources and functions of the organization. Each question requires qualitative answers by the respondents. Connected to the checklist, guidelines were developed by which it was possible to interpret how the organization will fulfill its safety tasks.

1. Design and change of the organization
2. Organizational philosophy and model
3. Acceptable technical standard
4. Experiences from other organizations
5. Economy and safety
6. Risk and safety analysis
7. Influence of contractors on organizational change
8. Distribution of responsibility
9. Working groups
10. Experiences of process operation
11. Distribution of tasks between corporate headquarter and plants
12. Acceptable safety
13. Safety awareness of personnel
14. Management of competence
15. Training programme
16. The formal and informal organization
17. Information management
18. Interaction between the organization and its members
19. Physical and social working enviroment
20. Planning and prognosis
21. Development of the organization

Figure 5: Examples of main topics of questioning in the checklist.

Figure 6 illustrates the design of relations between the documents used.

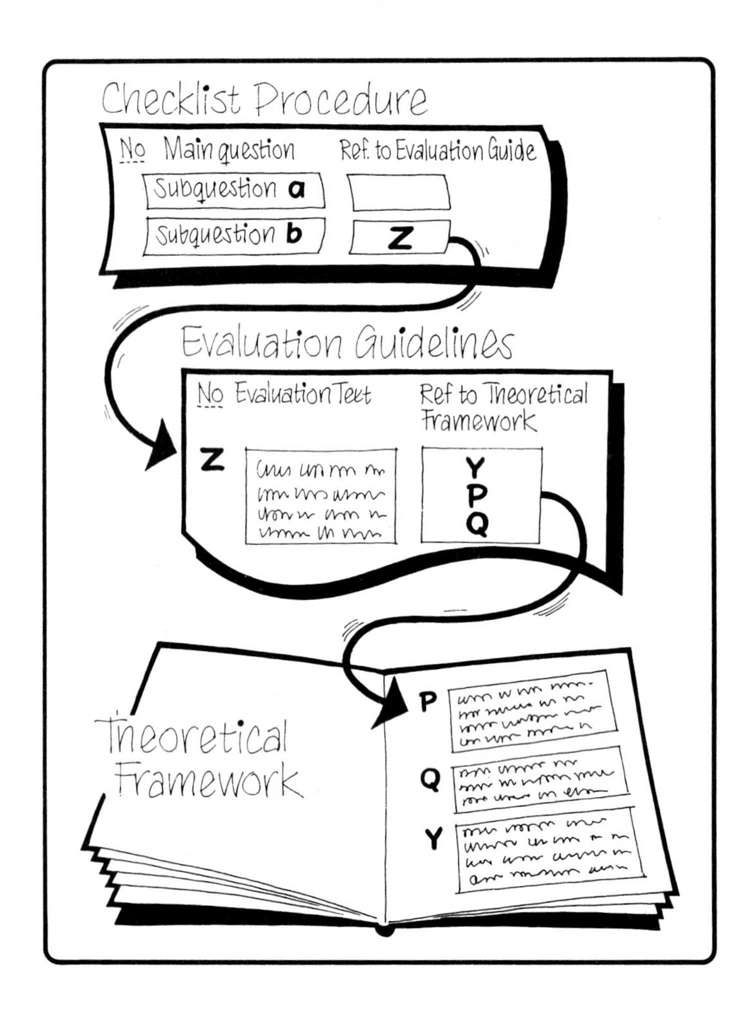

Figure 6: Document relationships.

CASE STUDIES

The checklist procedure was applied to three different organizations in different situations: one hydro plant dispatch center involved in a network breakdown a few years ago- (Example 1), one nuclear power plant building up the organization and staff approximately 18 months before startup-(Example 2), and one nuclear power plant facing symptoms of stress corrosion cracking in piping-(Example 3)

Example 1: One indirect reason for plant organization incapacity to avoid the breakdown was that the superior organization did not encourage plant management or personnel to develop efficient and safe procedures, routines and equipment. On the contrary, the plant personnel experienced signs of distrust of their capabilities at the same time feeling the responsibility (technical, economical and juridical) without adequate resources and support. This situation showed one part of a vicious circle consisting of one plant supervisor not relying on the competence of his personnel. Safety goals and safety performance were negatively affected resulting in the network breakdown.

Example 2: The plant organization consists of four departments: operation, maintenance (mechanical and electrical) and planning. The tasks and responsibilities of each must be fully known and accepted by the other departements, otherwise there is a risk of misunderstanding and conflict which in turn affects safety.

Applying the checklist procedure uncovered the operating staff considering their tasks, compared to the other departments, as more important for reaching plant goals including safety. This opinion was in full agreement with the plant manager, but not yet formally established. However, this opinion was not shared by the other departments.

Evidently, there was a lack of information and explanations necessary to reach consensus. Furthermore, the reasons for this stratification seemed to depend more on the tradition rather than careful analysis of operating experience and organizational perfomance.

Example 3: At one nuclear power plant the radiation protection unit (RPU) considered it more effective supervising the maintenance crews during works in radioactive areas than the maintenance management could do. This could be explained by the fact that the RPU had aquired a good experience and local knowledge of maintenance in plant environments. The responsibility of the unit is formally restricted to protection of every man in those parts of the plant where radiation can be released.

According to the RPU, the supervising initiative was taken when process state made it obvious that time for maintenance activities could be saved which means minimization of radiation to personnel and of economic losses. Implicit, judgements had been made that plant safety was not negatively affected.

However, from the perspective of a safety oriented organization, a few questions can be raised in view of these actions.

No doubts must exist about which unit has the responsibility and what constitutes that responsiblity. Safety implies that each member of the staff is fully aware of his tasks and his authority. Initiatives taken at one situation, which seem to be rational and also give positive results, can reinforce processes which in the long run negatively affect safety.

Plant management needs continous information of actual activities in the plant in such a way that evaluation is possible of how safety goals are achieved. In complex man-machine-systems safety risks can be introduced as a result of actions outside formal and informal procedures and routines.

DISCUSSION

The intended goal for the project was achieved. In defining the model, concentration was made on an organization in operation. Since nuclear power in Sweden, as a result of a referendum is to wind up by the year 2010 further development is needed for the corresponding organizational states.

The three case studies demonstrated that it was possible to point out organizational processes and phenomena of value for safety judgements in plant operations.

Safety is a true dynamic concept. Different values in society, economical realities together with operating experiences can change the prerequisites for todays safety goals. The operating organization needs to be prepared for changing conditions.

The checklist procedure is general in its nature. Adjustments are necessary for applications at specific plants or disturbance cases. Finally, safety evaluations of organizations have no tradition in the plants compared to technical analyses. Organizational reviews need to be justified and implemented by management in a very active way.

ACKNOWLEDGEMENTS

Gratefully we thank the Nordic Liason Committe for Atomic Energy, the Swedish Nuclear Power Inspectorate and the Swedish State Power Board for providing necessary funds in realizing this project. We also thank our comembers of the project in their efforts in inspiring us to write this paper.

REFERENCES

NRC Guideline (NUREG-0731). Guidelines for Utility Management Structure and Technical Resources. Wash DC, September 1980.

NRC Guideline (NUREG-CR-3215). Organizational Analysis and Safety for Utilities with Nuclear Power Plants. Wash DC, August 1983.

Osborn R.N., Sommers P.E., Nadel M.V. An Analysis of Existing Management and Organization Guidelines for Nuclear Operations, (BHARC-400/82/015), Batelle Seattle, Wash.

Kemeny J. et al. The Accident at Three Mile Island, Report of the President´s Commission, Wash DC, 1979.

Human Factors in Organizational Design and Management
H.W. Hendrick and O. Brown, Jr. (Editors)

A CASE STUDY IN REDUCED ORGANIZATIONAL ATTRITION

MICHAEL W. PEASE
LIEUTENANT COMMANDER, U.S. NAVY
ORGANIZATIONAL EFFECTIVENESS CENTER
PEARL HARBOR, HAWAII
USA

Human Resource Availabilities (HRAV's) have historically shown no significant long term effect on retention. This work examines the effect on retention of an emphasis shift to Organizational Effectiveness (OE) and three generations of technology which the OE Consultant has utilized. Two cases, both experiencing extremely poor retention are examined and analyzed against a systems model. Because of the method of reporting retention, six and twelve month summaries, and other factors significance testing of our intervention proved impossible.

INTRODUCTION

Studies have shown that the Navy Human Resource Availability (HRAV) based upon survey guided development produces no significant long term effect on first term retention. [1] The concept of the HRAV was to administer a Likert derivative survey to the entire crew of a Navy command, analyze the survey and feedback to the individual commanding officer. This was followed by a "waterfall/bubble-up" where the crew received the survey results top down and raised issues bottom up. Only those issues which could not be solved within the chain of command were addressed back to the commanding officer. Crew training was then to be given based upon assessed command need. A number of reasons are hypothesized for the failure of the HRAV to affect retention significantly. The HRAV, though mandated, was never rewarded or punished like other operational and administrative inspections. HRAV was in competition for time and energy with these other higher priority inspections. A common perception among Navy consultants is that HRAV became a "check in the box" for individual commanding officers and training degraded to a quick communications or time management workshop. Another factor may well have been the survey, a Likert system four instrument administered to a basically system one organization. The typical outcome was low numbers which a commanding officer, who has proven himself the best of the best through a long screening process, equated as "bad". The concept of this type of threat has influenced feedback design and will be explored further. The Navy rotation policy is also a negative factor. The entire crew changes within a three year period, including the commanding officer. In an aviation command, the commanding officer has changed twice in a thirty month period. Thus the command evaluated twelve to eighteen months down the road is a significantly different organization. A fourth possible explanation of the HRAV failure to affect long term retention, at least in recent years, is the failure of the feedback process. Waterfall/bubble-up has not occurred within Pearl Harbor for at least the past seventeen months. In some instances, feedback has not gotten below the Executive Department. An incident where this occurred will be briefed later.

An attempt to utilize a control group failed. Two groups were set up, group A which received OE services and group B which did not. Within group A, we tracked commands worked against those we did not. Within group A, we worked all but two commands within the year, a pleasant but unexpected occurrence. A closer comparison revealed group B to be not useable as a control group. Though sharing the same home port and basic mission they had different immediate superiors and significantly different hardware technologies. A perhaps more significant factor is the way retention is reported based upon six and twelve month totals. The implication is that retention trends are good or bad based upon a comparison between what happened this month and seven or thirteen months ago. Raw monthly data was not available.

The retention data still proved extremely valuable. By tracking the data, commonalities were observed. In almost all cases the beginning of an OE intervention arrested retention drops for a period of time fig (1). This might be equated to crew expectation that conditions were going to change. Beyond feedback the retention picture, at least from the consultants point of view, is a factor of how much the command did with the feedback. In all but one case, retention stabilized or improved. That one case is where the command, admittedly, totally disregarded the data and did not provide crew feedback. The group A commodore begin to use retention as a criteria for scheduling individual commands for OE services. This knowledge combined with a picture of retention gave the consultant some gauge of client sensitivity around the retention issue.

With an awareness that HRAV and the Navy survey was failing to fulfill the needs of the client the spring of 1983 saw interviews, generally open ended, being utilized on a rather consistent basis. The concept of an interview was extremely popular with clients. Focused information, no necessity for waterfall/bubble-up and less time invested were all strong marketing points. Many consultants also readily accepted the interview as the way to gather data. Interviews were not without their difficulties. At higher levels, there was some concern because the survey supported a very large Navy data base. Interviews do not data base very well for the upper echelons to make decisions from. For the consultant in the field, the problem was much more immediate. The issue was how to analyze the interview. Nothing in the Navy consultants training prepared them for interview analysis. In retrospect, what happened was individual consultants analyzed data against their own internal models of how the Navy works. This resulted in some differences based upon background. In conceptualizing these interal models it is not surprising to find a very large block labeled command or leadership. The inability to conceptualize the internal model of the Navy organization was a serious flaw. The result was feedback packages with numerous factors tied together. The product, in some respects, looked more like the consultant than the client because of the integration effort. In some cases the consultant lost a degree of objectivity. The result was a feedback package that was too big, with too much analysis, too many recommendations and a client with a bloody nose. As one Commanding Officer said, "I feel like swiss cheese." The bottom line was any follow on development was done without the consultant and it took fourteen months for the first client with one of these feedback packages to voluntarily return.

The realization that while interviews have better information, they also exacerbated the threat led to systems modeling. The various models utilized have yielded consistently positive results. Internally the consultants have much more consistent analysis. As implemented, the models require less writing skills allowing junior personnel to share a greater role in writing feedback packages. The feedback format has remained fairly constant while what is actually placed on the subsystems has varied, including: verbatim comments, comment synopsis, and even single words. The decision is based upon the consultants perceptions and knowledge of the individual commanding officer. Resultant consultant/client interpersonal dynamics have been positive. Much of the consultants energy goes into briefing the model, and objectivity is maintained. The client has a much better understanding of the analysis process, the feedback looks more like the command and the threat is significantly lessened. This issue has been addressed with numerous Commanding Officer's and Executive Officer's. An interesting point is that the Commanding Officers recognized the threat issue while the Executive Officers did not. The openness that has been fostered has allowed the consultant to point out holes in the data and for the commanding officer to invite the consultant back for a follow-on. Together the client and consultant have been able to agree on the key systematic issues and the follow on actions required. The most dramatic turnaround occurred with the command which, as mentioned earlier, had disregarded their data. Because of a disasterous retention picture they were required to again receive OE services. Based upon a systematic approach, the commanding officer was able to accept the new data as valid, and to develop an action plan. Both sets of data were tested against a systems model and found to yield essentially uniform results. Of special significance is the fact that this commanding officer invited the consultant back to evaluate the entire retention program. This case is one of two which will now be covered in detail.

The first systems model utilized in an analysis was the Kast Rosenzweig (K&R) Organizational Model. There were three interview questions asked of an entire crew, utilizing horizontal slicing. The questions were:

1. How effectively do people communicate through the chain of command? What could be done to improve communications?
2. What could the chain of command do to better manage routine operations?
3. What one thing could this command do to keep you in the Navy?

Additionally the Navy survey was administered to the entire crew.

The interview verbatim comments were placed within the appropriate subsystems and analyzed as positive or negative based upon their impact. This data was analyzed looking for cause/effect relationships and discontinuities. Though the commanding officer received a complete copy of the verbatims, only key items were presented visually and discussed. No written recommendations were provided at this point. The data was debriefed based upon cause/effect relationships and as the commanding officer addressed the issues, suggestions were made as to possible changes which might satisfy both command and crew needs.

The data from both survey and interview indicated that some levels of middle management were significantly negative compared to other commands, but the reasons were not clear. The consultant was invited back to address this issue to those concerned and to make recommendations. These interview responses were also analyzed utilizing the K&R Model with a focus on the environment and teamwork of the involved individuals. Two key systematic issues surfaced. Environmentally, a majority of the crew had not accepted, as important, the political reality of this command nor had the command addressed the matter. The other issue was role ambiguity. The entire command had a very strong "can do" spirit. The bottom line was that doing any task properly and promptly had driven each level of supervision to do the subordinates job. Upper levels were so harried that "crisis management" was the accepted norm. All feedback recommendations were presented and evaluated on a cost/benefit basis. The command responsed in a typical "can do" task like manner to all identified issues. The positive effect on retention (Fig 2) has continued for some months.

The second case was extremely sensitive. This command had totally disregarded their previous feedback and was not wanting any type of OE services. Due to continuing poor retention (Fig 3 T-2) the immediate superior in command forced the command to work with the consultant. This was an undesirable mode for the consultant to operate in. The consultant's initial visit with the Commanding Officer was low key stressing trust, data confidentiality and a sincere desire to help. An intervention design was agreed upon. A special purpose survey was administered based upon individual needs, program knowledge and the extent needs were being met. Two interview questions were asked of a selected group with less than one year remaining or less than six months on board. The questions were "what must the command do to meet your personal and professional goals for the next six months?" and "what are your expectations?"

The feedback presentation process was more important than content, considering the previous failure and present pressure on the commanding officer. It was decided to utilize a graphic representation of the survey with each need, as prioritized by the crew, graphed against programs and information questions which correlated to that need. The entire survey was based on a standard Likert response scale and presented without comment or recommendation. The consultant determined not to utilize the K&R Model for the interview data because the center (implied main) subsystem was leadership, a very sensitive issue. The McKinsey 7S Model was selected based upon shared goals and values being the central subsystem, an identified interest of the commanding officer. The key issues identified were environmental, reward, procedural and middle management. The data was presented within the subsystem as single word comments. The consultants aim was twofold: First, to minimize threat, and second, to force the commanding officer to ask questions and become involved with those comments which were of interest. The goal was to establish some level of client ownership of the data. Following feedback, the consultant withdrew to allow the command to deal with the data. The command response was positive. Reading the data package was mandated down through the first line supervisor and open throughout the entire command. All middle management was involved in action planning and program development. The consultant was invited back for a program assessment at a later date. The consultant opted not to do the assessment, with command concurrence. Retention figures demonstrated that retention has turned the corner (Fig 3 T-3). Due to the large amount of outside intervention into the command process it was felt that an assessment might signal to the crew that something was still wrong. Though the program might need some degree of fine tuning, the potential risks and expense out weighed the possible gains. The consultant proposed to do Middle Management Skills Training based upon both previous data gathering efforts indicating that middle management was somewhat uninvolved. It was felt that improving middle management might strengthen the retention program. This proposal is presently under negotiation.

The events of the past seventeen months have proven that the Navy consultant does indeed need a variety of tools to perform the OE mission. It also appears that the consultant can impact retention if feedback is presented of a quality and in a manner which is useful to the commanding officer.

(1) Thomas, E.D., Human Resource Management Cycle: Effect on First Term Retention. (NPRDC TR 80-19)

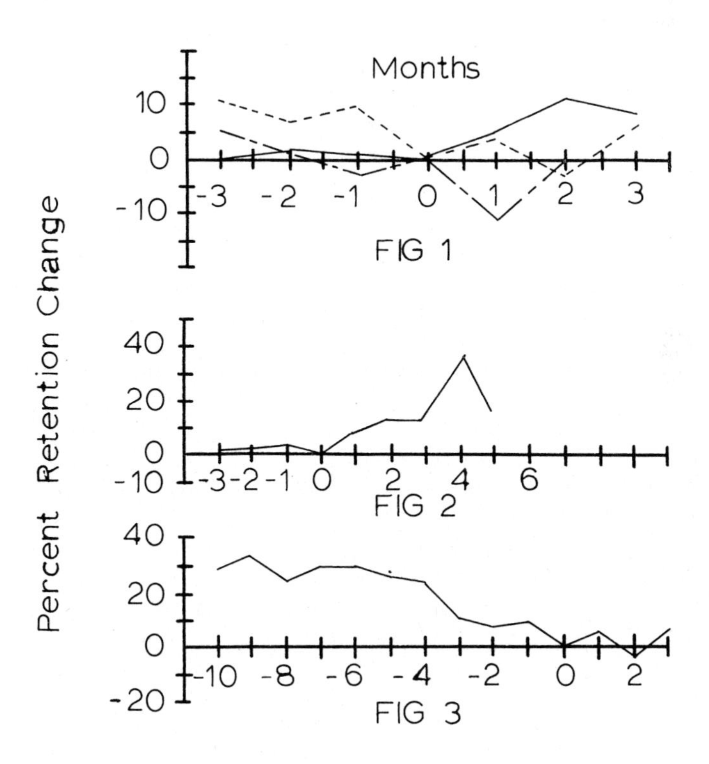

FIG 1

FIG 2

FIG 3

QUALITY OF WORK LIFE, ATTITUDES, AND VALUES

Human Factors in Organizational Design and Management
H.W. Hendrick and O. Brown, Jr. (Editors)
Elsevier Science Publishers B.V. (North-Holland), 1984

New technology and the effects on jobs

R G Sell
Work Research Unit
Department of Employment
Steel House
11 Tothill Street
London SW1H 9NF

Abstract

This paper looks at the effects on the jobs that people do of new technology and, in particular, that of developments in Information Technology and other computer advances. It makes the point that with the same technology change there are usually options which can either improve or degrade jobs: the degree of involvement of the employees in the change process can also affect the success with which the change is carried out.

Introduction

There is no stopping the development and introduction of new technology. There are, however, choices as to whether the changes which will result will benefit people at work or make jobs more de-personalised, routine and boring. The risk is that the latter situation will prevail, the aim of this paper is to show that it is not a necessary consequence. Whilst high levels of unemployment are likely to remain because the new methods make production per employee much higher than previously the concern of this paper and the Work Research Unit is with the still large proportion of the population who remain in employment (in the United Kingdom in 1984 some three million people were unemployed and some 23 million were employed). There are,of course, many agencies whose interests are with the unemployed.

The main technology change which is coming about is caused by the Information Technology explosion. This is reducing the costs and increasing the ease of obtaining information to a greater extent than ever before. The basic choice is whether this information is shared, leading to open participative styles of managing or kept for the use of a limited number of top managers leading to more authoritarian styles.

The views expressed in this paper are those of the author and do not necessarily represent those of the Work Research Unit or the Department of Employment.

The same technology change can be seen as either good or bad by those affected depending on the manner of introduction and the values of those responsible for managing the change and the new system. There is, in the UK, still a strong bias on the part of many managers to think of people as objects rather than as human beings and the technologists to want to pursue technology to its known limits.

With the development of IKBS there is a need to go back to consideration of the role of the operator and that of the machine as with the traditional allocation of function discussions of ergonomics. The technologist is likely to give the maximum to the machine whereas, from the point of view of the job design specialist, the man should be the master and not the slave.

There is no doubt that computers can improve processes like medical diagnosis because they can quickly evaluate a large number of alternatives. They cannot replace the doctor however and cannot cope with unforeseen and unprogrammed situations. Where a system cannot be made completely automatic, computers should not be used to monitor performance, only to aid it.

It is necessary to check the validity of any assumptions made. What makes for a good job depends on how the jobholder himself sees it and the observer may not see it in the same way. The author is conscious of a study he carried out on the London Underground Victoria line when it was first opened (Sell, 1970). The trains do not have a driver or a guard in conventional terms but one man who checks that the passengers are clear of the doors, closes them and only has to press a button for the train to go automatically to the next station.

Whilst this may be seen as de-skilling of the train driver's job the man himself sees it as changing him from an operator to a train manager. No longer does he have to keep the dead man's control pressed all the time; no longer does he have to wait for the guard to tell him when to start and with the improved communication system, he knows exactly what hold-ups there are over the whole line and whether he is likely to be held up or not. He has a job for which he feels responsible but, at the same time, he feels his job is well integrated into the whole organisation and the necessary knowledge has been shared with him.

The following sections look at the likely options and posible effects of various technological changes. They are not mutually exclusive.

The information explosion

As has been mentioned the main feature of the present technology development is the ease and cheapness with which inormation can be obtained, stored and processed and this is a feature which is still being developed. This information can be made available to a wide range of people so that they can understand what is happening throughout the organisation and can feel well integrated into it, as with the Victoria line operator.

On the other hand it can be kept to a select group of individuals and used as a management control device on the basis that 'information is power'. It can be used as a measure of performance in that, for instance, the number of key depressions on a visual display terminal can be recorded, sent to a central control position and the operator can receive feedback as the manager feels appropriate. A procedure which is more likely to alienate than integrate.

There is the risk that the information which is collected may be that which is easy to collect as with key depressions above but which may not be appropriate. VDT operations should only be a part of a person's job so any measure of performance should be based on the whole job, not just a part.

The author, in his work with the police, has heard vehicle location systems described by policemen as spy systems which can tell when they are at a cafe for a cup of tea and by management as a device for being able to send help if necessary without the driver having to say where he is. Vehicle location systems are likely to spread to other types of organisations and similar issues will arise.

Organisations will want to make use of the information they can have to analyse options and simulate alternatives so as to optimise their activities. However, they must become aware of the special knowledge which people have which they need to be able to use, eg routeing of delivery vehicles or salesmen's calls can be programmed by computer but the delivery driver or salesman will know himself better the up-to-date road conditions and should be able to take the computer print out as a recommendation which he can modify with his greater knowledge and experience. Similarly with production planning in a factory the production controller will know better the particular skills of different operator/machine combinations.

Systems designers will always need to be reminded of both the special abilities of people and the need to design systems optimise on both technical and social aspects.

Factory automation

The main development in factory automation is the introduction of the robot. This, as with the Victoria Line above, is seen by many people as de-skilling. It does, however, also mean that the operator is de-coupled from the machine; he does not have to have his 'hands on' all the time; he can organise his work himself; he is not fixed in one position; he is responsible for more output. He can in fact have a much more complete job.

There is though the situation that quality depends much more on the machine so the person who is highly skilled will not stand out so much as before as even the low skilled person will be able to turn out good work. What is still valued, however, is the ability to know when system performance is becoming degraded and what to do about it.

With some kinds of robots such as paint sprayers the operator has initially to 'teach' the robot the correct movements. As time goes on it might become very difficult for the human operator to maintain that skill without continuous practice.

With other kinds of systems controlled by tape modification to that tape may be required. Should the operator himself do it or should it be done by someone from the technical department? All the principles of job design say that it should be the man on the shop floor as it would help to maintain and lift up his skill level.

Whilst many routine jobs and those in bad physical conditions will be automated it is possible that, where it is difficult with the present level of technology to automate all jobs, some manual jobs will be included in a production line which are worse than those existing before. One Japanese VCR production line working on a nine-second cycle has 27 automated positions and one manual one. (The more recent line has no manual positions).

Non-factory automation

There is a risk that office jobs in the future will become more like the worse kind of factory jobs of the past with differentiated and monitored jobs like VDT data entry. As VDTs become even cheaper, however, it is likely that they will be used more widely and at all levels and that complete jobs will be created of which VDT operation is only a part.

Word processing is already becoming accepted as a tool available throughout organisations with their restricition to central pools becoming less. Again the skill aspect is important as authors see a perfect output, irrespective of the skill levels of the typist.

In some traditional organisations where the use of keyboards is seen as a lower level clerical skill some managers may be resistant to using them but the widespread use of home computers may change that. Should voice input become technically possible then great changes will happen.

The introduction of laser reading of prices in supermarkets, like automation in factories, should also not be seen as de-skilling but as allowing the check-out person opportunity for exercising human relations/sales person skills by actually talking to the customer and by ensuring that their needs have been met in the store.

Traditional ergonomic problems of posture, vision, lighting, glare, etc are now becoming more important in offices. In the UK the trade unions have become much more concerned than they ever did about factory problems of noise, posture, etc. This shows a developing awareness but it may be because they see the possible large-scale redundancies which could result and look at ergonomic problems as an acceptable displacement activity.

Introduction of complex equipment

One benefit of the reduction in cost of computer/electronic equipment is that complex inspection devices can be more cheaply supplied. This can mean that more operators can do their own inspection and then carry out their own rectification and re-testing signing off their own work thus increasing their level of responsibility.

It also means that, even though people may have less direct responsibility for the output, they can feel responsible for an important and complex piece of equipment.

One trend which has been developing for a number of years is the move towards more process production methods where the operator is separated from the actual product he is making and may not ever see it, feel it or smell it directly.

Improved communication

Better communication means that some people can work at computer terminals away from their normal place of work. When it means that they can stay at home there is obviously an advantage to those that would otherwise be unable to work because of physical handicaps or home ties. On the other hand, if others are forced to work at home then it will increase their social isolation. Similarly in factories it is possible that social isolation may be increased by people working terminals in rooms isolated from others.

Working at home will require a degree of trust over the amount of work which is done when on the spot supervision is not possible. It is also possible, however, there will be an increase in out-working on data preparation rewarded on a payment-by-results system which can be monitored automatically.

Changes in management structure

Logical developments in organisational structures as a result of the kind of technological change being discussed here is that they become flatter. With much more information handling being done by computer the basis for decisions becomes clearer and there is less need for as many levels of management as is usual at present. This can lead to more co-operation between the top and bottom with more responsibility being pushed to the operator level. Roles of people will need to be actively examined with management more of a supporting than controlling type.

Logic does not always come through however. If existing organisations do not change as indicated then problems of role conflict and role ambiguity will be prevalent with consequent stress for individuals. There will be a lack of trust with alienated managers whose real work has become redundant, engaged on checking and monitoring the work of others and so making them also more alienated.

Working hours

With the introduction of some complex equipment which may still be expensive by usual standards there will be pressure to have it working shifts throughout the day and night even when it is not technically essential. With systems requiring communications over long distances there will be pressure to work when telephone charges are cheaper as at night and at weekends.

This extension in shiftworking may be seen as beneficial by those who want to pursue leisure activities at times when most people are at work and to use scarce facilities but not by those who wish to pursue social activities with people working normal hours.

Occupational change

The changes in technology with the consequent increases in productivity will produce two sorts of occupational change. More people will have to be prepared to change their occupation and more people will be unemployed for larger proportions of their life unless very much shorter working hours become acceptable to both managements and workers.

What would be different in the future is that with the use of computers for more efficient decision-making, more management people will be affected and society has yet to come to terms with this.

Whilst this paper is not directly concerned with unemployment the effect that the fear of it has on people does affect their attitude to their job. In the UK culture any job is usually better than no job.

Change

The message that the experience of the Work Research Unit brings is that the more involvement and choice that people have in both the process and the result of change the more effective any change is likely to be. This is confirmed by the work that other organisations in the UK are carrying out such as the Human Sciences and Advanced Technology Group (HUSAT) at Loughborough and the Manchester Business School. (With its effective teaching and human implementation of computer systems: ETHICS).

A number of surveys have shown that many computer systems fail to work as specified. A recent survey of a computer-based package, the MRP II System for materials and resource planning, which requires a high degree of user commitment and involvement showed that 80% failed to work properly because not enough people had been involved to develop the required level of commitment (Woolcock, 1984). For people to change people from one kind of skill to another is difficult and takes time. When new systems are being instituted, especially if they involve new powerful computing techniques, users find it difficult to understand the potential of the system and they may find it difficult to see how they would like it to work.

Experience has shown that, if flexible systems can be provided which allow users to choose how to design their jobs when the equipment is on site or if they can choose when to move over as when word processors are provided without taking the typewriters away, then the changeover is more smoothly made. If choice is not offered then people are likely to avoid the system, not feeding information in and so the general level of trust will drop and it will fall into disrepute and be abandoned.

References

Sell, 1970 'The Victoria Line - operational aspects', Applied Ergonomics 1, pp 113-120

Woolcock, 1984 'The best laid plans of MRP II is .....' Technology 8, 9 April 1984, p 10.

Human Factors in Organizational Design and Management
H.W. Hendrick and O. Brown, Jr. (Editors)

IMPLICATIONS OF NEW TECHNOLOGY ON WORK ORGANISATION A CASE STUDY

EASON K D and GOWER J C

HUSAT Research Centre
Department of Human Sciences
University of Technology
LOUGHBOROUGH, Leics

A case study is reported of the effects of office automation on work organisation. Organisational change was not intended but the technology undermined the traditional demarcation of roles in the organisation and, in particular new strategies for using secretarial resources began to emerge. As a result, the technical change is seen to create an impetus for evolutionary organisational change.

INTRODUCTION

Many companies are now going through the pains of organisational change in the wake of introducing office automation. There have been many predictions of major organisational change as a result of this technology; the end of secretaries, a massive shift to working from home, etc (see Toffler 1980). It came as something as a surprise to us to find in a recent survey of job changes brought about by word processing (Pomfrett et al 1984) that many companies were introducing the technology without intending organisational change or recognising that it may occur. The aim in many cases appeared to be to replace the typewriter with a word processing system and leave everything else unchanged. Our evidence is that this is a pious hope; the technical change has indirect consequences which unsettle existing job structures and can create stress and ineffectiveness.

In this paper we present a case study of one organisation going through this transition. We present it primarily as an example of the indirect effects the technology can have on role definition. We did not undertake this case study as passive researchers; we were active in helping the organisation recognise the changes taking place and confront the policy implications for organisational structure. In the limited space available we will however concentrate upon describing an organisation undergoing change rather than focus on our role as change agents.

THE TECHNICAL CHANGE

The organisation is a city firm run by a partnership of forty six people. The firm is renowned for the high quality and speed of its work which we were to find are essential requirements in professional city firms. Because of this there is a substantial commitment to provide the best support possible to the technical fee-earning staff. They were more concerned to get good quality support from, for example, secretaries, and were not seeing information technology as a way of reducing support staff. An initial survey showed that the organisation of the support staff was as shown in figure 1.

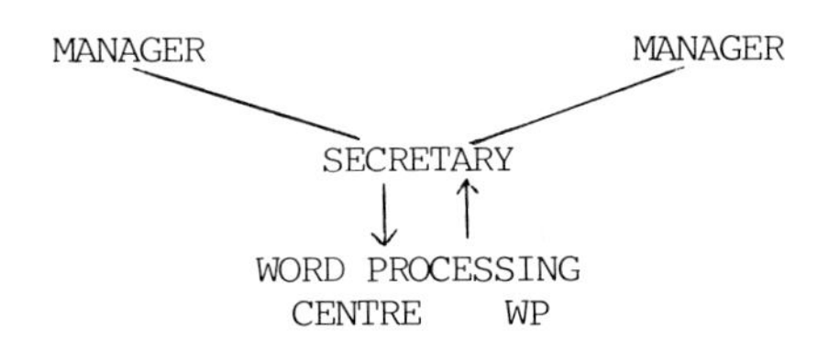

Key :

WP = Word Processing
Terminal

FIGURE 1
Organisation of support services before further introduction of technological support.

Such a structure creates a secretarial job profile which our investigations found consisted of four basic activities :

a) Secretarial Typing : Letters, memos, telexes and short documents

b) Long Documents : Lengthy reports and technical documents

c) Administration : Telephone, filing and personal assistance to technical staff such as managers

d) Technical Support : Assistance with routine, non-specialist technical activities.

With support services organised as above, long documents were sent to the centralised word processing unit which enabled the secretary to concentrate upon the other three activities. Also, as each secretary used a golfball typewriter, rather than equipment which allowed easy error correction, drafting of long documents was very difficult.

The new system to be introduced was a distributed 'administrative system' capable of word processing, communication and information search and retrieval. The aim was to provide each secretary with access to this system; it was not to change the jobs of any members of staff. We helped to create a high level steering committee to manage the introduction of this system, which we then advised. As a result of its deliberations, a new organisational structure, depicted in figure 2, emerged. The new addition is 'floor level' word processing centres as well as the main centre. These 'floor' centres enabled long documents to be handled locally so that authors could obtain a more personal and faster service than they could get from the main centre.

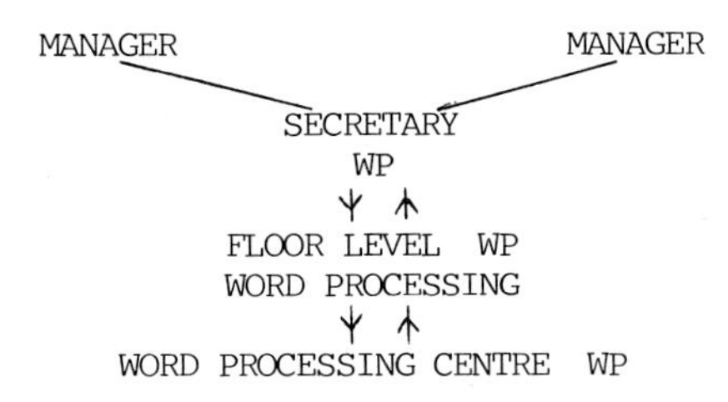

Key :

WP = Word Processing Terminal

FIGURE 2
Organisation after the Technical Change

Whilst this structure sustained existing jobs, it created a new layer in the structure and had important implications for job content and role relations.

ROLE IMPLICATIONS

Making available to many people in an organisation a very powerful tool for processing information changes people's perception of what can be done and who should do it. In this firm it changed the managers' perception of what secretaries and word processing staff could and should do. In this organisation there was no standard policy on the use of support staff and, as a result, we found a number of strategies emerging in the use of secretaries. The stimulus for the change lay in the fact that use of word processing equipment should, in theory, improve secretarial productivity, eg, it takes less time to correct errors, standard paragraphs can be used, etc. Thus, if all other things remain equal, the secretary should be left with some spare capacity. The strategies which developed were as follows :

a) Increased Secretarial Typing

The increased capacity of the secretaries could be used by authors who could use the system to create more drafts of the same document. They may start from less well-constructed drafts or become more particular about the finished document. Effectively, the secretarial workload is expanded without a corresponding increase in the volume of output. This situation was disliked by secretaries but some of the authors found it helpful to their style of producing documents. There was considerable evidence that this was happening.

b) Increased 'Spare' Time

Where the amount of time spent on secretarial typing is reduced because of the word processing equipment, the secretaries should have some extra capacity. Where there is no scope to increase administrative duties or other forms of support, the capacity may be surplus. One suggested use of such time was to require secretaries to work for more authors, thus increasing their workload. This possibility creates flexibility in the support to authors but was actively disliked by both secretaries and authors because it was a move towards a 'pool' concept.

c) Increased Administrative Time

Any time gained could be used to increase the time spent on administrative duties. This assumes that such duties are available and that the secretary is capable of performing them. This situation could lead to more support for authors and some secretaries would welcome this increased range of duties. However, there are some substantial negative implications. Some secretaries who want to do it may not have the work, some secretaries may be unable to do it. Also, role differentiation may lead to tension and increased use of the system for administrative duties may create greater author dependence on secretaries. It was a strategy that tended to be in use by senior managers and by technical staff who had administrative work appropriate for non-technical support staff to conduct.

d) Emergency Word Processing of Long Documents

Because secretaries have the potential to undertake word processing work which would previously have gone to the word processing centre, it is possible that the authors might utilise such potential in an emergency.

Such a strategy provides support for the authors in times of need and creates personal service where before it had been impersonal. There is also a greater involvement of the secretaries in the author's work. Conversely, such a situation would create occasional long hours and the related health problems of working in front of a visual disply unit for many hours. There may also be some resentment by some secretaries who may feel that they are being turned into word processing operators. It is also likely that these duties will take precedence over other duties and administrative duties will be curtailed. The secretary would also find it difficult to be fair to both the authors for whom she works. Finally, the role of the word processing centre would be diminished. This strategy appeared to be becoming prevalent.

e) All the Word Processing Needs of the Manager
The previous implicit 'rules' about sending long documents to the word processing centre are forgotten and the secretary does all the work. Such a practice would erode the administrative capacity of the secretary and would create a word processing operator/secretary. Such a dramatic change in the role of the secretary would probably lead to the loss of secretaries who do not wish to become word processor operators. The authors would lose their present administrative support and the question of fairness to both authors would arise again. There would also be a dramatic change in the role of the word processing centres. Few managers or technical staff had gone this far, but there was considerable fear amongst secretaries that it might develop that way in the future.

OTHER DEVELOPMENTS

In addition to the effects on the role of the secretary, effects were also beginning to be seen on the roles of the other staff, particularly the managers and the word processing staff.

i) The Managers An evaluation study was conducted to examine whether the managers felt the new system was performing satisfactorily. It yielded positive results; most managers reported faster turn round of documents, faster handling of amendments, better quality output and a more personal service. These were the effects of a powerful text processing tool in the hands of someone who gave them a personal service. The evaluation also displayed some changes in the managers' behaviour. Many, for example, were working through more drafts than previously and some were passing more administrative duties to their secretaries. A development which was just beginning was the use of the technology directly by the managers. This was driven by two different forces. First, some of the younger managers could see a value in using the system themselves and secondly many were becoming aware they were much more dependent on secretarial and word processing staff than hitherto. With documents and files stored electronically, managers were cut off from their work when the secretary was away or it was out of normal hours. As a result a debate was beginning about the appropriate type of training and access for managers to the technology with secretaries fearful of inexperienced hands inadvertently destroying files and junior managers fearful of expectations developing that they could do most of it themselves.

ii) The Word Processing Centre The main Centre began to change its role as the technology spread through the organisation. Instead of being the place where all the work was undertaken, it continued to do certain kinds of work, to provide an evening service and, above all, to be the operational centre for the system and a training and support centre for other less experienced staff. There was some evidence that in time they would move towards becoming what is now known as an 'information centre' (Eason 1984), a repository of knowledge about the information and information processing resources of the organisation:

CONCLUSIONS : THE EVOLUTIONARY DEVELOPMENT OF SOCIO-TECHNICAL SYSTEMS

The major implications of this case study are that this form of technology gradually calls into question the role definitions prevalent in an organisation as staff find appropriate ways of harnessing the new capability at their disposal. As a result jobs begin to change. This is not a planned, deliberate form of job design as often depicted in the job design literature and in our view cannot be because change has to be the result of individual & institutional learning about the role of the technology. At best organisational change becomes an evolutionary development as the institution recognises that new socio-technical structures are possible and desirable (Eason 1982). At worst it becomes a process of organisational drift as old structures are formally maintained whilst informal new practices are created and staff suffer partially recognised forms of role ambiguity and stress and fear for their task territory within the organisation. The challenge for the organisational change agent in these circumstances is, we feel, to help individuals and institutions perceive these symptoms of change and find effective ways of planning evolutionary development.

REFERENCES

(1) Eason K D. The Process of Introducing Information Technology, Behaviour and Information Technology 1.2 (1982) 197-213

(2) Eason K D. The Continuing Needs of End Users, in Firnberg D (ed) Information Centres (Pergamon - Infotech State of the Art Report, Maidenhead, 1984)

(3) Pomfrett S M, Olphert C W and Eason K D. Work Organisation Implications of Word Processing, in Shackel B (ed) 'Interact 84' (North-Holland, Amsterdam, 1984)

(4) Toffler A. The Third Wave (Collins, London, 1980).

Human Factors in Organizational Design and Management
H.W. Hendrick and O. Brown, Jr. (Editors)

HUMAN FACTORS REVIEW OF CHANGE IN PREFERENCE OF AUTOMOBILE USERS IN JAPAN

Takeshi KURABAYASHI
Hiroshi IKEDA
Japan Institute of Synthetic Technology,LTD
1-8 KOUJIMACHI CHIYODAKU JAPAN

Automabile users in Japan are diversifying with a sharp increase in the number of woman drivers. So are the determing factors for the purchase of cars. Cars are increasingly used by housewives and young women for "shopping and running errands."They account for 16% of the automobile users according to a survey by the Japan Automobile Manufactures' Association. Automakers' quality control rechnology and development capability have advanced to such a degree that users place total reliance on the performance and drivability of their cars. Appearance and style have assumed importance as deciding factors for the selection and purchase of cars. In 1975,15% of the users decided on the purchase of ergonomics in this situation is clarified, and automobile factors truly improtant for man are discussed in addition to factors important for car sales strategies.

CHANGE IN NUMBER OF AUTOMOBILES OWNED IN JAPAN

Since the oil crisis of 1974, Japan's economy has been growing stably at low rate to date, but the number of automobiles has been smoohtly increasing. The number of automobiles owned in Japan rose by about 18 million from 1973 to 1983, a figure equivalent to the number of automobiles ownid in Britain in 1982.[(1)] Ninety percent of the automobiles in Japan are passenger cars. The passenger cars increased at an average rate of 1 million per year from 1980 to 1983 and are expected to increase at this rate in the future, too. Japan has the world's second most automabiles at about 43 million, following the United States with some 160 million units.[(1)]

SPREAD OF AUTOMOBILES AND RESULTANT CHANGE IN LIVING ENVIRONMENT

As more and more people came to own automobiles between 1970 and 1975, Japan had frequent automobile pollution problems. In particular, the occurrence of photochemical smog prompted the public to turn critical eyes to automotive emissions. As a result, 1975 saw the establishment of automotive emission control standards and the start of work on the development of automotive engines to meet the emission control standards.

Consequently, automobile development technology has advanced further and users have come to place strong confidence in the hardware functions of cars.[2] Japan's automobile ownership density is one of the highest in the world. Accordingly, traffic noise is also one of the largest in the world.[3] Japan is slightly smaller in area than Sweden and Spain, but owns about 13 and 4 times as many automobiles as the two countries do, respectively. Despite the high automobile ownership density, traffic snarls and accidents are not so severe and frequent. This is attributable to the improvement of highways and other automobile related facilities and the provision of accurate traffic information to drivers. The increase in traffic in limited time zones only is another major reason. The fact that traffic snarls are not so extensive for the number of cars indicates that the Japanese have changed their use of cars.[4] Japan Automobile Manufacturers' Association surveys of 1979[5] and 1982[6] show that the households whose head drove the car and the households whose female members drove the car accounted for 68 and 10%, respectively, of all of the households that owned passenger cars in 1979 and that the corresponding figures changed to 56 and 16%, respectively, in 1982. Female drivers are increasing in this way and are predicted to exceed 20% by 1985. In view of these findings, the increase in the number of automobiles owned in Japan is largely ascribable to the women who have bugun to drive cars.[7] The woman drivers use their cars more frequently for shopping and running errands than for commuting and business. Therefore, they drive over short distances in the day time zones when traffic is relatively light.[8] This is one of the reasons why extreme traffic jams rarely occur.

EXPANSION IN USAGE OF AUTOMOBILES, MAINLY BY FEMALES

As women, particularly housewives, have started to drive cars, car usage has been diversified.[8] According to the results of 1981[9] and 1983[10] surveys by the Japan Automobile Manufacturers' Association, 54% of the housewives questioned drove their car for the purposes of shopping and running errands in 1981, and this percentage rose to 62% in 1983. Of the male drivers polled, a mere 12% used their car for shopping and running errands. Of all the people surveyed, 16% used their car for these purposes. This means that the number of passenger cars will increase further if more people own them for the purposes of shopping, running errands or leisure. Most of such drives are female. Japan's automakers should have woman drivers in mind when they forecast the demand for their passenger cars.[7],[8]

DIVERSIFICATION OF CRITERIA FOR SELECTING AUTOMOBILES

The Japanese place great reliance on their cars and especially absolute confidence in the mechanical performance and safety of the cars. When they select cars, therefore, the Japanese focus more attention on ergonomic aspects rather than mechanical reliability. The increase in the number of female drivers has mainly increased the car users and diversified the usage of cars. The carmakers are starting to produce and supply many models of automobiles to meet a variety of users' demands. The author's investigation shows that Japan's automakers produced 1,159 and 2,281 models of passenger cars by body type in 1974 and 1984, respectively, a twofold rise in the number of passenger car models in ten years. When commercial vehicles are included, 3,841 automobile models are produced in Japan now. When the automobile colors specified by users are taken into account, the user can select one passeger car from among 10,000 different models.

ROLE OF ERGONOMICS

Before 1985, the Japanese economy attained rapid growth. Then, most users emphasized engine running performance, size and style of cars when they decided to but one.[10] When the users replaced their current cars with a new one, they strongly tended to select passenger cars of larger dispalcement and better style. Since the oil crisis of 1975, interest in appearance and style has been growing yearly. Interest in drivability, ridability and fuel economy has been rising as well . The criteria by which people select the passenger cars to purchase are mainly ergonomic ones as described above, but the functions the users expect of passenger cars to purchase are mainly ergonomic ones as described above, but the functions the users expect of passenger cars have greatly diversified. A 1982[6] survey by the Japan Automobile Manufacturers' Association indicates this trend, as represented by the following data:

(1) Drivability is expected by 76.8% of the users.
(2) Ridability is expected by 61.1% of the users.
(3) Appearance and style are emphasized by 53.5% of the users and particularly strongly by 71.3% of the young users of 24 and under.
(4) Stability at high speed is emphasized by 50.7% of the users.

These items do not widely differ among users of different ages and sexes, but drivability is judged by different users according to different criteria. Automobiles equipped with automatic transmission and power steering seem ideal from a mechanical point of view, but only about 47% of the users want such cars.[10] Desigh parameters depend on how the automakers grasp this fact. They now highlight design features of their products as specific sales points. Ridability is also evaluated by users of different ages and sexes from different standpoints. In mechanical aspects, vibration and noise are chech items. Automakers have different criteria by which to judge what are comfortable vibration and noise, and there are no generalized indexes of these characteristics of passenger cars. Noting the increase in the number of female drivers, the author investigated the comfort of driver seats and the ease of driving sitting on specific seats. The results of the investigation show that male and female drivers have no large differences in the comfort of the driver seat, but complain many problems with the comfort of the seat and the ease of driving sitting on the seat. Such problems are concerned with differences between men and women in body dimensions. As it was strongly recognized in the past that automobiles were mainly intended for men, the driver seats are probably chiefly for men in terms of height, space and width. This explains the condition that many woman drivers place a cushion between their back and the seat. As they cannot comfortable reach the brake and accelerator pedals, many female drivers have the seat moved to the foward limit. Many women also drive their cars with their body almost touching the steering wheel. Today in Japan, there are no dependable data on differences between men and women in body dimensions. Accurate anthropometric data are available only on high school students and younger children and are concerned only with stature, weight, chest circumference

and sitting height. Therefore, each carmaker designs the driver seats of their cars on the basis of data abtained through individual efforts. According to a 1981 survey by the Tokyo Metropolitan Government, 16 years old male and female high school students differ about 13 cm in stature, 10 kg in weight and 5 cm in chest circumference. The difference in stature between men and women tends to decrease with age and the difference in stature between people of 40 and 16 years old is estimated at about 10 cm. In terms of stature, small and tall people differ 20 to 30 cm. Automobile manufacturers must more positively consider ergonomics to meet these varying body dimensions of drivers. At present, the automakers appear to sell their cars by images in terms of appearance and style, but not to have much interest in appearance and style from and ergonomic point of view. The items of a questionnaire by a Japanese carmaker to users are listed below for reference. The check items related to drivability are:
(1) Lightness of steering wheel
(2) Ease of backing and ease of garaging
(3) Operability of transmission
(4) Arrangement of pedals
(5) Ease of assuming driving posture
(6) Forward and backward visibility
(7) Readability of meters and instruments
The chek items related to ridability are:
(1) Comfort of driver seat
(2) Comfort of back seat
(3) Legroom of driver seat and back seat
(4) Comfort of ventilation and air conditioning
(5) Overall comfort
(6) Noise and vibration during running

It is delightful for professional ergonomists that automakers are showing increasing interest in ergonomic aspects when designing cars. At present, however, ergonomists do not play an important role in the design of automobiles, probably because data on body dimensions of drivers are meager. Ergonomists tend to turn their eyes to people in special conditions. They must pay more attention to general people in ordinary conditions. To put it more specifically, the ergonomists must be more interested in anthropometry and strive to gather authoritative data. They must also collect data on the body conditions of people in normal action. It is important for them to share these data and produce common and highly reliable data on human beings. If the ergonomists can reach common recognition from this standpoint, they should be able to play an effective role in the design, development and sale of automobiles.

References

(1) "Monthly Automobile Statistics (in Japanese)," Japan Automobile Manufacturers' Association, No. 3' 1984.
(2) "Automobile Industry (in Japanese)," ibid., Vol. 17, No. 3, 1983, pp.3-12.
(3) "Consideration of Environment and Pollution (in Japanese)," Tokyo Metropolitan Board of Education, 1983.
(4) "Traffic Yearbook (in Japanese)," Tokyo Metropolitan Traffic Safety Association, 1982.
(5) "Passenger Car Demand Trend Survey (in Japanese)," Japan Automobile Manufacturers' Association, 1979.
(6) ibid., 1982.
(7) "Automobile Industry (in Japanese)," ibid., Vol. 17, No. 1, 1983.

(8) "Women and Automobiles (in Japanese)," Toyota Motor Co., December 1983.
(9) "Passenger Car Demand Trend Survey (in Japanese)," Japan Automobile Manufacturers' Association, 1981.
(10) ibid., 1983.

Human Factors in Organizational Design and Management
H.W. Hendrick and O. Brown, Jr. (Editors)

Variables Influencing QWL and Job Performance in an Encoding Task

Paul T. Cornell

Human Factors Activity
Burroughs Corporation
Plymouth, Michigan
U.S.A.

A four factor model of the work environment was evaluated in a study of bank proofing operations. A total of 22 main effects and 50 interaction terms were used to represent the physical environment, organizational setting, job content and individual differences. Seven variables represented quality of work life and performance. The data were analyzed using a stepwise multiple regression. Fifty percent of the variance in satisfaction and 55% of the variance in performance was accounted for. The regression equations were qualitatively and quantitatively dissimilar, suggesting that the variables which contribute to satisfaction are different than those which contribute to productivity.

Two multidimensional categories of measurement are important when assessing work environments: the quality of work life (QWL) and job performance. It is often suggested that a causal relationship exists between these, i.e., high satisfaction leads to improved performance. Others have suggested the converse, i.e., superior performance contributes to satisfaction [6]. Unfortunately, empirical support for either contention is meager [1].

The practice of job design has survived a variety of theories and methodologies that have attempted to improve job performance and/or satisfaction. Each of these had, as its primary focus, a different aspect of the work context. Nonetheless, to varying extents all these approaches acknowledged the multidimensionality of work. Conversely, job design research has not always incorporated the concept of multidimensionality into dependent, independent or control variables. An exception is the work by Rousseau [9]. The piecemeal approach to the study of work is particularly noticable in recent research examining the impact of technology on jobs. A noted exception is the study by Sauter et al. [10].

Following the approach of Rousseau and Sauter, a model of the conditions which influence performance and QWL was developed and evaluated. This model contains four factors each of which is hypothesized to influence performance and QWL. The factors are labelled job content, physical/environmental, organizational/social and individual differences. The core of the model is based upon the job characteristics approach of Hackman and Oldham [4]. A major tenet of the model is that the four factors contribute both interactively and differentially to QWL and performance. Hence, all four factors and their potential interaction must be considered when designing or improving work environments.

A high productivity, data entry task was used as a test case for the model. It was hypothesized that the variables which contribute to performance would not be the same as those which contribute to QWL. It was also hypothesized that, using a multivariate statistical analysis, significant interactions would be found, suggesting that variables should not be viewed in isolation.

METHODS

Subjects. Proof encoding employees and supervisors from seven U.S. banks participated. Proof encoding is a data entry task which requires operators to encode numeric data on checks. The volumes of checks which must be processed daily puts a premium on productivity, or documents per hour. At the same time, accuracy must be near perfect. Participation was voluntary and agreed to in advance by bank management. A total of 163 employees completed questionnaires. The banks were located in Maryland, Ohio, Illinois and California.

Apparatus. The questionnaire consisted of the Job Characteristics Inventory [11] and Sections 3 - 7 of the Job Diagnostic Survey [4]. Context satisfaction was a summation of the four JDS context measures. The questionnaire also requested biographical and physical stress data. A list of the variables is given in Table 1.

A list of topics was covered in the semi-structured management interview, including Pugh et al.'s [8] formalization and centralization. Training was defined as the number of formal programs required of new employees. Span of control was defined as the ratio of the number of supervisors to the number of operators in the department.

An observational checklist was developed from a variety of ergonomic guidelines ([2], [3], [5] and [7]). These were grouped into the four variables listed in Table 1. Values for these variables ranged from 0 to 9.

Procedure. Approval to conduct the study was received from bank management prior to visiting the proofing departments. After approval, the department supervisors were contacted and sent copies of the questionnaire.

The banks were visited during normal working hours. Questionnaires were typically distributed first followed by the interview. Employees were briefed on the intent of the study and informed that their participation was voluntary. They were also given a one-page summary of the study. Participants usually completed the questionnaire at their workstation in 20 to 30 minutes. The researcher was available throughout this period to answer questions. The interview took place with an immediate supervisor from the department.

RESULTS

Variable Screening. To increase the power of the results, a preliminary analysis was conducted to determine whether any variables could be eliminated prior to the regression analysis. Seven were dropped due to near-zero variability. These included three of the physical/environmental and two of the organizational variables. This is not surprising given the similarity in department organization, function and equipment across banks.

Gender was dropped as a variable because 89% of the participants were female. Marital status was dropped for similar reasons. "Kneeroom" (one equals acceptable, zero equals poor) was substituted for physical accommodation as this turned out to be the only major difference between workstation design. Retained variables are denoted in the right-most column of Table 1. The analysis examined the effects of the variables on general satisfaction and productivity only.

Table 1
Variables Measured in the Study

Factor	Variable	Measured	Retained
Physical/ Environmental	movement accommodation	observation	no
	physical accommodation	observation	kneeroom
	chair design	observation	yes
	visual characteristics	observation	no
	noise	observation	no
Job Content	skill variety	questionnaire	yes
	autonomy	questionnaire	yes
	task identity	questionnaire	yes
	feedback	questionnaire	yes
Organizational/ Social	formalization	interview	no
	centralization	interview	no
	training	interview	yes
	span of control	interview	yes
	department size	interview	yes
Individual Differences	gender	questionnaire	no
	age	questionnaire	yes
	years of education	questionnaire	yes
	height	questionnaire	yes
	marital status	questionnaire	no
	experience	questionnaire	yes
	fulltime	questionnaire	yes
	context satisfaction	questionnaire	yes
	growth need strength (GNS)	questionnaire	yes
Quality of Worklife	general satisfaction	questionnaire	yes
	motivation	questionnaire	yes
	growth satisfaction	questionnaire	yes
	physical stress	questionnaire	yes
erformance	productivity	questionnaire	yes
	turnover	interview	no
	absenteeism	interview	no

Interactions. A number of interactions, computed as cross-product terms, were generated based upon the literature and intuition. From an ergonomic viewpoint, the physical/environmental variables should interact with the individual difference variables of height, age and fulltime, resulting in six terms. Since job content has been hypothesized to be moderated by individual differences, these variables were crossed with education, age, experience, GNS and context satisfaction to form 20 new variables. Organizational/social variables could interact with individual differences, and were combined with age, fulltime, education and context satisfaction, creating 12 terms. Job content and organizational/social interactions were also considered, adding another 12 terms. Given the particular variables measured in this study, it seemed unlikely that the physical/environmental variables would interact with the job content or organizational/social variables.

Stepwise Multiple Regression. General satisfaction and productivity were regressed on the sixteen main effect variables and the 50 interaction terms. The results are shown in Table 2.

Table 2
Multiple Regression Results

	General Satisfaction			Productivity		
Source	SS	df	MS	SS	df	MS
Regression	118.479	7	14.810	11750880.0	6	1678697.0
Error	105.489	132	0.805	8787330.0	116	76411.5
Total	223.968	139		20538210.0	122	
F			18.391			21.969
R^2 adjusted			0.500			0.546

Variable	t*	Variable	t*
age	2.896	age	-4.926
GNS	4.420	context	2.602
span of control	2.717	size	-4.141
fulltime	-3.671	education	1.934
size x context	8.839	training x age	8.017
size x GNS	-8.092	task ident x experience	6.122
size x skill variety	3.159	autonomy x education	3.095
training x fulltime	2.323		

* $p < 0.05$, df = 60, $t \geq 1.671$; $p < 0.025$, df = 60, $t \geq 2.000$

The satisfaction data suggest that older operators were generally more satisfied with their job. The data also indicated that departments with a higher management to employee ratio had more satisfied employees. Some of the interactions, however, were more difficult to interpret. A subgroup analysis of the three interactions with size indicated that the satisfaction relationship was different for smaller departments. The interaction with context indicated a stronger relationship between context and satisfaction for larger groups. The GNS results did not agree with the main effects, i.e., there was a strong negative correlation between GNS and satisfaction for larger units. Skill variety appeared to be more strongly linked to satisfaction in smaller departments. The fulltime and training interaction suggested that fulltime operators who receive more training are generally more satisfied.

The productivity results indicated that operators who are satisfied with the organizational context (i.e., co-workers, supervision, job security and compensation) are more productive. Combined with the main effect of age, the training/age interaction suggests that younger operators perform better, but that older operators respond to training more readily. A subgroup analysis of the identity/experience interaction revealed that identity contributed positively to performance for less experienced employees, but had little effect for the more experienced. A similar analysis of the autonomy/education data showed that job autonomy enhances performance for those with more than a high school education.

DISCUSSION

An important finding was that the variables which contributed to job satisfaction were different than those which led to productivity. This does not support the belief that satisfied employees are more productive. In fact, the correlation between satisfaction and productivity was -0.240. This finding may be the result of many factors; e.g., one could imagine that a satisfied employee may tend to socialize more, reducing the time spent encoding.

Variables representing three of the four factors in the model contributed to both satisfaction and productivity. This supports the contention that a multidimensional approach is necessary. In addition, there were many strong pairwise correlations in the data that did not appear in the regression. The number of the significant interactions supports the view that multivariate analyses are more appropriate in this kind of research.

The lack of statistical significance of the physical/environmental variables could be due to the small variability of these measures. More sensitive ergonomic measures may yield different results. However, it is also possible that the effects were outweighed by the presence of the other variables, i.e., the others were more important contributors. The correlation between kneeroom and productivity was 0.206, suggesting that viewed in isolation, this variable may have been significant.

REFERENCES

[1] Blumberg, M. & Pringle, C.D. The missing opportunity in organizational research: Some implications for a theory of work performance. **Acad. of Mgmt. Jour.**, 1982, **7**, 560-569.

[2] Cakir, A., Hart, D.J. & Stewart, T.F.M. **Visual display terminals.** New York: Wiley & Sons, 1980.

[3] Grandjean, E. **Fitting the task to the man, an ergonomic approach.** London: Taylor & Francis, 1980.

[4] Hackman, J.R. & Oldham, G.R. **Work redesign.** Reading: Addison-Wesley Publishing, 1980.

[5] Konz, S. **Work design.** Columbus: Grid Publishing, 1979.

[6] Locke, E. The nature and causes of job satisfaction. In M.D. Dunnette (ed.), **Handbook of Industrial and Organizational Psychology.** New York: Wiley & Sons, 1976.

[7] Niebel, B.W. **Motion and time study.** Homewood: Irwin, Inc., 1982.

[8] Pugh, D.S., Hickson, D.J., Hinings, C.R. & Turner, C. Dimensions of organization structure. **Admin. Sci. Quar.**, 1968, **13**, 65-105.

[9] Rousseau, D.M. Characteristics of departments, positions and individuals: Contexts for attitudes and behavior. **Admin. Sci. Quar.**, 1978, **23**, 521-540.

[10] Sauter, S.L., Gottlieb, M.S., Jones, K.C., Dodson, V.N. & Rohrer, K.M. Job and health implications of VDT use: Initial results of the Wisconsin-NIOSH study. **Commun. of the ACM**, 1983, **26**, 284-294.

[11] Sims, H.P., Szilagyi, A.D. & Keller, R.T. The measurement of job characteristics. **Acad. of Mgmt. Jour.**, 1976, **19**, 195-212.

Human Factors in Organizational Design and Management
H.W. Hendrick and O. Brown, Jr. (Editors)

INCREASING WHITE COLLAR PRODUCTIVITY BY HUMANIZING THE WORKPLACE: A DIAGNOSTIC APPROACH

Dr. James L. Freeley

Associate Professor of Management
Long Island University
Greenvale, New York
U.S.A.

Dr. Mary Ellen Freeley

Assistant Professor of Education
College of Mount St. Vincent
Riverdale, New York
U.S.A.

Faced with the challenge to increase white-collar productivity in a service dominated economy, more sophisticated approaches toward humanizing the workplace must be developed. Toward this end, an overview of an instrument known as the Productivity Environmental Preference Survey is presented which identifies the individual working styles of participants. By diagnosing an individual's preferred working characteristics, the workplace can be adapted to accommodate employee needs resulting in increased effectiveness and productivity.

INTRODUCTION

As the United States faces the challenges of the post-industrial society, there is an increasing need to develop new approaches toward managing human resources in a service dominated economy. The accelerating pace of technological change is having a substantial impact on the ability of organizations to adapt to a rapidly changing environment.

The decade of the 1970's ushered in the era of the post-industrial society in this country. For the first time ever in any industrialized nation, the role of service oriented enterprises became dominant in terms of employment of the work force. The post-industrial or service economy is not a new term, but its impact on the infrastructure of society has yet to be fully understood and dealt with in terms of increasing the productivity level of human resources.

Although the seventies did not emphasize the critical dichotomy between the manufacturing and service sectors and allowed for co-existence between them, the acceleration of the post-industrial economy accompanied by the substantial growth of the white-collar worker is forcing Management to re-evaluate and question the validity of the techniques used in the past. One of the most crucial problems to be addressed is that associated with improving white-collar productivity.

According to the most recent figures available from the Bureau of Labor Statistics, service oriented businesses now employ a record proportion of the work force - nearly three of every four workers on non-farm payrolls. Overall, more than 50% of all employees in the United States are now considered white-collar.

As a further indication of the shift away from a manufacturing or blue-collar economy toward a service or white-collar one, the March 8, 1982 issue of Business Week amply demonstrated this trend. In discussing computer innovations such as the use of robots on the assembly line, Raj Reddy of Carnegie-Mellon stated that "currently, around 25-28 million people are employed in manufacturing in America. I expect it to go down to less than 3 million by the year 2010." This trend is also reflected in other highly industrialized countries such as Japan, which uses robots extensively in its manufacturing sector.

Productivity Decline

Although the United States was successful in developing its economy to a point where it made the transition from a manufacturing to a service economy, it is ironic that this should be accompanied by a decline in productivity. In an article in the Journal of Methods and Time Management, Hilchey (1982) states that the U.S. annual productivity growth rate fell to about 1.5%. In 1980, the rate was negative. Not only was this incongruous, but it also caused the United States to fall behind most of its international competitors in terms of productivity growth.

As a possible explanation of this phenomenon, Christie (1983) pointed out, in the Journal of New England Business, that productivity improvement efforts have generally concentrated on the blue-collar worker while neglecting the white-collar worker. He further contended that experts generally agree that white-collar employees do about 4 hours of actual productive work in an 8 hour day.

Recent studies have confirmed the decline in white-collar productivity. An investigation of 99 firms conducted by the American Productivity Center (1983) attempted to determine the causes for the productivity slump. Documenting the shift to a service economy, the research evidenced that a great deal of work no longer fits the traditional manufacturing process. Therefore, white-collar work is intangible and not easily measured. The report suggested that productivity be approached from an integrated perspective.

With this in mind, the substantial number of white-collar workers employed in an office environment represents a critical component of any attempt to raise productivity levels. If ways could be found to nurture and accommodate these individuals, substantial productivity gains will result which will be mutually beneficial to both employer and employee.

Office Employee Attitudes

In 1978 and 1980, Steelcase Inc. sponsored pioneering studies of the attitudes of office workers and corporate office planners toward the offices of today and tomorrow. Those studies were based on in-home interviews with persons who worked in their office jobs for six months or more and who worked for organizations which employ 25 or more persons. Detailed survey questionnaires were designed using a combination of structured, unstructured, and projective questions.

The survey attempted to explore the subjective factors that contribute to people's attitudes toward their jobs and their resultant performance. It examined office workers' attitudes in the following major areas:

- ° Job satisfaction and job criteria
- ° Job performance and productivity, and particularly how the office is felt to contribute to these
- ° Perceptions of and satisfaction with tools, equipment, tasks and workspaces
- ° Participation in the office planning and design process
- ° Anticipation of changes on the job and in the workplace

It is these authors' opinion that a glaring omission from this comprehensive listing was the determination of each employees' individual working style characteristics. Although the participants were asked what physical changes they would make to improve their comfort in the office (they would improve the heating and air conditioning; obtain more space; get more privacy; reduce the noise level; obtain more comfortable office and lounge chairs; and, improve the lighting) the fact of the matter is that the majority of respondents do not have any actual say in determining office conditions. Yet, 53% of the office workers and 63% of the executives maintained that they would be able to do more work in a day if conditions in their offices were changed to make them more comfortable.

Furthermore, in a number of instances reported by Steelcase and the American Productivity Center, the redesign of the office environment has resulted in improved overall productivity and morale. In particular, case study summaries from major corporations such as Blue Cross and Blue Shield in Virginia and London Life Insurance in Ontario, Canada indicated that greater attention to facility planning and employee comfort has contributed to improved efficiency.

Going one step beyond environmental design, BEA Associates in New York recognized that some people need sufficient autonomy to determine how their work is organized and administered.

While the above research and case studies contribute to a great deal toward developing improvements in the office environment, it is critical that a diagnostic technique be utilized which identifies a white-collar worker's preferred working style. This would allow the office environment to be adapted to the human needs of the workers resulting in increased productivity.

In the following pages, an overview of an instrument known as the Productivity Environmental Preference Survey will be presented as it relates to identifying an individual's personal working style. Based on this approach, some implications for the white-collar employee's working environment will be discussed.

The Instrument

The Productivity Environmental Preference Survey (PEPS) (Dunn, Dunn, and Price, 1979), based on factor and content analysis, is a comprehensive approach to the identification of how adults prefer to function, learn, concentrate and perform in their occupational or educational activities in the following areas:

- ° Immediate environment (sound, light, temperature, and design).
- ° Emotionality (motivation, responsibility, persistence and structure).

° Sociological needs (self-oriented, peer oriented, authority oriented and combined ways).
° Physical needs (perceptual preferences, time of day, intake and mobility).

Within the first category concerned with the immediate environment, data on the individual's preference for light, sound, temperature and design is collected. Although these environmental elements are often taken for granted in an office environment, it has been evidenced that there are differences among people as to the levels of sound, light and temperature they prefer and work best in. This is particularly true with the element of design. Some individuals prefer to work in an informal environmental setting composed of carpeting and soft chairs, while others prefer the more traditional office environment with a hard desk and chair at which to work.

An article in the September 22, 1983 issue of the Wall Street Journal emphasized the critical importance of the environmental category in raising productivity levels. After extensively interviewing members of its software programming staff as to the type of environment they preferred, TRW found that there was a strong preference for quiet, privacy, and comfortable furniture. When the work environment was adapted to these needs, the productivity of their white-collar programming staff increased as much as 39% in the first year.

With reference to the second category of emotionality, the elements of motivation, responsibility, persistence and structure are analyzed. While much has been written about the area of motivation, it is also necessary to determine whether an individual has a high level of responsibility accompanied by persistence in seeing a project through to completion. After determining those three levels, it is critically important to ascertain whether the individual prefers a structured office environment in which he receives detailed instructions or prefers a relatively unstructured climate in which to accomplish objectives with little supervisory direction.

An analysis of the third category of sociological needs reveals the elements of self-oriented, peer oriented, authority oriented and combined ways. All too often, the design of office environments does not allow for the unique sociological working preferences of the employees. Individuals who are self-oriented prefer to work by themselves in a manner similar to that mentioned by BEA Associates with a great deal of autonomy. By comparison, authority oriented persons prefer to work with a supervisor and are more productive when reporting to an authority figure. Furthermore, some prefer to work in teams with peers in the office environment while others prefer more variety.

In terms of the fourth category relating to physical needs, the perceptual elements (including visual, auditory, tactual and kinesthetic preferences) are examined as well as time of day, intake and mobility. Of all the working style elements presented in the model, perhaps the most important is that of perceptual strength. It is vital to determine which modality a worker prefers - does he work more effectively by seeing (visual), listening (auditory), touching and manipulating (tactual), or actually doing and experiencing (kinesthetic). Enormous amounts of money are wasted in installing technological innovations without determining whether the worker can readily adapt to the new equipment.

Furthermore, planners of office environments should take into consideration the element of time of day and attempt to accommodate employees' peak energy levels by appropriately scheduling working hours which parallel an individual's needs.

Uses of the Instrument

The administration of the valid and reliable instrument takes approximately 30 minutes. The data collected from the PEPS will:

- Permit individuals to identify how they prefer to work.
- Provide a computerized profile of each individual's preferred style.
- Provide a basis for supervisor/employee interaction.
- Provide a computerized group summary enabling planners to design office environments based on similarities among employees' working styles.

Implications for the Working Environment

With the increasing need to raise white-collar productivity by humanizing the workplace, it is critical that the planning and financial resources committed to designing an office environment be carried out in an optimum manner. At the present time, despite all the research to date, the ability to diagnose an individual's preferred working style is lacking in the design of office environments.

Based on the aforementioned needs, we must develop a whole new approach toward designing future office environments which accommodate an individual's working style preferences. In this way, the resources allocated will result in maximum effectiveness. In specific terms, we propose that a new office environment model be utilized which consists of three stages: (a) employee diagnosis; (b) feedback; and, (c) participative discussion.

As currently practiced, most office environment decisions are concerned with two essential choices - design of a conventional office environment vs. the "open" office approach. In most instances, this basic decision is made by management without any involvement of the affected employee. A much more sophisticated approach is represented by the model outlined below.

Employee Diagnosis

Based on the previously reported research, no attention has been given to diagnosing an employee's unique working style characteristics. By utilizing the Productivity Environmental Preference Survey, it is possible to determine an employee's personal working style in relation to the 18 elements contained in the four categories of environmental, emotional, sociological and physical needs. These findings provide a solid foundation for developing an optimum office design.

Feedback

Having identified each employee's working style preferences, the findings should be shared with the individual worker. This sharing of information will not only enable the worker to know their unique working style characteristics, but also provide a basis for decision making as to the type of office environment most suitable for the organization. In most instances, it is likely that different types of feedback will result depending on the department and the individuals involved.

Participative Discussion

In far too many cases, the implementation of an office environment is carried out without any prior discussion with the employees. The announcement is made by Management that new equipment and technology will be arriving within a particular department. All too often, the arrival of this technology is greeted with a high degree of resistance by fearful and untrained employees.

A much better approach would be to involve the employees in the decision making as it relates to their office environment. By using the PEPS as a basis for opening lines of communication between management and employees, a reasonable compromise can be reached which will not only accommodate the working styles of the employees, but also raise white-collar productivity levels.

CONCLUSION

Faced with the challenge to raise white-collar productivity in a service dominated economy, more sophisticated approaches must be taken to optimize the office environment and thereby humanize the workplace. In a somewhat similar technique to that used in implementing quality circles, the employees need to have input into the final decision.

It is the authors' view that the Productivity Environmental Preference Survey represents a unique approach in addressing the problem of white-collar productivity. Its implementation will provide a useful tool for increasing the effectiveness of the human and financial resources committed to optimizing the office environment and thereby increasing white-collar productivity.

REFERENCES

[1] Christie, C. Pursuing productivity in the white collar world. New England Business 5 (2) 58 (1983) 16-21.

[2] Harris, L. The Steelcase national study of office environments: Do they work? (Steelcase, Inc., Grand Rapids, Michigan, 1978).

[3] Harris, L. The Steelcase national study of office environments, no. II: Comfort and productivity in the office of the 80s. (Steelcase, Inc., Grand Rapids, Michigan, 1980).

[4] Hilchey, R. World of the white collar: People, paper and productivity. Journal of Methods and Time Management 9 (4) (1982) 16-19.

[5] Steelcase and the American Productivity Center. Case Study Reports (Steelcase, Inc., Grand Rapids, Michigan, 1983).

Human Factors in Organizational Design and Management
H.W. Hendrick and O. Brown, Jr. (Editors)

A CASE STUDY OF WORKER PARTICIPATION IN WORK REDESIGN: SOME SUPPOSITIONS, RESULTS AND PITFALLS

Christof Baitsch & Felix Frei

Industrial Psychology Unit
Swiss Federal Institute of Technology
Zurich
Switzerland

Worker participation in work redesign is essential for the quality of working life. Some successes and possibilities of participation are illustrated in a case study example. But with the implementation of increasing automation, there are special problems for such participation: The relationship between automation and participation is discussed on the basis of two case studies which illustrate the characteristic mechanisms observed. Certain conclusions relating to automation and participation are drawn as a result of the observations made.

INTRODUCTION

In a research project[1] in West Germany, we investigated how work should be designed so that the workers can become better qualified and develop personally **in the process** of working. It became evident that perhaps the most important condition of all in this regard is that the workers themselves can **participate** in work redesign. To illustrate: by means of participation, it was possible in one of the companies studied - a meat factory - to extensively improve the outer work conditions in a certain department according to stipulations made by the workers. In a second step, the workers of this department taught one another all of the work activities which are performed in the department, so that complete polyvalence was established. And in a third step, the retiring foreman was replaced not by a new foreman, but simply by an additional worker; for now the activity of foreman was taken over by each of the workers in the department in monthly cycles.

The success of this project as well as other - our own and also unrelated - positive experiences made in this realm confirm the importance of participation for the quality of working life and at the same time demonstrate its feasibility. In terms of its **success**, we can draw the following conclusions:

1) Workers are generally willing to actively participate in the (re-)designing of their own work-systems. This can to a great extent be attributed to **the human need for control** over one's immediate environment.

2) Work-systems designed with worker participation tend to exhibit other qualities than those without worker participation. Such work systems are better adapted to the workers' needs and available qualifications and abilities are better utilized. Often workers are seeking opportunities for learning on-the-job which can lead to their becoming better qualified and thus more flexible; this is true both of the individual and of the group as a whole.
3) Participation in the process of re-design leads to effects which we consider to be beneficial for personality development. While involved with work design and in the process of cooperating with, or resisting management, workers are also learning, for example, to identify and express their own interests, or to recognize the possibilities and limits of realizing these interests under existing conditions.
4) In terms of content, such design processes tends to develop from an initial concern with contextual aspects to ever-in-creasing involvement with factors of work content and work organization. For the workers, being occupied with aspects of their own work environment, ergonomics and safety tends to foster the development or regaining of the social competence necessary for dealing with subsequent more complex problems.

This positive judgement of the possibilities of participation assumes, however, that several important **problems** will be handled carefully. We now would like to present three **examples**, which are appropriate for their general validity:

(1) Usually there is a substantial hierarchy of power among the different participants. Participatory work design, however, requires an anxiety-free atmosphere. If such a project is to succeed, groups (and individuals!) lacking in power must be guaranteed protection from negative consequences in whatever form.

(2) Participatory work design assumes awareness of the problems as well as the "suffering" they cause, and this from both sides, from the management as well as the workers: without awareness of the present unsatisfactory situation and without a desire for improvement there will be no readiness for change. Awareness of problems can be taken for granted at all levels of a hierarchy, although it often is covered by resignation, especially at "lower" levels.

(3) The payment system plays a crucial role. Accord system and other quantitative-oriented systems tend to undermine social support, mutual teaching, on the job learning and the development of content-related work motivation; moreover, the supervisor's task is rendered more difficult. In contrast, wage systems such as a polyvalence payment, or a system based on the qualitative results of the entire group will generate a different relationship to the joint task and its mastering.

However, new and specific problems for participation arise when work redesign involves extensive **automation** of already-existing work activities. With technological development being as it is today, however, just this complication is ever more the case. The problems which arise as a result of this strike us as so important that we now wish to discuss them in detail. We will now present some material from two studies conducted in the second company which served us in our researches.

CASE STUDIES

In this West German tire factory, our project was limited to one department of the company.

First Case

In the department under consideration, tires for small automobiles are manufactured. Some 350 people in two, in part, three, shifts produce about 10'000 tires per day; the company employs a total of 5'500 workers. This department consists of three sections: in the first, several parts of the tires are manufactured; in the second, a complete tire is built up out of these parts; and in the third section, the tires are vulcanized.

Between the scond and third sections, there is a stock room where the unvulcanized tires are stored, if necessary, until they can be sorted and sent via an automatic transport system to the vulcanization section. Normally four workers per shift endeavor to pick up each tire individually from a conveyor belt on the floor and place it either in a paternoster system or, if possible, in the proper order of types, directly on the corresponding wagon of the transport system. For the workers, this amounts to a heavy physical load and very modest job content. On the other hand, the workers can manage this job only if they first put an assortment of the arriving tires on the ground, so as to have a "buffer supply" between the irregularly arriving tires and the varying needs of the following vulcanization section. Having unvulcanized tires on the floor, however, leads to quality problems which are not immediately evident, but statistically certain.

We initiated our project in this stock room in line with a proposal made by company management. Participation of concerned workers was sought in the hope of finding a solution for the problems mentioned. It should be added that 12 out of the 13 workers involved are foreigners, principally of Turkish nationality and with very little knowledge of the German language. This was one of the handicaps for successful participation, but certainly not the most important one.

Ultimately, the participating group of workers was unable to find fundamental solutions to the problems. Subsequently, some technical experts of the company entered the scene: their proposal was brought up with the workers, who were able to identify weaknesses and disadvantages of the expert model, but they still could not produce viable alternatives.

After this second step, the manager of the department was discharged for reasons of health. The new manager rejected the experts' proposal and submitted a completely different model, which was not restricted to a re-designing of the stock room only, but also involved the immediate surroundings. It is important to add that, in the meantime, a crucial part of these surrounding aggregates had become redundant because of another technological advance in tire manufacturing unrelated to our project, and only this change rendered possible the solution proposed by the new manager. In subsequent discussions with the workers, this new model was appreciated: it proposes extensive automation of their work.

The consequences of this model - which is just about to be implemented - include the fact that only one person per shift will be required in the stock room, which means that three times three people must hope to find another job in the company. But this is a dubious proposition in a company which, for economic reasons, has been reducing the number of employees during the last three years.

Bearing this in mind, it is surprising that the participating workers supported the idea. But, due to errors made in the process of intended participation, several explanations can be given: First of all, at the time of the original questioning, the threatening consequences seemed far away. Second, these workers had had no experience with participation in questions of technological change: they were not accustomed to defending their own interests; on the contrary, they tended to share the views of a technician or manager. Third, and probably most important, they were not - throughout the project - genuinely supported by the works council, that are the elected representatives of the workers, in spite of the fact that two representatives of this council were involved in the project steering committee.

We can sum up the result of this first case as follows: although user participation was employed throughout three steps, it was not possible to avoid negative - in this case quantitative - consequences of technological change in the work situation.

Second Case

After the failure of the first step in the above-mentioned case, a parallel project was initiated in the same department. This time, the management had chosen a part of the first section of the department where the so-called beads of the tires are manufactured. In this group, two men work at winders, which semi-automatically wrap the beads. Some 14 women apply a rubber apex to the beads with mechanical help. They work on a piece-rate system and are required to complete about 1'300 pieces per shift.

Approximately half of the workers of each of the two shifts participated in monthly group discussions over a period of nine months.

The group discussions began with the collection of problems raised by the workers. The list of these problems extends from details of technical aspects over the question of a more optimal flow of material, to problems concerning the entire work organization. The process of finding solutions began with the detail problems and gradually became more and more complex. Implementation of the proposals of the workers - if accepted by management - followed with a certain delay in the same order as discussed.

Although the process of realization of the whole re-design of this part of the department is, at this time, not yet completed, we can anticipate the results. The following dilemma summarizes our experiences in this process in a simple manner: although each individual step of this participatory process satisfied not only the specific wishes of the involved workers, but also - at least to the extent of reducing physical load - psychological criteria of job design, the result of the entire process of re-designing the jobs was not satisfying, due to more and more shortened cycles, less job content, more monotonous actions, and, perhaps most significantly, a reduced number of required workers.

In fact, we must state that - as in the first case, too - worker acceptance of this final result was at least partly "caused" by the applied form of user participation.

Before drawing conclusions on automation and participatory job design, we should analyze the mechanisms of this dilemma and its final effects.

ANALYSIS OF THE MECHANISMS OF THE TWO CASES

Both the stock room case and the bead case are characterized by an important mechanism: in both cases, a simple "step-by-step participation" was employed. In the reported project, only user participation was used, but we suspect that the fundamental problems of a "step-by-step participation" would remain the same in the case of union participation, as well, if there were not a structurally different approach, as will be discussed below.

Figure 1 illustrates the effects of such a simple "step-by-step participation":

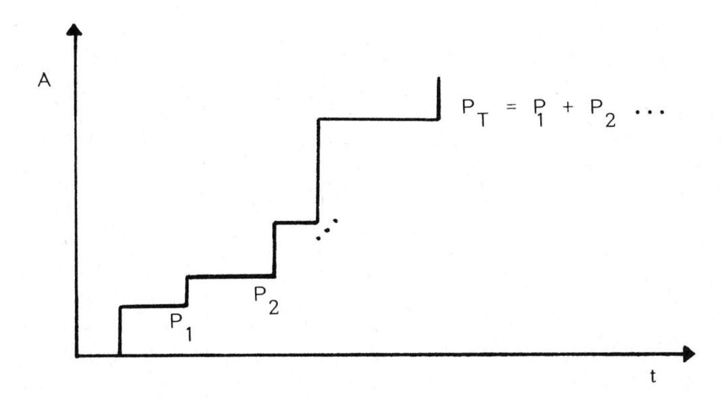

Fig. 1: "Simple step-by-step participation": Unknown quality of automation, and danger of unanticipated side effects.
(A = degree of automation; P = step of participation)

The total **degree** of automation, of course, is the sum of the effects of every single step of participation. But the degree of automation, per se, is not a sufficient criterion for evaluating a system. Much more important is the **form** of automation; that means the question of what has been automated, what has been left for the workers to do, or what has been primarily and consciously delegated for the people to do. In the case of a simple step-by-step strategy of participation, this latter question can neither be answered in advance, nor can one steer toward a principally defined goal, because the subject of participation in every step is restricted to the result of this concrete step only. We would guess that this problem amounts to the most crucial reason for the above-mentioned dilemma, which caused the failure of the discussed cases. Alternatives to this step-by-step strategy shall be discussed below. First we wish to show another fundamental problem of both cases.

From the point of view of a socio-technical system design, a joint optimization of both the social and the technical subsystem can only be successful if both the social and the technical aspects of the problem correspond to the area marked off for re-design. This criterion of correspondence has been injured in both cases, but in contrasting ways:

As fig. 2 shows, in the stock room case, the area marked off did not include all of the technical aspects of the problem. This means that the participating individuals did not represent the whole field of the problem. Thus they could not be **able** to solve the problem.

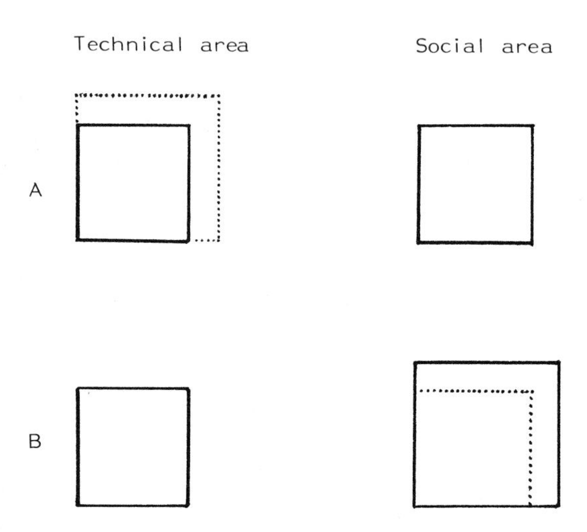

Fig. 2: Injuring the correspondence criterion of both technical and social area marked off: A = the stock room case; B = the bead case.

In the bead case, the facts are reversed: a joint optimization of both technical and social aspects of the problem would certainly concern less people than are working there now. This means that some - but it is unknown which - of the participating people should develop ideas which would render their own jobs obsolete. So they could not - or to say it more precisely: should not - be **willing** to solve the problem.

In fact, the latter way of injuring the correspondence criterion would have occured in the stock room case, as well, if the social area had been marked off in correspondence with the technical problem: this occurs in every case where, by means of automation, the number of workers is to be reduced.

This will be taken into consideration in the following discussion of some consequences of our experiences with automation and participation.

CONSEQUENCES

There would appear to be four consequences with regard to the question of automation and participation which can be drawn from these experiences:

First: To control not only the degree but also the quality of automation, it is an absolute necessity to orient participation from the beginning toward the complete re-design of the system. After outlining a desirable final state of the system, the required steps to attainment of this goal can be deduced. That means that participation should follow a "differentiated step-by-step strategy". While the technical realization of the developed solutions must move, of course, from details to the whole, the development of these solutions must move from the whole to the details.

Figure 3 might illustrate the difference. A not unimportant implication of this suggested differentation of the strategy is that one needs more time, because the process of participation and the process of technical realization cannot be overlapped that easily in time. But this must not be a severe disadvantage - it is simply the price that must be paid to avoid unanticipated and undesired side effects of automation in cases of participation.

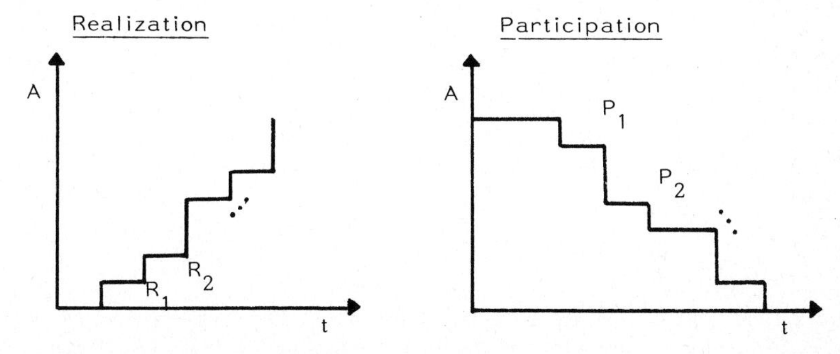

Fig. 3: "Differentiated step-by-step strategy of participation": Planability of both degree and quality of automation. (A = automation; P = steps of participation; R = steps of technical realization)

On the other hand, it is just the price, and no guarantee that such side effects will be avoided. There are still other consequences to be considered.

Second: Specifying the area of a system which is to be redesigned must be done in such a way as to meet the criterion of correspondence of social and technical aspects of the problem. In fact, it is one thing to mark off a technical area corresponding to the problem, and it is quite another to mark off the corresponding social area. In the first case, it is a question of system dependencies, regulation necessities, technical functions, and so forth. This is not always easy to do, but in principle, it is manageable. At least, in the process of planning the system re-design, a wrong or insufficient marking off of the technical area might be modified. It is much more difficult to define the appropirate social area. One reason for this derives from the fact that, in general, automation aims at reducing the number of actually required persons. A way out of this problem could be the following: Anticipating the possible outcomes of the participation process, one can, so to speak, "virtually" enlarge or reduce either the marked-off technical or the social sub-system. For example, one can virtually enlarge the technical sub-system by both defining the social sub-system as fixed - so that the expected rationalization effects are compensated by adding new duties to that group - and by defining the finding of such new and significant duties as an additional topic of the participation. Or, on the other hand, there is the opposite way out of the problem, which is much more problematic: one can reduce the social sub-system and involve only those people in participation who presumably can be certain of remaining in the particular technical sub-system in spite of proceeding automation.

At least at this point, it appears to be clear that user participation exclusively cannot guarantee the avoidance of unanticipated side effects of automation.

Third: If one aim of automation, or, at least, one of its possible outcomes, is reducing the required personnel, or even if people could fear such an effect, a participation approach definitely needs to provide a guarantee for the people involved that they will not lose their jobs, or at least that they can be assured of an equivalent job in another part of the company. Such a guarantee is a subject of negotiation, and it is neither probable nor desirable that such negotiations be the matter of user participation. Rather this should be a matter of the union, whether at the local level or at a higher level. Suitable examples of such negotiated union-employer agreements on technological development are known, e.g., from Norway. It should be added that such guarantees are a matter of course for the workers involved within the frequently-mentioned model of 'quality circles' - just one form of user participation - in Japan.

Fourth: At least one further conclusion must be mentioned: Constant technological innovation increasingly becomes the predominant characteristic of working life, be it with or without participation be it user or union participation. What lesson can we learn from our experiences of the relationship between automation and participation in a case where the above-mentioned three consequences are rejected by the management, although a certain kind of user participation is desired by this same management? We don't know the answer. In such a case, one should at the very least be careful and make certain not to promise anything to the participating workers beyond one's true intentions. Otherwise frustrating experiences could give participation the reputation of being nothing but a tricky instrument for, yet one time more, taking advantage of the workers.

FOOTNOTE

1) The project was supported by the Federal Ministry of Research and Technology, Bonn/FRG (Nr. 01 HA 059 - AA - TAP 0015).
The resonsibility for the contents of this paper, however, is carried by the authors.

Human Factors in Organizational Design and Management
H.W. Hendrick and O. Brown, Jr. (Editors)

MANAGING COMPUTER-BASED OFFICE INFORMATION TECHNOLOGY: A PROCESS MODEL FOR MANAGEMENT

Urs E. Gattiker

School of Management
The University of Lethbridge
Lethbridge, Alberta
Canada T1K 3M4

The introduction of computer-based office information technology has usually been technology-led, without consideration of potentially negative effects upon human resources and the quality of work life. This approach is forced on management by the fact that the effects of innovation on personnel are generally unknown at the time of its adoption. However, in order to make the most effective use of their technology, people need to feel comfortable with it. In this review, a model is offered which employs a longitudinal process approach to assess the perception by the workforce of any new technology, and also its influence on organizational profit and costs. The discussion centers on the model's application along with an attempt to outline the organizational constraints and contingencies affecting its successful introduction.

INTRODUCTION

In organizational research there appears to be general agreement that the term technology refers to the mechanical or intellectual processes by which an organization transforms inputs into outputs (Zey-Ferrell, 1979, pp. 106-118). In line with this assertion, a distinction must be made between machine technologies, which are part of the entire organization, and human technology, which consists of particular tasks performed by individuals (Zey-Ferrell, 1979, pp. 106-107; Miles, 1980, chap. 3). Thus the definition of technology can range from a very narrow perspective (one machine) to an extremely broad one including intellectual skills, such as computer languages and contemporary analytical and mathematical techniques. Pfeffer (1982, p. 152) stated that "most conceptualizations of technology have focused less on the operations or production technology and instead have conceptualized technology in terms of its complexity, analyzability, or routineness."

In this essay, the interrelationship of computer-based office information technology with human resources and a company's earnings will be discussed. Computer-based office information technology in this context is defined as a multifunction computer system which handles information and offers its users the possibility of interaction. This form of technology may include a wide variety of functions, from simple electronic mail to entire management information systems (Gutek, Bikson & Mankin, in press).

Organizations usually require multiple technologies to carry out their missions (cf. Woodward, 1965). Since a business often manufactures a variety of products, it may be impossible to assess overall inputs and outputs. Instead, a smaller unit (a plant, department or work-group) should be chosen for analysis. Another approach would be to examine how technology affects jobs and work features (Stolz, 1982). Several researchers have suggested that the technologies' effect upon the workforce should be evaluated so that its impact upon company effectiveness and profit can be measured in turn (e.g., Bauknecht, 1983; Mey, 1981).

Organizational research should be based on data that will allow generalizations for different organizations (Duncan, 1981). However, early instability in research results suggested that companies may have many technologies, highly divisible environments, and heterogeneous structures (Hickson, Pugh & Pheysey, 1969). Since technology is not just machinery but includes the intellectual capabilities of the workforce as well, it is usually safe to assume that organizations will be different from each other even though they may belong to the same industry, such as banking or retail. Considering that each company's product mix, workforce and human resource policies are unique, it would be fruitless to attempt a comparison of their transformation processes of input to output (cf. Gattiker & Larwood, 1984).

A model integrating organizational and behavorial variables will be developed here. Its major objective is to provide a theoretical framework for a conceptual analysis of the effect of computer-based office information technology upon human resources and company profit. Thus organizational researchers and management will be able to analyze a plant or department and detect benefits and liabilities created by technological innovations.

EFFECTS OF COMPUTER-BASED OFFICE INFORMATION TECHNOLOGY

The decision to integrate any kind of technology into the transformation process is usually made without consideration of its social impact (Kahn, 1981). Equipment is often installed before change-sensitive dimensions are defined and potential reactions assessed (Tydeman et al., 1982). An organization's wholesale negation of technology, on the other hand, could be interpreted as non-adaptive, and may be harmful to its future prospects (Gold, Rosegger & Boylan, 1980, chap. 3; Starbuck, 1983). Management could exhibit proactive behavior, that is, anticipate problems and work on their solutions, if the effects of a given technology were indeed known in advance. It is unfortunate that only a technology's diffusion can provide the necessary information about its impact upon the different company stakeholders, such as employees, customers, government agencies, competitors, suppliers, distributors, etc. (cf. Mitroff, 1983, chap. 2). There will always be a trade-off in the integration of any technology into an organization's transformation process. While technological advances may provide a competitive edge, the company will also have to cope with the "growing pains" common to innovators.

Due to space limitations, this article will not deal with the change process put into motion when a new technology is adopted by an organization, nor the development of the necessary new skills by employees. The reader is referred to detailed discussions of these issues by others (e.g., Eason, 1982; Turner, 1980).

Organizations must constantly analyze, develop and adjust their computer-based office information technologies, and also keep abreast of external changes in technology design. A conceptual model, which can be operationalized easily for the assessment of developments in computer-based office information technology, would improve the company's planning and adaptation capabilities. The level of analysis for such a model must include organizational (macro) and individual (micro) variables along with their effects upon profit and costs (cf. Katz, Kahn & Adams, 1980, pp. 7-8; Pfeffer, 1982, pp. 151-159). Obviously, the bottom line will always be dependent on the effective interaction of "man and machine." Management will arrive at a better fit between individual and organizational needs by employing data obtained in the application of the model proposed in this paper. As a result, system contingencies and human behaviors, which affect any adaptive action by the company, can be taken into consideration when contemplating changes in the technology system (cf. Turner, 1980).

DIMENSIONS OF THE PROCESS MODEL TO BE INTRODUCED

Organizational demography. Organizations are similar to small cities in that their populations fluctuate. Therefore, Stinchcombe, McDill & Walker (1968) contended that demography variables, such as age or sex, could be applied in organizational research. This concept provides a means by which individual and organizational attributes can be linked, reflecting the aggregation of personal characteristics within the unit (Pfeffer, 1983). Organizational demography is also a macro level property of the organization, incorporating global characteristics and relationships of its members (Wagner, Pfeffer & O'Reilly, 1984).

Stinchcombe et al. (1968) claimed that one reason organizational demography is ignored in industrial settings could be the belief that it is controlled by the organization. However, their research showed that several variables, such as geographical location and the net flow of the population in its territory, are beyond the company's influence. For instance, in a rural area it may be more difficult to find employees with special skills, and average organizational tenure of the workforce may be higher due to limited employment opportunities (Pfeffer, 1983).

Blau (1977, pp. 8-9) distinguished between two types of structural parameters to assess organizational demography: nominal parameters to define the horizontal differentiation or heterogeneity in groups (e.g., sex, marital status, place of work), and graduated parameters to describe the vertical differentiation or inequality of status distribution, such as age, tenure, education and income. Some of these parameters have been applied at the individual psychological level. Age can affect the perception of job environment; older employees have expectations different from their younger colleagues (cf. Schaie, 1983). An examination of age and tenure distribution in a company is important in analyzing micro variables (perception of office technology, organizational commitment, job involvement, etc.). Most of the research conducted in organizational demography to date has been limited to graduated parameters (e.g., Stinchcombe et al., 1968; Allison, 1978; Pfeffer & Moore, 1980).

Organizational commitment. This concept reflects a general affective response by the individual to the organization as a whole, including its goals and values (Mowday & Steers, 1979). Although day-to-day events may affect an employee's job involvement and satisfaction, transitory incidents usually will not cause a person to reevaluate his/her commitment to the organization (Mowday, Porter & Steers, 1982, chap. 2).

Two dimensions of organizational commitment, attitudinal and behavioral, have been identified in organizational research (Steers & Porter, 1983, pp. 425-430). While the former is based on the extent to which an individual identifies with a unit by accepting its goals and values, the latter reflects the employee's feeling of being bound to the organization by benefits which he/she cannot afford to give up. Research data shows that attitudinal commitment can affect turnover, absenteeism and job performance (e.g., Mowday, Steers & Porter, 1982). Behavioral commitment, on the other hand, will lead to psychological bolstering in the form of dissonance reduction and self-justification processes (Angle, 1983; Steers & Porter, 1983, pp. 425-430). Antecedents of organizational commitment are demographic variables, such as age, sex and marital status, which, in contrast to education, have been reported to be positively correlated with this type of commitment (Angle & Perry, 1981; Hrebiniak & Alutto, 1972).

Based on these discussions, one could hypothesize that organizational demography variables can influence the organizational commitment of a work group. However, the magnitude of its effect upon perception of computer-based office information technology is unknown at this time. Behavorial commitment may in fact be detrimental to the organization after a certain point because people might feel trapped into staying with the company by long-term or deferred benefits (Tokunaga & Staw, 1983). Thus, although behavioral commitment is relatively high, job involvement and employee perception of this technology may be unsatisfactory from the organization's point of view. For these reasons it is in the company's interest to create high attitudinal commitment.

Employee perception of computer-based office information technology. Much of the research in this area concentrated on the information technology's enhancement of productivity (Bikson, 1981; Lieberman, Selig & Walsh, 1982; Stolz, 1982). However, the aspect of user perception in the study of office information technology must not be neglected when trying to facilitate productivity and enhance job satisfaction (Bodmer, 1982; Bikson & Gutek, 1983). On the eve of a wide-sweeping office revolution, the inclusion of employee perception of computerized office technology in management decisions should be of prime importance (cf. Giuliano, 1982).

It has been shown that negative perception of a new system can nullify any potential benefits for the organization (Cheney & Dickson, 1982). Gattiker, Gutek & Berger (1984) conducted a study to assess user perception of various kinds of office technology, such as the telephone, typewriter and computer. The results indicate that all technologies were perceived as useful in increasing work effectiveness, but the telephone was also considered disruptive, while computers were thought to be less helpful for communication than telephones or typewriters.

Another issue to be studied is the effect of age, education and the influence of organizational cohorts on computer-based office information technology. Obviously, the availability of computer technology in education and recreation will have a profound influence on today's adolescents (Condry & Keith, 1983). They are much less likely to be intimidated by the technology than their middle-aged colleagues (cf. Tamir, 1982, pp. 117-122). Furthermore, organizational demography may greatly influence a group's perception of office information technology. For instance, a highly educated clerical workforce may want to prevent the introduction of technology to avoid routinization of their jobs through automation, which is a legitimate concern (Morgall, 1983).

Job involvement. Organizational commitment and demography as well as technology perception constitute the antecedent conditions for job involvement (cf. Turner, 1980), which can be classified as cognitive, or a belief state of psychological identification with one's job (Kanungo, 1982a). Psychologists have used a positive exploratory and empirical approach toward job involvement (Kanungo, 1982b, pp. 18-58). This phenomenon is limited to an analysis of behaviors at the individual level. Job involvement must be clearly distinguished from its antecedent conditions; it is in fact the result of these conditions (Kanungo, 1982b, pp. 74-93; Morrow, 1983).

Kanungo (1982b, pp. 118-119) states that people with extrinsic needs, such as desire for money, will be involved in their jobs as highly as peers with intrinsic needs (e.g., satisfying work which allows self-expression), provided that a job has the potential to satisfy one's prominent needs. He also cautions against using work involvement and job involvement interchangeably. While the former refers to a general normative belief and represents the moral value of work in one's life based on early socialization, the latter is a result of the potential of one's current job to satisfy personal needs. Therefore, depending on individual perception, job involvement may act as a motivational force.

Clearly, job involvement and organizational commitment are two concepts with different foci. One describes the degree of daily absorption in work activity, the other measures loyalty to one's employer. Organizational commitment delineates the level of desire to remain with, and the willingness to exert high efforts for, the organization. Job involvement attempts to define the depth of a person's psychological identification with his/her work. Moreover, if a person is job-involved it does not necessarily mean that he/she believes in and accepts the values and goals of the organization (Morrow, 1983).

Organizational profit and cost structure. Management's primary concern with any new technology lies with its impact upon organizational profit, not productivity (Gold, 1982). Consequently, it is imperative to assess the effect of human resource variables upon profit ratings. Most types of technology can indeed assist organizations in improving their effectiveness. However, it will be necessary to measure organizational demography, organizational commitment, job involvement and perception of computer-based office information technology to achieve this objective. Organizations have already shown that some areas of the transformation process (e.g., overhead costs, product mix and profit) can be positively affected by this technology (Credit Suisse, 1983), but the full use of any innovative technology requires an investigation of its effect upon the workforce as well.

The types of measures used to assess organizational profitability must also be considered. Generally, it is safe to assume that they will depend upon the management control system in effect in the organization to be surveyed (cf. Maciariello, 1984). When comparing bank branches, for example, it would be advisable to look at such cost ratios as total wages or credit servicing costs versus branch credit portfolio.

ORGANIZATIONAL PROFIT AND EMPLOYEE PERCEPTION OF COMPUTER-BASED OFFICE INFORMATION TECHNOLOGY

Figure 1 depicts a basic system of technology perception and organizational profit, designed to chart essential relationships and their impact. Job involvement, organizational profit, demography, commitment and employee perception of computer-based office information technology are the dimensions to be measured. This system is a conceptual process model which could be used as a basis for a generalizable analytical framework. The model should be employed to assess the magnitude of the population variance explained by using job involvement, perception of office technology and also organizational demography and commitment as predictors of the organizational profit and cost structure.

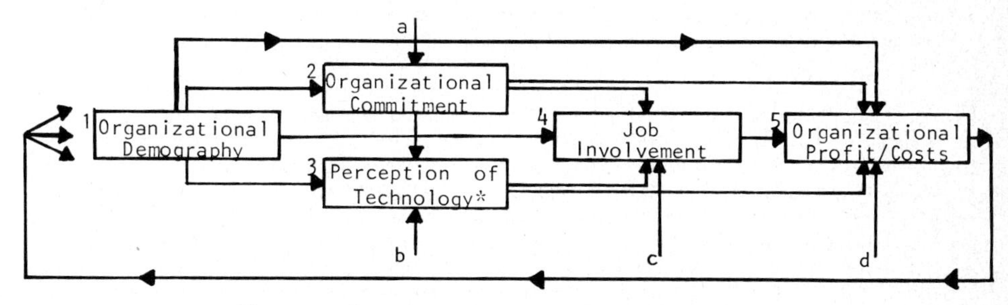

*Computer-based office information technology
a, b, c and d = residual variables

Figure 1
Computer-Based Office Information Technology: A Process Model

The model consists of a set of cause-effect relationships. It is unidirectional, containing one exogenous variable (1), whose variability is assumed to be determined by causes outside the model. As a result, the determination of variable #1 is not under consideration. The four remaining variables (2, 3, 4 and 5) are endogenous, that is, their variation is explained by either the exogenous variable (organizational demographics) and/or the endogenous variable (organizational commitment, perception of computer-based office information technology, job involvement and organizational costs and profit) and the residual variables a, b, c and d.

It appears that an organization's demography has a direct impact on organizational commitment and perception of computer-based office information technology as well as job involvement and organizational profit. In addition, organizational demography is shown as having an indirect effect on job involvement through the perception of this technology, organizational commitment and profit. In other words, the model consists of an independent variable and three mediator variables, all of which have an impact upon organizational profit. Furthermore, organizational demography and commitment affect the perception of computer-based office information technology. Since it will probably be impossible to account for the total variance of variable 2, 3 or 4, residual variables are introduced to indicate the effect of variables not included in this model.

A series of sequential occurrences over time are presented in the model to explain how the phenomena of organizational costs and profit happen. Every process model uses time ordering of the contributory events as a critical characteristic for the end result (Mohr, 1982, pp. 1-70). Such a model would measure the cycles of the effect of the antecedent conditions on the outcome. For example, it is assumed that the organizational profit and cost structure will start a reinforcing process. This process and its effect upon the field application of this model will be discussed in greater detail in the following section.

ORGANIZATIONAL APPLICATION OF THE MODEL

To obtain external validity of the model under discussion, a company should conduct a field study by surveying its workforce. Furthermore, longitudinal data is necessary to show management a pattern of developments. Individual variables, e.g., organizational commitment, perception of technology and job involvement, as well as system variables, such as organizational demography and profit/costs, may be measured with this model. Thus management and researchers will gain insights into the perceptions of individuals in organizational contexts by treating the company as an entity. In large corporations, individual departments, profit centers, plants or subunits may need to be analyzed separately to permit the effective use of this process model.

After all the dimensions discussed in the model have been measured in an organizational context, the back-loop process will be easily recognized. For example, in a well-established, healthy manufacturing company, profits will usually lead to growth which may require several new hires, affecting average organizational tenure. Furthermore, management may decide to offer additional benefits, which usually influence organizational commitment and have an impact on job involvement in turn. Naturally, company costs and profit will be affected as well, starting the entire cycle over again. Longitudinal, periodic assessment of these variables, whether it be semi-annually or anually, is of prime importance when employing the model as suggested. Strategic planning may be adjusted according to the results of the model's application, thereby improving management of human as well as capital resources.

CONCLUSION

The system described in Figure 1 and the subsequent discussion show that it is possible to use a quantitative approach which demonstrates the interplay of scientific theory with practice. This model employs an open-system approach by continuously monitoring all pertinent variables, permitting the introduction of changes where necessary (cf. Katz & Kahn, 1978, pp. 752-760). It also provides organizational researchers with an opportunity to do interdisciplinary work while being consistent with open-system theory, which regards the organization as a biological organism within a larger environment (Katz, Kahn & Adams, 1980, pp. 1-9). By using a longitudinal rather than a cross-sectional approach, the process model introduced in this essay achieves this consistency. Successful application enables management to refine the transformation process of technology from an external to an internal stage, which may increase the effectiveness of the human-technology interface.

In the field application of any model a careful balance must be struck between complexity and feasibility. For instance, an excessively long survey instrument may result in reluctance by management to administer the study. Other problems could be partially completed questionnaires and a poor response rate on the part of the workforce. Obviously, there are additional dimensions whose inclusion could easily be justified, such as turnover and job features, but if management feels threatened or intimidated by a model's complexity, the success of its implementation is questionable. Therefore, it is advisable to begin with a small-scale model which is simple yet sophisticated enough to increase our knowledge sufficiently and encourage further studies.

At this time, only the general four-stage pattern of the process illustrated in Figure 1 is recognized. Further answers must remain tentative until more research is conducted to determine the strength of each dimension. Longitudinal studies using this model will help to increase management's and organizational researchers' understanding of how changes in computer-based office information technology may affect individual and system variables. The process model introduced here is just one more step in moving toward better comprehension of the phenomena involved in "man and machine" interaction.

REFERENCES

1. Allison, P.D. (1978). Measures of inequality. American Sociological Review, 43, 865-880.
2. Angle, H.L. (1983, August). Correlates of instrumental and affective attachment to organizations. Paper presented at the annual meeting of the American Psychological Association, Anaheim, CA.
3. Angle, H.L., & Perry, J. (1981). An empirical assessment of organizational commitment and organizational effectiveness. Administrative Science Quarterly, 26, 1-14.
4. Bauknecht, K. (1983). Gedanken zur Bedeutung der Mikroelektronik. Die Unternehmung, 37, 128-131.
5. Bikson, T.K. (1981). Electronic information systems and user contexts: Emerging social science issues (P-6690). Santa Monica: Rand Corporation.
6. Bikson, T.K. & Gutek, B.A. (1983). Advanced office systems: An empirical look at utilization and satisfaction. In X.X. Furst (Ed.), Proceedings of the National Computer Conference, 52, (pp. 319-328). Arlington, VA: AFIPS Press.

7. Blau, P.M. (1977). Inequality and heterogeneity. New York: The Free Press.
8. Bodmer, W. (1982). Uberlegungen zur Anwendung der Mikroelektronik. Ergebnisse der Beratungen eines Gesprachskreises der SZF. Wirtschaftspolitische Mitteilungen, 38(5).
9. Cheney, P.H., & Dickson, G.W. (1982). Organizational characteristics and information systems: An exploratory investigation. Academy of Management Journal, 25, 170-184.
10. Condry, J. & Keith, D. (1983). Educational and recreational uses of computer technology. Youth & Society, 15, 87-112.
11. Credit Suisse. (July, 1983). Aktionarsbrief. Zurich: Author.
12. Duncan, W.J. (1981). Quasi experimental research in organizations: A critique and proposed typology. Human Relations, 34, 989-1000.
13. Eason, K.D. (1982). The process of introducing information technology. Behaviour and Information Technology, 1, 197-213.
14. Gattiker, U.E., Gutek, B.A., & Berger, D.E. (1984). Perceptions of office technology by employees. Manuscript submitted for publication.
15. Gattiker, U.E., & Larwood, L. (1984). Organizational ephemerals and the search for rationality. Manuscript submitted for publication.
16. Giuliano, V. (September, 1982). The mechanization of work. Scientific American, 149-164.
17. Gold, B. (November, 1982). Robotics, programmable automation and international competitiveness. Transactions in Engineering Management of the Institute of Electrical and Electronic Engineers, 29.
18. Gold, B., Rosegger, G., & Boylan, M. (1980). Evaluating technological innovations. Lexington, MA: Lexington Books.
19. Gutek, B.A., Bikson, T.K., & Mankin, D. (in press). Individual and organizational consequences of computer-based office information technology. Annual Review of Social Psychology.
20. Hickson, D.J., Pugh, D.S., & Pheysey, D.C. (1969). Operations technology and organization structure: An empirical reappraisal. Administrative Science Quarterly, 14, 378-397.
21. Hrebiniak, L.G., & Alutto, J.A. (1972). Personal and role-related factors in the development of organizational commitment.
22. Kahn, R.L. (1981). Work and health. Some psychosocial effects of advanced technology. In B. Gardell and G. Johansson (Eds.), Working life (pp. 17-37). Chichester, UK: John Wiley & Sons.
23. Kanungo, R.N. (1982a). Measurement of job and work involvement. Journal of Applied Psychology, 67, 341-349.
24. Kanungo, R.N. (1982b). Work alienation. New York: Praeger.
25. Katz, D., Kahn, R.L., & Adams, J.S. (1980). The study of organizations. San Francisco: Jossey-Bass.
26. Lieberman, M.A., Selig, G.J., & Walsh, J.J. (1982). Office automation. New York: John Wiley & Sons.
27. Maciariello, J.A. (1984). Management control systems. Englewood Cliffs, NJ: Prentice-Hall.
28. Mey, H. (1981). Mikroelektronik--Moglichkeiten, Gefahren und Grenzen. Wirtschaftsbulletin 30. Zurich: Zurcher Kantonalbank.
29. Miles, R.H. (1980). Macro organizational behavior. Glenview, IL: Scott, Foresman & Co.
30. Mitroff, I. (1983). Stakeholders of the organizational mind. San Francisco: Jossey-Bass.
31. Mohr, L.B. (1982). Explaining organizational behavior. San Francisco: Jossey-Bass.
32. Morgall, J. (1983). Typing our way to freedom. Is it true that new office technology can liberate women? Behaviour and Information Technology, 2, 215-226.

33. Morrow, P.C. (1983). Concept redundancy in organizational research: The case of work commitment. Academy of Management Review, 8, 486-500.
34. Mowday, R. T., Porter, L. W., & Steers, R. M. (1982). Employee organization linkages. New York: Academic Press.
35. Mowday, R.T., & Steers, R.M. (1979). The measurement of organizational commitment. Journal of Vocational Behavior, 14, 224-247.
36. Pfeffer, J. (1982). Organizations and organization theory. Boston: Pitman.
37. Pfeffer, J. (1983). Organizational demography. Research in Organizational Behavior, 5, 299-357.
38. Pfeffer, J., & Moore, W.L. (1980). Average tenure of academic department heads: The effects of paradigm, size, and departmental demography. Administrative Science Quarterly, 25, 387-406.
39. Schaie, K.W. (Ed.) (1983). Longitudinal studies of adult psychological development. New York: The Guilford Press.
40. Starbuck, W.H. (1983). Organizations as action generators. American Sociological Review, 48, 91-102.
41. Steers, R.M., & Porter, L.W. (1983). Employee commitment to organizations. In R.M. Steers and L.W. Porter (Eds.), Motivation and work behavior (3rd ed.) (pp. 441-451). New York: McGraw-Hill.
42. Stinchcombe, A., McDill, M., & Walker, D. (1968). Demography of organizations. The American Journal of Sociology, 74, 221-229.
43. Stolz, P. (1982). Technischer Wandel, Rationalisierung und Arbeitsmarkt. Die Unternehmung, 36, 229-246.
44. Tamir, L.M. (1982). Men in their forties. New York: Springer Press.
45. Tokunaga, H.T., & Staw, B.M. (1983, August). Organizational commitment: A review and critique of current theory and research. Paper presented at the annual meeting of the American Psychological Association, Anaheim, CA.
46. Turner, J.A. (1980). Computers in bank clerical functions: Implications for productivity and the quality of working life. Dissertation Abstracts International, 41, 1834B. (University Microfilms No. 80-23, 556).
47. Tydeman, J., Lipinski, H., Adler, R., Nyhan, M., & Zwimpfer, L. (1982). Teletext and Videotex in the United States. New York: McGraw-Hill.
48. Wagner, G. W., Pfeffer, J., & O'Reilly, Ch. A. (1984). Organizational demography and turnover in top-management groups. Administrative Science Quarterly, 29, 74-92.
49. Woodward, J. (1965). Industrial organization: Theory and practice. London: Oxford University Press.
50. Zey-Ferrell, M. (1981). Dimensions of organizations: Environment, context, structure process, and performance. Glenview, IL: Scott, Foresman & Co.

Human Factors in Organizational Design and Management
H.W. Hendrick and O. Brown, Jr. (Editors)
Elsevier Science Publishers B.V. (North-Holland), 1984

Job design and work organisation in the United Kingdom

Reg Sell,
Work Research Unit
Department of Employment
Steel House
11 Tothill Street
London SW1H 9NF

Abstract

This paper looks at the developments in the United Kingdom in the areas of job design and work organisation over the ten years since the Department of Employment's Work Research Unit was established. It concentrates on work carried out by the Unit and discusses the moves towards increasing involvement at all levels and the issues that raises within the contraints of British industry. Particular emphasis is placed on the need to take into account the values of managers.

Introduction

The Work Research Unit was set up in December 1974 within the Department of Employment because the Department was then concerned about the problems of job satisfaction, particularly of those people engaged on repetitive types of work such as car and electronics assembly. It had in 1970 commissioned a retiring Chief Psychologist from the Ministry of Defence to survey what was going on throughout Europe and the USA as in Volvo in Sweden, Bosch in West Germany, Philips in The Netherlands and AT & T in the USA.

Following his report (Wilson, 1973) the Unit was established within the headquarters of the Department and, around the same time, a Steering Group on Job Satisfaction was formed under the chairmanship of a Minister in the Department with representatives from the Employers (CBI), Trades Unions (TUC), and the Department. This Steering Group, which is still in operation, has the role of advising the Secretary of State for Employment on matters pertaining to job satisfaction.

The views expressed in this paper are those of the author and do not necessarily represent those of the Work Research Unit or the Department of Employment.

Activities of the Work Research Unit

For the first five years the Unit had four main activities:

Advice and assistance: working with companies to help them re-design jobs

Research: sponsoring research in universities and business schools to establish better the conditions under which job design changes can take place. In 1975 the Unit also started to take on the 'customer' responsibility for research on job-related stress being carried out by the Medical Research Council

Promotion: holding meetings, appreciation seminars, etc to encourage more organisations to consider changes in job design and work organisation to increase job satisfaction, etc

Information: an Information System was established to provide information for the staff of the Unit, managers, trade unions and others on actual projects being carried out,relevant research, etc.

In the ten years which have passed since the Unit was established it has developed its role to respond to changes which have taken place in the environment and as a result of its experience in working with organisations.

The emphasis has moved from job satisfaction which has a rather narrow connotation to the wider one of quality of working life. The range of jobs concerned has increased considerably and has covered all levels and types including supervisors, senior police officers (Sell 1984 (i)), office work as well as manual factory jobs. There has also been a greater emphasis on the process of change (Cuthbert, Smith and Sell 1984, White 1980).

In practice, the Unit has not been able to have much influence on the lower level kinds of factory jobs with their short cycle repetitive tasks. This is probably due to two special reasons; the expectations of people who do these jobs are limited in that their friends all have the same kinds of job and so they do not expect any better and are not therefore disappointed. Also the capital investment tied up in these kinds of jobs is very high and so change is not easy unless a major re-investment programme is possible.

Principles of job design

The general principles which the Unit seeks to apply have changed little can be listed as

Variety: people should be able to vary the tasks they do, work at different speeds and move about whilst carrying out their jobs. Repetitive short cycle tasks should be avoided as much as possible, especially if they are machine paced. Although it appears that people can tolerate jobs which are repetitive the evidence is that rigidly paced work can be very stressful and people do try to move away from this kind of work.

Autonomy: people should be involved in decisions surrounding their job such as deciding the way it should be done, the order in which tasks should be carried out and, in particular, they should have some say when changes are due to take place.

Identity: tasks should fit together to make a complete job, as when someone makes a complete article.

Feedback: it should be possible to know how well a person is performing without them having to be told. When this is not possible positive feedback and reward is better than negative as, apart from the obvious motivational aspects, it encourages the correct behaviour whereas only knowing when you are wrong does not actually indicate what is right.

Responsibility: people need to feel responsible for their work. If the above conditions of autonomy, identity and feedback are satisfied then responsibility normally follows. They have a complete job which they can own and this is more likely when they are able to do their own inspection and so sign off a product as theirs.

Social contact: most, but not all, peoppe desire to nave contact with other people as part of their job and not just at break times. They do not want to be forced to be isolated in one room without the possibility of choosing whether to have company or not. On the other hand, they do like to be able to choose to have some privacy sometime.

Achievement: they like to be able to go home at the end of the day feeling they have done a useful job and have achieved something preferably within the aims of the job rather than 'getting something over the boss'.

Opportunities for learning and development: although this might seem an unobtainable aim in any job it is possible to provide off-line possibilities for this even where the main job is of the routine type as with participating in Quality Circles or other problem-solving groups or to act as trainers of new staff.

Optimal loading: too much work can lead to strain and so can too little. A balance has to be aimed for which may be difficult in practice as the right levels vary very much from person to person.

Minimal role conflict: problems in role conflict can arise in many supervisory and managerial jobs as when organisations say they are concerned with safety but actually do not reward attention paid to safe working practices when they conflict with short-term production requirements.

Minimal role ambiguity: this is also a problem of many managerial jobs where the job-holder is uncertain of what he is actually responsible for and how much he will be called to account. The police are a good example of this in that the police constable who is usually the first person who has to deal with an incident is often uncertain as to how much he will be supported by his superiors who will be able to make a more leisurely account of what an appropriate action should have been (Sell 1984 (i)).

It must be accepted that these are guidelines and that they cannot be applied without further thought to all situations, especially where people are carrying out jobs which do not match up to the ideal criteria above. As an exponent of these principles to practicising managers one is always taken and shown people who are carrying outs jobs which are obviously repetitive and short-cycled and which, if they are not paced, are causing no problems of absenteeism, turnover, poor quality work, etc.

In applying the above principles in any particular real-life situation the desires and perceptions of those actually doing the jobs must be taken into account. For instance, it is often also suggested that it helps to make complete jobs if people are able to do the necessary paperwork, writing up of reports, etc. However, in practice, these activities are often seen by the job-holder as something extra which they wold prefer not to do. Completeness, as with the other design principles, is seen as meaning different things to different people. Using the police as an example, most police officers see the end of a case as when they have made an arrest; the writing up of the case for the prosecution they would prefer to pass to someone else. The writing up of a final research report also appears to be a chore which researchers would prefer not to do.

If someone is actually doing a job, no matter how bad it is in job design terms they have accepted it and it does their self-respect no good to be told it is bad. They have, in effect, a psychological contract to do the job in that way and contracts should not be broken unilaterally.

The Unit therefore has moved from the directive approach which tends to say 'we are the experts on job design; you have a badly-designed job so we will re-design it for you' to the more participative one 'how would you like us to help you re-design your job?'

Recent developments

The approach of the Unit is essentially practical. Its aim is to get more attention paid to quality of working life by all types of organisations in the United Kingdom taking account of the particular constraints which prevail. It is not in the business of academic research but in helping to create changes in practice.

Emphasis has shifted to trying to understand better the real life conditions which facilitate or hinder change towards better quality of working life conditions. The Unit has a remit to spread experience for the benefit of other organisations. Although it does not charge for its services it does expect the client companies will allow both successful and less successful changes to be publicised (anonymously if necessary). There is always a risk with units set up like the WRU in central departments that they will lose contact with reality and the WRU's advisory and assistance role in working with organisations on real problems is a way of avoiding this.

With the difficult economic situation of the 1980s it is obvious that quality of working life on its own is unlikely to be given much consideration by the managements of any kind of organisation unless it also contributes to effectiveness, efficiency, productivity, etc.

The shift towards the greater involvement of people illustrates that this is so. People have a lot to contribute to the success of their own job. Even at shop floor level the person who knows best how to improve performance is the person who is actually doing the job. If this knowledge can be tapped and acted upon there can be an increase in both the performance of the!job and the quality of life of the job-holder.

It is of interest to note that in the UK the present panacea for management is Quality Circles. These are consistent with the WRU's QWL approach to involvement but actually arrived via Japan where they had been developed as a means of harnessing the workers' knowledge as a way of improving performance.

The WRU's real concern now regarding Quality Circles in the UK is the degree to which managements will be actually able to develop invovlement whether it is to improve performance or QWL. The traditional management style is more inclined to authoritarianism than the more open participative ones which are being espoused by this paper. The risk is that Quality Circles and other approaches will be tried out by managements of the more traditional authoritarian kind but without the necessary !hanges to make them work. Effort will be put into training the Quality Circle members but it may not be put into training the various levels of management to listen and act upon the ideas arising from the groups. 'Not invented here' is a well known phenomenon.

Values

In the author's experience the main factor which leads to a successful change is the value system of the top manager of the establishment or organisation. This is shown by studies of actual cases which have looked at examples which have moved from a success to failure (eg Glaser, 1976). When the top manager is moved and is replaced with someone with different values the QWL work often stops and this is confirmed by situations where the WRU has experience.

Many of the case studies written up in the literature have been based on small operating units contained within larger factories where the level of support for the experimental activity has not spread to the factory as a whole. This has been so even when the experimental unit has been shown to be financially superior to the traditional way of operating.

Managers and engineers brought up in the more traditional way of working find it difficult to appreciate that new ways of working can be both effective and satisfying to the workforce. This can be illustrated by the common view amongst the traditionalists in the UK that the work carried out by Volvo, Sweden, has been unsuccessful in commercial terms. Nothing can be further from the truth. In the Kalmar factory, the most advanced in terms of job design, the man hours per vehicle is 25% lower than in

Torslanda (which is itself more advanced than most other car factories). And it has the lowest number of white collar working hours per vehicle produced. Volvo as a whole is commercially successful and is not subsidised by outside sources (Aguren 1984). The evidence on which the Unit bases its approach, as with Volvo above, is that there does not have to be a trade off between effectiveness and quality of working life. The Unit tries as much as is possible to establish an organisation-wide approach to reduce the effects of a change in management with different values taking over before the new culture has been established.

Management of change

When people are given the opportunity to contribute to decisions regarding their own job they need to be listened to by their superiors and their ideas acted upon or at least given real and positive consideration. This often means there is a need for a change in management style, away from the more common authoritarian style previously mentioned. One way of starting up a new project has been to interview individually all the members of the management team and to feed back to them as a group the issues which are necessary for them to deal with if a more participative approach is to succeed.

Once the commitment of the top management group has been established and some agreement obtained on the next step it is necessary to develop commitment to middle management levels (which is often the most difficult group to deal with). There are choices as to whether the work should be done across the whole organisation from the start or concentrated initially in certain areas. This decision must depend on the levels of commitment, understanding and competence of those at middle management level.

Developing participation is not easy and it is necessary to pick at least one area which is likely to be successful. It is likely to be one with a manager who shares the values of participation. If it is also decided to pick a more difficult area on the basis that if it succeeds there it will succeed anywhere it is even more important that that should also be one with a supportive manager.

Participation is two way and it is necessary to establish on what basis other people wish to participate. This may require some form of attitude survey, interviewing programme or just informal approaches by managers. If a sophisticated data collection approach is embarked upon it needs to be managed by some form of representative group covering the whole company to ensure that it meets the needs of employees at all levels. There is a tendency to take standard questionnaires off the shelf but these may not ask the questions which the respondents want to answer. One possible approach which also has a value as a management development exercise is to use staff from one part of the organisation to interview those in other parts. Although this does raise problems of confidentiality this is not usually as serious as is sometimes believed.

In the United Kingdom industrial relations situation it is necessary to work with the trade unions, if they are recognised in the organisation. This means getting the co-operation of both the in-company representatives and the local officials. This is both important and potentially difficult when job restructuring is being considered because it may well be that the tasks at present carried out by the members of one union should logically in the future go with jobs of members of another union.

The process of increasing involvement and job redesign is not a way of improving industrial relations. In fact it is more likely to work successfully if industrial relations are good from the start. In changes of the type being discussed it is necessary for there to be an element of trust right from thestart and this tends to go along with good industrial relations.

It is not possible to apply 'package' solutions to all types of companies. Each company has its own unique culture and what is right in one place is not acceptable in another. The Unit has established networks of companies in different parts of England which meet together at regular intervals when one of the companies will present the work it is doing. One can see from the questions which are asked at these meetings that, although they all have some belief in the QWL type of approaches which the WRU is supporting, they have very different attitudes to the way in which it can be appplied, depending on their own payment systems, industrial relations history, market situation and, most important of all, values of their top management.

A major problem which arises in changing to a more participative style of operation is that of competence. Participation is not something which can be taught in the traditional way. It might be learnt from experiential types of training courses. It is more likely to be practised if it is rewarded in some way by higher levels of management. There is the paradox of 'you will participate .

In one company in which the Unit has been working there is the aim of putting all those who wish to be trained through a one-week residential course of an experiential type. Each course has people from all levels from top management, which can include a director, to the shop floor. There is, of course, the problem with the first few courses that even if the course members come back believing in participation and better able to do it they return to a company of untrained and perhaps unbelieving people (Cuthbert, Smith and Sell, 1984).

People need to feel a degree of security and to expect to be treated fairly. Whilst it is not always possible to avoid some people being losers in a change the maximum possible involvement from all those affected will reduce the risk because the open discussions which take place will being the problems out into the open and, if it is within a supportive environment, allow them to be faced up to and resolved. Most people will accept a change which is not entirely to their own advantage if they feel that as much account as is practicable has been taken of their views, if they think that the organisation does care and if they trust and feel committed to the organisation and the changes which are taking place.

Greenfield sites

So far in this paper the emphasis has been on changing an existing organisation as this is the more usual situation with which the Unit is faced. In theory, a greenfield site, when one is starting from scratch with no in-built prejudices, should be a lot easier to influence. In practice, however, projects are controlled by engineers who are being judged by their ability to produce items from the completed factory by a specified date. Very rarely do they expect to be rewarded for installing a system which pays attention to quality of working life (unless the factory is also on time!).

There are exceptions to this where the ultimate client includes in his specification for the factory or organisation a concern for the people who have to work there. Volvo is the most outstanding example of this with its Managing Director's insistence on paying attention to the need of its workers (Gyllenhammar, 1977).

In the UK there are some examples of this happening. Lisl Klein of the Tavistock Institute was involved in the setting up of a new confectionary factory in which the operators were arranged into autonomous work groups with more say in how their jobs were organised and without the usual first line supervision. The management of this private family-owned company have, as with Volvo, shown a great interest in the application of behavioural science knowledge and a concern for their workforce and the kind of jobs they ask them to carry out. This project had been independently reviewed when it was found that the new organisation was working satisfactorily and was readily accepted by the workforce. (Kemp et al, 1983).

Carreras Rothmans in setting up a new cigarette factory, established it on the basis of task groups, each containing 25 people responsible for the making and packing of the product. Each group has flexibility within its own boundaries, can evaluate its own work and carry out its own routine maintenance. These have also been judged as successful by the employees and have allowed more responsibility to operators, greater commitment and a broadening of skills (Work Research Unit, 1982).

Courage, the brewers, wish to build a new brewery which will replace an existing and outdated one and be some miles away from it. They set up a number of teams including trade union officials to advise on all aspects of the Brewery, including harmonisation of conditions, pay and grading, job design and work organisation. Again in the new Brewery there has been a high degree of satisfaction of people to transfer from the old Brewery with the kinds of job they now have under the conditions of work in the new plant (Work Research Unit, 1982).

Problems with measurement

When attempting to look at changes in work organisation in a scientific way the problem of measurement arises. In practice the Unit has relied on the subjective assessments of the employees and the management as to whether they would want to go back to an older way of working.

There are so many other changes in terms of market pressures, etc which influence measures of profit, cost, productivity, labour absence and turnover that objective measures of performance, etc relevant to changes in quality of working life are almost impossible to obtain.

Standard attitude surveys repeated over periods of time are believed by some to give an indication of change but the frame of reference in which the replies are being given is changing also. If participation or job design is improved then this new level becomes the standard: expectations are increased and so the replies may not reflect the real increase level.

Standard measures of job design characteristics, etc are not usually seen as relevant by job holders in real jobs and reflect aspects chosen to fit a theoretical framework rather than practical change in industry. It is not possible in practice to say that the same sets of job characterstics are those which are the most important for that particular operation.

Pay

Within the UK payment systems can be seen as a problem. Increasing job involvement can be seen as a requirement for increasing pay if pay is judged by a job evaluation scheme based on knowledge, skill, responsibility, etc. Some operators express the view that they are not paid to think.

The importance of pay as an issue, however, depends very much on the type of payment systems in use and the degree to which pay levels are an issue. Where the employees at all levels feel they are fairly paid and treated they are more likely to take on additional tasks of a more interesting type without immediately agitating for more pay. In one with strict control of pay, individual payment by result schemes and continual negotiating for more money then it may be much more difficult.

One problem with traditional types of job enlargement/job enrichment on shopfloor jobs is that before any change the jobs may rate low on knowledge skill responsibility but collect premium payments for bad conditions. Any increase by job enrichment in the pay entitlement may be much less than that which is lost by the removal of the bad conditions premium and so may not be acceptable to the jobholders who are quite likely to be people who have an instrumental appraoch to work.

Divisions in the workforce

Within the UK there are many divisions within the workforce which can cause problems. A major one which is under the control of management is that between hourly paid workers and the staff who are monthly paid. With the development of more participation hourly paid people are taking on tasks which have traditionally been the prerogative of monthly paid and they are starting to question the many privileges which staff people have which they do not.

There are also problems between unions and, especially with the introduction of new technology, areas of work which have always been carried out by members of one union are now becoming more open to question. For instance,with computer controlled machine tools should adjustments to tapes, etc be carried out by people who come from the shop floor who are on the spot or from the planning staff who produced the tapes in the first place?

The future

In the future the major problems are likely to arise in offices with the present development of VDUs and computer terminals spreading to all levels and with extra problems likely if voice input to computers becomes possible (Sell, 1984 (ii)).

The contact that the Unit has with industry is continuing to develop, even with the difficult conditions of 1984. So far, however, the balance has been towards manufacturing rather than towards offices as it appears that those responsible for the development of offices have not yet realised the importance of work organisation problems.

References

Aguren S et al, 1984 'Volvo/Kalmar revisited: Ten Years' experience'. Report prepared by Development Council of Sweden

Cuthbert C, Smith A and Sell, R G, 1984 'Forming the future together'. Employment Gazette, Vol 92 (also issued as WRU Occasional Paper No 30, March 1984)

Glaser D, 1976 'Productivity gains through work life improvement'. New York: Harcourt Bruce Jovenovich

Grayson D, 1982 'Job evaluation and changing technology' WRU Occasional Paper No 23

Gyllenhammar P G, 1977 'How Volvo adopts work to people' Harvard Business Review 55, pp 102-113

Kemp et al, 1983 'Autonomous work groups in a greenfield site: a comparative study' Journal of Occupational Psychology 56, pp 271-288

Sell R G, 1984 (i) 'Work organisation in the police' Proceedings of International Conference on Occupational Ergonomics, Toronto

Sell R G, 1984 (ii) 'New technology and the design of jobs' Proceedings of First International Symposium on Human Factors in Organisational Design and Management, Hawaii

White G, 1980 'Job satisfaction and motivation - The development of practical strategies for their enhancement' in Duncan K, Gruneberg M, Wallis D(Eds) 'Changes in work life'. J Wiley & Sons

Wilson N A B, 1973 'On the quality of working life' Manpower Paper No 7, London, HMSO

Work Research Unit, 1982 'Meeting the challenge of change - Case studies'. Obtainable from the Work Research Unit

Human Factors in Organizational Design and Management
H.W. Hendrick and O. Brown, Jr. (Editors)

THE BASIC LIFE ORIENTATION CONCEPT
UNDERSTANDING AND MEASURING THE WORK ETHIC: AN EMERGING & SEMINOLE FACTOR IN ORGANIZATIONAL DESIGN & MANAGEMENT PRACTICE

Gary L. Benson

Director, Business & Management Programs
University of Wyoming-Casper
Casper, Wyoming
U.S.A.

The human factor in the work environment has become a focal point for both management theorists and practitioners. Increasing concern for the productivity and performance of people has resulted in an attempt by both academicians and practicing managers to understand and predict human behavior at work. The Basic Life Orientation Concept is a construct that provides both a theoretical and a pragmatic framework through which the impact of work ethic on organizational design and management practices can be examined.

The Basic Life Orientation Concept begins with the fundamental proposition that before human behavior at work can be understood and managed effectively two things about each individual employee must be examined.

1. The relative strength (the presence or absence) of a work ethic, that is to say, the position or importance of "work" as a life activity, and,
2. The position or importance of the "job" in the total life space of the employee.

WORK ETHIC

Max Weber, well-known German sociologist, popularized the concept of a work ethic in his book, The Protestant Ethic and the Spirit of Capitalism. He characterized the growth and development of capitalism by thrift, hard work, and the capacity for deferring gratification. According to Weber, the fundamental basis of capitalist culture is

> . . .an obligation which the individual is supposed to feel towards the content of his professional activity (work), no matter in what it consists. . .in particular no matter in what form it appears. . .the spirit of hard work, of progress or whatever else it may be called.[1]

In fact, the concept and presence of a strong work ethic is so critical to dynamic growth and economic well being, on both the macro economic system level and the micro firm level, that Tilgher in his book, Work: What It Has Meant to Man through the Ages, concluded that the value and position of "work" in a society is in large measure concommitant with its rise and fall.

If Weber, Tilgher, and others are right (and history seems to have affirmed their conclusions) the question then becomes: Do people in general, in the United States and elsewhere, still possess a strong work ethic, a sort of reverence and worship of hard work?

Many authors have concluded that the work ethic is either dead or dying in the United States as compared to the work ethic present in other countries such as Japan and West Germany. Unfortunately, much of this research and writing is highly speculative, theoretical and suspect in methodology. The Basic Life Orientation Concept was developed to help fill the need for empirical investigation of the impact of work ethic on organizational design and management practice worldwide.

THE BASIC LIFE ORIENTATION CONCEPT

The Basic Life Orientation (BLO) Concept is an operationalization of the concept of a work ethic. Encompassed in the BLO Concept are a Basic Life Orientation (BLO) Index and a Basic Life Orientation (BLO) Grid.

THE BLO INDEX

The BLO Index is the measurement instrument used to gather information from respondents (employees) concerning their attitudes toward work, their jobs, and other life activities. Responses to the questions in the BLO Index are tabulated to produce a respondent plot or profile on the BLO Grid in one of eight octants as shown in Figure 1 below.

FIGURE 1
BASIC LIFE ORIENTATION GRID

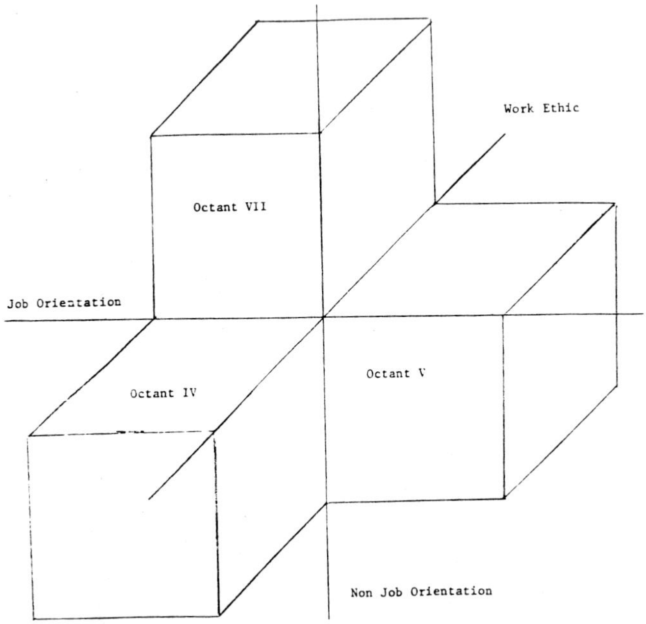

THE BLO GRID

The BLO Grid is a three dimensional model, with work ethic plotted on one axis, job orientation on another axis, and non-job orientation on the third axis. The Grid provides for eight octant plots or Basic Life Orientation profiles as shown below in Table 1.

TABLE 1
BLO PROFILES

Type	Work Ethic	Job Orientation	Non-Job Orientation
I	High	High	High
II	High	Low	High
III	High	Low	Low
IV	High	High	Low
V	Low	High	High
VI	Low	High	Low
VII	Low	Low	High
VIII	Low	Low	Low

Critically important in the application of the BLO Concept to the practice of management and organizational design is the realization that "work ethic" and "job orientation" are different constructs. It is possible, and in fact occurs frequently, that a person may have a strong work ethic and choose to express that ethic in a non-job, rather than a job environment, thus having a low job orientation while at the same time having a strong work ethic.

APPLICATIONS OF THE BLO CONCEPT IN MANAGEMENT AND ORGANIZATION DESIGN

The BLO Concept has many applications in the practice of management and organizational design. In the largest study conducted on the BLO Concept to date a sample of 734 people employed in five different companies (an insurance company, a savings and loan, an electronics manufacturing and assembly plant, a large retialer, and a government agency) was drawn and the respondent's BLO profiles plotted on the BLO Grid. The 734 participants plotted as follows by grid octant.

TABLE 2

Octant/Type	Number of Plots
I	129
II	151
III	77
IV	185
V	17
VI	64
VII	88
VIII	23

From these results it is possible to conclude that people do have and exhibit different Basic Life Orientation profiles with different work ethics and job orientations.

DECAY VERSUS TRANSFER

From this research, and other related BLO research projects, it is possible to conclude that in many cases and with many people the problem is no so much one of a decaying work ethic, but rather that of a transferring or shifting work ethic where people choose to express their work ethic in non-job related, as opposed to job related environments at an ever-increasing rate. The next question becomes: Why?

From the mid-1800's to the mid-1900's society moved from an average of 72 hour work week to a 40 hour work week (a 12 hour, 6 day week to an 8 hour, 5 day week). In recent years that "normal" 40 hour (8 hour, 5 day) work week has often been reduced even further by some combination of 4-10 hour days, 3-12 hour days, flex time, job sharing or other methods of work scheduling.

In short, society has increased its non-job (leisure time), that time required beyond eating, sleeping and miscellaneous necessities, from an average of 2 hours per day in the mid-1850's to an average of 8 to 10 hours per day in the mid-1980's.

Many people, if not the majority of all workers, have found that non-job institutional commitments in the life space are more interesting, challenging, and rewarding than their jobs. Hence, many workers have chosen, either consciously or unconsciously, to transfer the expression of their work ethic to non-job commitments and activities in life, thus relegating their jobs to a position of secondary importance in life and viewing their jobs as only a means to an end rather than an end in and of themselves.

This massive transfer of work ethic in the job market, particularly evident in the United States during the past two to three decades, has a very real impact on management practice and organizational design. The dilemma lies in the fact that management now finds itself in the position of literally having to compete for the interest and attention of the employee. Organizations find that they must take alternative approaches to organizational and job design and structure in order to facilitate the transfer of the work ethic back to the job environment.

The Basic Life Orientation Concept can thus be seen as a vehicle that management and organizations can use to evaluate the presence and/or absence of work ethic expression within the job environment. Such an evaluation can then be used as the basis for the justification, development and implementation of alternative management practices and job and orgranizational designs needed to enhance management and organizational performance and productivity.

ADDITIONAL APPLICATIONS

The BLO Concept also has additional applications in the area of management practice and organizational design. Below are brief descriptions of two such applications.

PERSONNEL MANAGEMENT

In a small study designed to evaluate the usefullness of the BLO Concept in screening and placing job applicants in a restaurant in Casper, Wyoming, the BLO Index and Grid proved to be very helpful in narrowing a field of 53 applicants for 4 positions to a field of 15 applicants based upon work ethic and job orientation. Four of the 15 finalists were hired and 3 of the 4 achieved higher levels of job performance than any of the existing employees based upon end of probation and annual performance reviews. The BLO Concept may be seen, then, to have application in the area of personnel management selection, placement, and promotion practices.

MANAGEMENT DEVELOPMENT/SUPERVISORY TRAINING PROGRAMS

The BLO Concept has been used a number of times in management development and supervisory training programs to assist managers in self-evaluation of management practices and career planning. In the context of such programs, the BLO Concept, Index, and Grid have been found to be very useful in assisting managers and supervisors in achieving the kind of introspection required to improve career planning and both short term and long term value to employers. There are, in fact, many other potential applications for the BLO Concept in management development/supervisory training and employee/organizational development programs.

In addition to personnel management and management/organizational development programs, the BLO Concept has potential application(s) for management practice and organizational design in areas such as productivity improvement and cost reduction programs, job design and motivation, and compensation practices.

CONCLUSION

If Weber, Tilgher and others have been right in their assertions that "work ethic" is the key to the economic well being of individual firms, companies and organizations, as well as societies, cultures and nations, then the Basic Life Orientation Concept may well be an important contribution to understanding and managing the impact of the human factor on management practice and organizational design.

[1]Max Weber, The Protestant Ethic and the Spirit of Capitalism, Allen & Irwin Ltd.: London, 1930, p. 139.

Human Factors in Organizational Design and Management
H.W. Hendrick and O. Brown, Jr. (Editors)

RETURN TO WORK PROGRAM: A Benefit to Both Employee and Employer

Gene L. Dent

Return to Work Coordinator
Lawrence Livermore National Laboratory
Livermore, California
U. S. A.

Increasing costs and a trend toward self-insurance/self-administration of benefits have led corporations to initiate health care cost control measures. One approach is a return to work program aimed at employees experiencing difficulty returning to work after injury, illness, or disability. Helping employees return to work requires an appreciation of the complexities that tend to prolong disability as well as an understanding of all the factors that contribute to return to work. Vocational rehabilitation practices, employee assistance counseling, communication of all parties represent the foundations of a return to work program. While these practices may not be new, their systematic application within corporations represents a new approach toward remediating the human and financial costs of disability to both the employee and employer.

What is a Return-to Work Program?

The program to be described is one predicated on vocational rehabilitation techniques, supplemented by employee assistance and risk management practices. Vocational rehabiliation is not an unfamiliar term to employers. In some states, worker's compensation law requires provision of vocational rehabilitation services in selected cases. Some employers may regard vocational rehabilitation as an additional costly benefit within the legal proscriptions of worker's compensation law. Rather, vocational rehabilitation, applied willingly and proactively, contributes to the employer's cost effective administration of benefits because 1) it focuses attention and effort on difficult cases, 2) it defuses problem cases from growing in complexity, and 3) all its effort is directed toward returning employees to work.

However, the program to be described here will be a new application of vocational rehabilitation. A few large corporations have operated vocational rehabilitation or disability management programs for several years. Titles such as "disability manager" or "health care cost containment specialist" are appearing within the organization charts of an increasing number of corporate human resources and benefits departments. Large corporations, particularly self-insured entities, have much to gain by formalizing a return-to-work program. A formal program adds visibility to the company's effort, insures consistent treatment of employees, and can help in the early and certain identification of cases that need attention.

The return-to-work program must be directed toward and for the benefit of all employees -- whether or not the disability is work-related. To reserve this type of assistance for only work-related or only the most serious cases is to require cases to become work-related or "serious" before help is made available. Consequently, the program's emphasis should be directed toward preventing problems. While the term "preventive rehabilitation" may, at first, appear to be a self-contradictory term, it expresses the main purpose of an internal return-to-work program. Even though injury and illness cannot always be prevented, even though disability may not be prevented, any resultant handicap can be prevented. That is, a return-to-work program can prevent disability from being an obstacle to the employee's ability to work.

Therefore, in keeping with the desired preventive perspective, the return-to-work program may be most advantageously situated within an employee assistance function. At their best, employee assistance programs are preventive in nature -- directly affecting the cost of health-related benefits. Disability often provokes the kinds of problems with which employee assistance programs customarily deal. And vocational rehabilitation provides employee assistance a problem-solving methodology and a much needed means for directly measuring cost-effectiveness. Other natural locations for a return-to-work program would include Occupational Health or Benefits departments.

What Does a Return-to-Work Program Do?

Counseling: One of the first considerations for the return-to-work coordinator should be to determine the motivation of the injured or ill employee. Motivation is not a static thing, and it can be difficult to determine. But it is always critical to successful return to work. A disability never happens in a vacuum -- figuratively speaking. Disability happens to a person. Though disability certainly causes new and difficult problems, it may solve pre-existing problems. For example, convalescence may be regarded as a satisfying respite from a disliked job or supervisor. Or the employee may see little economic incentive to return to work quickly because of benefit payments and personal circumstances. Consequently, thorough vocational rehabilitation must take into account the entire person -- including emotional, financial, and interpersonal as well as medical needs. The physical convalescence of the employee is important, but equally important are all the other factors which influence the employee's attitude toward disability and return to work. The best rehabilitation plan is certain to be frustrated if the return-to-work coordinator and disabled employee are working at cross purposes. Consequently, it is important to help disabled employees explore their own values and motivation early in return-to-work planning.

Job analysis/modification: Traditionally, vocational rehabilitation counseling has taken a systematic approach -- fitting all the pieces of the "disability puzzle" together. Often, one piece of the puzzle is an analysis of the employee's job. Industrial engineers have used job analysis techniques to make jobs safer and workers more productive. Vocational rehabilitation uses similiar techniques to find a match of disabled employee to the job. Job analysis is the systematic identification of those physical and psychological requirements essential to a particular job. An advantage of the written job analysis is that it enables the treating physician to participate in return-to-work planning. Without a clear statement of the job's requirements, the doctor is guessing. But with a job analysis, the employee's physician is often willing to help determine to what degree recovery and return to work can contribute to each other, and recommend temporary job modifications.

The other half of job analysis is job modification. For some disabilities, job modification may entail changed or reduced work hours, temporary reassignment of heavier work, a modified work site, or simply the opportunity to take periodic rest or exercise breaks. If the disabled worker is assigned to a new job, that job, too, may require analysis. Sometimes a job only seems easier; disabling conditions can be quite selective as to the kinds of activities they allow. Normally effortless movements can take on a radically new character during a period of temporary disability. The prevalence of job analysis and job modification within vocational rehabilitation highlights a strong bias toward simple solutions. Prompt return to work is usually best served by return to the same job, building upon established skills and relationships.

Adaptive devices: Adaptive devices can enable disabled persons to return to work and to be more productive. Examples might include simple grasping aids, orthopedic supports, or special tools. In addition, advances in electronics are making possible futuristic communication and mobility aids. The most far-reaching adaptive device of all time may be the soon-to-be-ubiquitous computer terminal. Most good adaptive devices are the product of cleverness more than cost, but even expensive items represent a reasonable investment when compared with the alternative of employees being off for extended periods of time or going out on disability retirement. Furthermore, the devices may increase productivity and safety on the work-site and be available if a similar need should arise in the future. Though adaptive devices can sometimes be more trouble than they are worth, when an aid works it can be like slipping a key into a lock. Selection of the proper device should include careful analysis and the thorough involvement of the disabled employee. Another source of help for difficult cases can be community resources. State Departments of Rehabilitation, private rehabilitation centers and hospitals, and non-profit organizations have skills and experience that should be utilized.

Re-Training: The least common but occasionally necessary step to return disabled employees to work is re-assignment and even re-training for a new position. To compound the difficulty of adjusting to return to work by complicating it with re-training, a new job, and new relationships is to increase the risks for all concerned. Since employees can get as lost in retraining as they were in disability before, try to build on existing skills and relationships. If an alternate job is necessary, on-the-job training and continual monitoring of performance should be emphasized.

How the costs of adaptive devices or retraining are borne will depend on the employer's situation. In those states with mandatory rehabilitation as part of worker's compensation, reserving requirements may dictate how those expenses are covered. In non-work related cases, many short-term disability carriers express ready financial support for plans that will help their policy-holders return to work. Insurance companies have been quick to develop rehabilitation programming. If you decide to discuss financial support for rehabilitation with your temporary disability insurer, be sure to have specific cases and specific proposals ready. Self-insured or permissibly uninsured organizations are in the best position of all to realize cost savings through return-to-work planning.

Other considerations: A rehabilitation specialist can help the employee understand and cooperate with prescribed treatment. Moreover, it may be important to place the employee-patient as the central actor in convalescence and rehabilitation. Too often, a patient feels like things are being done to him or her. Many things about disability foster feelings of dependency. The patient is acted upon by a succession of health care personnel, coming to feel like little more than an inanimate object. Consequently, if recovery is delayed or pains continue, it is natural to blame the doctor or the treatment. Therefore, it may be to everyone's benefit to expect the patient to be in charge, to take responsibility for convalescence and return to work. The chances of success are maximized if the patient/employee can be brought to center stage as the principal actor and encouraged to use health-care personnel as resources. Though this may represent only a change in the patient's attitude, the outcome can be remarkably transformed if the patient has taken responsibility for recovery and return to work. This re-framing of the recovery process can be accomplished if an early intervention is made.

Several techniques help the employee take an active and constructive role in rehabilitation. One example may be to encourage the employee to regard the disability payments, not as compensation for being hurt or payments for being off work, but as surrogate salary for an important and usually more difficult new job -- convalescence and return to work. Then, when recovery has been completed, the employee can rightfully take satisfaction and pride in his or her accomplishment. Whatever intervention is used to counsel the disabled employee, it is predicated in large part upon information obtained in earlier counseling and motivational assessment discussions. An appropriate framing of return-to-work planning will be based upon careful attention to the patient's statements, ideas, and feelings about self and others.

While there may be no observable difference between patient-centered convalescence and treatment-centered convalescence, having the patient own the responsibility gives them control and can remarkably affect return to work. Techniques of behavioral medicine, popularized by W. E. Fordyce among others, deserve greater consideration within medical and rehabilitation practice. The formal study of behavioral approaches toward rehabilitation is in its infancy and holds considerable promise for influencing patients' attitudes toward convalescence and chronic conditions.

During convalescence, it is important for the injured employee to keep in touch with his or her supervisor because the single most important element in expeditious return to work is good communication between principals. If job modification is to be necessary, the supervisor will be the one to provide it. Efforts to keep the supervisor informed of return-to-work measures are rewarded when a return to light duty or a modified job is appropriate. As in so many other instances, the burdens of reasonable accommodation fall upon the supervisor. Once the employee has returned to work, a vocational rehabilitation program can help reduce those burdens. Supervisors support vocational rehabilitation efforts because it helps them to resolve unfamiliar and difficult problems. As with all employee assistance practices, it allows supervisors to be supervisors and to focus on work requirements.

Why Operate a Return-to-Work Program?

The human and financial costs of extended absence from work are great to both the employee and employer. From the employee's perspective, disability benefits rarely equal lost salary. The stresses of enforced idleness, reduced income, and irregular communication with the employer combine to compound the "disabling" effects of illness and injury, and may have the effect of prolonging the period of inactivity and making return to work more difficult. Careers and earning power are jeopardized by disability.

From the employer's perspective, the productivity of a trained employee is lost, and the supervisor and co-workers are affected. Though most employees return to work without difficulty, the costs generated by difficult cases warrant attention.

Consequently, both employee and employer benefit from a return-to-work program.

Employee benefits:

1. Return to work means resumption of salary. Disability benefits rarely equal full salary.
2. Return to work represents a genuine expansion of the employee's disability benefit. The disability benefit comes to mean more than simply a check in the mail.
3. Disability causes concern about earning power and career prospects. A return-to-work program responds directly to those needs, resolving the uncertainty and worry surrounding career prospects.
4. Disability provokes stress throughout all areas of one's life. In considering the whole person, return to work addresses these necessary concerns, too.

Employer benefits:

1. Return to work represents recognized Affirmative Action toward a particularly important group of disabled persons -- one's own employees.
2. Return to work helps supervisors deal with the difficult and unusual problems that illness and injury cause. Prompt return to work eases the burden on supervisors and co-workers.
3. One of the not-so-hidden causes of lowered productivity is lost work-time. Return to work contributes to maximum human resource utilization.
4. Return-to-work programs save money. The costs associated with disability benefits provide the means to measure cost-effectiveness.

As much as has been made of this last benefit, an example seems in order. In California, a common worker's compensation temporary disability payment would be $196/week (sometimes supplemented by the employer). Therefore, every month of absence due to industrial injury costs $800/month, not counting supplemental payments and medical costs. Return to work saves temporary disability costs. Alternatively, other arenas for potential savings are short-term and long-term disability programs and disability retirement channels. If an employee can be returned to work rather than leaving for disability retirement, the savings can amount to thousands of dollars each year before normal retirement.

ROBOTICS AND AUTOMATION

Human Factors in Organizational Design and Management
H.W. Hendrick and O. Brown, Jr. (Editors)

SAFETY DESIGN OF ROBOT WORKPLACES

Martin Helander, Ph.D.

Department of Industrial and Management Systems Engineering
University of South Florida
Tampa, FL 33620

This paper provides a summary of design principles which may be used to improve the safety of industrial environments. A model for safety interaction and safety communications between humans and robots is given. There are several options for safety communication of warnings between humans and robots including the use of visual and auditory signals, synthesized speech, and computer voice recognition. In the future, technological development of robots' sensory capabilities will make it possible to utilize the robots' sensing to improve the safety at the workplace. Several different types of robot sensors may be used, for example: floormats, automatic gates, computer vision, sonar devices, force devices, force sensing, and touch sensitive skin. The type of safety system chosen depends largely on the size of the robot, and the types of human and robot tasks being performed. The last part of the paper gives examples of some ergonomic design improvements of robots that may improve comfort and productivity as well as safety.

INTRODUCTION

Robotic safety has been an issue for many years, actually long before robots were introduced in the industrial environment. In 1940-50, Isaac Asimov published several science fiction-type stories about robots (Asimov, 1950). His three laws of robotics provided an ethic for the interaction with robot:

1. A robot must not harm a human being, nor through inaction allow one to come to harm.

2. A robot must always obey human beings, unless that is in conflict with the first law.

3. A robot must protect itself from harm, unless that is in conflict with the first or second law.

This paper focuses primarily on the safety of human beings and less on the safety of the robots themselves. Often, however, the issues go hand-in-hand.

There has not yet been published much statistics on accidents with robots. It is therefore difficult to evaluate the safety of robot workplaces as compared to other industrial environments. A study conducted in Sweden (Carlson, 1980) analyzed 15 accidents which were reported over a 30-month period from January 1976 to June 1978. Common causes of accidents were "pushing the wrong control buttons," and "being cut by tools grasped by robots". From this study, it was concluded that the accident rate was about one accident per 45 robot years. This may be compared to industrial presses, the most hazardous industrial machines, which have an accident rate of one accident per 50 machine years. Sugimoto and Kawaguchi (1983) analyzed 18 near-accidents caused by industrial robots that occurred in Japan. Summarizing the results of this and the Swedish study, the authors stated that:

-Most manufacturing companies are aware of the dangers of robots.

-The accident rate of robots is significantly higher than for other automated machines.

-A majority of accidents occur during teaching, testing, and maintenance.

As a result of their investigation, Sugimoto and Kawaguchi (1983) concluded that only when robots themselves are able to detect the approach of humans and perform appropriate actions to avoid accidents, will safety in the workplace be assured.

It may seem that Asimov's concerns were legitimate: robots have to watch out so that they do not harm human beings. Accordingly, this paper will describe several different types of sensors that make it possible for robots to sense human activity in the workplace, and supply the human operator with adequate warning signals.

A MODEL FOR HUMAN-ROBOT SAFETY INTERACTION

Figure 1 presents a model for human-robot safety interaction. Robots may sense human activities in several ways. The information from the safety sensors is then used for modifying the robot's arm movement (motor response) and for communicating safety messages to the human operator. There are several modulating factors which influence both the robot sensing strategy and the motor response. These factors include type of sensing and motor equipment and software support.

Humans, on the other hand, evaluate the ongoing activity using perceptual and cognitive rules. As with robots, there are several modulating factors, for example, previous training and experience, and the condition of the sensory equipment (vision and hearing). As a result of the human sensing, there are human motor responses and human safety communication to the robot. The details of this model is further elaborated upon below.

ROBOT SAFETY

The sensing capabilities of robots are limited to sensing of human proximity.

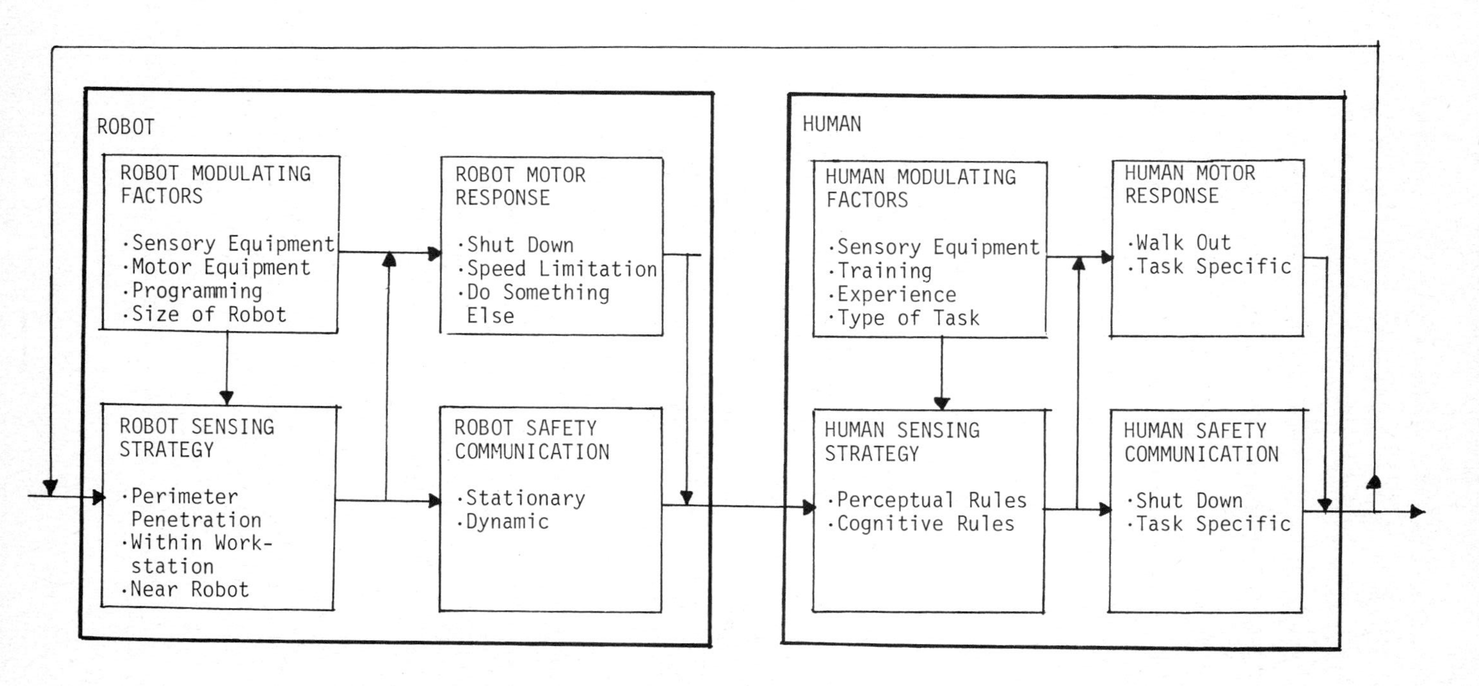

Figure 1
Human-Robot Safety Interaction

(In the future it is feasible that expert systems may be used for evaluating other aspects such as human decision making and intent.) Several different types of sensor devices have been tested by the National Bureau of Standards (Kilmer, 1982), see Table 1.

There are three different principles for sensing human proximity:

- -Perimeter Penetration
- -Within Workstation
- -Near or On Robot

For sensing of perimeter penetration, there are four types of sensor devices: automatic gate, floormats, photoelectric beam, and detection of motion by the use of ultrasonic, microwave or infrared sensors. The automatic gate, floormats, and the photoelectric beam are methods presently used in industry, whereas the motion sensing devices are still at an experimental stage.

A photoelectric beam or an automatic gate are usually mounted on poles that surround the workplace. Opening the gate or walking through the beams will initiate a safety response. Typically, the robot is shut down, but there are other alternatives such as speed limitation of the robot arm and/or alerting the intruder by visual or auditory warnings or by the use of spoken messages derived through computer speech synthesis.

However, the use of poles, chain-link fences or cages introduce safety risk to human operators since there is an increased probability of being pinned between the robot and the fixed obstacle.

For detection within the workstation there are two principles: boundary detection and point detection (much of this is still at the experimental stage). Motion sensors such as ultrasonic, microwave or infrared sensors may be used to sense intrusion across boundaries within the workstation. Alternatively, sensing devices may be used to detect human proximity to critical points, for example, the elbows of a robot. Obviously, touch sensitive floormats may be used both for perimeter detection and detection within the workstation.

There are several principles for robot motor response. The first choice in Table 1, shut down, has time-consuming consequences, since the robot must be started up again. There are other alternatives, such as restricting the speed of the robot arm, instructing the robot arm to use alternative motion paths or to perform other tasks until the intruder disappears. The latter option is possible for flexible manufacturing systems or other workplaces where the robot serves several machines.

The robot may communicate the hazardness of a situation to the human operator in four different ways: stationary warning signs, visual or auditory signals, or by the use of speech synthesis. Stationary warning signals, such as "Hazard-Pinch Point", are always present in the working environment. As a result, workers do not always pay attention to them. Much better are adaptable safety communications that take into consideration the present state of the system. Visual signals, auditory signals or speech synthesis may be used. Of these, speech synthesis has the greatest versatility since depending upon the situation, several different safety messages may be used.

Table 1. Principles for Robot Sensing of Human Proximity

Human Proximity	Type of Sensor	Motor Response	Safety Communication
1. Perimeter Detection	a. Automatic Gate b. Photoelectric Beam c. Motion Sensors ·ultrasonic ·microwave ·infrared d. Floormats	a. Shut Down b. Speed Limitation	a. Stationary Warning Signs b. Visual Signal c. Auditory Signal d. Speech Synthesis
2. Detection Within Workstation a. Boundary Detection b. Point Detection	a. Motion Sensors ·ultrasonic ·microwave ·infrared b. Floormats c. Computer Vision	a. Shut Down b. Speed Limitation c. Use Alternative Motion Path d. Perform Other Task Until Intruder Disappears	a. Visual Signal b. Auditory Signal c. Speech Synthesis
3. Tactile Detection	a. Touch Sensitive Surface (Safety Skin b. Pressure Sensing ·strain gauge ·hydraulic pressure	a. Shut Down b. Speed Limitation	a. Speech Synthesis

Tactile sensors respond to contact forces between the robot and solid objects. Two types of tactile sensors are presently being researched (Weintraub, 1981): touch sensors and force sensors. The touch sensor allows the robot to determine if it is in contact with an object or person. This is accomplished by providing the robot with a skin of two rubber sheets lined with wire. The top layer is electrically charged. When the sheets press together, the wires contact which changes the voltages. The force sensor, on the other hand, allows the robot to determine the magnitude of a contact force. Weintraub (1981) described an end effector equipped with a bracelet on its wrist that shines a light into three light detectors on the hand. The hand is connected to the wrist by springy cylinders of metal and rubber. When the hand exerts a force, the cylinder moves, causing the light to move across the detector.

Computer vision systems may also be used to recognize objects in the workplace. Cameras may be positioned on the robot arm or in the vicinity of the robot. A major problem with this type of system is the requirement for fairly large sized computers in order to process the great amount of information generated. These systems will become more common in five to ten years.

Other types of proximity sensors include eddy-current sensors and magnetic field sensors. The eddy-current sensors may detect changes in flux density on or very close to the robot surface. The magnetic-field sensors are sensitive to the capacitance bewteen a person's body and the surroundings.

Many of these options require extensive development of software, which should be taken into account before the decision to invest in the system. A cost-benefit analysis must be made. This should include benefits due to increased safety and improved productivity.

As mentioned in Figure 1, the size of the robot is important for the selection of safety systems. It is unlikely that large, medium-size and small robots would require similar types of safety systems. For example, a small Seiko robot used for assembly of watches handles only small objects and is not much of a threat to a human being. In contrast, a Cincinnati Milicron T3 robot is larger than human beings and poses much more of a threat. Therefore, depending upon the size of the robot, different safety devices may be used.

Some researchers (e.g., Macek, 1981) are of the opinion that it may not be desirable to cage a robot, since it may create the impression that the robot is a menace. If so, other protection is necessary. A chain-link fence, for example, would protect operators from flying objects that may accidentally be released from the robot's grip.

Engelberger (1980) proposed that robots should be human sized, see Figure 2. Human sized robots are more comfortable than large robots, since it is easier to get a visual overview of both the robot and the rest of the working environment. Operators working with medium-size robots, therefore, have a better sense of control of the environment.

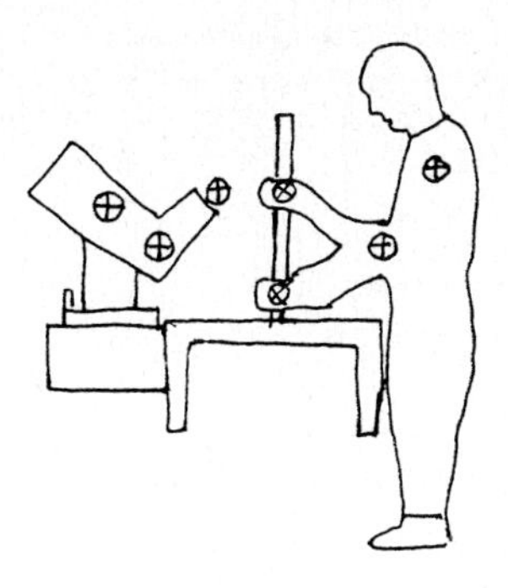

Figure 2
Robots Should Not Be Larger Than Humans
(Source: Engelberger, 1980).

HUMAN SAFETY

As indicated in Figure 1, humans and robots have certain similarities in sensing information, communicating about safety, and type of motor response. For evaluation of information the human being utilizes perceptual and cognitive rules. As with robots, there are several factors which modulate the information processing and the strategies for decision making. These include the status of the sensory equipment (vision and hearing), training, experience, and type of task performed.

Obviously, several safety communication devices are inappropriate if the human operator has hearing or visual deficiencies, or if the industrial environment is such that the noise level or the type of illumination makes it difficult to perceive auditory or visual warning signals.

Previous experience with robots is of great importance. Most people like to observe robots working, thereby exposing themselves to hazards. Until robots become less conspicuous, it is likely that there will be more exposure of people who are not familiar with them. For this reason, it might be necessary to provide additional safety guards around the robot.

As pointed out previously, individuals who program and/or maintain a robot expose themselves to many safety hazards because they have to work close to the robot. Another potential danger is that those working around the robot for some time begin to take its movements for granted. However, power interruptions, which might bring the robot arm to its resting position, or malfunctions which send the robot arm to its limit of travel, could be hazardous. These dangers can be minimized through in-teaching by using slow operating speeds which allow the human instructor to get out of the way if the wrong button is pushed. Another option is to always have two people teaching the robot thereby insuring that there is easy access to emergency stop controls.

Sometimes people have difficulty in perceiving what is going on. When a robot is working, its power is evident and the on-lookers keep their distance. But when it stops, they may move in for a closer look. This may pose a great danger, since the robot may simply be stopping as a natural sequence of the program and could take off again with great force a moment later. Sugimoto and Kawaguchi (1983) distinguished between several different halt conditions, see Table 2.

Table 2. Causes for Robot Stoppage (Source: Sugimoto & Kawaguchi, 1983).

Emergency Halt With Emergency Stop Button
Temporary Halt With The Pause Button
Malfunction Halt Due To Abnormality
Runaway Halts for Machine Failure
Condition Halt for Machine Recycling
Halt Due To Work Termination
Apparent Halt Due To Fixed-Point Position Control

Emergency and temporary halts may be induced by the robot technician durinn the work. Malfunction and runaway halts are due to machine failure, and condition halt for machinery cycling and apparent halt are induced by software. Both for the experienced operator and the inexperienced observer, it may be difficult to understand why the robot stopped and if it is safe to approach the robot. Under such circumstances, a warning signal such as a flashing light or a synthesized speech message could be used to indicate the status of the robot and its potential danger.

The human motor response to a safety hazard depends on the task. Mostly, the human operator will simply walk out of the work area. Occasionally, however, he/she might try to defeat the robot by exerting pressure on the robot arm or by ignoring a safety warning.

To communicate safety actions to the robot, the human operator might choose to shut down the robot. Depending upon the situational characteristics and the design of the robot, there is a wide difference in appropriate control actions:

- -Cutting off the power
- -Cutting off the drive and oil pressure pump power supply
- -Cutting off the power supply for several controls
- -Use the force button

Maybe standardizing the control layout and control functioning of robots would be help for operating the robot.

ERGONOMIC DESIGN IMPROVEMENTS

Ergonomics, or human factors, design philosophy focuses attention on the "user-friendliness" of the human-machine interface. Although this can be justified on moral and ethical grounds, practical reasons are the usual motive. Ergonomics considers that people come in all sizes and shapes and have a wide variety of needs, capabilities, limitations and handicaps. The previous discussion on different types of robot safety devices and how to utilize them, is largely an ergonomic issue. However, the interest in ergonomics is broader; there are several other possible design improvements, relating both to hardware and software design. These may increase the operator comfort and productivity, and may only indirectly have consequences for the safety of the system.

Before human factors design is attempted, it is common to perform a task analysis of the work. Since there may be many different users and many types of tasks performed, the needs are usually different. For example, in a manufacturing plant, there are usually four classes of users: the programmer who programs the motions of the robot arm, the plant engineer who primarily uses a keyboard and a display for program editing and system configuration, the operator who monitors the system, and the field maintenance engineer who may interact with several systems components. Each of these individuals have different needs which might be satisfied with different (hopefully not conflicting) design improvements.

Very few manufacturers of robotics systems have yet addressed ergonomic design of robots. One exception is a study by Olex and Shulman (1983), which analyzed the design of the Nordson's spray paint robot. Several recommendations were made, and almost all of them were incorporated in the new design of the robot. The following list provides examples of several improvements.

1. Control panel lights for critical functions, such as hydraulic power, should be visible from 20-30 feet.
2. The control panel should be designed so that visual displays and indicators are positioned on a vertical surface slightly below operator eye-height. Function keys and the alphanumeric keyboard should be positioned on a horizontal surface at a comfortable height for keying, (Van Cott and Kinkade, 1972).
3. The labeling of controls for robot axis movement is often confusing. Since the location of the control panel might differ from one installation to another, it is not meaningful to use "left" and "right" or arrows indicating the direction of movements or an x-y-z coordinate system. More appropriate is to use numbers for indicating direction of movement, for example 1-4, 2-5, and 3-6. The same consideration is also important for teach pendants which the operator carries around the workstation.
4. Controls on the keyboard or teach pendant should be positioned in functionally related groups according to subtask. Preferably, it should be possible to operate the controls sequentially (from left to right) within each group.
5. The teach pendant should be designed so that it can be held comfortably with either hand and operated with the other hand.

These are examples of design improvements that would improve both operator comfort and productivity, and maybe indirectly also the safety.

Few manufacturers of robots utilize ergonomics for design improvements. However, it is interesting to note that German robot safety standards, recently submitted to the International Standards Organization (1983) address some ergonomic design issues. Similarly, the ANSI standardization committee on Industrial Automation which was formed in 1983, will address both ergonomics and safety of robots. These activities might take several manufacturers by surprise.

REFERENCES

Asimov, I. 1950. I, Robot. Doubleday, Page & Co., New York, N.Y.

Carlsson, J. 1980. Industrial Robots and Accidents at Work. TRITA-AOG-0004. Stockholm, Sweden.

Corlett, E. N. 1983. Robots, Safety, Fault Analysis and Ergonomics. Unpublished Manuscript. University of Nottingham, U.K.

Engelberger, J. F. 1980. Robotics in Practice - Management and Application of Industrial Robots. ANACOM, New York, N.Y.

International Standards Organization. 1983. German Proposal on Safety Requirements Relating to the Construction, Equipment and Operation of Industrial Robots and Allied Devices. ISO/TC97/SC8/WG2/N32. Geneva.

Kilmer, R. D. 1982. Safety Sensor Systems for Industrial Robots. Proceedings of Robots VI: 479-491.

Leipold, F. P. 1982. Consideration for Robot Users and Manufacturers: Safety First. Proceedings of Robots VI: 149-158.

Macek, A. J. 1981. Human Factors Facilitating the Implementation of Automation. Journal of Manufacturering Systems, 1(2), 195-206.

Olex, M. B. and H. G. Shulman. 1983. Human Factors Effect in Robotic System Design. Proceedings of Robots VII: 9-29 - 9-40.

Sugimoto, N. and K. Kawaguchi. 1983. Fault Free Analysis of Hazards Created by Robots. Proceedings of Robots VII: 9-13 - 9-28.

VanCott, H. P. and R. G. Kinkade. 1972. Human Engineering Guide to Equipment Design. U. S. Government Printing Office, Washington, D. C.

Weintraub, P. 1980. Raising the Robot's IQ. Discover, 2, 66-70.

ACKNOWLEDGEMENT

I much appreciate the assistance of Ms. Debbie Chandler for typing the manuscript.

Human Factors in Organizational Design and Management
H.W. Hendrick and O. Brown, Jr. (Editors)

EDUCATION AND TRAINING IN RELATIONSHIP BETWEEN MAN AND ROBOT

K. Noro

University of Occupational and Environmental Health, Japan
1-1 Iseigaoka, Yahatanishi-ku
Kitakyushu, 807
Japan

Industrial robots are expected to be used in ever increasing numbers. College students must be educated in robots to prepare them for the coming robot age. This report introduces the education and training of medical students in robots.

OBJECTIVE OF EDUCATION IN ROBOT MEDICINE:

Oshima et al. (1980) pointed out that many Japanese ergonomists have educational backgrounds in the medical field as compared with other countries. Before 1980, none of Japan's medical schools had courses on ergonomics (systematic education in ergonomics for six months or one year) in their curriculum.

In other words, graduates of the medical schools who acquired the knowledge of ergonomics after graduation used to work as professional ergonomists in Japan. The year 1972 saw the establishment of the Occupational Health Law that obligates companies employing 50 or more people to appoint full-time industrial doctors. The University of Occupational and Environmental Health, Japan was founded to educate industrial doctors as dictated by the new law. The objective of the university's medical school is, of course, to nurture doctors who can solve and improve human-related problems in industry under cooperation with engineers and managers of the firms.

The remarkable development of industrial robots in Japan is based on a wide range of studies conducted in many fields and places. Industrial doctors are required to approach robotics from a standpoint entirely different from these studies. The functions required of industrial doctors with respect to robots are summarized in Table 1. Based on the table, robot medicine was chosen as a subject of education for the medical students of the university.

Robot medicine is defined as a branch of medicine that studies the mental and physical effects of robots on human beings in the society where human beings coexist with robots and pushes forth the health control of human beings. Another reason for selecting robot medicine as a course in the curriculum is the appearance of robots not only at the workplace but also at the home in the future society. Future doctors will have to work in such an environment. In the society where people coexist with robots, the problem of stress the people suffer cannot be ignored. When a new factory is to be designed or a new robot is to be developed, the doctor should be capable of raising the problems involved with the project.

Doctors used to assume an indifferent attitude to science and technology. In the coming age of science and technology, doctors will have to do more than simply alleviate the condition of patients. That is, they will have to perform the new tasks as listed in Table 1.

Table 1. Roles required of present and future industrial doctors with respect to robots.

Present	Clarification and elimination of mental and physical problems of workers at factories where robots are already introduced
Future	Medical assessment and elimination of adverse effect of man-robot interface on humans
	Advice on design of robots
	Advice on design of automated factories, including robots

The above description may be concerned with the field of preventive medicine. Robot medicine can be considered as part of preventive medicine. Present medical students who will later work as doctors in such a society must now be educated in robotics. The earlier, the better.

CURRICULUM ON ROBOT MEDICINE:

In 1983, education in robot medicine was started for the university's medical students in their first and third years of study. This education is designed to develop doctors who can successfully solve human problems at the factories where robots are in operation and is thus composed of two flows.

<u>One flow: education in the principles of robotics</u>;
One is a robot with a human form as described by Asimov (1950) and others, while the other is a robot that functions like a human. The development of robots of the former type has a very long history. This course teaches the students how humans are treated in this history and how humans differ from robots.

The freshmen study the principles of robotics. The sophomores are educated in human-related subjects like anatomy and physiology. The juniors learn ergonomics in general and industrial robots in particular.

<u>Other flow: industrial robot</u>;
As they aspire to become industrial doctors, the students are taught the classification, operation and construction of the industrial robots as compared with human beings. (See Figures 1.)

The curriculum according to which the medical students study robot in comparison with humans is presented in Table 2. Practical training is provided in parallel with the aforementioned classroom education. In one of practical exercises, a student is assigned as a worker to supply parts to a robot repeating a simple task, and another student measure, analyze and evaluate the physiological change, mental stress and other conditions

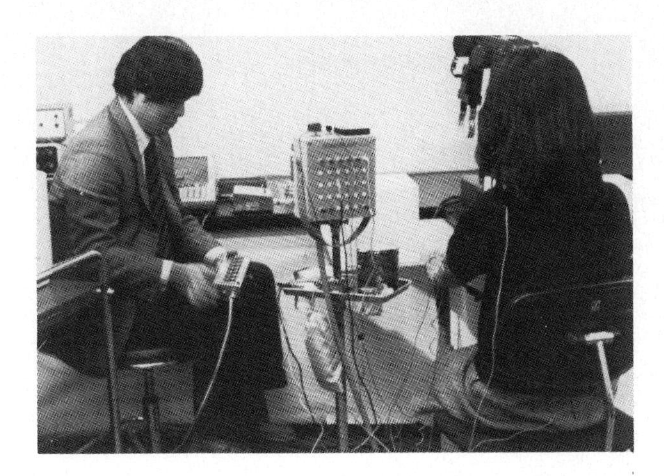

Figure 1. First exercise: Worker (right) feeding parts to repeatable robot and observer (left).

Table 2. Curriculum of "Introduction to Robot Medicine" for freshmen in last half of 1983 academic year.

Time	Topics
1	Present situation of microelectronics and robotics
2	Advantages and disadvantages of robot (1)
3	Advantages and disadvantages of robot (2)
4	Characteristics of man and acceptance of robot in society
5	Future of system in which man and robot coexist
6	Training in operation of robot (1)
7	Training in operation of robot (2)
8	Medical care and technology
9	Role of medicine
10	Discussion

First period of every Monday: 8:50 to 10:20

of the former at work from a medical viewpoint (Figure 2). In another practical exercise, the robot is programmed to perform an intelligent task. The student is then instructed to carry out the same task in front of the robot and to report the experience. In the report, the student has to describe to what extent he thinks the robots will advance and what jobs must be left to humans to perform. Through this experience, the students consider what jobs truly befit humans.

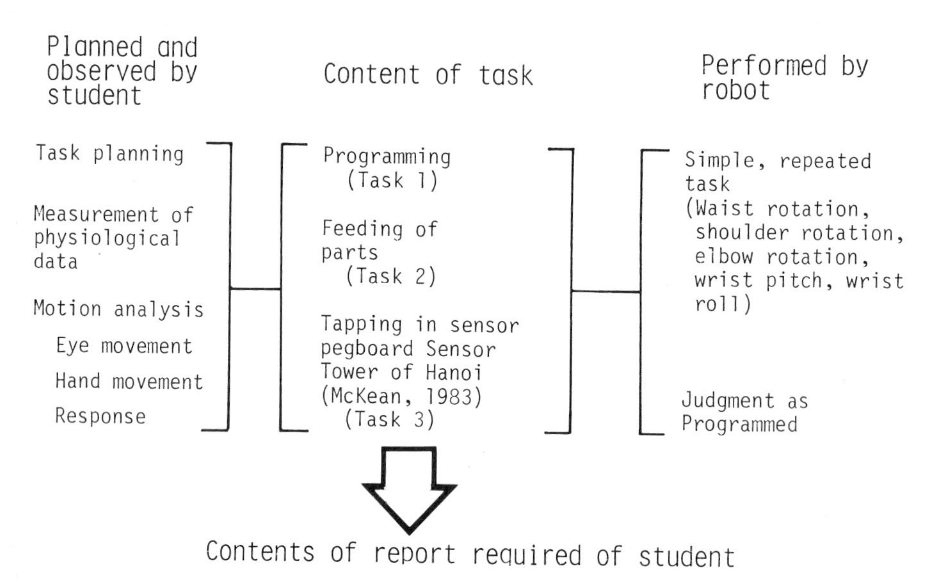

Figure 2. Practical exercises in robot-related tasks.

RESULTS OF EDUCATION:

Typical results of practical exercises:
Table 3 summarizes the report made by a student about the practical exercises shown in Figure 2.

Evaluation:
In general, it is difficult to evaluate the outcome of education. This example is no exception. The points the author noticed in the study are described below.

(1) The effectiveness of this robot medicine education in solving the problems encountered by doctors in industry is suggested by a report of David Lammers (1984) of the Associated Press and an article in the Wall Street Journal (1984).
(2) In Japan, industrial robots are classified into six types from a manual manipulator to an intelligent robot (Japanese Industrial Standards Committee, 1982). Experiments with robots in the respective categories will be more effectively conducted.
(3) Similar education should be given to engineering students, too. Not only robotics but also the problems created by robotization for humans

Table 3. Summary of report made by one student about practical exercises in robot medicine.

Task 1	Programming There is no large change in heart rate. Heart rate is 70 to 80 contractions/ min before work and slightly rises to 90 contractions/min after work. Sharp increase in working time This may be ascribable not only to nature of work, but also to surrounding environment, unsettlement due to mistakes and unfamiliarity with keyboard. Comparison of two subjects Subject K takes longer working time than Subject H but has smaller change in heart rate.
Task 2	Feeding of parts to robot Heart rate does not largely change (80 to 90 contractions/ min). When robot moves fast, heart rate is large on average and eye movements become constant.
Task 3	Tapping in sensor pegboard test Sensor Tower of Hanoi test (Mckean, 1983) Comparison of two subjects There is practically no change in heart rate in series of operations, but heart rate of Subject K seems to be gradually increasing.
Summary	Above three sets of results suggest that task 1 and task 3 contributes more to increase in heart rate than task 3. Heart rate alone was used as index in present exercises. It appears necessary to observe workers according to various other indices and objectively view various changes on basis of observations.

and the impact of robotization on society should be explained to them.

REFERENCES:

(1) Asimov., I., I, Robot (Doubleday, Page & Co., New York, N.Y., 1950).
(2) Japanese Industrial Standards Committee, Types of Industrial Robot (Author, Tokyo, 1982).
(3) Lammers, D., Robot production to grow 25 percent in 1984, AP-TK-01-27 (Associated Press, 1984).
(4) McKean, K., Memory, Discover (November 1983) 18-27.
(5) Oshima, M., Hayashi, Y., and Noro, K., Human Factors Which Have Helped Japanese Industrialization, Human Factors 22(1) (1980) 3-13.
(6) Wall Street Journal, Japan School Explains Robot-Related Stress (January 24, 1984).

Human Factors in Organizational Design and Management
H.W. Hendrick and O. Brown, Jr. (Editors)

ROBOTICS IN THE WORKPLACE: ROBOT FACTORS, HUMAN FACTORS, AND HUMANE FACTORS

Olov Ostberg
Telecommunications Administration
12386 Farsta, Sweden

Jan Enqvist
Working Environment Journal
11220 Stockholm, Sweden

The short history of robots indicates that they are actually no different from the more conventional automation devices. The primary goal of robots has been to boost production and not to humanize the workplace. Emphasis has been on "robot factors" at the expense of "humane factors". With the advent of more advanced robots and greater potential for far-reaching, robotic-based production, human factors professionals can and must pioneer a more balanced approach. However, recent European experience indicates that it is difficult to create robotic production systems in which the workers are content with both the mental and physical working environments.

ROBOTS AS PART OF A SYSTEM

The term robotics downplays the focus on robots per se in favor of a systems approach that sees them as integral parts of production automation. This approach has important implications for any discussion of industrial robots in society and the workplace, since their introduction affects the entire work process and workforce. Essential to any approach to robotics is the definition of an industrial robot. To the Japanese, industrial robots include manually operated manipulators, simple pick-and-place material handling devices, and fixed-sequence automatic equipment (Noro and Okada, 1983). European and U.S. definitions of robots are narrower, however, and indicate that a robot should be a reprogrammable, multifunctional manipulator designed to move materials, parts, tools, or other specialized devices through variable programmed motions for the performance of a variety of tasks (RIA, 1981).

The latter definition may or may not include the traditional notion of a robot--the humanoid mechanical worker, often known as "Robbie the Robot". Whether or not this science fiction creature qualifies depends on how it will be used--either as a simple pick-and-place device, or as a reinstructable multifunctional tool handler equipped with various feedback sensors and artificial intelligence. Paul and Nof (1979) found industrial robots to be better modeled as human operators than as machines. Engelberger (1980), on the other hand, found that in the interest of production efficiency, many industrial workers have been reduced to the level of a robot by being put to grossly subhuman activities. Is the worker a robot or is the robot a worker? And just what is a robot? The International Labour Office describes the influence of robots on workers' jobs in its recent Encyclopedia of Occupational Health and Safety:

> "The development of the robot was logical. The idea underlying mass production was that a job broken down into simple steps could be done more rapidly if each worker was confined to repeating one simple process. Later, as most of the processes were taken over by machinery, the worker was left with the function of a loader, positioner, operator or unloader. Without going much further it was possible to replace him in these functions, so the robot was born./---/ When a robot or an automatic production machine is inserted in a line with human operators, it must be realized that the worker to whom the robot passes the work is paced by the machine. This can be very difficult for the less agile, the very small, or those with arthritis or other joint troubles." (ILO, 1983)

A recent description from a roboticized welding line at a major U.S. corporation drew a similar picture:

> "The small, stocky man in blue overalls picked up the welding torch, turned to the unfinished transmission casing on the workbench, and, with a slight nod, dropped the protective hood over his face. With a searing white light, the torch sputtered liquid metal a foot in all directions. After a few minutes, the worker passed the casing on to his colleague--a robot. The 4-foot mechanical arm clicked into action, tracing the outline of the transmission casing with another torch. After 12 minutes the part was finished./---/ The man had done the intricate work, the robot had followed with the easy./---/ The human welder worked on incentive, meaning he got more money if he kept the robot busy. He now welds more parts than he did before." (Guterl, 1983)

These two quotations illustrate the dilemma surrounding industrial robots. Do we consider them humanoid machines designed to replace workers already reduced to machinoid humans? If that is the case, then galloping technology is a social disease (Frank, 1983) and a crisis in human dignity (Proshansky, 1983). A systems concept is no solution per se, if the systems are approached from a narrow production engineering perspective. As discussed by Helander (1983), when adapting products and production methods to the available robotics technology, the robotics streamlining may eventually make it economical to keep the workers and forego robots. Helander pointed out that systems resulting from such a process cannot really be adapted to humans. Ideally, products and production methods should be analyzed from a human worker's perspective; then systems can be designed to be both economical and worker-friendly.

OPTIMISTIC AND PESSIMISTIC VIEWS

The earlier quotation from Guterl was part of an editorial in a professional engineering journal that concluded that little has been done and few plans have been made to ease the transition in developed industrial nations to a more highly automated society. The welding operation described by Guterl will be truly automated when robots have been developed sufficiently to do even the intricate work without the help of a human. According to the U.S. Society of Manufacturing Engineers, such will be the case well before the year 1990 (Hegland,

1983). The concerted national robotics development project recently launched in Japan gives strong support to this projection (Togai, 1984). A telling example of the speed of the technological development is a robotic meat cutting system introduced in 1984 by Meat Industry Technique in Sweden. This firm probably did not know that 2 years earlier this application had been categorized as something no robot would ever be likely to do (Ayres and Miller, 1982).

According to a study at Carnegie-Mellon University, the number of U.S. workers that could be affected by roboticized production during the next two decades may amount to 7 million, mostly machine operatives and assemblers (OECD, 1982). The investigators at Carnegie-Mellon hoped that as robot use increased, sufficiently enlightened, humane, and far-sighted policies would emerge so that the reduction of skills, displacement, and unemployment of workers would cause no social problems (Ayres and Miller, 1982). If such policies were not developed, these investigators foresaw possible severe deterioration of industrial labor relations, increased union militancy, and a rash of ill-advised legislative remedies. Even worse, they believed that an anti-technological backlash could occur, accompanied by social pathologies from anomie to organized industrial sabotage. Later these authors concluded that the potential for social unrest could not be taken lightly--especially not among unskilled and semiskilled women and minorities in the metal working industries in the Great Lakes States (Ayres and Miller, 1983). As an example of the potential impact on the workforce in this region, the General Motors Corporation had fewer than 2,000 robots in 1982, but they have indicated that they will have more than 20,000 by 1990. This figure may be substantially larger if the unions do not take an active role in managing the implementation of this technology. In late 1981, GM's Chairman, Roger B. Smith noted: "Every time the cost of labor goes up $1 an hour, 1,000 more robots become economical" (Foulkes and Hirsch, 1984).

Noble (1983), a specialist of the history of automation, concluded that there is no hope for any enlightened, humane and far-sighted policies to emerge in the area of robotics, and he actually recommended that workers smash robots and that intellectuals smash the mental machinery of domination to bring about the necessary beneficial changes to society. Dewey (1983) pointed out that industry is to blame for such reactions since it is discouraging for workers to hear industry spokesmen repeatedly declaring that it depends on the workers if industry "shall automate, emigrate or evaporate", that the manufacturing world "must become a friendly place for a robot to work", and that "the salvation of the world is in the hands of technologists".

Salvendy (1983) asserted that the current robotics technology is developed to a level that should enable U.S. industries to use more than 1 million robots. He believed that the present use of only 1 percent of that figure was largely attributable to insufficient considerations of human factors in the analysis, design, implementation, and operation of robotics systems. Another believer in the benefits of robotics is James S. Albus, Chief of the U.S. National Bureau of Standards' Automated Manufacturing Research facility. He has projected that the robotics revolution will free the humans from the regimentation and mechanization of manual labor and human decision-making in factories and offices, that millions of exciting new jobs will be created, and that the robot industry will eventually employ at least as many workers as the computer and automobile industries do today (Albus, 1982; 1983).

The labor market actually seems to be developing in the opposite direction, however. In the United States, more workers are currently employed in fast-food services (hamburger restaurants) than in the auto and steel industries together. Similar figures are reported from Japan, where employment in manufacturing grew by only 70,000 workers from 1970 to 1981, but employees in retail, wholesale, and service industries increased by 5.2 million (Foulkes and Hirsch, 1984). Regarding the millions of exciting new jobs promised by Albus, the U.S. labor statistics strongly indicated a gross national reduction in skills required by jobs as a result of new technology (Levin and Rumberger, 1983). Furthermore, of the 20 occupations expected to generate the most jobs in the 1980's, not one is related to new technology, and only two require a college degree.

THE WORKER-ROBOT INTERFACE

The history of automation, current political-economic debate, and forecasts of future automated production are all important elements governing the design, implementation, and operation of robot-based production. Yet these factors are elusive and even irrelevant to the worker's pragmatic, here-and-now point of view. In the United States, spokespersons from unions representing auto workers (Unterweger, 1983), machine operatives (Nulty, 1982), electrical workers (Pillard, 1983), and the federation of unions (AFL-CIO, 1983), fear that unless management is forced, it will ignore safety and health, the quality of working life, and the job security of the workers affected by the introduction of robots. According to Shaiken (1984), this fear is a realistic perception. Using four case studies, Shaiken found that robot-based U.S. manufacturing was characterized by such negative aspects as paced work, monotony, and stress. He concluded that serious discrepancies exist between potential workplace benefits and the current use of industrial robots.

Ayres and Miller (1982) used a questionnaire to poll U.S. industries on the motivating factors behind the introduction of robots. Reduced labor costs, increased output rate, and improved product quality were the most important factors cited, but the possibility of relieving workers of tedious and/or dangerous jobs were also frequently mentioned. Similarly, an improved working environment is alleged by management to be the primary motivation for introducing robots in Sweden (Sten, 1977), Germany (Kalmbach et al., 1982), and Japan (JISHA, 1983).

Though industry may claim that their investment in robots is motivated by concern for the worker, others hold a different view. Engelberger (1980), representing a robot manufacturer, asserted that the primary factor in justifying a robot is labor displacement. He had found that industrialists are mildly interested in shielding workers from hazardous working conditions, but that their key motivator is to reduce costs by supplanting a human with a robot. Mickler et al. (1981) made an extensive investigation of the use of robots within the Volkswagen Corporation. They found that (1) every single robot had been introduced with a pure profit motive, (2) the use of robots had resulted in decreased physical work, job skills, and decision latitude, and in increased machine-pacing and isolation, and (3) contrary to the actual results, the robots had been promoted as a technology designed to humanize working life. The same group of researchers later cited three other German studies that strongly challenged the conclusion that industrial robot application results in improved working conditions, reduced physical labor and liberation of workers from monotonous and environmentally stressful jobs (Kalmbach et al., 1982).

It is particularly disturbing that what industry considers to be a successful robot installation often involves a decrease in worker skill level (Fleck, 1983). Even when management's intentions are good, shorter cycle time and more monotonous work may result simply because of current engineering difficulties in meeting the requirements of both humans and robots (Langmoen, 1982). In representing a U.S. robot manufacturer, Rosato (1983) went as far as to make the following recommendations for modern robots: (1) Robot operators should only start, monitor, and stop the robot, (2) maintenance personnel should complete the rest of the operation and maintenance, and (3) all programming should be done by technicians or engineers. Such a division of labor naturally meets resistance from the workers, especially since only a marginal portion of the U.S. workforce affected by the robot revolution is covered by policies with provisions for worker training (Workers and automation, 1983).

Asendorf and Schultz-Wild (1983) found that workers with the rather average qualifications and experience, typical of traditional high-volume production can be successfully prepared for more demanding jobs. They argued that current workers should be trained to assume new duties, and that it is neither necessary nor preferable to skim the cream from the workforce when staffing robot-based production. However, Foulkes and Hirsch (1984) found that although retraining provisions reduce employee anxiety and fear, they restrict management flexibility and are unlikely to gain acceptance in the near future. The same authors also reported that though management understands the benefits of having well-trained robot operators who are knowledgeable in routine programming and maintenance, they have nevertheless chosen to use the division of labor as a means of preventing the unions from getting a strong bargaining position. On the other hand, a growing body of literature shows that unions (Hirsch and Link, 1984) and indeed workplace co-determination (Gardell, 1982) have positive effects on productivity.

EMERGENCE OF FLEXIBLE MANUFACTURING SYSTEMS

Robots, numerically controlled machine tools, and transportation systems may be coordinated through a computer network controlled by computerized management and design systems. These systems make it possible to organize batch production based on customer orders without sacrificing the production flow benefits of large-volume production. In the past, robots were used primarily in volume production. In the near future, they will increasingly be used for batch production of 20 to 20,000 units per year. Batch production constitutes almost 40 percent of the manufacturing volume and value in the United States.

To achieve flexibility, many different types of human skills will have to be available or production will not run smoothly. Sheridan (1976) noted that we automate and roboticize what we understand and can predict, and we hope that the human supervisor will take care of what we do not understand and cannot predict. Production workers are often superior to supervisors in this respect. Aron (1982) reported on a study at a Nissan automobile plant. During a 19-month period, 197 production problems occurred. Of these, 100 percent were correctly diagnosed by the workers, 79 percent by the foremen, and only 4 percent by upper management. The importance of having multi-skilled and motivated workers seems to have been considered in the robot-based Nissan plant in Kentucky, U.S.A., where the employees are no longer called workers but production technologists (Van Blois, 1983).

Seppala and Salvendy (1982) investigated centralized versus decentralized supervisory control of flexible manufacturing systems (FMS). They found overwhelming support of decentralized control, since such control significantly decreased both the response time and error rates of supervisors. Most FMS operate on the basis of a hierarchical, centralized, and highly specialized job structure. This traditional approach may pose serious long-term problems. Kohler and Schultz-Wild (1983) reported that apart from creating skill deficiencies, hierarchical work organizations in FMS tend to cause worker dissatisfaction, which can cause production problems. Preventive maintenance, monitoring machines, the speed and accuracy of repairs, inspection, etc., largely depend on the motivation of the workers. In an article on computerized manufacturing (Gerwin, 1982), a pressing need was identified for redesigning jobs and work organizations. One suggested possibility is a joint work group that eliminates separate jobs for operators and loaders. Because an integrated system is composed of self-contained tasks, the group can as a whole be responsible for loading, monitoring, routine repairs, and tool setting. Then motivation is likely to increase, and dissatisfaction should decline.

A similar challenge came from Rosenbrock (1981, 1982), who expressed concern about robots being used in ways that further fragment and trivialize existing jobs. Rosenbrock is currently involved in designing a model automated cell for FMS production. Here technology is seen as a tool that the worker can use, rather than the reverse. The philosophy behind this project is described in the book New Technology (1981). In working on the design of the automated steelworks of the future, Hedberg (1984) found it necessary to expand this philosophy to accommodate the workers' career paths (sequences of jobs). Such systems are at variance with the philosophy of the predominant production engineering school, which is described as follows:

> "Engineers exist in that part of the system that is insulated from the consequences of their designs. In the economist's language, they escape the externalities of their decisions--the external costs (borne by those outside their operation). Such costs include excessive fatigue, boredom, excessive work load, isolation, frustration..../---/ For want of a robot, an operator is used. This perspective, ingrained in students by engineering schools and common in top management, pervades the culture of the design engineer. It leads to a social structure of incentives, punishments, physical layouts, output measures, etc., that reinforces the perspective of designing out the 'man' in the loop." (Perrow, 1983)

In the near future there may not be any need for workers to perform the downgraded jobs referred to by Perrow. This view has been taken by the Swedish Employers' Confederation, which argues that the intermediate use of new technology may cause problems, but that full use would allow more positive solutions. In a summary of the first 7 years of its comprehensive research project (Aguren and Edgren, 1980), a full use of new automation technology was said to be able to fulfill the following benefits and criteria of the working environment:

(1) The individual is less tied to the job. A key criterion for good job design is that the individual should be able to work at his own pace.
(2) The worker enjoys longer task cycles and complete jobs. Opportunities exist to manufacture a complete product or a logically limited part of a product. Thus the worker takes responsibility for quality in the work and can more efficiently combine responsibility for quality and quantity.
(3) Both group work and individual work are possible. Cohesiveness and feelings of commonality in a parallel group can be strengthened if the group members can distribute the work themselves and bear collective responsibility for supplementary tasks such as quality control, adjustments, and material supply.
(4) Better ergonomics. New technology can contribute to more comfortable working positions and better work environments as a whole. Manual handling of heavy components can be eliminated.

The following section describes the race for new factories between the Swedish auto makers Volvo and Saab. Note, however, that Salvendy (1982) recommends extreme caution in the cross-cultural and cross-national transfer of methodologies requiring the interaction of humans with other systems.

ANALYSIS OF DEVELOPMENTS IN THE SWEDISH AUTO INDUSTRY

With a ratio of 500 industrial workers per industrial robot, Sweden has relatively speaking, three times as many robots as Japan and ten times that of the United States. Sweden also has proportionately more numerically controlled machines and automated machining cells than any other nation. Sweden is thoroughly unionized, with factory workers, salaried employees, and professionals all belonging to unions. This situation has engendered a long tradition of tripartite policy-setting with regard to workplace-related issues between unions, employers, and government. Thus an international audience should be interested in the gradual development of the production process in Sweden, which has led to the introduction of robots.

Examples leading to the current level of robot use will be presented from the Swedish auto industry (Volvo and Saab). All workers in the auto industry belong to the Union of Metal Workers, which has long been positive toward the introduction of new technology. In the early phase of robotization, a union spokesperson indicated that working environment arguments alone would justify replacing every tenth industry job with a robot (Jonsson, 1975). This positive attitude toward modernization has prevailed, and in 1982 the president of the Union of Metal Workers rebuked groups asking for contract language containing the right to veto new technology. Rather he suggested they seek a veto right on old technology. Behind such confidence is the knowledge that individual workers largely have employment security, and collectively they have considerable influence in the design, implementation, and operation of new production systems. This influence is achieved by means of a network of laws and agreements in which the introduction of new technology is approached as part of a larger set of worker and union rights to constant improvement of the workplace.

Volvo's Biotechnology (1961)

Ergonomics on a large scale was introduced in Sweden through the biotechnology concept implemented at the Volvo Penta plant for engine assembly in 1961. The harsh moving belt production, based on piecework and MTM-designed jobs, was modified according to man's biological functions. The driving force for biotechnology was the plant's safety and health unit, which came to serve as a model for industrial ergonomics in Sweden. In a very ambitious training program, production engineers were taught biotechnology principles and were encouraged to bring in outside expertise in ergonomics/biotechnology during major development projects. The basic goals were to aim for an optimal area between the risk zone and the luxury zone, and create physically safe and healthy workplaces. These old Volvo concepts can be recognized on the cover of a recent book on machine pacing and occupational stress. Thus to cushion unrest and possible ill-health, researchers were said to aim at enhancing the human values in the paced-work environment without hindering productivity (Salvendy and Smith, 1981).

Saab's Sociotechnology (1972)

Volvo's biotechnology was very successful, but during the 1960's, was realized that physically safe and healthy workplaces were not enough. These alone could not prevent worker monotony and alienation, which in turn resulted in high turnover, absenteeism, and low productivity. The labor movement argued that it was not enough to have shorter working hours, longer annual leave, more sick compensation, and higher pay; the empty working time could not be compensated by a full leisure time. In fact, it became generally accepted that empty work resulted in empty leisure (Johansson, 1971).

The Saab-Scania assembly plant in Sodertalje pioneered a sociotechnical approach as a means of enriching the working conditions. "Socio" refers to a work organization that makes room for job enlargement, job rotation, social contacts, and (to some degree) self-determination within smaller production units. "Technical" refers to bringing in new technology to boost productivity and developing the new work organization to uncouple the workers from the line speed.

At Saab-Scania, socitechnology meant that parts of the production were highly mechanized in traditional line fashion, and the manual assembly line for building engines was broken up into parallel flow lines in the shape of loops on the main line (Aguren and Edgren, 1980; Goldmann, 1979). This arrangement was achieved by means of ergonomically designed assembly trolleys that could be individually pulled into the loops and assigned to one of the workers within the loop unit. With these trolleys, an engine block could easily be positioned at the correct working height for the worker, who could then carry out a complete assembly cycle. Most of the workers were women, and thanks to the loop system (which also served as an in and out buffer), they could help each other during difficult or heavy work tasks.

This system was highly productive and met some of the desire for quality of working life. But shortcomings existed with regard to both the "socio" and the "technical" subsystems. Though no salary penalties were imposed if the production quota could not be met, the workers were asked to discuss and find solutions to the problems. This approach resulted in group pressure on workers who fell behind in production; it also caused the workers in a loop to voluntarily speedup the production to obtain a period of nonwork of up to 2 hours at the end of the shift. In turn, even greater group pressure was exerted for greater production demands than required by management. Another shortcoming of this approach related to the fact that the workers represented a great many nationalities and language barriers prohibited social interaction. A serious technical drawback was that the loops were still lines within which it was not possible to change position between individual assembly trolleys.

Volvo's Teamwork (1974)

The Volvo Skovde plant for engine assembly, which opened in 1974, amplified both the biotechnology and sociotechnology concepts. The production had vertical job enlargement, self-propelled assembly carts, ergonomically designed workstations and handtools, and a flow layout and plant architecture that supported the teamwork concept. The Volvo Kalmar plant for final car assembly is internationally better known but based on the same principles. Union and management representatives from all over the world have made study visits to this plan during the last decade. Gyllenhammar (1977), the Volvo chief executive officer, was rightly proud of the Kalmar plant and its embodiment of the 'New Factories' ideas of the Swedish Employers' Confederation (see Aguren and Edgren, 1980). In 1984, Gyllenhammar declared that this plant was the most productive within the Volvo family. Though the workers are positive about the basic ideas, they maintain that the plant has received more praise than it deserves. Their main criticism is that the production is still based on a moving-belt technology, even though the robot carts make the belt invisible. They also feel that there is not enough buffer capacity in the loops, and that it takes young, strong workers to manage the production speed.

Saab's Self-Paced Work (1983)

While the world is still admiring the teamwork in Swedish auto assembly plants, the Saab-Scania plant for engine assembly in Sodertalje has gone further and has actually abandoned the moving belt and teamwork concepts. The work design principles of biotechnology and ergonomics are further amplified here, as is the drive toward truly parallel production flow lines. A computer-aided assembly carrier has been developed that surpasses Volvo's robot cart.

In this system, a number of heavy and/or short-cycle operations have been assigned to industrial robots. Their tasks include mounting valve springs, sealing valves, mounting flywheels, and threading and torquing bolts. Great care is taken to separate the robot and the manual assembly operations. The manual assembly is carried out in individual, ergonomically well-designed workstations. As the workers complete their tasks, they send away completed products and order new assembly carriers according to their individual workpaces. The production layout is shown in Figure 1.

Purely ergonomic considerations aside, the new assembly system was launched to boost output capacity. In particular, it was designed to handle custom-order production. Thus about 30 different engine versions are being assembled at any given time. For the workers, this self-paced work is an improvement over the previous group-paced work, with its concomitant, self-inflicted high workload. Even if the overall production is not lower in the new system, self-paced work is obviously less stressful than that set by the group of workers.

The workload under the group-paced assembly production was witnessed by a group of U.S. auto workers testing the previous Saab-Scania system. As reported by Goldmann (1979), the U.S. workers found the plant superior to U.S. plants with regard to ergonomics and environmental factors, but they were troubled because the women they worked with "worked harder than anyone in Detroit." This hard work showed up as sick-leave increase. In addition, the company physiotherapists were busy treating neck, shoulder, and arm ailments among the assembly workers. Though it is still too early to draw firm conclusions from the new self-paced work, the prevailing impression has been described as "Good-bye to group assembly. No more stress and load injuries--but a less interesting job." (Enqvist, 1984).

CONCLUDING REMARKS

The history of automation and industrial robots indicates that there is a growing production sphere in which increasingly skilled robots can successfully compete with humans. Workers' jobs have been gradually reduced in skill level because of management determination to maintain production control and to increase productivity by means of traditional production engineering techniques. In Sweden, where the Employers' Confederation officially has taken issue with these techniques, efforts have been made to create worker-friendly environments without reducing productivity. Typically these attempts have taken the form of a production design process in which human values play an important role and unions have input to the development projects. In the resulting systems, robots are integral parts of a larger production concept and do not operate to eliminate skills or jobs. But even in Sweden it has not been possible to create large-volume production in which the workers are content with both the mental and the physical working environments.

Acknowledgment: This paper was prepared while the senior author was a Foreign Visiting Research Scientist with the U.S. National Institute for Occupational Safety and Health, Cincinnati, OH. Comments on the manuscript by B. C. Amick, M. G. Helander, L. E. Schleifer, and M. J. Smith, are gratefully acknowledged.

FIGURE 1. Saab-Scania's engine assembly plant i Sodertalje. At the automated section, robots mount valve springs, collets, flywheels, and (top) thread and torque bolts. At the manual section (bottom), the final assembly is carried out in parallel, individual workstations, to which computer-aided assembly carriers are brought in on the workers' demand. (Photos by Saab-Scania).

REFERENCES

AFL-CIO Committee on The Evolution of Work, The Future of Work. Washington, D.C.: AFL-CIO, 1983.
Aguren, S. & Edgren, J., New Factories: Job Design through Factory Planning, Stockholm: Swedish Employers' Confederation, 1980.
Albus, J.S., Industrial robot technology and productivity improvement. In Exploratory Workshop on the Social Impacts of Robotics. Washington, D.C.: Office of Technology Assessment, U.S. Congress, 1982. (Pp. 62-89).
Albus, J.S., Socio-economic implications or robotics. Proceedings of the 1983 World Congress on the Human Aspects of Automation. (MM83-479).
Aron, P.H., Testimony before the Committee on Science and Technology, U.S. House of Representatives, June 2, 1982.
Asendorf, I. & Schultz-Wild, R., Work organization and training in a flexible manufacturing system - An alternative approach. Proceedings of the IFAC Workshop on Design of Work in Automated Manufacturing Systems, Karlsruhe, F.R.G., November 7-9, 1983.
Ayres, R.U. & Miller, S.M., Robotics: Applications and Social Implications. Cambridge, MA: Ballinger, 1982.
Ayres, R.U. & Miller, S.M., Robotic realities: Near-term prospects and problems. Annals of the American Academy of Political and Social Science, 28-55, 470, 1983.
Dewey, D., Robots reach out. United, 92-99, 28(7), 1983. (United Airline).
Engelberger, J.F., Robotics in Practice. New York, NY: American Management Association, 1980.
Enqvist, J., Good-bye to group assembly. Working Environment 1984 in Sweden, 14-17, 1984.
Fleck, J., The adoption of robots. Proceedings of the 13th International Symposium on Industrial Robots and Robots 7, 1:41-1:51, 1983.
Foulkes, F.K. & Hirsch, J.L., People make robots work. Harvard Business Review, 94-102, Jan/Feb., 1984.
Frank, J.D., Galloping technology, a new social disease. Journal of Social Issues, 193-206, 39, 1983.
Gardell, B., Worker participation and autonomy: A multi-level approach to democracy at the work place. Stockholm: Stockholm University Research Unit for Social Psychology at Work, 1982. (Report No. 31).
Gerwin, B., Do's and don'ts of computerized manufacturing. Harvard Business Reivew, 107-116, Mar./Apr., 1982.
Goldmann, R., Six automobile workers in Sweden. In R. Schrank (Ed.), American Workers Abroad. Cambridge, MA: MIT Press, 1979. (Pp. 15-55).
Guterl, F., An unanswered question: Automation's effect on society. IEEE Spectrum, 89-92, 20(5), 1983.
Gyllenhammar, P.G., People at Work. Reading, MA: Addison Wesley, 1977.
Hedberg, B., Career dynamics in steelworks of the future. Journal of Occupational Behaviour, 53-69, 5, 1984.
Hegland, D.E., No end in sight. Production Engineering, 47-51, 30(4), 1983.
Helander, M.G., Human factors aspects of computer-integrated manufacturing. Documentation of the Workshop on Human Factors in Automated Manufacturing held at the 27th Annual Meeting of the Human Factors Society, Norfolk, VA, October 10-14, 1983.

Hirsch, B.T. & Link, A.N., Unions, productivity, and productivity growth. Journal of Labor Research, 29-37, 5(1), 1984.
ILO's Encyclopaedia of Occupational Health and Safety. (On "Robots and automatic production machinery"). Geneva: International Labour Office, 1983. (Third and revised edition).
JISHA: Prevention of Industrial Accidents due to Industrial Robots. Tokyo: Japan Industrial Safety and Health Association, 1983.
Johansson, S., Investigation of the standard of living. (In Swedish). Stockholm: Allmanna Forlaget, 1971.
Jonsson, B., 1975. Quoted in Ostberg, O., Production with limited manpower. How? What are the social effects? (In Swedish). University of Lulea Technical Reports, No. 50T, 1978.
Kalmbach, P., Kasiske, R., Manske, F., Mickler, O., Pelull, W. & Wobbe-Ohlenburg, W., Robots effect on production, work and employment. Industrial Robot, 42-45, March, 1982.
Kohler, C. & Schultz-Wild, R., Flexible manufacturing systems - manpower problems and policies. Proceedings of the 1983 World Congress on the Human Aspects of Office Automation. (MM83-470).
Langmoen, R., Automatic assembly of electric heaters. Proceedings of the 12th International Symposium on Industrial Robots, Paris, June, 1982.
Levin, H.M. & Rumberger, R.W., The educational implications of high technology. Stanford University School of Education, CA, 1983. (IFG 83-A4).
Mickler, O., Pelull, W., Wobbe-Ohlenburg, W., Kalmbach, P., Kasiske, R. & Manske, F., Industrial Robots. Conditions and Social Consequences of the Introduction of New Technology in Auto Industry. (In German). Frankfurt/Main: Campus-Verlag, 1981.
New Technology: Society, Employment and Skill. London: Council for Science and Society, 1981.
Noble, D.F., Present tense technology, Part III. Democracy, 71-93, 3(4), 1983.
Noro, K. & Okada, Y., Robotization and human factors. Ergonomics, 985-1000, 26, 1983.
Nulty, L.E., Case studies of IAM local experiences with the introduction of new technology. In D. Kennedy, C. Craypo & M. Lehman (Eds.), Labor and Technology: Union Response to Changing Environments. College Park, PA: Penn State University Department of Labor Studies, 1982. (Pp. 115-139).
OECD synopsis report on The Impact of Automated Manufacturing Equipment on the Manufacturing Industries of Member Countries. Paris: Organization for Economic Co-operation and Development, 1982. (Annex to report DSTI/IND/82.2).
Paul, R.P. & Nof, S.Y., Work measurement - a comparison between robot and human task performance. International Journal of Production Research, 227-303, 17, 1979.
Perrow, C., The organizational context of human factors engineering. Administrative Science Quarterly, 521-541, 28, 1983.
Pillard, C.H., Remarks to the IUD Conference on Technology, September 16, 1983. Washington, D.C.: AFL-CIO's Industrial Union Department, 1983.
Proshansky, H.M., The environmental crisis in human dignity. Journal of Social Issues, 207-224, 39, 1983.
RIA Worldwide Survey and Directory on Industrial Robots. Dearborn, MI: Robot Institute of America, 1981.

Rosato, P.J., Robotic implementation - do it right. Proceedings of the 13th International Symposium on Industrial Robots and Robots 7, 4:32-4:49, 1983.
Rosenbrock, H.H., Engineers and the work that people do. IEEE Control Systems Magazine, 4-8, 1, 1981.
Rosenbrock, H.H., Engineers, robots and people. Chemistry and Industry, 756-759, 19, 1982.
Salvendy, G., Human-computer communications in computer integrated manufacturing systems. Proceedings of International Conference on Productivity and Quality Improvement, Tokyo, October 20-22, 1982. (C-3-1).
Salvendy, G., Review and appraisal of human aspects in planning robotic systems. Behaviour and Information Technology, 263-287, 2, 1983.
Salvendy, G. & Smith, M.J., Machine Pacing and Occupational Stress. London: Taylor & Francis, 1981.
Seppala, P. & Salvendy, G., Decision making styles and productivity of first-line managers in computerized flexible manufacturing systems. West Lafayette, IN: Purdue University School of Industrial Engineering, 1982. (Unpublished Human Factors Report).
Shaiken, H., Automation in the workplace. Case studies in the manufacturing process. In Computerized Manufacturing Automation. Employment, Education, and the Workplace II. Washington, DC: Office of Technology Assessment, U.S. Congress, 1984.
Sheridan, T.B., Modeling supervisory control of robots. In A. Morecki & K. Kedzior (Eds.), Theory and Practices of Robots and Manipulators. Proceedings of an International Symposium, Warsaw, 1976. (Amsterdam: Elsevier, 1977).
Sten, H., Inventory of industrial robots in Sweden in June 1977. (Stencile in Swedish). Stockholm: Sveriges Mekanforbund, 1977.
Togai, A., Japan's next generation of robots. IEEE Computer, 19-25, 17(3), 1984.
Unterweger, P., Work, automation, and the economy. Proceedings of the Annual Meeting of the American Association for the Advancement of Science, Detroit, MI, May 1983.
Van Blois, J.P., Strategic robot justification: A fresh approach. Robotics Today, 44, 45, 49 p., April, 1983.
Workers and automation. Technology Review, 83, 86(2), 1983.

Human Factors in Organizational Design and Management
H.W. Hendrick and O. Brown, Jr. (Editors)

OPERATION RATE OF ROBOTIZED SYSTEMS : THE CONTRIBUTION OF ERGONOMIC WORK ANALYSIS

Alain Wisner, François Daniellou

Laboratoire d'Ergonomie
Conservatoire National des Arts et Métiers
41 rue Gay-Lussac, 75005 PARIS
FRANCE

The ergonomic work analysis (E.W.A.) of the real behaviour of operators working on various automated systems in the automobile industry shows that a significant proportion of breakdowns can be attributed to the fact that the organization of the automated process has only taken into account the simplified activity that the methods specialists have imagined to be the real activity. This means that the many adjustement and corrective activities that enable non-robotized systems to function properly must still be practised by the operators of robotized systems. The results of the E.W.A. before the introduction of automatic systems gives the possibility of knowing with accuracy what makes up the real work being done, thus ensuring a more complete and efficient automation. Analyzing operator activities on automatic systems when these systems are functioning in a degraded operating mode can enable recommendations to be drafted that will reduce stoppage time.

INTRODUCTION

There is now a high rate of stoppages and time loss due to operating incidents on a number of automated sequential units that have been set up within the last few years (Sugimoto et al., 1983). This is deliterious both to the company, since it lowers the system's time of full operation, and to the operators for whom such conditions will involve more periods of increased work activity with its attendant time pressures.

What is the contribution that ergonomics can make to understanding the causes of these dysfunctions and identifying possible solutions?

The following points have been taken from research on automated systems in the French automobile industry (Daniellou, 1982). This has been obtained by means of ergonomic analysis on the work really performed by the operators and maintenance workers of these systems.

Ergonomic Work Analysis (E.W.A.) takes into account all the activities of an operator, since there are always many activities an operator has to do on an automated system. It therefore covers the following behavioural activities :

-actions on the system itself,
-uptake of visual, auditory and olfactive information,
-verbal communications and signals.

The results of this behavioural activity analysis were then put to the operators and to their supervisors, who would then often give most interesting explanations about the technical difficulties that induced the behaviour described.

Two basic characteristics of human interventions will be stressed : the taking into account of variability and early diagnosis. We will then deal with the way these characteristics can be considered when designing an automated system.

TAKING VARIABILITY INTO ACCOUNT

An anaysis of work performed by operators of robotized welding lines showed that frequent physical interventions had to be made on the unit for the positioning of parts, for restarting an interrupted welding cycle, or for checking out oil leaks etc.

Generally speaking, interventions affecting the control of automated systems can be grouped into two categories :

-Those which are intended to overcome technical deficiencies in the system itself; they are usually easily indentifiable and proper attention will be given to such breakdowns by the relevant technical departments:

-Those which are intended to resolve situations that automation can't deal with by itself: in such cases, work analysis before and after automation can contribute tools for understanding the difficulties being experienced.

Work analysis has made it possible to make the following statements in non-automated situations, and these points are now well-known: (Teiger, 1978).

-In real mass-production conditions, the usual situation is not one of stability but rather one of variablity : metallurgical sizes and characteristics often drift, tools wear down, machinery goes off its ajustment;

-Because of this variability, a production can only be considered ultimately suitable both in terms of quality and quantity because of the amount of regulating operations undertaken by the operators in charge. Behind the purely "manual" and repetitive aspect of work on a production line, there is ceaseless mental activity going on in the minds of the operators in order for them to be able to produce whatever they are making within the alotted time in spite of the incidents that arise.

An analysis of several automated systems has shown that in many cases the automation has not covered the real work, but a mere theoretical view of what the work should be, with the accent on stability and repetitiveness. The variable components which are still in the system are simply not dealt with by the automat, and this provokes dysfunctionings which then require human intervention to overcome them.

EARLY DIAGNOSIS

The fact that the role of the operators is so essential does not necessarily mean that one would recommend automatic detection and processing of all foreseeable mechanical incidents. On the contrary, work analysis enables one to stress the specific aspects of the way an operator discovers that something is wrong.

It can be seen that the surveillance being undertaken by the operators is designed to detect an "ongoing" incident, that is to find it out before it has any serious consequences on the parts being made or on the production unit itself.

Very often, automatic sensors will detect an incident at a stage when more harm has been done, thus involving more time to put things right.

This early detection of incidents by the operators themselves is based on the use of varying types of indications : the "official" indications that are provided by the screens, dials, and lights on the machine, along with the informal indications that come from experience, i.e. the way a part looks, the way a tool is vibrating etc.

Prevention of any stoppages in the unit presupposes the

operators'easy access to the following sources of information :

- design and placement of formal indications,
- the possibility of close contact with production operations, thus facilitating the intake of informal indications.

WORK ANALYSIS AND DESIGN

The aforementioned aspects show that any analysis of breakdowns in automated systems must do more than just take into account the technical dependability of the unit's components.

The contribution of such analysis can serve to demonstrate :

- How important it is to have a knowledge of the real work before automation takes place, and particularly the components of its variability;

- The role of early human intervention on automated units. The difficulties that production operators have in obtaining formal and informal information at all times throughout the ongoing cycle. This contributes to delay in diagnosis, and in many cases, it increases the seriousness of the incident, and hence the stoppage time;

- The importance of the conditions in which the maintenance operators intervene. This aspect has not been developed here (Daniellou op. cit.) but it involves the need not only to analyse the number of stoppages but also the time taken up by each one, because this is often of significant importance to the maintenance operators since it indicates the difficulties they have when establishing a diagnosis or when intervening.

This approach enables us to define the following causes of stoppages :

- Those which involve starting up a new unit, which will always entail a certain amount of debugging;

- The more deeply-lain causes that result from the intrinsic variability in the operations and machinery being used. In many cases, if an attempt is made to radically eliminate these variability-linked incidents by increasing the automation, the cost becomes prohibitive.

At that point, it becomes absolutely necessary to take the specifically human needs and interventions into account.

A promising approach for improving the stoppage rate of

production layout and the operators' working conditions appears to be to systematically undertake the automated system design using degraded operating modes analysis and considering the human activities needed to overcome them. The results of work analysis have shown that such an approach entails two conditions :

. A systemic approach : dysfuctions in a unit will be the outcome of human and technical factors that will include operations that are upstream from the unit, as well as associated services.

. The identification and recording of the operator's competence in the design process.

As has been mentioned above, all production, whether automated or not, will depend on ajustment activites being carried on by the operators. Adapting the means of production to such activities would entail all the operators being associated with all the different design phases.

Defining the technical and social conditions under which the designers' and operators' know-how can be confronted appears to be one of the greatest challenges in formulating the organizational design of automated systems.

REFERENCES

Daniellou F. (1982)
The Impact of New Technologies on Shift Work in the Automobile Industry. European Foundation for the Improvement of Living and Working Conditions, Dublin, Ed.

Sugimoto N., Kawaguchi K. (1983)
Fault Tree Analysis of Hazards Created by Robots. XIII th International Symposium on Industrial Robots.
Conference Proceedîngs 9 - 13-28.

Teiger C. (1978)
Regulation of Activity : An Analytical Tool for Studying Work Load in Perceptual Motor Tasks. Ergonomics, London 21, 3, 203-253.

Human Factors in Organizational Design and Management
H.W. Hendrick and O. Brown, Jr. (Editors)

EVALUATING THE EFFECTS OF AUTOMATION ON INTERACTIONS BETWEEN ORGANIZATIONS

Anna M. Wichansky*

AT&T - Bell Laboratories
Whippany, New Jersey

Interorganizational interactions and productivity of organizations indirectly affected by new automation were assessed in 22 operations centers working with a Bell Operating Company Maintenance Center, before and after trial installation of a mechanized telephone line testing system. Specific recommendations were implemented for changes in systems features, methods of use, and streamlining the interorganizational process required to perform telephone line maintenance.

1. INTRODUCTION

Automated systems are typically introduced into business operations to reduce the need for human performance of repetitive, low-level functions, thereby saving time, labor, paper, and money. Human factors evaluations of such systems are usually concerned with the effects of automation on the performance, health, and job satisfaction of individual system users. An issue which is rarely addressed is the effect of automation on interactions between organizations within a larger corporate unit. This is clearly an important factor in the performance of the company as a whole, given some newly automated part. It also reflects how well a particular computer system fits the application for which it was designed or selected within the larger context of total organizational performance.

Most organizations are made up of functional units, which are responsible for performing particular tasks in a highly coordinated way with other units in order to accomplish a corporate mission. A computer is often introduced to automate or mechanize some or all of the tasks of one particular unit. Computerizing one unit's functions, however, may have a "ripple effect" throughout the entire organization. The content and format of information exchanged between units to get the job done may have to change. Established workflows may become obsolete. Interactions between units may also be affected by changes in personnel skills and experience accompanying automation.

These changes in interorganizational dynamics may be used as a barometer of an organization's general level of adjustment to new automation. Observation over time of interactions between functional units directly and indirectly affected by the computer may show the need to adjust systems features or work methods to get the maximum benefit from the new system. Otherwise, automating one unit may improve its efficiency while jeopardizing performance of the organization as a whole.

In 1982, a fully mechanized system developed at AT&T-Bell Laboratories for testing telephone lines (called loops) was installed for an operational trial in a Bell Operating Company Maintenance Center. Use of this system allowed reduction of skill and knowledge levels of test personnel, cutting costs for the operating company. This paper describes the pre- and post-conversion evaluation of interactions between this Maintenance Center and other organizations coordinating with it to maintain records, interact with customers, and repair broken telephone plant facilities. The operations centers working with the Maintenance Center were categorized as Outside Craft Organizations; Facilities, Service, and Records Control Centers; Customer Interface Centers; and Central Office Organizations. The parameters of interest included information exchange, coordination of activities, and productivity of interacting organizations, all of which affect customer perceptions of telephone company service.

2. METHOD

2.1 Pre-Conversion Study

In the 9 months before systems conversion, a study was conducted to describe the interactions between the Maintenance Center and associated operations centers, as a baseline for later comparison with fully mechanized operations. Six Outside Craft, ten Facilities Control, four Customer Interface, and two Central Office Organizations were identified as subject groups. Within the 22 centers, 35 managers were interviewed and 122 craft and clerical personnel were surveyed using a paper and pencil questionnaire, concerning the purpose, frequency, content, method, persons contacted, and subjective perceptions of interactions with the Maintenance Center. Respondents' identities were recorded to allow for post-trial within-subjects comparisons. As a validity check, interviews of Maintenance Center personnel were conducted using a parallel form of the survey concerning their interactions with the operations centers.

The organizations experiencing the highest levels of interaction with the Maintenance Center, and therefore most likely to show any effects of the new system on their own productivity, were the Outside Craft and the Central Offices. Individual productivity reports routinely kept by the company were obtained for a representative 3 month period before the trial for 36 outside repair technicians interacting with the

Maintenance Center. Time logs showing frequency and duration of Maintenance Center calls to Central Office technicians were also acquired, as well as counts of false troubles dispatched to the Central Offices. All data were coded and stored on-line on a per respondent and per center basis.

2.2 Post-Conversion Study

In the follow-up study conducted 3-6 months after systems conversion, an attempt was made to gather the same data from the same personnel in the same operations centers. Of the original 157 survey/interview respondents, 105 personnel or their staff replacements still interacted with the Maintenance Center in the same capacity at the time of the post-conversion study. Productivity indices for 32 of the original 36 outside repair technicians were obtained for a corresponding 3-month post-conversion interval. Central Office activity logs were obtained for the 6 months following the start of the trial. Within-subjects before and after comparisons were conducted using Statistical Package for the Social Sciences (SPSS).

3. RESULTS AND DISCUSSION

Personnel in certain operations centers but not others reported changes in patterns of interactions with the Maintenance Center following conversion to the new testing system. Statistically significant interactions between organization type (Outside Craft, Facilities Control, Customer Interface, or Central Office) and time (before or after conversion) were found on survey measures of operations center-initiated contact frequency (Figure 1; $F = 3.72$, $df = 3, 92$, $p < .014$), and Maintenance Center response time (Figure 2; $F = 17.7$, $df = 3, 86$, $p < .0001$). Interactions of this nature indicated that analysis of the data by organization type would be useful in discerning the true effects of conversion on interorganizational performance.

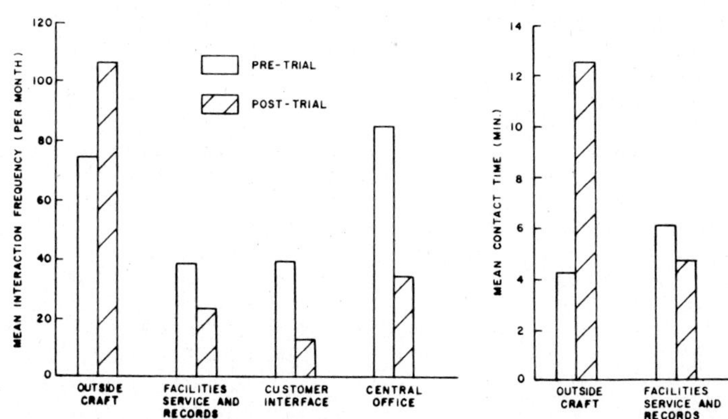

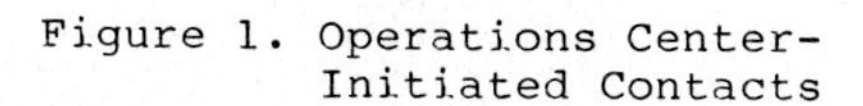

Figure 1. Operations Center-Initiated Contacts

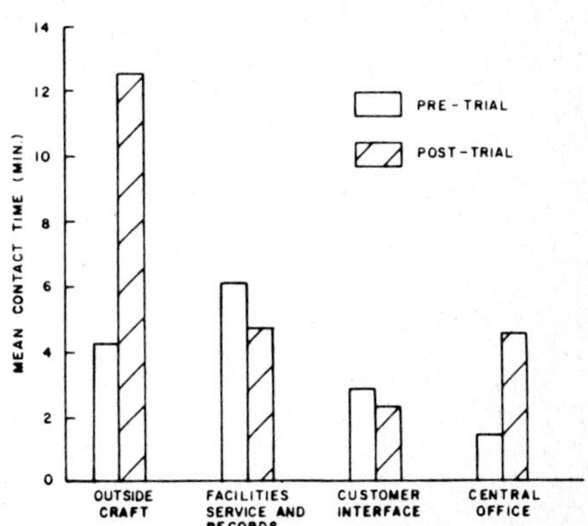

Figure 2. Maintenance Center Response Times

3.1 Changes in Interactions with Outside Craft

Outside craft survey responses following trial conversion showed significant increases in estimates of time to contact the Maintenance Center for test support ($t = -5.71$, $df = 44$, $p < .003$) and frequency of attempts to make contact ($t = -3.09$, $df = 44$, $p < .003$). The survey results were validated by independent telephone traffic studies showing longer hold times on craft calls to the Maintenance Center in the post-trial environment. Despite these problems, outside craft productivity was not significantly affected by the systems conversion.

Since the trial, improvements in transaction times have been made in the Maintenance Center by running abbreviated versions of tests. New call handling procedures were developed to better serve outside craft needs. Specific tests requested by craft personnel on the survey that were unavailable in the trial are currently under development as software enhancements.

3.2 Changes in Interactions with Central Offices

Analyses of Central Office records of time spent helping Maintenance Center personnel with testing showed some transient productivity changes in the first five months of the trial. Calls per day to the Central Offices from the Maintenance Center increased 12 - 25% from pre-trial levels. Transaction time per call increased 5 - 28%, depending upon the Central Office. The increases in calls and transaction times were largely due to the mechanized systems users' inexperience in working with the Central Offices at the start of the trial.

The total number of troubles per month referred to the Central Offices by the Maintenance Center increased significantly following system conversion ($t = -4.71$, $df = 10$, $p < .01$). The percentage of false dispatches to the Central Offices also increased by 17% during this period. This resulted from mechanized systems users' lack of familiarity with tests for sectionalizing troubles, and from the need to calibrate test trunks in the trial site.

Several steps were taken to mitigate the effects of system conversion on the Central Offices. Methods for screening Central Office troubles were reviewed with Maintenance Center personnel. System users were encouraged to run shorter versions of certain tests and be ready to test before calling the Central Offices, which reduced transaction times. Test trunks were calibrated to provide more accurate sectionalization. The effects of system conversion began to taper off 6 - 9 months after the start of the trial.

3.3 Changes in Interactions with Other Organizations

In general, most organizations interacting with the Maintenance Center were unaware of the trial conversion and experienced few effects of it. There was a trend toward less overall contact with the Maintenance Center and greater concentration on internal operations mandated by company reorganization.

Opportunities were identified within two Facilities Control centers to directly access the fully mechanized testing system, using terminal hardware and data communication facilities already in place in their organizations. This virtually eliminated the need to interact with the Maintenance Center for these organizations, thus streamlining the maintenance process in the company as a whole.

4. CONCLUSIONS

The results of interorganizational analysis before and after automated systems conversion provide insights concerning the usefulness of the new system in a wider context of total organizational performance. Customers may be better able to determine the costs and benefits of automation in a particular unit, relative to the effects of that change on organizational performance as a whole. Requirements for new systems or enhancements and methods to improve manual and automated operations workflows may also be generated. Customer needs for systems modifications and new features which are likely to surface in the study will be of interest to systems developers and vendors. Minor software changes can be made before widespread installations take place. New features can be planned and future software releases scheduled early in the product introduction phase of the marketing cycle. New opportunities for systems implementation in customer organizations interacting with the original target unit for automation could lead to greater sales. Finally, ideas for new technology which would streamline complex interactions between customer organizations may be discovered, leading to future product development.

5. REFERENCES

[1] SPSS Release 9.0. (SPSS, Inc., Chicago, 1981).

* Currently with Hewlett Packard, Cupertino, California.

SAFETY FACTORS, WORKLOAD AND PERFORMANCE FACTORS

Human Factors in Organizational Design and Management
H.W. Hendrick and O. Brown, Jr. (Editors)

EVOLUTION OF SAFETY CONCEPTS IN THE PULP AND PAPER INDUSTRY IN CANADA, 1903-1983

Monique Lortie

Industrial Engineering
Ecole Polytechnique
Montréal, Québec
Canada

The object of this paper is to present an analysis of the evolution over the last 80 years of the approach and concepts in safety in the pulp and paper industry in Canada. This analysis is based on papers on safety published in the "Pulp and Paper Magazine of Canada" from 1903 until 1982. Basic ideas prevalent on safety and their implications and the organization of prevention will be received.

INTRODUCTION

The object of this paper is to present an analysis of the evolution over the last 80 years of the approach and concepts in safety in the pulp and paper industry in Canada. This look into the past will allow a critical reflection on our way to deal safety problems, namely on their management. Basic ideas prevalent on safety will be viewed in part I of the paper, and the organization of prevention will be presented in the second part. A large number of citations are included to stress some of the concepts, ideas, beliefs and strong assumptions that were prevalent during all the years covered by the study.

MATERIAL AND METHOD USED

Papers on safety published in the "Pulp and Paper Magazine of Canada" from 1903 (its first year of publication) until 1983 have been systematically censed and analyzed: this material covers about 600 pages of papers, chronics..., written by managers, safety directors, safety supervisors... and foremen. In addition further material which will not be presented here, has also been examined in order to gain a general understanding of that sector: namely papers dealing with industrial relation, working conditions, technological evolution, federal and provincial inquiry commissions and archives of a large Union.

PART I: BASIC PHILOSOPHY AND CONCEPTS

In this part, we will see what is the basic perception of the safety problem and its ramifications and implications. This will allow us to understand the organization of the accidents prevention analyzed in the second part of the paper.

As soon as 1915, it is said that "no more than 1/3 of the reduction (of accidents)...have been accomplished through the use of safeguards or mechanical equipement while 2/3...through organization and through education, in short, through reaching the workmen and securing their cooperation"(1). Causes of accidents are then classified in two categories: physical conditions, i.e. mechanical hazards and human factors such as human failure, and unsafe practices. The first category is believed to explain 10-20% of accidents, while the second explains 80-90% of them. This ratio, first specified in 1917 (2) will be quoted and always assumed as an inescapable truth until the 1980's. At the beginning, mechanical hazard was quite synonymous with machine hazard: dangerous conditions as differing from hazard will not be clearly introduced as a concept before 1955 (3), although some timid relations between general aspects of working conditions and safety had been previously established. In the 1930's, the understanding of human factors or human failure had evolved from the simple notion of carelessness to the one of lack of knowledge of the jobs and their hazards: "Careleness...has been used to cover up a lack of understanding of the real cause of accidents... (which is) poor training, ignorance" (4). Once training was introduced, the problem was reduced to a question of attitude. This basic classification had four main consequences as outlined below.

First, it was translated in a double sharing of responsibilities: the employer had to take care of mechanical risks, while the employee had to take care of the unsafe practices which are "beyond the power of employer..."(5). This latter responsibility was later shared, from 1930's onivard by foremen: "safety is a method of working, a method which is...under the control of the foremen. We put the safety responsibility squarely up to foreman... And we do everything to encourage that (accident are their fault) feeling" (6). It was then clearly said that safety staff and director "acting only as advisor...does not assume the responsibility for an accident" (7).

Second, safety being a question of attitude and training great emphasis was put on educational campaigns to make worker "safety minder". As a matter of fact, prevention is seen to be "largely a matter of mental attitude" (8). A classification of minds is even proposed: "unguarded mind, puzzled mind, misguided mind, stubborn mind, diverted mind, troubled mind... tired mind" (9). Therefore emphasis was "placed upon making men's mind safe for the machinery rather than making the machinery safe for men's minds, since the latter job will presently be achieved" (10). The problem of the man-machine relationship began to be treated only on of 1970. Emphasis was placed mainly on training: safety training at first and job training later, in the 1940's. It was then up to recently, strongly believed, that a "good" worker doesn't have accidents: "accident prevention is the logical and inescapable result of good workmanship" (11). "If you teach a man how to do a job properly, he knows how to do it safety" (12). Such a notion was prevalent up to the 1980's, and even in 1982, accidents were said to be "simply a matter of not doing our jobs properly" (13).

Third, this oversimplification of the safely problem brough about a total lack of intellectual effort to develop analytical tool for the accident phenomena, to try to understand the mechanisms of accidents and to identify the factors involved: "accident prevention...is just a matter of education" (14), "let us preach and teach our employees a religion...that of accident prevention" (15). "A prevention is simply being careful all the time" (16). "Safety is such a simple story" (17), "It's just plain common sense" (18). Over the last 80 years only one example of systematic analysis of a type of accident in relation to a dimensionnal characteristic of an equipment, namely the calendar, was censed [19]. The other exception is one underlining the needs of a scientific study of the characteristics of axes in relation to accidents [20]. However, many improvements were brought forward, quite often under pressure from the industries specialized in the design of safety products. But in most cases these technical solutions helped decrease accidents primarily attributed to human failure. The following example concerning slip and fall accidents is typical: "we know that errors of judgment ...inattention and poor discipline cause injuries as evidenced by slips and falls" [21]. Solutions, however will be technical.

Finaly, we may understand the astonishment of some safety directors when faced with accidents which in "spite of so good working conditions, proper sanitation and a splendid safety organization" [22], continue to occur. Psychology has brought some answer: accidents are due to a little group of accident-prone people "who resent authority in any form, have a fatalistic attitude and believe accidents have to happen" [23]: (in this particular case, the subject of the paper was the accidents related to block pile, which was an especially dangerous job). This led to the conclusion that accident would occur to either the permanently by prone (or accident addict) people or to those which were temporarily accident prone [24]. Having an accident will be squarely defined as a mental illness: "it's always puzzled me to realize that intelligent men working in a mill...will again and again expose themselves to all kinds of hazards even though they know perfectly well that they are running risks...perhaps...accidents are not accidents at all...perhaps they are some kind of subconscious rebellion..." [25]. "Accident...much better to call it a selfinflected wound" [26]. "The accident maker is suffering from a form of mental illness. He is a troubled person who unconsciously hurts himself" [27]. Overall such a basic philosophy had led to place more emphasis on the identification of a human culprit than on the hazard itself and its elimination. This was admitted as the object or advantage of a program of selections (to identify the accident prone and the "good material"), the constests, the recording of accident data. To create an atmosphere of safety "employees are encouraged to report... dangerous working habits by fellow employees" [28] and no paper will be more often quoted than the one entitled "to fix responsability" which mainly concern foremen [29].

PART II: ORGANIZATION OF PREVENTION

In this section we will examine the personnel involved in safety together with their respective functions. The organization of committees and the most privileged tools of prevention will be reviewed.

a) Staff

Foreman

A foreman is principally identified as a link between workers and management. He receive the mandate to "unite safety and work" [30]. He is strongly and in a continuous way identified as the "key man" in safety, especially from the 1930's onward. His functions are to teach safety job, to supervise and to inspect: "who can teach the workers better than the foreman...he comes to this position usually from the ranks and so knows how the work should be done well and safely" [31]. He "must learn to recognize... things which may lead to accidents" [32]. As key man, he will be pointed as responsible for accidents: "Put the responsability for the safety on each man directly on the shoulder of the foreman... tell the employees that their foreman is responsible" (33). This picture will almost not evoluate from 30's to 80's, with two exceptions: his needs of training are admitted at the end of the 1950's, and specific training departments are developed.

"Safety man":

As "technical man", with usually a background in engineering or mechanics the "safety man" has a very specific function: designing safety equipment or acting as an adviser. His function as safety director is most often defined as a function of the foreman. Allusions to conflics between safety staff and foreman are however quite frequent. He must only "advice or council... not to administer the program" (34), "The safety director should not be allowed to act as he were foreman" (35) and "interfere in the operation of a department" (36). Consequently, the worker is encouraged to report safety problems to the foreman rather than to the safety department (37). The foreman is supposed "to correct... without refering them (problems) to other departments or to persons of higher authority" (38). In exchange, the "safety man" who "must act only as an advisor... does not assume the responsability for an accident" (39).

After the second world war, we observe trends to group together everything dealing with employees, such as employment, safety, services... The safety director will then be often also director of personnel (or under his order). His background in industrial relations will tend to emphasize the trend toward viewing safety problems as those of human relations and communications.

b) Organization of committees

Two general trends may be identified: the multiplication of committees and the strong involvement of workers. Typically there will be a central or main committee involving high level staff who will define the program, a safety committee comprised of a "safety man", a physician, superintendants and eventually of foremen and workers, both a foremen and workmen committees with elected members, and finally an inspection committee. A great number of meetings are devoted to safety: few specialized staff however are hired to study these problems. Workers identified as strongly responsible for their safety are frequently invited to participate in these committees. It is worthwhile to note that these committees involving management and workers may represent the first sharing of management rights with workers. Later, from 1942 onward unions were also represented in these committees.

c) Prevention tools

A result of the basic philosophy as explained in the first part of this paper, was to place a great emphasis on publicity: posters, slogans, bulletin boards, movies, entertainment, contests...: "we felt that safety should be a game of good sport an so, it was added to several other sport games..." (40). In particular safety contests brought conflicts and a tendency not to report accidents. Critical analyses of such an approach did not appear until 1956. The idea to offer more specific job training, with a first objective to learn how to identify hazards, and second, to develop skills will gain importance only from the 1950's. The other most important prevention tools used is the protective personal equipement. The generalized use of goggles, safety shoes, hard hats..., began during the 1950's as well.

Most papers of the time deal then with the convincing of workers to wear this equipment: "Don't worry about the gadget may be not perfect... Make it a condition of employment" (41). From one advertisement to another, we learn however of that weakness were corrected and that resistance may have been not just "blind prejudice" (42). Selection of workers ("we have to start with good materials") (43) and detection of the accident prone worker will also be an important tool of prevention. Overall, prevention by the removal of hazard, improvement of working conditions analysis of working procedures and safety analysis are varely emphasized even though it is known, after the fact, that a better lighting, ventilation and housekeeping had beneficial effects. The preventions by a better relationship between worker and machine will begin to be stressed just at the end of the 1970's.

CONCLUSION

Frequency ratio and severity of injuries have been greatly cut down during the years. Even though pulp and paper industry continues to have a bad record compared to other industries. This quotation extracted from annual meeting summarizes particularly what happened: "while we are pleased to see that attention was been paid to elimination of hazards, most of the discussion and the meeting was based on human factor in accident prevention work" (44). As a matter of fact, technical evolution had brought more safety in general, particularly concerning the mechanization of operations. The decreasing, for example of accidents related to handling tasks, falls and sliding... is due above all to technical improvements. Although dangerous conditions leading to accidents have indeeed decreased over the years, the idea that accidents are essentially due to human factors still appears to prevail. No other concept brought more misunderstandings and inefficiency. Up to only very recently the prevailing notion was that it was the man that was not fit to the machine rather than the other way around. Such a notion seemed to be a straight extension of old common-law doctrine whereby employees assumed the risk of accident upon taking a job. With the workmen's compensation act, the concept of fault related to the right to receive a compensation for an accident has disappeared, but the concept that a worker assume the professionnal risk has not.

FOOTNOTES:

(1) 1915, April, p. 202; (2) 1917, May, p. 534; (3) 1955, April
(4) 1931, Oct. p. 1165; (5) 1915, April, p. 205
(6) 1929, May, p. 610 ; (7) 1960, June, p. 211
(8) 1925, Oct. p. 1123; (9) 1925, Oct., p. 1123
(10) 1925, Oct., p. 1123; (11) 1947, June, p. 118 (12) 1967, March, p. 105
(13) 1981, Sept. p. 95; (14) 1936, Feb., p. 135 (15) 1955, April, p. 238
(16) 1944, May, p. 496; (17) 1967, March, (18) 1956, Nov. p. 262
(19) 1958, Feb. (20) 1945, Sept., p. 784 (21) 1929, Jan., p. 158
(22) 1948, Jan., p. 125; (23) 1953, May, p. 289
(24) 1953, August, p. 192; (25) 1955, August, p. 209;
(26) 1956, Sept., p. 255; (27) 1961, June, p. 248; (28) 1961, Sept., p.220
(29) 1953, July, p. 198; (30) 1929, p. 610; (31) 1944, Oct., p. 810
(32) 1947, Oct., p. 141; (33) 1954, Feb., p. 111;
(34) 1951, May, p. 177; (35) 1945, May, p. 453;
(36) 1959, June, p. 212; (37) 1961, Sept., p. 220
(38) 1949, Feb., p. 50; (39) 1960, June, p. 211;
(40) 1938, Nov., p. 789; (41) 1959, Feb., p. 98;
(42) 1947, Aug., p. 131; (43) 1938, Nov., p. 792;
(44) 1958, Oct., p. 432;

Human Factors in Organizational Design and Management
H.W. Hendrick and O. Brown, Jr. (Editors)

THE IMPLICATIONS OF TECHNOLOGICAL CHANGE FOR WORKER HEALTH: THE ROLE OF THE JOB REDESIGN MODEL IN PUBLIC HEALTH

Benjamin C. Amick, III and Carol S. Weisman, PhD.
The Johns Hopkins University
Baltimore, MD

Michael J. Smith, PhD.
National Institute of Occupational Safety and Health
Cincinnati, OH

The purpose of this paper is to present a model relating technology, which is one characteristic of the work organization, to an individual's quality of working life. A study of U.S. postal workers was used to examine the question of whether technology, mediated by job dimensions, has an effect on individual attitudes and health. What makes the model unique is the extension of quality of working life beyond job attitudes to an individual's health. Extending the model has implications for organizational productivity, since workers today more than ever exercise their right to paid sick leave. The inclusion of job dimensions in the model as mediating factors creates a flexible framework for management to propose a variety of modifications to ameliorate the potential adverse affects of technological change. Furthermore, this model allows occupational health specialists to predict the acute and chronic health consequences to the worker following technological shifts in the organization.

Examining the effects of technology on the worker has a long tradition in organizational research. Recently, Rousseau [1] has proposed an integration of the socio-technical systems approach with the job redesign approach. Rousseau showed that different forms of operations technology are associated with different constellations of job dimensions. Using Woodward's classification of technology, she found long-linked technologies (e.g. mass-production technologies) were negatively associated with most job dimensions and job satisfaction, while being positively associated with absences, propensity to leave and experienced stress. Rousseau then showed that this observed relationship between technology and the individual attitude or behavior was mediated by job dimensions.

The above research does not typically examine health outcomes. On the other hand, occupational stress research in public health has focused on these outcomes relating them to job attitudes (e.g. job satisfaction) and job dimensions (e.g. job demands and autonomy). However, rarely are the individual health outcomes related to a particular technology or any characteristic of the organizational structure. (The exception to this is the recent research on video display terminals.)

As a result of this focus on individuals in occupational stress research, rather than the relationship of the organization to the individual, stress management programs have been the only scientifically defined intervention available to management. Yet, in these times of rapidly changing technologies, it may be a more effective strategy to propose organizational changes which affect jobs, than to spend numerous person-hours in stress management clinics.

Combining the above two research traditions provides the logic for the following heuristic model:

TECHNOLOGY ⟶ JOB DIMENSIONS ⟶ JOB ATTITUDES ⟶ HEALTH OUTCOMES

Although it is recognized from the organizational design and human factors research areas that technology has both direct and indirect effects, this paper only explores the effects of technology on job attitudes and worker's health, while also considering the effect of job dimensions. This model also addresses a central drawback in occupational stress research by relating an organizational characteristic of the workplace to an individual's health.

Data from a 1979 National Institute of Occupational Safety and Health study of postal workers were used to explore the research question. A cross-sectional research design was used with mailed questionnaires to collect information about the employees' job and health. The overall response rate was 50%. The sample of postal workers used for this analysis was those who sort the mail. By focusing on a single job, sorting mail, inter-job differences were controlled.

Three types of technology for sorting mail are included in the analysis: a machine-paced technology, a self-paced technology, and an intermediate technology. In the machine-paced technology, postal clerks (N=3225) sort mail using a multiple position letter sorting machine (MPLSM). Clerks have a letter fed into the machine every second and, they then have a second to scan the address so they can key the letter to the appropriate bin on the other side of the machine. Distribution clerks (N=1528) are slf-paced. They choose a letter, read the address and manually place the letter in the appropriate bin at their own pace. People who work at the MPLSM have no control over when the task is initiated or it's duration. Distribution clerks have control over both the duration and initiation of the task. The single position letter sorting machine (SPLSM) clerks (N=150), although small in number, are an important comparison group on the technology continuum, because these clerks perform the same task as the MPLSM clerks. The SPLSM clerks have control over the initiation of the task, but not the duration of the task. The above classification of technology is based on an article by Dainoff [2].

Many characteristics of the job have been identified, but not all are influenced by the type of technology used in the work process. Table 1 shows the job dimensions used in our model. Job demands were measured as workload pressures; as the technology becomes more machine-paced, job demands increase. Autonomy was operationalized as an individual's control over the job; as the technology becomes more machine-paced, individual autonomy decreases. Co-worker support can be seen conceptually as both friendship opportunities and assistance in the completion of tasks; as the technology becomes more machine-paced, co-worker support decreases. Supervisor support is one aspect of the feedback a worker gets about work performance; as the technology becomes more machine-paced, supervisor support decreases. Each of these job dimensions has been identified in the job redesign field [3]. All of these dimensions have been associated with mental and/or physical health outcomes. The final job variable included is pay satisfaction which is used as a surrogate fcr pay in this analysis. This characteristic varies significantly between postal employees working at the MPLSM and Distribution clerks.

One job attitude, job satisfaction, was included in the model. The measure is Quinn and Shepard's [4] content-free scale. It can be hypothesized that the more machine-paced the technology, the less satisfied the worker is with the job. Job satisfaction should be positively affected by autonomy, co-worker support, supervisor support and pay satisfaction. It should be negatively related to job demands.

The health outcome defined in this study is a psychosomatic symptoms index, composed of self-reported symptoms which have occurred during the past year. Considering these symptoms as acute responses to working conditions, it can be proposed that postal workers who experience these symptoms to a greater degree will use more sick leave and be absent from work more often. In accordance with earlier bivariate hypotheses, the more machine-paced the work, the more symptoms will be reported. Among job dimensions, the more autonomy, co-worker and supervisor support, and pay satisfaction, the fewer symptoms reported. Increased job demands should lead to increased reporting of symptoms. Finally, increased job satisfaction should lead to decreased symptoms.

Table 1
PEARSON PRODUCT-MOMENT CORRELATIONS AMONG ALL MODEL VARIABLES
(N=4318; all p-values less than 0.001)

Model Variables	1	2	3	4	5	6	7	8
1. Technology*		-.32	.46	.14	.06	.04	.16	-.16
2. Job Demands			-.51	-.10	-.18	-.13	-.26	.22
3. Autonomy				.16	.22	.10	.32	-.18
4. Co-Worker Support					.24	.08	.13	-.11
5. Supervisor Support						.10	.38	-.16
6. Pay Satisfaction							.15	-.05
7. Job Satisfaction								-.29
8. Psychosomatic Symptoms Index**								

* Technology is coded 1,2,3 with machine-paced technology 1.
** The index is the negative reciprocal transform of the index.

The unit of analysis in all tables is the individual. Using the zero-order correlations in table 1, each of the bivariate hypotheses is explored. To explore the validity of the model, hierarchical regressions are carried out. In table 2, job satisfaction is the outcome. If the job dimensions account for the relationship between technology and job satisfaction, then the standardized regression coefficient for technology should diminish or become non-significant when the job dimensions are added to the model. The psychosomatic symptoms index is the outcome in table 3. The logic for analysis is the same as in the first hierarchical regressions, except now job satisfaction should account for the relationship of the job dimensions to individual health. Therefore, job satisfaction should have the largest beta, and the betas for the job dimensions as a group should diminish.

Table 1 shows the correlations among all independent and dependent variables used in the regression models (excluding covariates). Reading across row 1, the direction of the relationship between technology and the other model variable follows the bivariate hypotheses. This is also true for every other hypothesis stated concerning relationships of all the independent variables with job satisfaction (column 7) and the psychosomatic symptoms index (column 8). These trends agree with findings in the literature. Of particular interest are the strong correlations between technology and two job dimensions, autonomy and job demands. These two job dimensions seem more closely tied to the particular technology than the others.

Job satisfaction is the outcome in table 2. In each model the covariates are entered first, since the variance they explain in the outcome is exogenous to the model. The beta for technology is (0.13) when first entered into the regressions. As technology becomes more machine-paced, the worker is less satisfied with the job. However, when the job dimensions are entered into the model, technology's effect diminishes to zero. Concommitant with this decrease is an increase in the variance of job satisfaction explained by the independent variables as a group.

All the job dimensions vary with job satisfaction in the manner predicted. Comparing the rank of each dimension in table 2 with their relative rank in table 1 shows the same pattern, with supervisor support most strongly related to job satisfaction. The strong zero-order correlations of technology with job demands and autonomy may account for the majority of the change in the beta for technology. Overall, these hierarcical models suggest that job dimensions account for the relationship between technology and job satisfaction.

Table 3 extends the traditional organizational design model to explore how technology can impact on a person's health, given job dimensions and job attitudes. Again, when technology is first introduced into the regression it has a beta of -0.13. The negative sign indicates that the more machine-paced the technology, the more psychosomatic symptoms reported. Introducing the job dimensions causes the coefficient for technology to diminish to near zero. The largest coefficient is for job demands and not supervisor support as in table 2. The most intriguing relationship is the small beta for autonomy, which in table 2 was ranked second. These hierarchical regressions follow the same pattern demonstrated in table 2. The only difference is in how the job dimensions relate to the psychosomatic symptoms index.

Table 2
REGRESSIONS OF JOB SATISFACTION ONTO TECHNOLOGY, JOB DIMENSIONS AND COVARIATES
(Standardized Partial Regression Coefficients)

	Hierarchical Regression Models		
Model Variables	Covariates Only	Covariates and Technology	Covariates, Technology, Job Dimensions
Years on the Job	0.11*	0.04*	0.04*
Education	-0.16*	-0.15*	-0.13*
Technology**	-	0.13*	-0.007
Job Demands	-	-	-0.10*
Autonomy	-	-	0.17*
Co-Worker Support	-	-	0.03
Supervisor Support	-	-	0.29*
Pay Satisfaction	-	-	0.07*
R-Squared	0.04	0.06	0.23
F-Test	82.52	72.68	123.81
P-value (less than)	0.0001	0.0001	0.0001
df	2;3689	3;3691	8;3400

* P-value less than 0.05
** Technology is coded so that the lower the score, the more machine-paced.

By introducing job satisfaction into the model the coefficient for technology does not change much. As expected , the largest coefficient is for job satisfaction, indicating that as job satisfaction increases, there is a decrease in symptoms reported. All the job dimensions change slightly except for supervisor support which changes from the second largest beta to near zero. Perhaps the most interesting effect of introducing job satisfaction into the model is the minimal effect it had on job demands, which continues to have a relatively strong impact on symptoms.

Table 3
REGRESSIONS OF THE PSYCHOSOMATIC SYMPTOMS INDEX ONTO TECHNOLOGY, JOB DIMENSIONS, JOB SATISFACTION AND COVARIATES
(Standardized Partial Regression Coefficients)

	Hierarchical Regression Models		
Model Variables	Covariates & Technology	Covariates, Technology, Job Dimensions	Covariates, Technology, Job Dimensions, Job Satisfaction
Years on the Job	-0.004	-0.003	0.007
Sex**	0.22*	0.22*	0.22*
Education	-0.03	-0.04*	-0.07*
Technology***	-0.13*	-0.04	-0.05*
Job Demands	-	0.14*	0.11*
Autonomy	-	-0.06*	-0.02
Co-Worker Support	-	-0.04*	-0.03
Supervisor Support	-	-0.11*	-0.04*
Pay Satisfaction	-	-0.05*	-0.03
Job Satisfaction	-	-	-0.24*
R-Squared	0.07	0.13	0.17
F-Test	69.47	51.50	64.86
P-value(less than)	0.0001	0.0001	0.0001
df	4;3508	9;3249	10;3167

* P-value less than 0.05
** Sex is scored as a dummy variable; 0 = male, 1 = female.
*** Technology is coded so that the lower the score, the more machine-paced

The implications of these findings for public health are two-fold. First, they show that it is not technology per se that effects job satisfaction and the psychosomatic symptoms index, but the constellation of job characteristics associated with a particular technology. Second, they suggest that not all workplace intervention and prevention programs must focus on the individual. Instead, by changing characteristics of jobs, management may broadly impact workers' health. These findings are based on cross-sectional data which do not permit us to explore whether different technologies cause different patterns among the job characteristics, and which in turn cause changes in job satisfaction or a worker's health. Further research of an experimental or quasi-experimental nature is needed, to more clearly show how changing job characteristics by organizational change effects a worker's health.

References:

[1] Rousseau, D., Characteristics of departments, positions and individuals: Contexts for attitude and behavior, Administrative Science Quarterly 23 (1978) 521-540.

[2] Dainoff, M.J., Hurrell, J.J. and Happ A., A taxonomic framework for the description and evaluation of paced work, in Salvendy, G. and Smith, M.J. (eds.), Machine-Pacing and Occupational Stress (Taylor & Francis, London, 1981).

[3] Oldham, G.R. and Hackman, J.R., Relationships between organizational structure and employee reactions: Comparing alternative frameworks, Administrative Science Quarterly 26 (1981) 66-83.

[4] Quinn, R.P. and Shepard, L.J., The 1972-73 Quality of Employment Survey (ISR, Ann-Arbor, 1974).

Human Factors in Organizational Design and Management
H.W. Hendrick and O. Brown, Jr. (Editors)

ACCIDENT PRONENESS AND INTERVENTION STRATEGIES FOR SAFETY

Kenneth R. Laughery

Department of Psychology
Rice University
Houston, Texas
U.S.A.

An analysis of industrial accidents showed the observed distribution across employees differed from a Poisson distribution, supporting the notion of individual differences in susceptibility to accidents (accident proneness). Detailed analyses of scenarios for a sample of employees who incurred many accidents showed variability in the accident patterns. Management procedures should consider individual factors in developing strategies to improve safety.

INTRODUCTION

This paper presents some preliminary results of an ongoing project concerning industrial accident analysis. The specific issue of interest here is the distribution of accidents across employees. This issue has received considerable attention in the accident literature over the years and has usually been addressed under the label _accident proneness_.

The concept of accident proneness refers to the notion that people who are exposed to equal risk do not have equal numbers of accidents. It is generally agreed that the first scientific treatment of accident proneness was performed by Greenwood, Woods and Yule (1919) in their analyses of the frequencies of accidents among individual female workers in a British munitions factory during World War I. The question was formulated into a statistical problem in which the actual distribution of accidents was compared to a chance distribution in which it was assumed that accidents are a random occurrence with respect to the people employed. Greenwood et al. found significant differences in the distributions and noted that the observed distributions resembled an unequal liability distribution. In other words, their analyses provided evidence for accident proneness. In another "classic" study, Newbold (1926) examined accidents in 13 factories representing a variety of industries. She also found statistically significant differences between the observed and chance distributions. These studies represented important contributions to safety and influenced subsequent thinking through the 1930's and early 1940's--particularly in the United States. The thinking included the idea that an important approach to reducing industrial accidents was to identify and eliminate or change accident prone workers.

This point of view did not prevail, however, and in the past 30 years or so the dominant view has been that the accident proneness concept is of minor importance. Shaw and Sichel (1971) attribute the turnaround in part to earlier excessive claims about the utility of the concept. Also, while some studies have replicated the findings of Greenwood et al. and Newbold, others failed to find differences between observed and chance accident distributions (Mintz and Blum, 1949). Another important development in the shift away from accident proneness as a useful concept was the post World War II emphasis on the design of safer man-machine systems.

The emphasis on the design of safer equipment, environments and jobs is well founded and should be at the forefront of safety concerns and efforts. But have we gone too far in disregarding individual differences in susceptibility to accidents (accident proneness)? Is it possible to make significant contributions to safety by identifying individuals with high accident rates and by developing intervention programs based on such information? These questions, of course, are fundamental issues of management safety policy; but such policy should be based on good data about the various factors that contribute to accidents.

Shaw and Sichel (1971) have argued that the accident proneness concept does have utility. Their long-term study of traffic accidents among South African bus drivers showed improved accident rates on the basis of intervention programs that took into account individual differences. Cameron (1975) has criticized the Shaw and Sichel work because they did not demonstrate that proneness is an identifiable characteristic in the population generally nor that it is an enduring characteristic. Cameron's comment raises an issue as to what the accident proneness concept means. If indeed it is intended to refer to enduring general traits that account for individual differences in accident rates across a wide variety of work settings, then its limited utility is not surprising. If, on the other hand, the term is intended primarily to describe a statistical phenomenon, and if these statistical results lead to more productive accident intervention strategies, then the concept may have sufficient utility to warrant pursuing it further. To clarify, the issue is not one of enduring traits, but why are there individual differences in accident rates.

A potentially useful approach to understanding these individual differences is to examine in detail the types of accident events that occur. The logic is that if an individual has several accidents that follow a very similar pattern, one might focus on the equipment/environment/job factors. On the other hand, if an individual experiences multiple accidents with varying patterns, causes may be associated more with the individual. An example of the former would be a machinist who has several mechanical injuries while operating a particular piece of equipment, while the latter might be a machinist who has different types of injuries while performing a variety of tasks. In part, the work reported here is a preliminary exploration of the accident patterns (scenarios) of employees in a major manufacturing complex who had a significantly large number of accidents. The objective is to determine the extent to which the scenarios for a given individual are similar or different across accidents.

THE DATA BASE

The data include all accidents during 1981 and 1982 at Shell Oil Company's Deer Park Manufacturing Complex in Deer Park, Texas. During the 1981-82 period there were approximately 3300 employees including 1850 in the non-staff category. The total number of accidents during the two years was 3526 of which 3333 involved non-staff employees. Details of the type of information recorded about each accident and the coding scheme for getting it into a computer have been described in detail by Laughery et al. (1983). Of particular interest here is the scenario information which describes the dynamics of the accident. Four variables describe the accident event by breaking it down into time-sequenced elements. These variables are labelled prior activity, accident event, resulting event and injury event.

RESULTS

Three types of analyses are presented here. First, the test for accident proneness is a standard comparison of the observed and chance distributions. Secondly, accident rates are presented by job category. Thirdly, the scenarios for a sample of employees who were involved in a large number of incidents will be discussed. Data for staff employees were not included in these analyses, because people in this category had very few accidents.

The accident proneness issue is whether the accident distribution across employees can be attributed to chance. In order to answer this question the observed distribution is compared to a distribution predicted by the Poisson model. This model has been the standard for testing the accident proneness hypothesis. The procedures are described in Sachs (1982). The observed and expected distributions are presented in Table 1. A chi-square test showed the difference between the distributions to be statistically significant ($p < .0001$). More people had multiple accidents than would be expected by chance.

Accident rates (number of accidents/year/person) were computed for nine job categories. The eight jobs with the highest rates were carpenter (3.2), welder (2.1), mechanic (1.8), machinist (1.6), pipefitter (1.5), boilermaker (1.4), instrument man (1.1) and electrician (1.0). The operator job category (0.5) was included because it contained the largest number of employees (819). Of the 114 employees who had six or more accidents, 108 were in these nine job categories.

A scenario consists of a combination of values of the four scenario variables mentioned earlier. The objective of these analyses was to examine the patterns of accidents incurred by specific individuals. The accidents of ten employees were selected. These ten included the five people who had 12-16 accidents and five of the nine employees who had 11 accidents. The job categories and accident counts of these ten were mechanic (16), welder (15), instrument man and boilermaker (14), operator and instrument man (13), pipefitter (12), electrician (11), electrician (11), boilermaker (11), welder (11) and carpenter (11).

Space does not permit a detailed presentation of the scenarios associated with this large number of accidents. It is striking, however, to note the variability in the patterns for seven of the ten employees. An example is the boilermaker involved in 14 accidents. The injuries included thermal burns, various mechanical injuries, eye irritations, strains/sprains, a chemical burn, and one multiple injuries. The prior activities were handling materials, assembling equipment, in transit and climbing. Some of the accident events were impacting an object, missing a proper step, sudden movement of materials and overexertion. In short, the worker was involved in a large number of accidents that happened in a variety of ways. A second example is the operator-instrument man who had 13 accidents including mechanical injuries, strains/sprains, irritations and thermal burns. These incidents occured while the employee was operating valves, disassembling equipment, handling materials, climbing and doing housekeeping work.

Accident events were caught between, tool slipped, impacting an object, person slipped and noncontainment of materials.

Table 1

Actual and Expected Distributions of Number of Employees Having a Given Number of Accidents

Number of Incidents	Number of Employees Having This Many Incidents	Expected Number of Employees Having This Many Incidents
0	553	307
1	519	552
2	296	497
3	182	298
4	121	134
5	65	48
6	50	14.5
7	22	3.7
8	13	0.83
9	10	0.17
10	5	0.03
11	9	0.005
12-16	5	0.0007

The scenarios for another five employees in the sample were similarly variable. There were three people in this group whose accident scenarios were more homogeneous. These three were one of the electricians with 11 incidents, and the boilermaker and carpenter who also had 11. A good example of this group is the carpenter. The prior activities were all either handling materials or assembling equipment, while the accident events were either impact with an object or a tool slipping. All but one accident involved mechanical injuries to the hand or fingers.

DISCUSSION

The checkered history of the concept of accident proneness seems to have revolved around a few issues or questions. The search for enduring general traits that would identify individuals as prone to accidents has proved to be generally unrewarding. This is not to suggest that cognitive, personality or motivational factors do not contribute to accidents; undoubtedly they do. But it would be naive to believe that such factors account for large portions of the variance in accident events--which raises a second point concerning the history of the concept. An apparent reason for the concept's declining stock has been that eliminating "susceptible" people from the workforce does not have a major effect on accident rates. This may well be true, but it is not a reason for ignoring individual differences in accident rates. The thesis underlying the work presented in this paper is familiar; namely, the analysis of accidents and the development of intervention strategies requires a systems approach. One must take into account the environment, the equipment, the job, and the person as well as their interactions.

The analyses of the Shell data showed that the distribution of accidents across employees was not attributable to chance. Many people had more accidents than expected on this basis. The analyses further showed that most of these people were in job categories where the accident rates were high. Unfortunately, sufficient data does not yet exist to test the proneness hypothesis within job categories. But even if such analyses show statistically significant differences within job categories, the questions still remain as to how the environment, equipment, job and person are interacting when accidents occur. This paper has presented one approach to addressing such questions. Much can be learned by analyzing the dynamics of the accident events, the scenarios, for it is in the event itself that the interactions of the system components can be observed.

All of this logic is, of course, an argument for a particular management strategy in dealing with safety problems. Managers need good accident data, particularly information about how accidents are happening. Traditionally, information is available about the who, when and where of accidents, but it is the how and why that counts.

REFERENCES

(1) Cameron, C. Accident proneness, Accident analysis and prevention, 7, (1975) 49-53.
(2) Greenwood, M., Woods, H.M. and Yule, G.U. A report on the incidence of industrial accidents upon individuals with special reference to multiple accidents, in: Haddon, W., Suchman, E.A. and Klein, D. (eds.), Accident Research (Harper & Row, New York, 1964).
(3) Laughery, K.R., Petree, B.L., Schmidt, J.K., Schwartz, D.R., Walsh, M.T. and Imig, R.C. Scenario analyses of industrial accidents, Proceedings of the Sixth International System Safety Conference, Houston, 1983.
(4) Mintz, A. and Blum, M.L. A re-examination of the accident proneness concept, Journal of Applied Psychology, 33, (1949) 195-211.
(5) Newbold, E.M. A contribution to the study of the human factor in the causation of accidents, in: Haddon, W., Suchman, E.A. and Klein, D. (eds.), Accident Research (Harper & Row, New York, 1964).
(6) Sachs, L. Applied Statistics: A Handbook of Techniques (Springer-Verlag, New York, 1982).

Human Factors in Organizational Design and Management
H.W. Hendrick and O. Brown, Jr. (Editors)

Why the difference in Actual vs. Reported Accident Causes? Phase II

Dr. Harv Gregoire

Institute of Safety & Systems Management
University of Southern California

Abstract

This is phase II of initial research developed to uncover specific areas of weakness in organizational safety. It was postulated that data concerning insurance claims, medical treatment, and other recorded accident cause variables may be contaminated since certain accident causes are either not reported or erroneously reported for a variety of issues such as fear of blame, misperceptions, and other intentional and unintentional reasons. This research delves into the reasons for intentional misinformation in reporting accident causes.

Introduction

Background

The initial phase I research reported earlier (Gregoire, 1984) described an attempt at developing an approach to uncover actual causes of accidents or near accidents in which causes were not reported accurately or not reported at all. (1) The fact that many insurance forms and accident reporting forms are perceived as efforts to fix blame or assign culpability in accidents appears to bias some of the very data that they are designed to derive. While there is an undeniable need to use the multitude of such forms which exist for individual accidents, there is a separate issue of identifying generic causes of accidents in organizations throughout industry and government.

It has been postulated that in questionnaire surveys, respondent anonymity will result in more objective data gathering than those surveys which require respondent identification. In responding to a questionnaire, most people do not want to appear significantly "different" from fellow respondents.

(1)Gregoire, H., An Identification of Actual vs. Reported Accident Causes, Proceedings of International Conference on Occupational Ergonomics, 1984.

Most importantly, respondents generally do not want to appear any less careful, responsible, or worthwhile than contemporaries whether the roles involved are that of employee, associate, or insurance client. It may be a combination of social pressure and human pride that often causes respondents to answer questions in terms of what they perceive the interrogators want as data or what is socially acceptable rather that what may be objective truth as they know it. We will probably never discern how much questionnaire data has been biased in the desire to answer questions in a socially appropriate manner rather than a completely objective manner. As this research indicated, information concerning accident causes is particularly biased when respondents perceive the possibility of direct negative consequences resulting from cause culpability which may result from reported accident information.

Method

First, an Accident Cause Evaluation (ACE) consisting of an interview and questionnaire protocol was administered to assess the issue of incorrectly reported accidents. The ACE device was administered to 40 men and women in an overseas military vehicle maintenance organization.

Secondly, information concerning the reasons for discrepancies between actual and reported accident causes was obtained from each respondent in very carefully designed interviews. The structure of the interviews emphasized (guaranteed) anonymity.

Discussion

Selected portions of the trial research questionnaire dealing with cause reporting disparities and resultant data are reproduced and discussed below. It should be emphasized that the author's post-item comments pertain only to this particular research administration and are not meant to be generalized across organizations.

Accident Cause Evaluation Questionnaire

In each question, the term "the accident" refers to an accident or near accident at work you are describing which was incorrectly reported as to its cause. (Note: Answers stated in percentage of respondents)

1. Rate how certain you are on a scale from 1 to 10 that you know what caused the accident (1 = uncertain, 10 = positive). - 10 = 80%; 8 = 10%; 1 = 10%.

2. Rate how certain you are on a scale from 1 to 10 that the item reported as causing the accident did not cause it. (1 = uncertain, 10 = positive) 10 = 80%; 8 = 10%; 1 = 10%. Both questions 1 and 2

indicate a high reported certainty of actual versus reported causes.

3. Which of the following categories of causes best describes the reported cause of accident? A = material failure - 40%; B = procedural error - 5%; C = communication problem - 10%; D = equipment design - 15%; E = inadequate training - 5%; I = negligence - 5%; J = fatigue - 5%; K = unknown cause - 10%; L = other (if L is selected, describe briefly) - 5%. Compare the two most frequently reported causes, material failure and equipment design, with those causes reported as actually causing the accident in question 4.

4. Which category in question 3 (above), actually caused the accident? (If L is selected, please describe briefly). If you list more than one answer, please list them in order of importance. A = 10%; B = 30%; C = 13%; D = 10%; E = 5%; I = 25%; J = 13%; K = 10%; L = 5%. These answers totaled over 100% of the respondents sice many insisted on identifying two "actual" categories of cause (procedural error and negligence). In subsequent interviews, it was determined that attempts at preventing these two causes would be more effective and less expensive than action formerly planned by management.

5. In your opinion, which of the following answers describes the reason why the cause of the accident was not accurately reported. If you list more than one answer, please list them in order of importance. A = person reporting cause didn't know - 5%; B = embarrassment - 35%; C = wrong cause reported due to incorrect or missing data - 5%; D = fear of blame and consequences - 30%; E = insurance would not pay damages or medical compensation - 15%; F = reporting the actual cause wouldn't result in any improvement - 0%; G = wrong cause or no cause reported to avoid involvement in accident investigation - 10%; H = other (describe) - 0%. Inaccurate reporting was identified as being caused primarily due to embarrassment and fear of consequences due to blame. It appears that the vast majority of cases where incorrect cause reporting occurred was intentional. If this trial research had been limited to organization - specific accidents, one could speculate that a major safety challenge to management would be to devise an acceptable, no-fault, accurate reporting procedure in order to identify and prevent actual accident causes. This issue of "no-fault" accurate reporting is the subject of this second phase of research.

Post-survey interviews were conducted to determine what measures might reduce the disparity between actual and reported accident causes. As might be expected, respondents were very sensitive to admitting that they may have misreported accident cause data, either verbally or via documentation for whatever reason. Several measures were taken during post-survey interviews to guarantee respondent anonymity. These measures included: (1) a letter from the commanding officer (Colonel) of the organizational unit commander (Captain) endorsed by the unit commander stating that (A) the "safety" interviews were strictly of a research nature to compile statistical data for future accident prevention; (B) participants were not to divulge their name or any form of identification; (C) no effort would ever be made to determine participant identity; (D) maximum honesty and cooperation in furnishing interview data was requested to enhance future accident prevention; (2) a standard set of interview forms were used which emphasized data collection accuracy rather than culpability; (3) stress was placed on clarifying accident cause identification as a preventative measure.

Despite these measures, of the 40 respondents, one refused to participate in the post-survey interviews allegedly due to mistrust of anonymity guarantees. Another original survey participant was not available (had returned to U.S.). Thus 38 subjects participated in the post-survey interviews.

Portions of the post-survey interview data are discussed below. The data is a sampling of the organization researched and does not necessarily generalize across other organizations. Answers are condensed into brief summary forms indicating percentage of respondents.

Question

1. Why do you think 35% of the safety survey respondents said that embarrassment was why actual accident causes were not reported in accidents which occurred in this organization?

   Answers - "Didn't want anyone to know" = 45%. This explanation only redefines the word embarrassment. Much of the anecdotal verbage however suggested the type of behavior causing the accidents in question involved complacency I.E. the operator had either occasionally, frequently or consistently engaged in the practice which caused the accident such as not following established procedures, using improper tools etc. It is extremely noteworthy that generally first-line-supervisors and occasionally shift leaders were cited as being aware of and tolerant of the potentially dangerous practices unless or until something accidental happened as a result. There were even indications from the post-analysis interviews that some supervisors covered up worker-caused accidents to the degree that supervisors might be implicated in permitting the potentially dangerous practices in the name of expediency. A total of 61% of the respondents described specific improper procedural shortcuts as "normally" done during their day-to-day tasks in order to "get the job done".

2. Why do you think 30% of the safety survey respondents said the causes of certain accidents were not accurately reported because of the fear of blame or consequences? Answers - Didn't want to get: in trouble, demoted, court martialed, charged for damages, reprimanded = 82%. Twenty-one (21) % of the interviewees indicated surprise that only 30% of the survey respondents selected the fear of blame/consequence answer as the reason for faulty accident cause reporting. A typical reaction of this 21% of the interviewees was that this reason was probably the primary cause for intentional inaccurate accident reporting in a far greater percentage of cases, particularly in those instances where accidents were of a serious rather than a minor nature.

The most noteworthy issue which stood out in analyzing the interview factors derived from this question was the typical respondent over-estimation of potentially resulting consequences which they feared would occur if they were implicated as the one responsible for the accident. Over 70% of the interviewees perceived that the consequences for "guilt" in causing certain accidents (we will term them "category B" accidents) would possibly range from; court-martial, dishonorable-discharge or paying for equipment damage. When discussing infraction and punishment correlations with those who administered such disciplinary measures it was discovered that the consequences for causing such "category B" accidents was usually a reprimand, extra training or occasionally restriction of privileges if gross neglect was involved. Thus, actual consequences were far less severe than perceived or anticipated consequences of "guilt" establishment. This phenomenon was especially true of younger interviewees (I.E., below the age of approximately 23).

The third most cited reason for inaccuracy in not reporting actual accident causes concerned the non-payment of damages or medical compensation via insurance, 15% of the survey respondents selected this answer. It was initially surprising that even as many as 15% selected this answer due to non-personal insurance nature of the organizational accidents researched I.E. within a military unit. The post-survey interviews clarified this somewhat. A majority of the interviewees (68%) indicated this answer was selected not so much as a concern over medical compensation or insurance payment of damages, relative to money exchanging hands; rather it was an expression of reciprocal concern that those found responsible for accidents would have to pay the "personal, job-based" costs mentioned in the previous answer concerning fear of blame, I.E. job loss, demotion, etc.

The last issue cited as contributing to non-reporting or inaccurate reporting was the desire to avoid involvement in the accident investigation (10% of survey respondents). Those interviewees who were aware of the reasons for selecting this response described instances of taking (or avoiding) accident reporting actions primarily to protect fellow workers from anticipated potential consequences of accident responsibility.

Conclusion

Initial (Phase I) research developed an Accident Cause Evaluation (ACE) protocol which featured respondent anonymity to discern any generic differences which might exist between actual and reported causes in organizational accidents. When the ACE was administered to a Military vehicle maintenance organization and extensive discrepancies were discovered between actual and reported causes, Phase II research attempted to determine what prompted the discrepancies.

Structured interviews which guaranteed anonymity were conducted with the survey respondents. The major factors which were identified in this trial research were the following summaries of why accident causes were not accurately reported when the causes reported were known to be different from the actual causes.

1. The reluctance to admit to complacency and supervisor - condoned or sanctioned complacent on-the-job procedures.

2. The predominately mistaken perception that consequences for accident responsibility were or would be far greater than what had traditionally been the case in reality.

3. The "social pressure" of protecting peers from potential anticipated consequences of accident responsibility.

The primary goal of developing a more objective device such as an improved form (document) for organizational accident investigation to uncover generic causes sometimes hidden from management was not achieved in this research. What resulted rather was a lesson in what should be accomplished from a management perspective in identifying and isolating actual accident causes.

First, recognition of and measures against potentially hazardous complacency in any form during daily work tasks. Complacency in terms of procedural short cuts, tool substitutions, and lack of detailed planning was the most frequently mentioned generic cause of accidents in the organization surveyed.

The consistent application of measures to prevent complacency is especially important at middle and foremen levels of management.

Secondly, the organizational atmosphere must be clarified and objectively understood by everyone down to the lowest level or newest employee relative to the non-punitive nature and need for accurate accident reporting. Employees must be made to understand that mistakes which cause accidents are tolerable and to some degree acceptable up to a point and that one of the best preventions of subsequent occurrences is accurate, open, honest identification of causes, whatever the reason. Management should devote necessary effort to whatever training or incentive programs are necessary to foster total "no-fault" honesty and thoroughness in accident reporting.

Hopefully, the two measures described above would remove a great deal of whatever constitutes the "social pressure" which operates in some instances to cover up or ignore actual accident causes in order to "protect" ones self or peers from the perceived potential of accident responsibility. Adequate management attention to issues identified in this research may have the potential of improving organizational safety via accurate identification of generic variables which cause organization-specific accidents.

Human Factors in Organizational Design and Management
H.W. Hendrick and O. Brown, Jr. (Editors)

THE RELEVANCE OF MORNINGNESS-EVENINGNESS TYPOLOGY IN HUMAN FACTOR RESEARCH: A REVIEW

Luciano Mecacci, Raffaello Misiti and Alberto Zani
Istituto di Psicologia del CNR
Roma, Italy

The review analyses the recent research on the individual differences in circadian rhythms (the so-called morningness-eveningness dimension). Variables like age, profession, personality and tolerance to shift-work have been particularly investigated. The relevance of the circadian typology in organizational design and management systems is discussed.

Recently there has been an increasing interest in the systematic investigation of inter-individual differences in circadian variations of physiological and behavioral responses, particularly in relation to the factors underlying the observed differences in tolerance to shift-work. Inter-individual differences in times of optimal performance and subjective alertness have already been known for long time. Kleitman (1963), for example, suggested the existence of "morning" and "evening" individuals, with intermediate types between the two extremes, according to an "early" and a "late" peak in body temperature and efficiency curves over the day.

A self-assessment questionnaire has been constructed to distinguish between the two extreme diurnal types. Öquist (1970) produced a Swedish language questionnaire. Östberg (1973a) modified the questionnaire to be used in a circadian rhythm study of food intake and oral temperature. Later on Östberg (1973b) redesigned the questionnaire for a shift-work study and claimed that this last version could point out the individual differences for suitability to shift-work: morning types do not adapt their habits to the needs of shift-work as readily as evening types and morning types have a more autonomous circadian rhythm. Finally, using the former questionnaires as a basis, Horne and Östberg (1976) designed an English language version, now universally used and translated in many other languages as Dutch (Kerkhof, 1981), German (Moog, 1981), French (Royant-Parola et al., 1980), and Italian (Mecacci and Zani, 1983). An other questionnaire has been constructed by Torsvall and Åkerstedt (1980).

The Horne and Östberg questionnaire consists of 19 questions on habitual bed and rising times, preferred times of physical and mental performance, and subjective fatigue after arising and before going to bed. The scores obtained in each single item are added together and the sum converted into a five point morningness-eveningness scale:

Type	Score
Definitely evening type	16-30
Moderately evening type	31-41
Neither type	42-58
Moderately morning type	59-69
Definitely morning type	70-86

Extreme morning and evening types represent in the whole the 5 per cent of the population (Posey and Ford, 1981; Mecacci and Zani, 1983).

The biological clocks of the two diurnal types have been shown to keep time in a different way producing a phase difference in their physiological indices and performance levels in motor and cognitive tasks. For example, in comparison to evening types, morning types have a phase advance in body temperature (Horne and Östberg, 1976), sleep behavior (Horne and Östberg, 1976, 1977; Webb and Bonnet, 1978), subjective alertness and concentration (Pátkai, 1971), subjective alertness and fatigue (Åkerstedt and Fröberg, 1976; Fröberg, 1977), and reaction times over the day (Kleitman, 1963; Fröberg, 1977; Kerkhof et al., 1981). Also the brain evoked potentials change over the day in relation to the diurnal type (Kerkhof, 1982). The Figure 1 shows how oral temperature and simple reaction times are affected

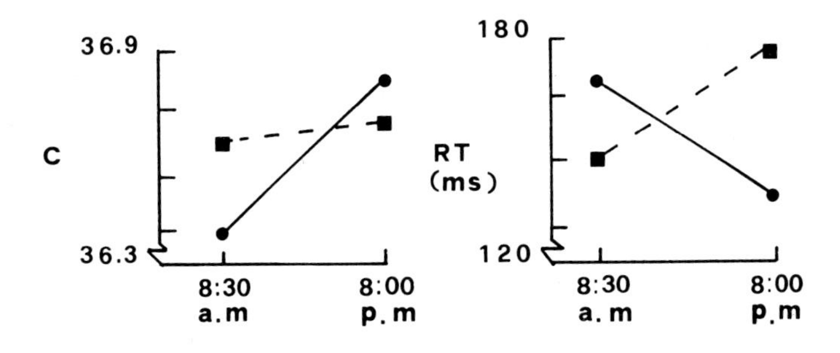

Figure 1
Diurnal variations of oral temperature (left) and reaction time (right) in morning (dashed lines) and evening types (solid lines). Redrawn from Kerkhof et al. (1981).

by the time of day and the diurnal type. In comparison to the evening types, morning types have a higher temperature and a shorter reaction time in the morning, whereas in the evening a delayed peak of temperature and shorter reaction time are achieved by the evening types.

The most of research on the morningness-eveningness dimension has been carried out on student samples. We wondered whether the acquisition of a regular job had some effects on the circadian typology (Mecacci and Zani, 1983). On the basis of a comparison of a sample of universitary students with a sample of white collars having a morning working schedule and the same mean age, a shifting towards morningness scores turned out in the worker sample. When morning and evening types are examined within a student sample, usually the following obvious difference in sleep-wake behavior is observed: morning types are the first to go to bed and the first to rise. In our worker sample, however, it was found that the difference between morning and evening types still remained, but with a remarkable attenuation. Since evening workers have a more delayed bedtime than morning workers and an early rise in the morning, they might suffer from a slight sleep deficiency. So in the evening individuals the acquisition of a regular job induces an advance in the preferred and actual sleep-wake behavior, whereas in morning individuals the working occupation would not have any significant effect on their sleep-wake habits, since they are fitted to the social working-rest schedule.

The influence of the occupation and life style on the circadian typology was investigated in athletes practicing several different sport activities at high levels of performance. We expected that the time of day when the match is usually played might be a significant factor in the circadian typology of the athletes. We compared 7 sport activities, from golf - a "morning" sport - to water-polo - an "evening" sport (Rossi, Zani and Mecacci, 1983; Zani, Rossi, Borriello and Mecacci, in press). A significant relation between the sport activities and the morningness-eveningness scores was found. A decrease in morningness scores was evident going from golf and shooting to volley-ball and water-polo (Figure 2). Answering to the question whether the phenomenon is due to the fact that practicing hard at different times of day, these athletes might develope different habitual activity patterns or, viceversa, it is due to inborn biological differences in the diurnal trends of arousal and fatigue, it is still difficult because of the few available data. We are now studying the relation between the diurnal type and the actual performance of the athletes in the competition.

The circadian typology seems to be dependent not only on the life-style and occupation, but also on the age. It is known that the sleep-wake habits change with aging: aged people are inclined to rise early in the morning and to distribute their sleep throughout the day (Tune, 1969; Webb, 1982). In a research on the morningness-eveningness preferences in aged people (Mecacci, Rocchetti, Zani and Lucioli, in preparation), it has been found a significant shifting towards the morningness scores (Figure 3). It is worth of noting that evening types were absent in our sample of aged people and in general they might be very rare.

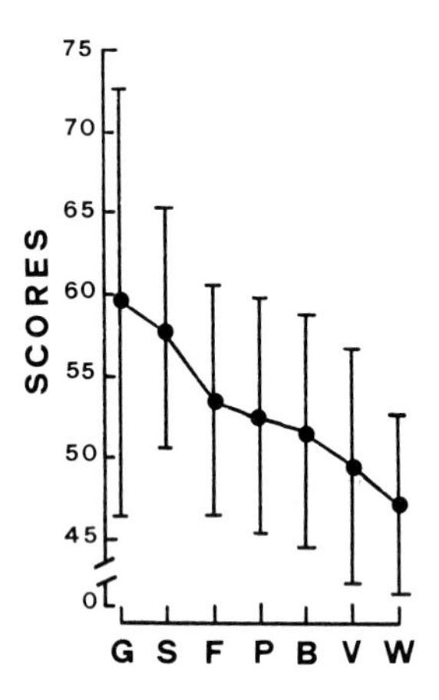

Figure 2
Morningness-eveningness scores in top level athletes.
G: golf. S: shooting. F: fencing. P: pentathlon. B: Basket-ball. V: volley-ball. W: water-polo.

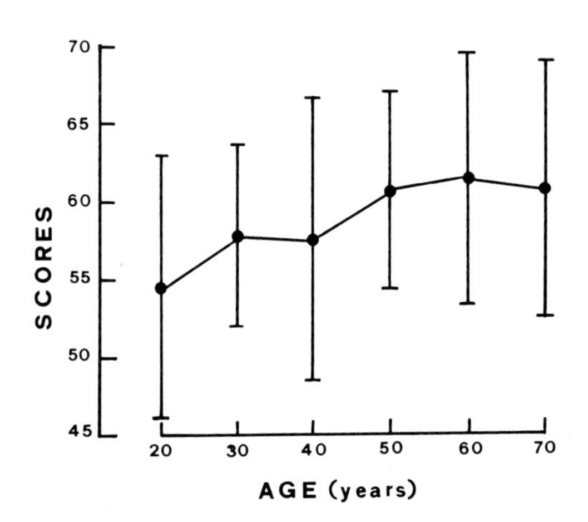

Figure 3
Morningness-eveningness scores in different age samples.
N= 435.

Summarising the main conclusions of our research on the morning and evening types, we may point out that the morningness-eveningness dimension is normally distributed only among student samples. When one considers a special sample in respect to the job or the habitual activity, the typology is usually skewed towards morningness scores (except in the case when an evening habitual activity is requested). The age is also a strong factor that should be carefully considered in shift-work schedules and planning.

We have finally to point out that the circadian typology has a relation to personality variables. Especially the introversion-extraversion parameters have been studied. The results are not completely in agreement. Introverts and morning types are higher in arousal in the morning in respect to extraverts and evening types whose arousal is higher in the late afternoon. If one correlates the scores on the two typologies, the correlation is not always significant. Some authors consider the circadian typology, in comparison to the introversion-extraversion typology, to be more predictive for that concerns the individual's diurnal trends (Östberg, 1973a, 1973b; Horne and Östberg, 1976, 1977; Kerkhof et al., 1981). In a pilot study on the personality profiles of morning and evening types in a worker sample (working schedule: 8-14 hours), we have found that no differentiation turned out in the introversion-extraversion parameters. However evening types, according to a psychosomatic inventory, showed a higher frequency of medical complaints than morning types.

Today it is well established that the shift-work causes a signficant reduction of well-being, particularly in connection with the night-shift. However it is obvious that this reduction is considerably different among the individuals. Recent research has shown that the morningness-eveningness dimension plays an important role in the speed of adjustment to the shift changes. In particular, Östberg (1973a, 1973b) found that the morning types had the most pronounced difficulty in adapting to the shift system. Hildebrandt and Stattman (1979) have shown that after about a couple of weeks of night work, the morning types hyperreacted when they were exposed to the inverse life pattern, having a greater deformation of pulse rate and body temperature than the evening types. These data are in agreement with the empirical observation that the most of the shift-workers who drop out the shift system are morning types (Aanonsson, 1964; Torsvall and Åkerstedt, 1979).

One aspect of the morningness-eveningness dimension remains completely unexplored. Except for one study on roommate relationships (Watts, 1982), no systematic investigation has been devoted to the question of the synchronization among individuals in the rhythms of their social and physical environments. Obviously, the quality of work life is affected by a not suitable agreement between the individual diurnal type and the working and activity schedules. However it should be stressed that this agreement is also related to the "circadian compatibility" of an individual with the other people acting in the same working and social environment.

Organizational and managerial systems should consider the factor of both the individual circadian typology and the inter-individual compatibility.

In conclusion, the morningness-eveningness dimension has emerged as an important factor in determining the individual efficiency in working and habitual activity. Some of the psychological and medical disturbances observed in workers seem to be dependent on the synchronization of the personal rhythms with the social schedules over the day. We may expect that the circadian typology will turn out to be a more crucial variable in affecting the quality of work life when one will consider the interpersonal relationships in the organizational and management systems.

References

(1) Aanonsson, A., Shift Work and Health (Universiteitsforlaget, Oslo, 1964).

(2) Åkerstedt, T. and Fröberg, J.E., Interindividual differences in circadian patterns of catecholamine excretion, body temperature, performance, and subjective arousal,Biol. Psychol. 4 (1976) 277-292.

(3) Fröberg, J.E., Twenty-four-hour patterns in human performance, subjective and physiological variables and differences between morning and evening active subjects, Biol. Psychol 5 (1977) 119-134.

(4) Hildebrandt, G. and Strattman, I.,Circadian system response to night work in relation to the individual circadian phase position, Intern. Arch. Occup. Environm. Health 3 (1979) 73-83.

(5) Horne, J.A. and Östberg, O., A self-assessment questionnaire to determine morningness-eveningness in human circadian rhythms, Intern. Journ. Chronobiol. 4 (1976) 97-110.

(6) Horne, J.A. and Östberg, O., Individual differences in human circadian rhythms, Biol. Psychol. 5 (1977) 179-190.

(7) Kerkhof, G.A., Brain potentials at different times of day for morning-type and evening-type subjects (Doctoral Thesis, University of Leiden, 1981).

(8) Kerkhof, G.A.,Event-related potentials and auditory signal detection: Their diurnal variation for morning-type and evening-type subjects, Psychophysiol. 19 (1982) 94-103.

(9) Kerkhof, G.A., Willemse-v.d. Geest H.M.M., Korving H.J. and Rietveld, W.J., Diurnal differences between morning-type and evening-type subjects in some indices of central and autonomous nervous activity, in: Reinberg, A., Vieux, N. and Andlauer, P. (eds.), Night and Shift Work: Biological and Social Aspects (Pergamon Press, Oxford, 1981).

(10) Kleitman, N., Sleep and Wakefulness (University of Chicago Press, Chicago).

(11) Mecacci, L., Rocchetti, G., Zani, A. and Lucioli, R., Morningness-eveningness preferences and aging (in preparation).

(12) Mecacci, L. and Zani, A., Morningness-eveningness preferences and sleep-waking diary data of morning and evening types in student and worker samples, Ergonomics 26 (1983) 1147-1153.
(13) Moog, R., Morning-evening types and shift work. A questionnaire study, in Reinberg, A., Vieux, N. and Andlauer, P. (eds.), Night and Shift Work: Biological and Social Aspects (Pergamon Press, Oxford, 1981).
(14) Öquist, O., Kartlaggning av individuella dygnsrytmer (Thesis at the Department of Psychology, University of Goteborg, Sweden, 1970).
(15) Östberg, O., Circadian rhythms of food intake and oral temperature in "morning" and "evening" groups of individuals, Ergonomics 16 (1973a) 203-209
(16) Östberg, O., Interindividual differences in circadian fatigue patterns of shift workers, Brit. Journ. Industr. Med. 30 (1973b) 341-351.
(17) Pátkai, P., Interindividual differences in diurnal variation in alertness, performance, and adrenaline excretion, Acta Physiol. Scand. 81 (1971) 35-46.
(18) Posey, T.B. and Ford, J.A., The morningness-eveningness preference of college students as measured by the Horne and Ostberg questionnaire, Intern. Journ. Chronobiol. 7 (1981) 141-144.
(19) Rossi, B., Zani, A. and Mecacci, L., Diurnal individual differences and performance levels in some sport activities, Perc. Motor Skills 57 (1983) 27-30.
(20) Royant-Parola, S., Benoit, O. and Foret, J., Diurnal variation of body temperature and "morningness", in Proceedings of the Fifth European Sleep Congress of the European Sleep Research Society (1980, abstract).
(21) Torsvall, L. and Åkerstedt, T., A diurnal type scale. Construction, consistency and validation in shift work, Scand. Journ. Work Environm. Health 6 (1980) 283-290.
(22) Tune, G.S., The influence of age and temperament on the adult human sleep-wakefulness pattern, Brit. Journ. Psychol. 60 (1969) 431-441.
(23) Watts, B.L., Individual differences in circadian activity rhythms and their effects on rommate relationships, Journ. Person. 50 (1982) 374-384.
(24) Webb, W.B., Sleep in older persones: Sleep structures of 50- to 60-eyar-old men and women, Journ. Gerontol. 37 (1982) 581-586.
(25) Webb, W.B. and Bonnet, M.H., The sleep pf "morning" and "evening" types. Biological Psychology 7 (1978) 29-35.
(26) Zani, A., Rossi, B., Borriello, A. and Mecacci, L., Morningness-eveningness preferences in athletes, Journ. Sport Med. (in press).

Human Factors in Organizational Design and Management
H.W. Hendrick and O. Brown, Jr. (Editors)

LABORATORY EXPERIMENTS ON VIGILANCE DECREMENT AND INTRADIAN CHANGES IN VISUAL RESPONSIVENESS

Lucia R. Ronchi
National Institute of Optics
6, Largo Fermi, Florence
50125, Italy

Antonina Serra
Chair of Physiopathological Optics
University of Cagliari
09100, Italy

Some data recorded by us in a "visual laboratory" are reviewed. We consider the effect of "sampling time" on a"secular" scale in a set of experiments ideally aiming at determining the"power spectrum" of operator's output. The concept of a "stable" human operator, whose visual responsiveness undergoes a complex ineluctable fluctuation could serve as a guide for the evaluation of the degree of intrinsic versatility of organizational design and management.

INTRODUCTION

A visual experiment, in general, is based on rigid rules aiming at minimizing the effects of some observer-related factors, which may obscure the stimulus dependencies of responses recorded at various outputs of the chain of stages of the visual pathway. In principle, the plot of a given response index versus time (by making abstraction from transient adaptive effects), for a "reliable" observer, is expected to consist of a set of data points, randomly scattered around a flat baseline. The smaller the entity of the scatter, the more skilled the observer is. Sometimes, to cope with day-today variability, "ratios" are taken, by assuming that changes in response level may be quantified through a multiplicative factor. During the past three decades (Ronchi, 1969; 1978, 1977) (Ronchi and Brancato, 1971) (Ronchi and Serra, 1979) we have been devoting some attention to a problem scarcely considered in the area of "pure" Physiological Optics: the effect of time, in a sense different from the duration of a light pulse or of a pre-adapting stimulus. Rather, we considered the influence of the duration of a session, its time distribution of stimulus presentations, as well as the time-of-day. Indeed, we noticed that an effort was needed to transfer the visual data gathered in a laboratory, under "aseptic" conditions, to the practice of offices. The attempt to fill-in a gap has been ecouraged by the technical progress in the tools at disposal of the operator. The advanced offices are becoming more and more similar to a visual laboratory, so that the optimization of working conditions becomes a part of a specialized organizational design. Now, in our opinion there is a point of basic importance: we do not expect that the observer is "stable"in the sense tacitly assumed in some kind of "pure visual research". Stability should be conceived in the broader sense of the word, being preserved even if sensitivity exhibits a steady-state time oscillation.

From literature reviews (Ronchi, 1972), it emerges that during the course of years, the existence of a wide gamut of periodicities of visual responsiveness have timidly suggested by a number of authors. The advent of Chronobiology (Halberg, 1969) has oriented towards the possible existence of a "circadian" rhythmicity. Some more-or-less fine intradian periodicities as well as the slower ultradian ones (Frazier et al. 1968; Globus et al.1971 1973; Klinker et al. 1972; Naitoh, 1973) could be regarded as the outcome of the combination of a set of (non linear!) interacting self-sustained oscillators, so that the resulting output may be interpreted in terms of carierrs and beats. However, a word of caution is needed, since biological rhythms and rhythms of performance are not necessarily monitored by the same "centers" (Ronchi and Barca, 1977). The old problem "from the single cell to the body as a whole" (Harker, 1964) will probably find an analytical acceptable solution in the studies on synergistic and cooperative phenomena.

Now, the ultimate experimenter's goal is still that of obtaining a power spectrum across a frequency range as large as possible. For this, responses are to be recorded at various successive instants. The problem consists in choosing the appropriate sampling interval. If too large, there is the risk of aliasing. If too small, there is the risk of an inter-stimulus interaction.

Some authors prefer to refer to a multiface concept like that of "fatigue". For istance, the decrement of performance at the end of the day, from another point of view, may be regarded as a branch of an intradian time dependence of performance. One of the puzzling aspects of this interpretation, however, is that it may even occur that no decrement in visual capability is reported, even under conditions of severe general weariness, subjectively reported.

Other aspects of "visual fatigue" are met, on different time scales. For istance, there is a very short-time aspect, related to the fading in the perception of a peripheral stimulus presented for a long time (Troxler's effect). Let us also recall that a colorimetric match persists only for a few seconds of fixation, and soon "Umstimmung" takes place (Verriest and Popescu 1974).

On a wider time scale, one finds the counterpart of "performance decrement" in situations of prolonged, uninterrupted, monotonous cyclical visual stimulation with near threshold light pulses.

In any case, the leading idea is that "fatigue" is a highly vulnerable feature, and it is a management task to monitor it.

EXPERIMENTAL FINDINGS

a) - Vigilance decrement for near-threshold signals.

Time-related phenomena in perceptual performances, in watchkeeping tasks, have been widely investigated during the past four decades (Mackworth, 1948; Colquhoun et Hockey, 1972). Briefly, the performance of even the most highly motivated observer will decline in time, if the situation is such as to involve brief, near threshold, irregularly occurring, infrequent signals with no feedback to the observer regarding performance. The above phenomenon has been investigated also in scotopic vision (Ronchi, 1970; Ronchi and Salvi, 1973). The data shown in Figure 1 refer to an extremely long, uninterrupted session (lasting 2 hours). Note that the "false" responses, for the considered observer are too few to warrant or even to permit analysis.

It is as if visual threshold would increase during the first, say, 50 minutes, while an oscillation is the main feature during the following hour.

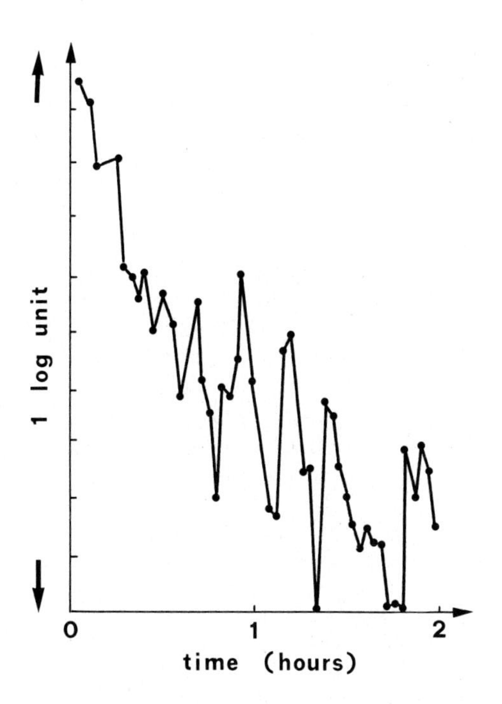

Fig. 1

Abscissae: time across the session, fixation being steadily maintained. Ordinates: filter density estimate, at absolute threshold. Pulse duration, 100 ms. Interpulse interval, 2 sec; retinal eccentricity stimulated: 10° nasally. Test-spot size: 10 min. of arc. Obs: CE, fully dark-adapted.

b) - Quantification of hits rate decay, and its diurnal variation.

The experimental situation considered in the present section is somehow similar to that of the previous one, although referred to more realistic conditions: the background is kept at a photopic level (100 nit), and the fixation is maintained for ten minutes only. Hit rate is defined as the proportion of signals detected across each block of, say, 15 presentations (covering 30 sec each). Hit rate decay is quantified by the slope of least-square justified line of best fit, to predict hit rates as a function of time in the test phase lasting 10 min. Now, time-of-day appears to be an important variable (Figure 2), the decay being greatest at the lowest point of circadian rhythm and zero in the middle of the day. It cannot be excluded that lighting factors may be important to ensure that comparisons of environmental conditions control for such effects as it is the case of circadian rhythms (Colquhoun,1970; Colquhoun and Hockey, 1972; Mackie, 1977).

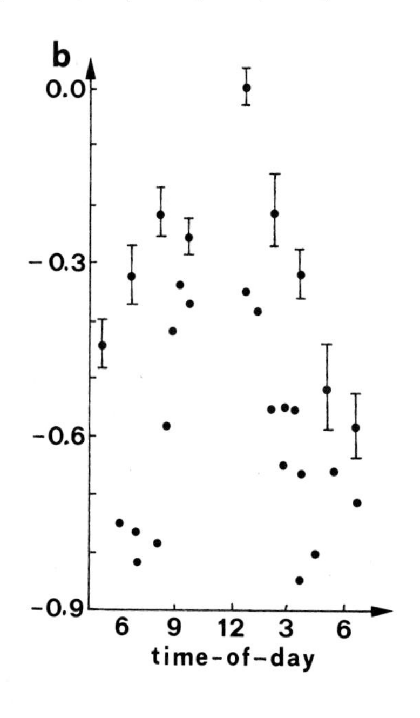

Fig. 2

Abscissae: time of day (hours)
Ordinates: slope (pulses per min), used to quantify hit-rate decay during prolonged watching.

Crosses: C-illumination.
Full points: low pressure Sodium illumination (of matched luminance)

Bars: Standard Deviation, for a sample of six observers.

c) - Inter-task and inter-individual differences.

The intradian pattern shown in Figure 2 is easily accepted since, as is said above, it conforms somehow to something basiç,like the circadian rhythm. However, the generality of the above finding is rather limited because of both the complexity of the intradian pattern and of individual differences (Ostberg,1973; Folkard,1975). For istance, in experiments where suprathreshold stimuli are used (like those concerning Figures 3 and 4),intradian multimodal patterns are obtained. We do not know whether this is bound to the peculiar inter-trial intervals there adopted, or else.

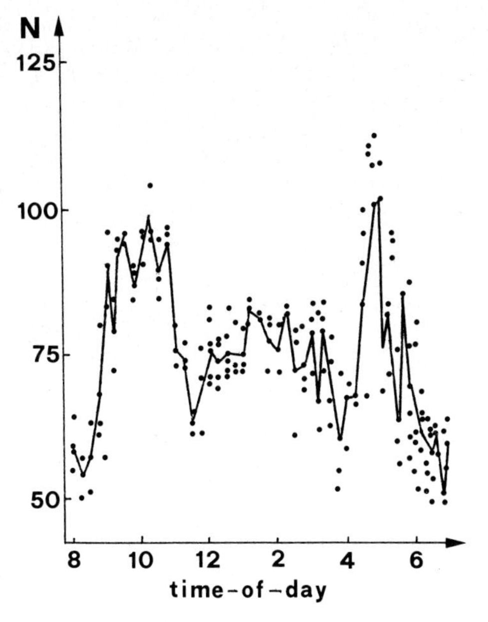

Fig. 3

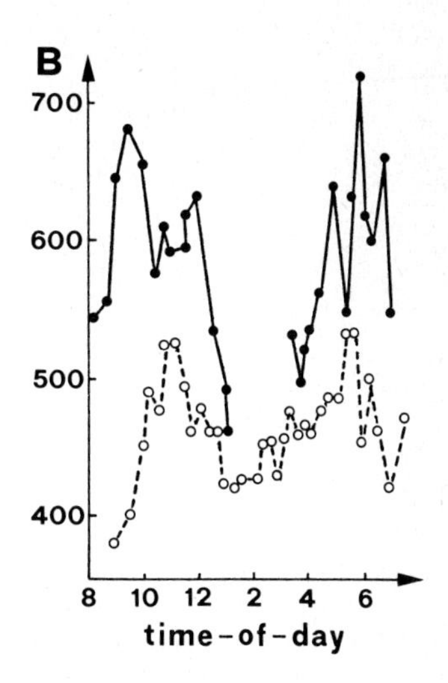

Fig. 4

Target: matrix of 10x15 Landolt rings, with suprathreshold break, in eigth different orientations.
Task: to identify and count rings with a pre-determined break orientation, across a given exposure time.
Abscissae: time-of-day, in hours.
Ordinates: number of correctly identified rings per presentation.
Intertrial interval: 15 minutes.
Obs.: MM. Data recorded in three different days are averaged (full line).

Stimulus: set of ten solitary flashes of given intensity.
Abscissae: time-of-day (hours)
Ordinates: amplitude of electroretinographic b-wave (uV), averaged across the set of ten flashes. Standard Dev. + 14 uV
Intertrial interval: no less than 10 minutes.
Obs.: LR (full line); AME, dotted line.

One of the basic pre-requisites upon which the reliability of the above data relies upon, is the rigorous control of retinal adaptation across the day. The ideal reference environment is a windowless building.

Now, the data shown in Figure 3 concern a problem historically related to the names of Dressler and Marsh, at the turn of the century, through Kleitman (1933), to Aschoff (1970), Folkard (1975) and Colquhoun (1971).The data shown in Figure 4 have been discussed for a long time. Although the diurnal change in electrical resistance (and hence conductance) of the body tissues on which the electrodes are applied does not seem the primary factor (Ronchi,1977),possible retinal changes are questionable (Ronchi, 1969).

Let us consider now an experiment where visuo-motor reaction time was recorded. The target was a line, printed on a paper running at a speed of 20 mm/sec. From time to time (at random intervals), a step-like jump occurred. The task consisted in depressing a lever as soon as the jump was perceived. The duration of each session was 40 minutes. The plot of reaction time (or response latency) versus time, across the uninterrupted session, shows a rising trend. Being fitted by an exponential law, the decrease in response speed is quantified in terms of a "time constant". Two subjects exhibit two opposite time patterns

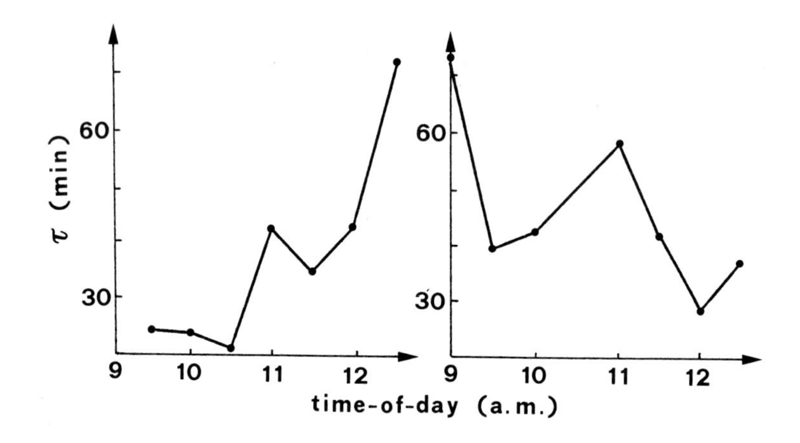

Fig. 5

Abscissae: time across the morning, in hours. Ordinates: the above defined time constant, in minutes. Data recorded in different days are put together. Obs: left plot, AZ; right plot, DD.

d) - The effect of an intertrial rest interval.

In an experiment where a near threshold stimulus is cyclically delivered for a long time, as is said above, the plot of perception probability versus time exhibits a decreasing trend, on which a damped oscillation is superimposed. If the observer is presented with a temporal pattern of stimulation, where blocks of flashes are separated by suitable intertrial intervals, the decay in performance is no longer expected. Let us recall that the problem of recommended pauses across a monotonous work has known industrial implications (Ostberg, 1973; Grandjean, 1979).
Now, some individual, in the absence of the day trend, persist in exhibiting a rhythmical change in their plots of response index versus time (see, for instance, Figures 6 and 7). For others, the rhythm does not concern the response index "amplitude", but, rather, the scatter of data points. In other words, the degree of observer's reliability varies more-or-less cyclically with time.

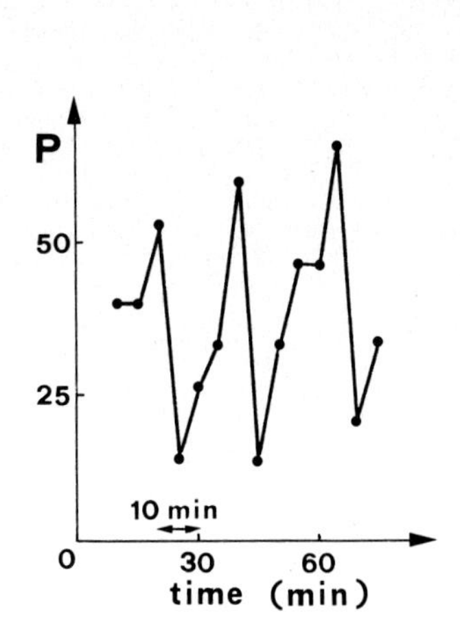

Fig. 6

Task: count the perceived flashes (of given intensity, close to absolute threshold), delivered at an eccentricity of 40°) Each block of 15 presentations (interpulse interval 2 sec), is followed by some minutes of rest.
Abscissae: time across the session, in minutes.
Ordinates: proportion of seen pulses.
Test spot size: 6 min. of arc.
Obs. O.N.

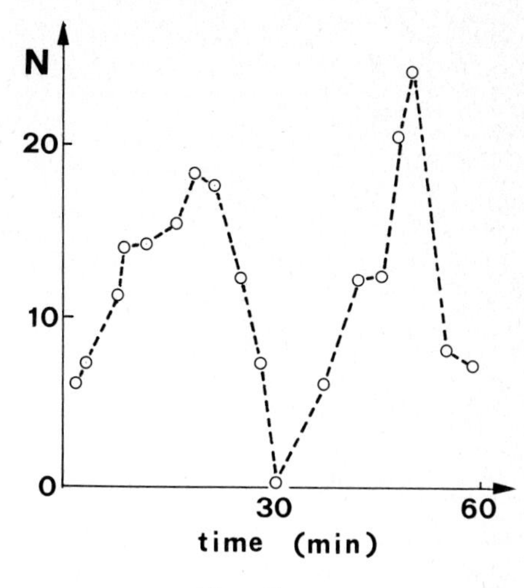

Fig. 7

Task: to arrange the caps of third box of F-M 100-Hue Test, the exposure time being 2.5 min., intertrial interval 2.5 min.
Ordinates: number of errors (score).
Illumination: C-, 100 lux.
Obs.: A.S. (binocular vision)
Ref. Ronchi, Stefanacci and Serra, 1981

e) - Experiments on refraction, by the use of a new Badal's Optometer.

The ergonomical solution of some problems concerning sight, labelled "Working Spectacles" (Ostberg, 1977), based on the correction of refraction at the working place, finds its support, amongst others, in some dependencies of accommodation and its meridional variability, on the spectral composition of illumination (Ronchi and Stefanacci, 1977; Ronchi et al. 1978;Ronchi et al., 1982), as well as in the correlation with task demands (Charman and Tucker, 1977), Illuminating Engineering being consequently involved (Wever, 1969; Faulkner and Murphy, 1973).
The experimental data obtained by us (Figures 8 through 11) aim at emphasizing the intradian dependencies of refraction estimates, including accommodation capability (through accommodative astigmatism) and "fatigue" (reduction in near point estimate after prolonged close work, for references see Ostberg, 1977).

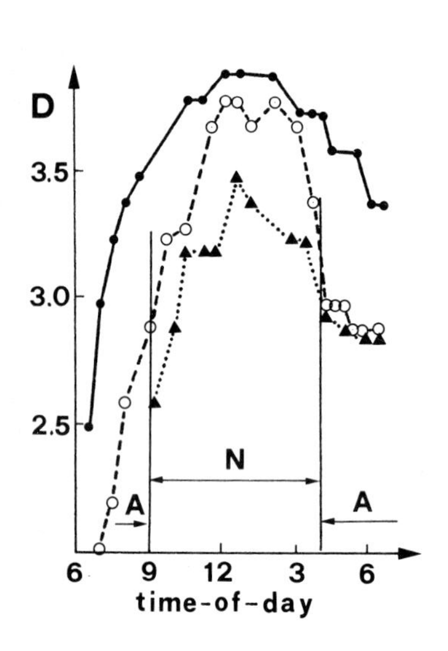

Fig. 8

Abscissae: time-of-day (hrs)
Ordinates: near point (dioptres)
Full line: constant lighting, 100 nit, 2.5 mm pupil. Dotted line, natural lighting, cloudy day. Broken line, natural lighting, clear day. Labels A and N stay for artificial and natural, respectively. Obs. LR.

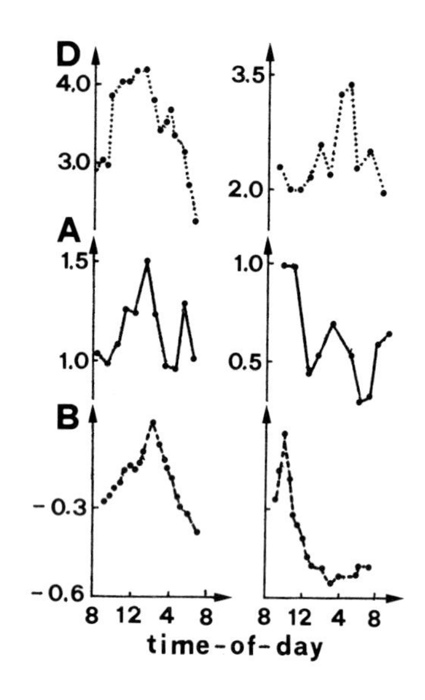

Fig. 10

Abscissae: time-of-day (hrs)
Ordinates: D, accommodation amplitude, in dioptres; A, accommodative astigmatism.
B (third plot), hit rate decay in a watch-keeping task.
Obs. LR (right); AS (left)

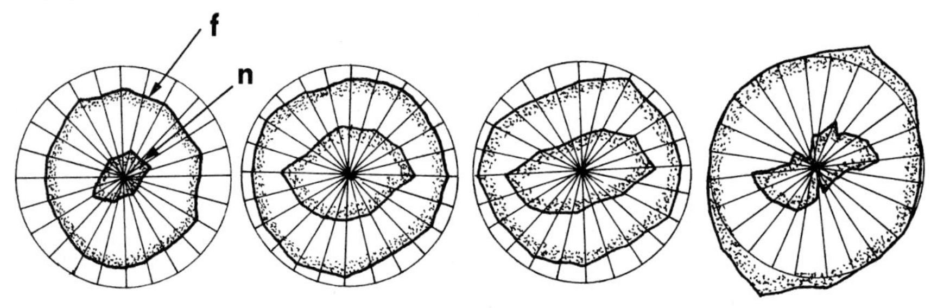

Fig. 9

Polar diagram showing the meridional dependencies of near point(n) and far point(f) estimates, at different times of day (4 hr interval), for a young adult, normal, individual (CM, 30). Note, amongst others, the intradian change in accommodative astigmatism, by comparing the shape of the n plot to that of the f one.

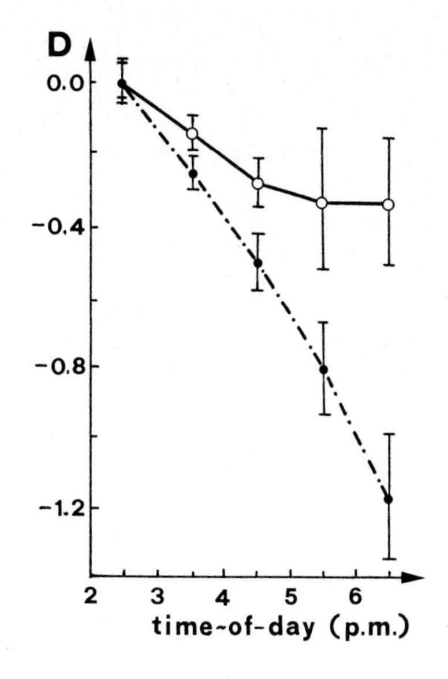

Fig. 11

Abscissae: time-of-day (hr)
Ordinates: loss of maximum accommodation power, expressed as a difference of near point estimates after and before a prolonged exposure to a cyclical sequence of near threshold pulses.
Duration of "fatiguing" exposure:
full line, 15 min.
broken line, 40 min.
Bars: Standard Deviation for repeat variability.
Obs. SIM

CONCLUDING REMARKS

An human operator, exhibiting a rigorously constant (visual) performance across every task, whatever the time of day, is unrealistic. The optimization rules should not concern only lighting and the viewing distance-correction of refraction paradigm: a proper time patterning of stimulus distribution is also needed. In the case of "fading", visual research has suggested some cyclical interruptions, to cope with the effects of some (probably distal) "time constants". The research on human factors has suggested intertrial rest pauses to cope with performance decrement, probably due to central factors. When this is done, there may remain a time oscillation in responsiveness, probably due to the (unvoluntary) shift of fixation, mediated by eye servosystems, calling into play, in succession, nearby retinal areas with slightly different response characteristics. Let us consider, now, the changes in performance across the day (cannot be excluded that they share some mechanisms with the physiological changes considered by chronobiologists). The question arises whether they are to be accepted as an ineluctable feature, or they should be counteracted, say, by improving correspondigly the viewing conditions. A general solution of this problem seems to be impossible, by considering both inter-individual and inter-task differences. "Averaging" visual data across a large sample of observes, unavoidably flattens out every plot. One might suggest that, whenever possible, once "fitted the task to the man" (as he is viewed from the designer), the "actual man" should let free to fit the task to himself, through the local versatility of proper tools. The recommendable ranges of this self-paced manipulation still wait to be standardized.

REFERENCES

Aschoff, J., Poeppel, E. and Wever, R.: J.Interd. Cycle Res. 1 (1970)105-10.
Charman, W.N. and Tucker, J.: Vision Res. 17 (1977) 129-39.
Colquhoun, W.P.: Human Factors, 12 (1970) 537-45.
Colquhoun, W.P. (Ed), Biological Rhythms and Human Performance (Ac.Press, New York, 1971).
Colquhoun, W.P. and Hockey, G.R. in:Colquhoun W.P. (Ed),Aspects of Human Efficiency, Diurnal Rhythms and Loss of Sleep (English University Press, London, 1972).
Faulkner, T.W. and Murphy, T.J.: Human Factors, 15 (1973) 149-54.
Folkard, S., in:Colquhoun W.P. et al. (Eds),Experimental Studies on Night and Shift Work (Westdeutscher Verlag, Opladen, 1975).
Frazier, T.W., Rummel, J.A. and Lipscomb, H.S.: Aerospace Med. 39 (1968) 383-7.
Frome, F.S., MacLeod, D.I.A., and Buck, S.L.: Vision Res. 21 (1981)1323-8.
Globus, G.R., Drury, E.A.,D. Phebus, E.: Perc. Mot. Skills,33 (1971)1171-6.
Globus, G.G., Phebus, E.C. and Humphries, J.:Aerospace Med. 44 (1973)882-6.
Grandjean, E.: Arbeitmedizinische und ergonomische Probleme der Arbeit an Bildschirmgeraeten (ETH Zentrum, Zuerich, Report 30 Maerz 1979).
Haeberlin, H., Jenni, A. and Fankhauser, F.: Int. Ophthal. 2 (1980) 1-9.
Halberg, F.: Ann. Rev. Physiol. 31 (1969) 675-85.
Harker, J.E., The Physiology of Diurnal Rhythms (Cambridge Un. Press,1964)
Kleitman, N.: Amer. J. Physiol. 104 (1933) 449-59.
Klinker, L., Kunkel, S. and Weiss, D.: Cycle Res. 3 (1972) 225-30.
Mackie, R.T. (Ed), Vigilace: Theory, Operational Performance and Physiological Correlates (Plenum Press, New York , 1977).
Mackworth, N.H.: Quart. J. Exptl. Psychol. 1 (1948) 6-16.
Naitoh, P.: Int. J. Chronobiol. 1 (1973) 223-8.
Ostberg, O.: Brit. J. Industr. Med. 30 (1973) 341-8.
Ostberg, O., in: Proc. 2nd World Congr. Ergophthal. (Stockholm, 13-16 June, 1977).
Ronchi, L., in: Santamaria L. (Ed), Research Progress in Organic, Biological and Medicinal Chemistry (North-Holland, Amsterdam, 1969).
Ronchi, L.: Vision Res. 10 (1970) 605-7:Space Life Sci. 4 (1973) 231-9.
Ronchi, L. and Brancato, R.: Ophthalmologica, 163 (1971) 189-98.
Ronchi, L., An Annotated Bibliography on Variability and Periodicities of Visual Responsiveness (Fond. G.Ronchi,Vol. 17, Florence, Italy, 1972).
Ronchi, L.: Rev.Sens.Disabil.29 (1977) 1-5;Med.Res.Engng.12 (1977) 20-4.
Ronchi, L. and Barca, L.,Biological Rhythms and Rhythms of Performance: An Annotated Bibliography (Fond.G. Ronchi, Vol.40, Florence, 1977).
Ronchi, L. and Salvi, G.: Ophthalmic Res. 5 (1973) 113-8.
Ronchi,L. and Serra,A., in:Proc. CIE 19th Session (Kyoto, 1979, P79-32).
Ronchi,L., Serra,A. and Meucci,R., in: Proc. Intern. Symp. Physiol. Optics (Tokyo, May 7-9, 1978).
Ronchi,L. and Stefanacci,S. in: Proc. 3rd LUX EUROPA (AIDI, Florence,1977)
Ronchi, L. et al.:Proc. AIC Meet.(Berlin,1981); Proc. Colour Dynamics 1982.
Verriest, G. and Popescu, M.P.: Mod. Probl. Ophthal. 13 (1974) 26-30.
Wever, R.: Pflueg. Arch. 306 (1969) 71-9.

Human Factors in Organizational Design and Management
H.W. Hendrick and O. Brown, Jr. (Editors)

THE MEASUREMENT OF HUMAN MENTAL WORKLOAD IN DYNAMIC ORGANIZATIONAL SYSTEMS: AN EFFECTIVE GUIDE FOR JOB DESIGN

N.Meshkati,[1] P.A.Hancock,[2] and M.M.Robertson.[3]

1. Department of Industrial and Systems Engineering
2. Department of Safety Science
3. Human Factors Department

University of Southern California
Los Angeles, California
U.S.A.

This paper reviews current methods of assessing human mental workload. The advantages and disadvantages of each approach are presented. Understanding and application of appropriate methodology is an integral component of job design. Further, workload assessment may help alleviate acute and chronic problems of work stress related illness. Also, the maintainence of a balance between workload requirements and an individual's capability can enhance system efficiency.

INTRODUCTION

Present day managers face the problem of distributing workload among their personnel in order to maximize system efficiency. Failure to match an imposed level of workload with an individual's capabilities may result in short-term acute effects of frustration and anxiety for the worker combined with an overall reduction in unit performance and product quality. If this imbalance between demand and ability persists for a longer period the chronic effects of this overload may be manifested in disorders such as ulcers of the gastrointestinal tract, hypertension and other heart ailments (cf., Welford, 1978). These in turn cause job burn-out, claims for work related stress injury (Yates, 1979) and a prolonged depression of system productivity.

The immediate questions of interest to managers are; How can an individual's reaction to workload be measured? and; How can we tell when an operator is either under or overloaded? Furthermore, the knowledge derived from mental workload analysis can be of value to management in the areas of job and system design, task allocation, the study of existing process operating methods, operator specification and selection, the development of system operating instructions, operator training and the planning of operator deployment in advance of system development (Kitchin and Graham, 1961).

When workload is defined in terms of physical demand there are several methods which satisfactorily index human response (cf., Astrand and Rodahl,

1977). However, when the load involved requires some form of mental or cognitive effort, assessment is considerably more difficult and it is this non-trivial problem which forms the focus of the present work. This study reviews four principal mental workload (MWL) measurement techniques and evaluates each of them from the view point of practicality and their utility for managers in job design.

PRIMARY TASK MEASURES

Primary task performance may be the most obvious method of workload assessment. If one wants to know how driving is affected by different loading characteristics e.g., traffic, fatigue or lane width, one should be able to utilize the driving performance as a criterion (Hicks and Wierwille, 1979). There are several methodological approaches to the measurement of performance or system output measures, which were discussed in detail by Chiles and Alluisi (1979). From the practical stand point the analytical approach appears most appropriate. Welford's (1978) concept of the analytical approach looks in detail at the actual performance of the task to be assessed, examining not only overall achievement but also the way in which it is attained. The advantage of this method is that the various decisions and other processes that make up performance are considered in the context in which they normally occur, so that the full complexities of any interaction between different elements in the task can be observed. For instance, it has been shown that the presence in a cycle of operations, of one element which has to be carried out more deliberately than the rest, slows the performance of all the elements in the cycle, so that a prediction of the time taken made on the basis of the time required to carry out each element in isolation would be too low.

This approach requires that several scores be taken of any one performance. For example, the component parts of a complex cycle of operations should be measured separately, errors may be recorded as well as time taken, and different types of errors need to be distinguished. Welford (1978) has argued that the greatest value of this approach is probably that it enables the more subtle effects of workload to be examined by showing the strategies in use, such as the balance between the speed and accuracy, and between errors of omission and commission, and methods of operation which in various ways seek to increase efficiency and to reduce excessive load. This approach has two major difficulties, first, the detailed scores required may be difficult to obtain for tasks such as process monitoring in which most of the decisions made do not result in any overt action, and second, that even where there is sufficient observable action, recording may have to be elaborate and analysis of results laborious.

Synthetic methods comprise another major approach to performance measurement as a MWL assessment technique. According to Chiles and Alluisi (1979) these methods start with a task analysis of the system in which the proposed operating profile is broken down into segments or phases that are relative homogenous with respect to the way the system is expected to operate. For each phase, then the specific performance demands placed on the operator are identified through task-analytic procedures. Performance times and operator reliabilities are assigned to the individual tasks and sub-tasks on the basis available or derived data. The information on performance times is then accumulated for a given phase and the total is compared with the predicted duration of phase. This comparison of time required with available can be employed as an index of workload.

Welford's (1978) approach to this method is relatively similar, he considered loads imposed by task demands (e.g., data gathering, choices, actions) which are separate in terms of time taken or other measures either in the laboratory of in artificially simplified work conditions. The total load is then assessed by adding the components together. This is the approach on which standard time analysis of manual work is based. For the assessment of mental workload based on this approach the work of Kitchin and Graham (1961) is considered of seminal importance. They questioned the applicability of conventional work measurement techniques to mental workload. They were able to show that those techniques could give a satisfactory quantitative expression of three types of mental activity on a time basis. First, mental activity involved in directing and coordinating muscular activity i.e., during all defined physical movements of the body, including highly manipulative work. Second, mental activity in perception (taken to mean the actual receipt of information by any of the senses) and finally, mental activity of the senses in searching for a random stimulus; in particular, alertness to pick up a random (but likely) signal demanding instant action. In summary, work-measurement, although predominantly a technique for measuring work which can be observed as being carried out physically, satisfactorily recognizes for practical purposes those mental activities which are an integral part of some physical activity and which are defined by the physical activity which they accompany in time.

The third major approach to MWL assessment via performance analysis is the multiple measures of primary task performance which is a composite technique. The technique might be considered useful for workload assessment, when individual measures of primary task performance do not exhibit adequate sensitivity to operator workload, who because of operational adaptivity due to perceptual style and strategy can compensate to variation. According to Williges and Wierwille (1979), using multiple measures in a combined analysis has the beneficial effect of reducing the likelihood that important strategy changes will go unnoticed. There are numerous studies that report both positive and negative results for the application multiple measures, which can be found in Williges and Weirwille (1979).

Although use of the multiple measures approach is potentially advantageous, it may have a detrimental effect similar to noise amplification. By measuring a large number of variables, it becomes more likely that some will not change reliably as a function of workload. Another area of concern is the differential sensitivities of the individual members of the multiple measures to the different aspects of the task. For instance, there might be two or more different measures, appearing almost equal in ability to discriminate change in operator workload, but may in fact have large differences in sensitivities. Consequently, this might cause disarray in scaling of the different workloads.

The lack of sensitivity of performance measures to the changes in mental workload levels is one of the major problems of these methods. This issue has been addressed by many authors such as Gaume and White (1975) and Gartner and Murphy (1976). Gaume and White argue that the level of mental workload may increase while the performance is unchanged, so that the performance may not be a valid measure of workload. Gartner and Murphy proposed that an operator may show equal performance for two different configurations, but in reality effort on one system may greatly exceed the effort on the other. Generalization to the different task situations poses an additional problem in the application of performance measures, since for each experimental situation a unique measure must be developed (Hicks and Wierwille, 1979). Williges and Wierwille (1979) also refer to this point and argued that the measures of performance of the primary task are task-specific. Each time a new situation is examined, new measures must be developed and tested. In several other techniques of workload assessment, the same measures can be used regardless of the application.

Williges and Wierwille (1979) cited numerous studies which support the same concept; namely that, no substantial change occured in the primary task as a function of workload. In general, the studies cited appear to have been performed at workload levels where the operator had sufficient reserve capacity to adapt to the increased load. Rouse (1979) also referred to long-term performance measures as the indicator of relative workload, although this would seem to provide at best an ordinal scale of workload measurement. Further, unless one is willing to assume that humans always operate to capacity and that all humans have the same capacity, the performance based workload scales would only reflect the states of particular individuals for which the data were collected. In other words, inter-individuals comparisons may not be valid for this measure.

SECONDARY TASK MEASURES

The concept of using a secondary task as a measure of MWL is grounded on the fundamental assumption of the limited channel capacity of human information processing system (Welford 1959, Kalsbeek 1964, 1973). This approach assumes that an upper limit exists on the ability of a human operator to gather and process information. The secondary task is a task which the operator is asked to do in addition to the primary task. There are two types of secondary tasks; 'loading' and 'subsidiary' or 'non-loading'. If the subject is instructed to aim for error-free performance on the secondary task at the expense of the primary task, the secondary task is called a loading task (Rolfe, 1973). In this case the operator must always attend to the secondary task so as to cause performance degradation on the primary task (Sheridan and Johannsen, 1976). If the subject is instructed to avoid making errors on the primary task, the secondary task is called a non-loading or subsidiary task. In this case the operator attends to the secondary task when time is available.

According to Knowles (1963), one of the best ways of measuring operator load is to have the operator perform an auxiliary or secondary task at the same time s/he is performing the primary task under evaluation. If the operator is able to perform well on the secondary task, this is taken to indicate that the primary task is relatively easy. If s/he is unable to perform the secondary task, and at the same time, maintain primary task performance, this is taken to indicate that the primary task is more demanding. The difference between the performances obtained under the two conditions is taken as a measure or index of the workload imposed by the primary task. There are two related but different reasons for using the secondary loading task. The first one, according to Knowles (1963), is to compensate for any deficiency in the loading of the primary task and to stimulate aspects of the total job that may be missing. Therefore, the secondary task is used simply to bring pressure on the primary task with the idea that as the operator becomes more heavily stressed, performance on difficult tasks will deteriorate more than performance on easy tasks. In this first application of the secondary task, the emphasis is upon stressing the primary task. Differences in operator workload are indicated by differences in the primary task performance measures taken under the stress induced by the auxiliary task. In the second application of the secondary task as a subsidiary task, the auxiliary task is used not so much with the intention of stressing the primary task as with the intention of finding out how much additional work the operator can undertake while still performing the primary task to meet satisfactorily the system criteria. Some examples of secondary tasks are arithmetic addition, repetitive tapping, choice reaction time, critical tracking tasks and cross coupled dual tasks. Varying combinations of these tasks have been employed by many investigators in their measurement of operator workload. Ogden, Levine and Eisner (1978) reviewed 144 experimental studies which used secondary task technique to measure, describe, or characterize operator workload. In addition to the secondary task techniques that have been reported by Ogden, his colleagues and Williges and Wierwille (1979), there are two other related (secondary task) techniques. The first of these is occlusion and the second is handwriting analysis.

The occlusion method of workload estimation is actually a time sharing technique. However, in occlusion, the time sharing is forced rather than voluntary. The operator is given time samples of visual information in performing the primary task, in other words, the time sharing is accomplished by suppressing the informational input (Hicks and Wierwille, 1979). The usual method is to block the operator's visual input from the display. Senders, Kristofferson, Levison, Dietrich and Ward (1967) and Farber and Gallagher (1972) used this technique in measuring the attentional demand of driving an automobile and reported some positive results. However, according to Hicks and Wierwille (1979) the occlusion method is not particularly sensitive and is more intrusive compared to other techniques in a simulated automobile driving situation. Handwriting analysis is another potential measure of MWL because of its deterioration due to distraction of the individual by other tasks. Kalsbeek and Sykes (1967) utilized handwriting as a secondary task while the primary task was to respond, via pedal depression, to a random series of binary choice signals which were presented to either the auditory or visual sense. They were able to show a step-by-step disintegration of writing performance provoked by the increasing number of binary choices.

The secondary task as a mental workload measurement technique has many shortcomings. Perhaps the most difficult aspect of secondary task methodology for assessing workload is intrusion. When the secondary task is introduced, performance on the primary task is known to be modified and usually degraged (Williges and Wierwille, 1979). This problem has been addressed by other authors such as Welford (1978), who regarded the extra load imposed by the secondary task as a factor that might produce a change of strategy in dealing with the primary task and consequently distort any assessment of the load imposed by the primary task alone. Brown (1978) argued that since the dual task method is essentially a resource-limiting device (e.g., human processing resources are limited), interference should occur within the processing mechanisms, rather than at the sensory input or motor output. He claimed there is empirical support for the idea that interference is maximal at the level of response selection. Brown indicated that the dual task interference is greater when the tasks share the same response modality than where responses occupy different modalities.

The nature of the primary task and its informational load and structural characteristics can cause problems of efficiency and a reduction of the utility of the secondary task. Workload may be largely a function of the structural characteristics of a task rather than of the informational load imposed by its component parts (Brown, 1978). Therefore, the more interesting tasks may be relatively inaccessible to study by the dual task methodology. This fact and other expected and unexpected interactions between certain tasks, drove Ogden and co-workers (1978) to point out that the choice of a secondary task is problematic (McCormick and Sanders, 1982).

The question of individual differences in secondary task performance has been addressed by some investigators through association with personality constructs. The introduction of an additional task may increase arousal which has been shown to affect differing personality types in contrasting ways (Gibson and Curran, 1974; Huddleston, 1974). Motivation is another related factor which may play an important role in secondary task performance (Kalsbeek and Sykes, 1967). Knowles (1963) attempted to provide a set of criteria against which to judge the desirability of a secondary task. These criteria included non-interference to primary task, ease of learning, self-pacing (in order for the secondary task to be neglected in maintainence of primary task performance) and compatibility with the primary task. Ogden and others (1978) added sensitivity and representativeness to the above set in order to address a wide range of human abilities and functions. Kalsbeek (1971) suggested the dual task method for use in two ways. First, in the traditional way, measuring the so-called spare mental capacity, and second, in the experiments where the main task, to which preference has to be given, is a simple or repetitive one. For instance, a binary choice task can be regarded as a stress condition in the performance of a secondary task.

The various problems outlined above, point to Brown's (1978) conclusion that; "the dual task method should be used only for the study of individual difference in processing resources available to handle workload... If it is so used, it should probably be in the form of an additional, secondary task, presenting discrete stimuli of constant load, on a forced paced schedule, and competing for processing resources only".

SUBJECTIVE RATING MEASURES

The subjective measures of MWL involve direct or indirect queries of the individual for his opinion of the workload involved in a task. The easiest way to estimate the mental workload of a person who performs a certain task, is to ask what s/he subjectively feels about the load of the task. Sometimes a list of key words or definitions describing different levels of load will be given. The subject then has to rate the load with reference to these levels (Sheridan and Stassen, 1979). Later Sheridan (1980) argued that MWL should be defined in terms of subjective experience and he continued; "subjective scaling is the most direct measure of such subjective experience".

The subjective estimates of the load have often been obtained through either the use of rating scales or interviews/questionnaires. A widely used rating scale in system evaluation is the Cooper and Harper (1969) scale. This scale was originally developed to measure the handling characteristics of aircraft by using the subjective reports of test pilots. Recently,

Wierwille and Casali (1983) introduced modifications to this scale in order to devise it for the purpose of assessing workload in systems other than those where the human operator performs motor tasks. Especially for systems which load perceptual, mediational and communication activities. The Cooper-Harper scale was designed primarily to assess flight characteristics and the descriptors of this scale pertain to "flyability" of an aircraft. Therefore, it can be applied to manual control tasks (Moray, 1982). However, according to Williges and Wierwille (1979), if this scale was used for workload assessment, the assumption must be made that handling difficulty and workload are directly related. The assumption is that if a pilot states that an aircraft is difficult (or impossible) to fly. This is equivalent to saying that the task of flying it imposes a very heavy or unsupportable load (Moray, 1982).

Wewerinke (1974) has confirmed the validity of Cooper-Harper scale as an indicator of MWL. He was able to report an extremely high correlation coefficient (0.8) between subjective difficulty rating and objective workload level in his study. Gartner and Murphy (1975) echoed the above idea and referred to the positive qualities of the Cooper-Harper scale as operational relevance, convenience, and unobtrusiveness. Some of the other examples of application of rating scales can be found in the work of Williges and Wierwille (1979).

Application of subjective rating techniques in strictly cognitive tasks has been addressed by Borg (1978) who was able to achieve a high correlation between subjective and objective measures of difficulty. The strictly cognitive tasks, unlike the manual control tasks and time stressed tasks (e.g. some signals must be processed before the processing of its predecessors is completed), are single-trial tasks. The subject is not under pressure associated with a continuous or arbitrary stream of signals that may arrive before s/he has finished dealing with an earlier signal. However, Phillip, Reiche and Kirchner (1971) argued, in the time stress task, it is not possible to differentiate unambiguously between the criteria of stress time and difficulty of the control task, based upon subjective rating methods. Therefore, the subjective feeling of difficulty in work-processing seems to be essentially dependent on the time stress involved in performing the task. Gaume and White (1975) in their study of mental workload, evaluated the subjective estimates of stress levels obtained during the experiment and they considered them as potentially valid and reliable indicators of MWL.

The second approach to the subjective rating is through the application of interviews/questionnaires. The procedures used in this approach are not as structured as rating scales. They range from completely open-ended debriefing sessions to self-reporting logs of stressful activities from carefully chosen questionnaire items (Williges and Wierwille, 1979). Usually, interviews and questionnaires have been used primarily as supplemental measures of other techniques. For example, Sherman (1973) demonstrated a high correlation between subjective measures of workload and various physiological measures. The advantages of this approach stem from its unobtrusiveness and extreme ease of application.

Since the subjective rating of the difficulty of a task is primarily a function of the raters' perception, the concept of 'perceived difficulty' has to be given importance and analyzed directly. Audley, Rouse, Senders and Sheridan (1979), proposed that the perceived difficulty of meeting the task demands should be the primary consideration and attempts should be made to dissociate this from other facets of subjective aspects of workload.

The perceived difficulty of a task might alter the human operator's attitude to it. This in turn could affect the time operators would be prepared to spend and the level of confidence in their decisions (Moray, 1982). The perceived difficulty of the individual is influenced by at least three groups of factors. The first group deals with the content of long term memory storage including both general experience and memories of similar tasks. The second group is the background factors such as personality traits, habits and general attitudes including likes and dislikes, aspiration and expectation levels. The third group of factors represents momentary conditions, e.g., one's emotional state, general fatigue, motivation and the importance ascribed to the task as well as by actual anticipated success or failure (Borg, Bratfisch and Dorine, 1971).

The subjective rating of task difficulty could also be affected by the situation and job as a whole rather than by only the task induced or individual rater's factors. Borg (1978) proposed that it is necessary to point out the set of factors which seem to cause the experience of the difficulty in one job which may be different from those in another job. Finally, it should be noted that individual differences in general and adaptivity of the operator to the system, the task and the resultant impression of the task as viewed by the operator, causes higher than normal rating (Williges and Wierwille, 1979).

PHYSIOLOGICAL MEASURES

Individuals who are subjected to a high degree of mental workload commonly exhibit changes in a variety of physiological functions. As a result, several researchers have advocated the measurement of these changes to provide an estimate of the level of workload experienced. The limit to the number of physiological mechanisms which have been used for such assessment has only been restricted by the ingenuity of individual investigators and in consequence, responses as varied as bodily fluid composition and muscular tension have been employed with greater or lesser success. In previous work, we (Hancock, Meshkati and Robertson, 1984) have attempted to differentiate the current physiological approaches on the basis of their validity as a measure and their practical utility in the applied work setting. From this analysis, the most practical method to emerge is a value for heart rate or it's derivatives (e.g., Kalsbeek, 1968) and the most valid measures are changes in brain activity associated with the presentation of specific stimuli, known as event related potentials or ERP's (e.g., Wickens, 1979). Further, we have suggested that a tympanic temperature measure, taken in the ear canal, can provide a potentially useful compromise in the trade-off between the concern for validity and practicality. The interested reader can find specific details of differing physiological measures in the useful review articles of Ursin and Ursin (1979), Weirwille and Williges (1979) and Meshkati (1983).

As with each of the previous methods of assessing response to imposed mental workload, physiological responses have certain inherent advantages and limitations. Principal among the advantages is the unobtrusiveness of the measure with respect to the primary task at hand. This means that the individual is not disturbed during normal activity. However, physiological measures require the imposition of some recording devices which themselves can range from complex and intrusive equipment to relatively simple and comfortable meters. An additional advantage is that physiological responses monitor an individuals involuntary reactions and provide somewhat superior quantification compared to subjective estimates. Despite these positive aspects, the relationship between mental workload and most physiological variations is tenuous.

The application of the physiological measures as indicators of MWL is influenced by a combination of several factors, such as the cost of both hardware and software to operate the equipment, the training level of the personnel who administer the physiological tests, environmental conditions of the workplace and willingness of the employees to be connected to a physiological recording mechanism. Due to problems caused by various combinations of these factors, there is at the present time the indication that physiological measures are amonged the least preferred methods of MWL assessment for managerial usage in complex person-machine systems. It should be noted, however, that technological innovations such as telemetric monitoring, has reduced some of the problems associated with these measures and use of these techniques in the working environment may be realized in the very near future.

DISCUSSION

In this brief review, four major MWL assessment methods, their applications, advantages, disadvsantages and feasibility from a scientific as well as managerial perspective have been presented. The subjective rating measure, due to it's apparent simplicity, ease of administration and interpretation is currently the most widely used technique for the measurement of non-physical workload of white and blue collar workers. However, as mentioned previously, the problem of perceived difficulty and it's resultant artifacts inherent in employee's responses are as yet unresolved issues (cf., Meshkati and Driver, this volume). Until these questions are answered, the subjective ratings of task difficulty should be analysed with extreme caution and it should include the use of multiple scaling techniques which are validated through cross-checking procedures.

One of the challanges to the present day manager is to be aware of and informed about the structure, content and interface of each job under his or her jurisdiction. Allocating tasks to the person and to the machine, the level of interaction between each component and the imposed MWL on the human operator adds a new dimension to the mangerial list of functions. In order to assist the managers to tackle these problems, there should be a close working relationship and regular dialogue between the manager and the Human Factors specialist who acts as a reference and provides technical expertise on MWL. Subsequently, the manager may effectively and efficiently use the knowledge concerning MWL in the design and redesign of job functions. This cooperation would lead to an increase in productivity, quality of work life and alleviate sources of stress which are of current concern in the managerial community.

REFERENCES

[1] Astran, P. and Rodahl, K. A text of work physicology (McGraw Hill, New York, 1977).
[2] Audley, R.J., Rouse, W., Senders, J., and Sheridan, T. Final report of mathematical modeling group, In: Moray, N. (ed.), Mental Workload, its theory and measurement (Plenum Press, New York 1979).
[3] Borg, G. Subjective aspects of physical work, Ergonomics, 21 (1978) 215-220.
[4] Borg, G., Bratfisch, O., and Dorine, S. On the problem of perceived difficulty, Scandinavian Journal of Psychology, 12 (1971) 249-260.
[5] Brown, I.D. Dual task methods of assessing workload, Ergonomics, 21 (1978) 221-224.
[6] Chiles, W. D. and Alluisi E. A., On the specification of operator or occupational workload with performance - measurement methods, Human Factors, 21 (1979) 515-528.
[7] Cooper, G.E. and Harper, R.P., Jr. The use of pilot rating in the evaluation of aircraft handling qualities. Moffett Field, California, National Aeronautics and Space Administration, Ames Reseach Center, NASA Tn-D5153, (April 1969).
[8] Farber, E. and Gallagher, V. Attentional demands as a measure of the influence of visibility conditions on driving task difficulty, Highway Research Record, 414 (1972) 1-5.
[9] Gartner, W.B. and Murphy M.R., Pilot workload and fatique: A critical survey of concepts and assessment techniques. Washington, D.C.: National Aeronautics and Space Administration, Report No. ASD-TR-76-19, (October 1975).
[10] Gaume, J.G. and White, R.T., Mental workload assessment, III, Laboratory evaluation of one subjective and two physiological measures of mental workload. Long Beach, California, McDonnell Douglas Corporation Report MDC J7024101, (December 1975).
[11] Gibson, H.B., and Curran, J.D. The effect of distraction on a psychomotor task studied with reference to personality, Irish Journal of Psychology, 2 (1974) 148-158.
[12] Hancock, P. A., Meshkati, N., and Robertson, M. M., Physiological reflection of mental workload. Proceeding of the 1984 International Conference on Occupational Ergonomics, Toronto, Canada, (May 1984).
[13] Hicks, T.G. and Wierwille, W.W., Comparison of five mental workload assessment procedures in a moving-base driving simulator, Human Factors, 21 (1979) 129-143.
[14] Huddleston, H. F., Personality and apparent operator capacity, Perceptual and Motor Skills, 38 (1974) 1189-1190.
[15] Kalsbeek, J.W.H., Measurement of of mental workload and of acceptable load: possible applications in industry, The International Journal of Production Research, 7 (1968) 33-45.
[16] Kalsbeek, J.W.H., Standards of acceptable load in ATC

tasks, Ergonomics, 14 (1971) 641-650.
[17] Kalsbeek, J.W.H., Do you believe in Sinus Arrhythmia? Ergonomics, 16 (1973) 99-104.
[18] Kalsbeek, J.W.H., and Sykes, R.N., Objective measurement of mental load, Acta Psychologica, 27 (1967) 253-261.
[19] Kitchin, J.B. and Graham, A., Mental loading of process operations: An attempt to devise a method of analysis and assessment, Ergonomics, 4 (1961) 1-15.
[20] Knowles, W. B., Operator loading tasks, Human Factors, 5 (1963) 155-161.
[21] McCormick, E.J. and Sanders, M.S., Human factors in engineering and design, (McGraw-Hill, 5th edition, 1982).
[22] Meshkati, N., A conceptual model for the assessment of mental workload based upon individual decision styles, Unpublished Ph.D. Disseration, University of Southern California (1983).
[23] Meshkati, N. and Driver, M.H., Individual information processing behavior in perceived job difficulties: A decision style and job design approach to coping with human mental workload. Proceedings of the First International Symposium on Human Factors in Organizational Design and Management, (Honolulu, Hawaii, August 1984).
[24] Moray, N., Subjective mental workload, Human Factors, 24 (1982) 25-40.
[25] Ogden, G.D., Levine, J.M., and Eisner, E.J., Measurement of workload by secondary tasks, a review and annotated bibliography, Advance Research Resources Organization, prepared under contract no. NAS2-9637 to the National Aeronautical and Space Administration, Ames Research Center, Washington D.C. (January 1978).
[26] Phillip, V., Reiche, D., and Kirchner, J. The use of subjective ratings, Ergonomics, 14 (1971) 611-616.
[27] Rolfe, J. M. The secondary task as a measure of mental load, in: Singleton, W.T., Fox, J.G. and Whitfield, D. (eds.), Measurement of man at work (Taylor and Francis, London, 1973).
[28] Rouse, W.B., Approaches to mental workload, in: Moray, N. (ed.), Mental workload, its theory and measurement (Plenum Press, New York, 1979).
[29] Senders, J. W. Kristofferson, A.B., Levison, W.H., Dietrich, C.W., and Ward, J.L., the attentional demand of automobile driving, Highway Research Record, 195 (1967) 15-33.
[30] Sheridan, T.B. Mental workload - what is it? Why bother with it? Human Factors Society Bulletin, 23 (1980).
[31] Sheridan, T. B. and Johannsen, G. (eds.), Monitoring behavior and supervisory control (Plenum Press, New York, 1976).
[32] Sheridan, T.B. and Stassen, H.G., Definitions, models and measures of human workload, in: Moray, N. (ed.), Mental workload, its theory and measurement (Plenum Press, New York, 1979).
[33] Sherman, M.R., The relationship of eye behavior, cardiac activity and electromyographic responses to subjective reports of mental fatigue and performance on a doppler

identification task, Master's thesis, Naval Postgraduate School, Monterey, California (September 1973).
[34] Ursin, H., and Ursin, R. Physiological indicators of mental workload, in: Moray, N. (ed.), Mental workload, its theory and measurement (Plenum Press, New York, 1979).
[35] Welford, A. T., Evidence of a single-channel decision mechanism limiting performance in a serial reaction task, Quarterly Journal of Experimental Psychology, 11 (1959).
[36] Welford, A.T., Mental workload as a function of demand, capacity, strategy and skill, Ergonomics, 21 (1978) 157-167.
[37] Wewerinke, P. H., Human operator workload for various control conitions, Proceedings of the 10th Annual NASA Conference on Manual Control. Wright-Patterson AFB, Ohio, (1974) 167-192.
[38] Wickens, C., Measures of workload, stress and secondary tasks, in: Moray, N. (ed.), Mental workload, its theory and measurement (Plenum Press, New York, 1979).
[39] Wierwille, W.W. and Casali, J.G., A validated rating scale for global mental workload measurement applications, in: Proceedings of the 27th Annual Meeting of the Human Factors Society, Santa Monica, California, Human Factors Society, (1983).
[40] Williges, R.C. and Wierwille, W.W., Behavioral measures of aircrew mental workload, Human Factors, 21 (1979) 549-574.
[41] Yates, J. E., Managing stress (AMACOM, a division of American Management Association, New York, 1979).

Human Factors in Organizational Design and Management
H.W. Hendrick and O. Brown, Jr. (Editors)

INDIVIDUAL INFORMATION PROCESSING BEHAVIOR IN PERCEIVED JOB DIFFICULTY: A DECISION STYLE AND JOB DESIGN APPROACH TO COPING WITH HUMAN MENTAL WORKLOAD

N. Meshkati[1] and M.J. Driver[2]

[1]Department of Industrial and Systems Engineering
[2]Graduate School of Business Administration

University of Southern California
Los Angeles, California
U.S.A.

Individuals with different information processing behavior react in predictably different manners to the imposed informational and environmental load of the job and organization. Their perceived (subjective) difficulty of the non-physical task which is a function of the above factors and their Decision Styles, could be estimated and determined. It has been demonstrated that unifocus subjects' perceived difficulty is significantly different from multifocus ones. Also, different dominant and back up Decision Style patterns exhibited completely distinct responses to the various levels of the mental workload.

INTRODUCTION

Numerous studies assert that excessive mental workload results in negative effects on human operators both in terms of performance and physical and psychological well-being. Welford (1978) considered two different kinds of symptoms as the result of an unbalanced mental load. First is the 'chronic overload' which appears to be implicated in many psychosomatic disorders such as ulcers of the alimentary canal, various neurotic symptoms, hypertension and other heart ailments. The second one is the complaints of the incumbents that after a few years the job 'gets on top of you' and results in a change of job. This is mainly due to the loads which become unbearable for reasons that are essentially emotional. Unbalanced workload is a potential source of stress and excessive stress tends to disrupt performance. Tikhomirov (1971) has cited some of the negative effects of stress as; narrowing the span of attention, forgetting the proper sequence of actions, incorrect evaluations of situations, slowness in arriving at decisions, and failure to carry out decisions made. Hyndman (1980) also considered stress-induced hypertension due to the mentally demanding occupational task as its potential hazard to health.

Since according to Strasser (1979) man is not, as we would like to assume, a factor which can be used (employed) and coupled with men and technical devices like a machine, without taking into account his wishes, necessities and individual requirements to communicate with colleagues, etc. Therefore, in order to keep the safety and performance at the optimal level, job demands and task characteristics should be determined. This knowledge contributes to better personnel selection and to a more efficient system design (Chiles and Alluisi, 1979).

This study tries to demonstrate three important facts. First; Information Processing Behavior (IPB) or Decision Style (DS) which stems from individual differences predictably influences perceived task difficulty and subjective rating. Second, each DS exhibits distinct psychophysiological response toward Mental Work Load (MWL). Third, at least one of the dimension of each job that should be taken into account, analyzed, and considered is the required IPB or DS of the ideal incumbent, in order to have the optimal job-person match.

THE DECISION STYLE MODEL

In terms of mental workload assessment, the human operator can be considered as an input-output system of information processing (Kalsbeek, 1968). This concept is also supported by Rohmert (1979) who described non-physical work as sensing, information processing, decision making and action generation. Therefore, MWL, information processing and decision making are closely related to each other.

The whole process of receiving data and converting it to an action is analogous to the conversion of raw material into a finished product in a production process. All the operations between the point of sensing the stimuli (data) and the point of taking a specific action can be regarded as information processing and decision making. However, since different individuals have different levels of conceptual structures (i.e., the way an individual receives, stores, processes, and transmits information) for processing information, then this process is fairly individualistic and is referred to as IPB (Schroder, Driver, and Streufert; 1967). The complexity of IPB of each person portraits the manner by which the individual decision maker seeks, acquires, evaluates, integrates and uses information in order to make a decision. The complexity of the IPB of an individual is primarily determined by his/her cognitive style (Alawi, 1973). Cognitive styles are defined as learned thinking habits. Intelligence tries to capture the upper limit of a person's thinking capacity; style tries to measure a person's typical method of thinking in a given situation. Cognitive styles are, therefore, not absolute in any sense. They can be modified by further learning, and there is no absolute best style (Driver, 1983). Cognitive style is an index of total personality systems and its functioning and development. Through the cognitive style, which is the representative of the individual's conceptual structure, behavioral variations and individual differences of decision makers can be explicitly identified and analyzed.

By employing this concept Schroder et al. (1967) and Driver and Streufert (1969) developed a human information processing model. This model suggests that environmental pressures (or load) systematically affect the complexity of information processing in persons following an inverted U-shaped function. Each individual can be considered to have a unique and consistent curvilinear information processing pattern (c.f., Driver, 1979 and Meshkati, 1983). Environmental Load which determines the amount of processed information is defined as the sum of the effects of four basic environmental factors: (a) information complexity in the environment, (b) Noxity or negative input, (c) Eucity or positive input, and (d) uncertainty (Driver, 1979).

Informational complexity is defined as the input to the individual which changes either probabilities or utilities perceived by him. Operationally, it is linked to such input aspects as number of message per time unit, complexity of message content, time pressure, number of people supervised, amount of reading, and a vast array of similar variables. Noxity is defined as the amount of negative input reaching to an individual; operationally, it has been linked to threat, failure, criticism, assault, and similar variables. Under too much or too little noxity, information processing systems close and become simpler (Schroder et al., 1967). Eucity is defined as the amount of positive input and has been operationally defined as praise, success, support and like. Although there are interpersonal variables which can enhance or diminish noxity and eucity, it is argued that particular values of these factors can be viewed as having widely held common impact on individuals (Driver, 1979). Finally, uncertainty refers to unpredictability of the situation.

The decision style model which is based upon the foregoing concepts has two basic dimensions; information use and focus. Information use refers to the amount and complexity of information actually used in thinking. Number of focus is defined as the number of alternative which are contained in the final solution. Focus is a continuous dimension ranging from unifocus to multifocus (Driver, 1979 and 1983). Both of the dimensions; information use and focus can be partitioned into categories for descriptive purposes. The information use dimension can be split at some point between two extremes. At one extreme are those individual who habitually use every piece of available and as much information as is compatible with non redundancy (termed maximizers). At the other end are those people who use just enough information to generate one or two useful alternatives (termed satisficers). The maximizer/satisficer dimension suggests low vs. high amount of integration, or the type and amount of connections between information units during analysis. The focus dimension has two extremes; the unifocus in which a single alternative forms the outcome, and multifocus in which many different options are included in the final answer. The unifocus style takes a given amount of data and connects it around a single solution or decision alternative, whereas the multifocus style takes the same amount of data and integrates it to several outcomes simultaneously or within a very short time.

By combining the dimension of information use and focus dimension the model seen in the following table can be generated.

Five Decision Styles

	Amount of Information Use		
Focus	**Satisficer**	**Maximizer**	
Uni-	Decisive	Hierarchic	Systemic
Multi-	Flexible	Integrative	

The Decisive Style (D) is defined as using only just enough data to generate a good enough answer which is irrevocably adhered to. Once a decision is made, it is final. There is no going back and reanalyzing data. This style is expected to be very concerned with speed, efficiency, consistency, and achievement of results. Although some people consider this style too rigid and simplistic, it is dynamic, strong, and reliable. The Flexible (F) Style also employs a minimum or sufficient amount of data and it is seen as one engendering use of enough data to reach one or two alternative conclusions which are constantly open to absorb new data and its re-evaluation, and generates new solution as needed. This style is associated with speed, adaptability, and a certain intuitiveness. The Hierarchic Style (H) presents a sharp contrast to the other two. This style shows a very high use of all available information to meticulously generate a single optimal plan of action - the "best solution". Then the best solution is often implemented using an elaborate contingency plan, but it is basically resistant to change. This style is seen as rigorous, analytic, precise, and even perfectionistic. The Integrative Style (I) also uses a large amount of information, but simultaneously generates a number of possible solutions for implementation. There is also a greater tendency to rely on creative synthesis rather than pure logic. This style is high inventive, empathic, and cooperative. It seems to no-Integratives that Integratives don't make decisions, yet in fact they do -- it is just that Integrative decisions are multipronged, often requiring many simultaneous actions. Systemic decision Style (S) combines features of Integrative and Hierarchic orientation. This style seems to operate at first as an Integrative-exploring all options; then they shift to a higher order schema to prioritize options more like a Hierarchic. They appear to be more methodical and careful than the Integrative yet more open than the Hierarchic style.

In conclusion the model proposes that there are five basic styles. Each person has acquired at least one basic or 'dominant' style that normally shows up under moderate amount of environmental load. For most individuals, a second or 'backup' style emerges in extreme load conditions, i.e., under- or overload. The overload or underload can be due to information load which refer to the total informational pressure emanating from the job environment which the job incumbent experiences during a period of time.

THE EXPERIMENT AND IT'S RESULTS

Experimental Procedures

A two factor experimental design with repeated measure on one factor was used. The 64 young male adult subjects were nested according to their dominant decision styles under this factor. The independent variables were MWL and DS. The dependent variables were Sinus Arrhythmia (SA) and Subjective Rating (SA) of task difficulty. The SA was defined according to Kalsbeek (1968), as the variability of the instantaneous heart rate and was quantified based upon his proposed method. The SR was obtained by employing the Constant Sum Method (Metfessel, 1947). The experimental procedure consisted of a rest period followed by three different randomized order loading tasks where each one was a descriptive case composed of different number of items of information and corresponding questions. The difficulty level of each task was controlled by the case complexity, description, number of items of information and their combinations/integrations in making decisions, number of questions, and time (pressure).

It should be noted that suppression of SA is taken as the indication of experiencing MWL and/or stress by the subjects (cf., Kalsbeek, 1968 and Meshkati, 1983).

Results

The Sinus Arrhythmia (SA) measure was able to detect a significant difference between rest on one hand and all mental load levels on the other hand. However, H's and D's decision styles did not show any significant difference in their SA, whereas F's and I's did. Also the D's started the experiment with a significantly lower (i.e., suppressed) SA in the rest condition and maintained it throughout the remaining parts of the experiment (Meshkati, 1983).

The subjective rating (SR) as an indicator of task difficulty was able to detect and exhibit significant differences among different mental load levels for H's and D's decision styles, whereas for I's and F's it failed to do so.

CONCLUSION AND DISCUSSION

The first and foremost conclusion that can be drawn is IPB or decision style of the individual plays a significant role in experiencing the burden of MWL (and under extreme conditions, feeling the effects of stress), as demonstrated by the behavior of the SA and SR dependent variables. Furthermore, since decision style model is an inclusive concept for relating personality to decision process, then by employing it, personality traits could be classified and incorporated in the psychophysiological and other MWL assessment methods in order to adjust them to the effects or artifacts, caused by individual differences.

The differences in subjective rating of unifocus vs. multifocus styles stemmed from their perceived difficulty of the mental tasks. Perceived difficulty of a strictly cognitive task is a partial function of the 'subjective complexity' of the subject, and also is influenced by the components of the Environmental Load factor (Borg, 1978 and Meshkati, Hancock, and Robertson, 1984). Since each subject, depending upon his/her decision styles, had different perception of the environmental load, then the different style-dependent subjective ratings were predictable. For instance, the SR of the H's and D's (i.e., unifocus) decision styles were fairly conclusive, whereas I's and F's (i.e., multifocus) decision styles ' SR were incoherent and associated with a large variance. The latter one might be attributed to several factors such as the inherent nature a multifocus decision style and being forced to report only a choice regarding the task difficulty. One could hypothesize that it is done through a random selection among all other generated choices. The other factor is caused by the scaling behavior of multifocus which might be multivariate, whereas their unifocus counterparts' linear scale.

The practical implications of these findings to the present day manager is twofold. First, since every sample of incumbents in any organization might contain some multifocus decision styles, then the reliability of self-reported aspects and characteristics of the task pertaining to it's difficulty is questionable and should be interpreted with extreme caution. Second, due to the fact of presence of the unifocus decision styles in the foregoing sample, utilization of (psycho-) physiological methods of MWL assessment should be based upon a thorough analysis of the composing decision styles of the individual members.

REFERENCES

[1] Alawi, H., Cognitive, task and organizational complexities in relation to information processing behavior in business managers, Unpublished Ph.D. Dissertation, University of Southern California (1973).

[2] Borg, G., Subjective aspects of physical work, Ergonomics, 21 (1978) 215–220.

[3] Chiles, W.D. and Alluisi, E.A., On the specification of operator or occupational workload with performance-measurement methods, Human Factors, 21 (1979) 515–528.

[4] Driver, M.J., Individual decision making and creativity, in: Kerr, S. (ed.), Organizational Behavior (Grid Publishing Inc., Ohio, 1979).

[5] Driver, M.J., Decision style and the organizational behavior: Implications for academia, The Review of Higher Education, 6 (1983) 387–406.

[6] Driver, M.J. and Streufert, S., Integrative complexity: An approach to individuals and groups as information processing systems, Administrative Science Quarterly, 14 (1969) 272–285.

[7] Hyndman, B.W., Cardiovascular recovery to psychological stress: A means to diagnose man and task? in: Kitney, R.I. and Rompelman, O. (eds.), The Study of Heart Rate Variability (Clarendon Press, Oxford, 1980).

[8] Kalsbeek, J.W.H., Measurement of mental workload and of acceptable load: Possible applications in industry, The International Journal of Production Research, 7 (1968) 33–45.

[9] Meshkati, N., A conceptual model for the assessment of mental workload based upon individual decision styles, Unpublished Ph.D. Dissertation, University of Southern California (1983).

[10] Meshkati, N., Hancock, P.A., and Robertson, M.M., The measurement of human mental workload in dynamic organizational systems: An effective guide for job design. Proceedings of the First International Symposium on Human Factors in Organizational Design and Management, (Honolulu, Hawaii, August 1983).

[11] Metfessel, M., A proposal for quantitative reporting of comparative judgments, Journal of Psychology, 24 (1947) 229–235.

[12] Rohmert, W., Human work taxonomy for system applications, in: Moray, N. (ed.), Mental Workload, Its Theory and Measurement (Plenum Press, New York, 1979).

[13] Schroder, H., Driver, M., and Streufert, S., Human Information Processing (Holt, Rinehart; and Winston, New York, 1967).

[14] Strasser, H., Measurement of mental workload, in: Moray, N. (ed.), Mental Workload, Its Theory and Measurement (Plenum Press, New York, 1979).

[15] Tikhomirov, O.K., The Structure of Human Thinking and Activity, Translated from Russian-language book, Moscow University Printing House, 1969; J.P.R.S. 52199 (January 1971).

[16] Welford, A.T., Mental workload as a function of demand, capacity, strategy and skill, Ergonomics, 21 (1978) 157–167.

Human Factors in Organizational Design and Management
H.W. Hendrick and O. Brown, Jr. (Editors)

THE EFFECT OF BLOOD ALCOHOL ON PERFORMANCE IN INDUSTRIAL TASKS OR WHEN THE RIGHT TO A PROFIT EXCEEDS THE RIGHT TO PRIVACY

Dennis L. Price
Department of Industrial Engineering and Operations Research
Virginia Polytechnic Institute and State University
Blacksburg, Virginia
U.S.A.

A series of seven studies are presented. These studies explore the effect of blood alcohol concentration on factors of interest in industrial work. The factors include incentives, task difficulty, work pacing, productivity and quality of performance. These studies contribute to the development of a data base to support conclusions about the right of management to monitor blood alcohol concentrations in employees and to require employee participation in Employee Assistance Programs.

INTRODUCTION

It may seem trite to some to perform research on the effects of Blood Alcohol Concentrations (BACs) on the performance of industrial tasks. There is, after all a considerable body of literature (Price, in press) which has examined the effects of BAC levels on performance of a large number of psychophysical elements. However, the research in the literature generally is on isolated elements rather than integrated tasks, except for the research performed on the task of driving a vehicle. Most tasks in industry integrate cognitive, psychomotor, information processing, and other factors in a complex way. The literature does not provide an adequate data base for management's present decision needs. Management is faced with a current crisis in the workplace in the use of alcohol and other drugs. The contention of management is that drugs in the bloodstreams of workers results in impaired performance that effects profitability. In a highly completitive marketplace, impaired performance is a serious matter. Some feel that management should have the right to require employees to submit to the sampling necessary to ensure that performance is not impaired and to require participation in Employee Assistance Programs where adviseable.

The problem is that management has little data that supports that there is a threat to profitability from the worker on the job with blood alcohol concentrations. The large body of literature on elemental responses effects may not provide convincing support for the need for management to supercede the employee's right to privacy in order to preserve the right of the company to a profit. Neither are the data on the alcoholic's increased absenteeism, insurance costs, mortality and illness rates directly applicable to the issue of the right of management to monitor employees physical state for drugs on the job. What is needed are data on the effects of drugs on the performance of industrial tasks.

It would be ideal to obtain such data directly from tasks performed in industry. This author's efforts to obtain such data have been thwarted by organizations and persons who see their right to privacy invaded by such research. It is necessary, therefore, to obtain the data in the laboratory by simulated industrial tasks. The decision to monitor workers in the workplace should be made with the assistance of a data base of support. The series of studies discussed below are designed to contribute to such a data base. Even without sufficient data on on-the-job performance, estimates of costs to industry run as high as 50 billion dollars annually. Blood Alcohol Concentrations for management and workers sufficient to impair performance on the job may involve as much as 20% of employees (Price, in print). There could be cause for concern, but relevant data are needed.

The following summarizes the research completed to date. The research uses Blood Alcohol Concentrations (BACs) of .09% or less, because it is assumed that workers with levels in excess of this amount would not remain on the job.

SUMMARY OF RESEARCH

The research conducted includes seven studies using tasks which on face validity are similar to tasks conducted in industry. Each study was a repeated measures design which includes a placebo level. Counterbalancing techniques were used.

The first study (Price and Hicks (1979)) utilized micromotion analysis of twist, grasp, and return motion elements to evaluate the effects of trace through .09% BAC levels during the assembly of six water faucets. Phase I results showed significant treatment effects at $p < 0.05$ using the Page nonparametric test for the overall elapsed time scores and for twist, grasp, and return micromotion work elements. There was a 2.9% increase in overall elapsed time with a mean blood alcohol content of .023%. Phase II results were significant at the $p < 0.05$ for increased assembly time and for increased times to perform twist motions. The assembly time at .04% BAC increased by 3.5% and at .09% BAC at 19% compared to the placebo level. Quality errors were too few in number to consider.

The second study (Price and Liddle (1982)) was of welding performance. The dependent variable of greatest sensitivity in this task was electric current. The task was arc welding. This task is especially common in certain Virginia industries. The flow of electric current is determined by the size of the gap maintained. The steady maintenance of the proper gap is considered essential for a quality weld. The measure used was a count of current reversals which increased by 3.11% at the .06% BAC level and 15.76% at the .09% BAC level.

The third study (Price and Flax (1982)) was a study of drill press operation which used both incentive and task difficulty as independent variables in addition to BAC. The factorial design included four levels of BAC, eight levels of task difficulty, and two incentive conditions. The incentive conditions were monetary rewards for speed while maintaining quality and for quality while maintaining speed of performance. These were examined for their effects hits and misses. Hits were defined as inserting the drill probe within a hole in a template without touching the probe to the template. Misses were the number of probe strikes against the template. Task difficulty was defined by an adaptation of Fitts Law. The number of hits decreased significantly ($p < 0.01$ using Tukey's HSD Test) by 12% and 19% at the .06% and .09% BAC levels respectively. There were no interaction effects between BAC levels and task difficulty or incentive conditions.

The fourth study (Barbre and Price (1982)) used a punch press simulator. This study maintained the subjects at a given BAC for either 27 or 97 minutes before the experimental trials. Performance under the longer duration was a higher productivity by 4.53% overall ($p < 0.03$). There was a 32% loss in productivity at the .09% BAC level, a 26% loss at .07%, and a 16% loss at .05% BAC when compared with the placebo level.

The fifth and sixth studies were conducted simultaneously using the same subjects and counterbalancing the presentation of the two studies' trials. The fifth study (Radwan, Price, and Tergou (unpublished manuscript)) was a simulated electronic assembly and test task. It involved selecting the correct resistors from a number of resistors and placing them on a grid board in specific locations. The task was paced by a conveyor belt which provided materials to the subject at either 100% of their predetermined speed capacity for the task, or 75%, or 50%. The assembled board was then placed on a holder and an analog meter was adjusted to a specified value. The average loss in productivity at .05% BAC was 12%, at .07% BAC was 22%, and at .09% BAC was 33%. These differences were significant at $p < 0.05$ for the Student Newman Keuls test. Quality of performance also degraded as BAC increased. At the .09% BAC level resistors placed out of position increased by five fold, resistors placed out of orientation increased by three fold, improper selection of resistors increased by more than ten fold, and incomplete units by a factor of four compared to the placebo. Gender made a difference for all variables and there were gender by BAC interactions for all dependent variables. In general, females performed faster and their performances were more sensitive to alcohol levels. There was an alcohol by pacing interaction. Performances were more sensitive to alcohol at the 50% pacing level than at either the 75% or the 100% level.

The sixth study (Price, Radwan, and Tergou (unpublished manuscript)) used the same task as that described above, but was unpaced, and two incentive conditions were used, one emphasizing speed and the other accuracy of work. In general, productivity declined by 41% at the .09% BAC, by 24% at .07, and by 15% at .05% BAC compared to the placebo. The quality errors increased by a factor of two to three compared to the placebo condition. There was an incentive by alcohol condition interaction in which the number of correctly completed units and the number of improperly selected resistors showed greater impaired performance under the speed incentive than under the accuracy incentive.

The seventh and last study (Barbre and Price (1983)) completed at this time used a touch-entry equipped CRT to present alphanumeric arrays in which a target character was embedded. Participants visually searched randomly generated 108 character arrays for the predefined target, touching the CRT surface at the target locations where discovered. Half of the search trials used arrays containing no target; therefore, the subject had to decide to give up the search at some point, if no target was located. A monetary incentive/penalty system was used to define low- and high-criticality search trials. Search time, touch accuracy, the number of trials completed, the percent of incorrect "give-ups", and hand travel time were all significantly degraded by the alcohol dosages used. An alcohol by criticality interaction was observed for percent give-ups, and an alcohol by target presence interaction was significant for mean search times. Unexpectedly, mean search time decreased with higher BACs ($p < 0.0001$) by 31 percent. The percent hits, or accurate touches, decreased by 12.3% at the .09% BAC level ($p <$

0.0006) compared with the placebo. The mean distance of touch location from target location more than doubled at the .09% BAC level ($p < 0.0002$). Percent incorrect give ups more than doubled (from 3% to 7.5%) at the .09% level compared with the placebo condition of ($p < 0.0001$). The number of search trials completed was reduced by 10.5% at the .09% BAC condition compared with the placebo ($p < 0.0005$). Increased BAC had a greater degrading effect on percent incorrect give ups when the error criticality was high ($p < 0.0006$). Suprisingly, higher BACs had little effect on search times when a target was in the scene, but resulted in a significant decrease in search time when no target was present.

CONCLUSIONS

Even through the research is still in progress, the evidence indicates that significant losses in productivity and quality of performance do exist. Productivity losses may be as great as 50% for some tasks at levels of BAC which might occur in the workplace. Errors which degrade quality might increase as much as ten fold. Incentives might not be as effective as they would otherwise be. The pace of the work might increase the effects of BAC impairment. Employees might take greater risks and make more mental errors. The evidence generally supports the idea that BACs above normal will impact performance to such an extent that profits could be substantially effected.

REFERENCES

[1] Barbre, W. E. and Price, D. L., Effects of maintained blood alcohol concentrations and task complexity on the operation of a punch press simulator, in: Proceedings of the Human Factors Society 26th Annual Meeting (1982) 916-919.

[2] Barbre, W. E. and Price, D. L., Effects of alcohol and error criticality on alphanumeric target acquisition, in: Proceedings of the Human Factors Society 27th Annual Meeting (1983) 468-471.

[3] Price, D. L., Effects of alcohol and drugs on human performance, chapter in: Human Factors in Accidents and Litigation (Garland Publishing, Inc., New York, in press)

[4] Price, D. L. and Hicks, T. G., The effects of alcohol on performance of a production assembly task, Ergonomics 22(1) (1979) 37-41.

[5] Price, D. L. and Liddle, R. J., The effect of alcohol on a manual arc welding task, Welding Journal 61(7) (1982) 15-19.

[6] Price, D. L. and Flax, R. A., Alcohol, task difficulty, and incentives in drill press operation, Human Factors 24(5) (1982) 573-579.

[7] Price, D. L., Radwan, M. A. E., and Tergou, D., Gender, Incentive, and Alcohol Effects on an Unpaced Manual Assembly Task, unpublished manuscript.

[8] Radwan, M. A. E., Price, D. L., and Tergou, D., Gender, Alcohol, and Pacing Effects on an Electronics Assembly Task, unpublished manuscript.

THEORETICAL FACTORS

Human Factors in Organizational Design and Management
H.W. Hendrick and O. Brown, Jr. (Editors)

THE ORGANIZATION AND MANAGEMENT OF HUMAN FACTORS AS AN INFLUENCE ON SYSTEM DESIGN

Thomas B. Malone, Ph.D.

Carlow Associates, Inc.
8315 Lee Highway
Fairfax, Virginia 22031

This paper describes the process of applying human factors in a system development effort. It begins with a discussion of the system acquisition process, focusing on the characteristics of the systems approach. This leads to a description of the role of human factors in system acquisition, centering on the relationships between human factors technology and system personnel readiness. The human factors design process in system acquisition is then described and the organization and management of a human factors program within a system acquisition effort is discussed. Finally, the benefits of applying human factors within a system acquisition effort are listed.

INTRODUCTION

A major problem with the application of human factors today is the fact that, all too often, human factors specialists focus on the design of equipment and software interfaces, and ignore the application of human factors to systems. Even if a human factors effort is mounted in the acquisition of a system, it is usually relegated to a supporting role. In such a role, human factors people put finishing touches in the system, and ensure that it's design does not violate human factors principles.

If a system includes people, human factors must be applied to its acquisition. This application must place human factors in a directive rather than a supporting role. In such a role, human factors influences design. By influencing design, human factors controls the determination of the roles and requirements of people in the system, and it influences the directions that they overall design concept will take. When fundamental design decisions are made, and the design philosophy for the system is established, then human factors can support the conduct of the design of equipment and software interfaces.

The effort to ensure that human factors does influence the design of systems requires the organization and management of a human factors program. Such a program includes the process for applying human factors, the data base, the personnel, and the resources required to integrate human factors with system engineering.

OVERVIEW OF SYSTEM ACQUISITION

The primary characteristics of system acquisition are that 1) a system is acquired through application of the systems approach, 2) that the acquisition is accomplished in a phased process, and that 3) specific development milestones are associated with the termination of each phase. To state that a system is acquired by application of the systems approach seems to be immediately obvious. Yet, it is not always apparent just what the systems approach means, what it involves, what it demands, and how it is applied.

In an application of the systems approach in a system design effort, a design process must possess certain specific characteristics. First of all, it must be interdisciplinary, drawing from the expertise and methods of a variety of scientific and engineering disciplines. A system, comprising an array of different components, interfaces, capabilities, and operations, will be optimal when designed by a team of professionals with interests and expertise spanning these system elements. The system becomes the synthesis of these multifaceted points of view.

Second, the systems approach requires consistent and constant attention to designing in terms of requirements. The comprehensive identification, analysis and integration of requirements drives the development of system concepts, and the selection of criteria to evaluate the effectiveness of alternate concepts. In its fixation on requirements, the system approach guarantees that system objectives will be met, that system constraints will be accommodated, that mission requirements will be satisfied, and that the system will, in final form, do what it was intended to do. System requirements of primary importance are: 1) information requirements, what the system must know; 2) performance requirements, what it must be able to do; 3) decision requirements, what decisions it must make; support requirements, how it must be supported; and 4) interface requirements, how the components and subsystems must be connected.

This brings us to the third of our characteristics of the systems approach, the concern for interfaces. A system is, by definition, an orderly combination or arrangement of parts or elements into a coherent whole, such combination following a rational principle. The rational principle is usually that the system output will progress directly from system input according to defined laws or rules. The basis for the orderly combination or arrangement of parts or elements (subsystems and components) is the design of interfaces or links among the elements. Design of these interfaces is a critical step in the development of a system.

A fourth characteristic of the systems approach is that it is integrative; that it synthesizes and aggragates the requirements associated with the parts into the whole. It is through the integration of system requirements that conflicting and competing demands are resolved, and that parts or elements are combined into the coherent system.

The fifth characteristic of the application of the systems approach is that it is iterative, or characterized by repetition at successfully greater levels of detail. A system analysis effort always begins at the most general level of specificity, and then proceeds iteratively to the most specific. Thus, missions are iterated into phases, which are broken into functions, each of which is analyzed into subfunctions, which lead to tasks, and then to subtasks, to task elements, to activities, etc. In the same way, the concept for the system is iterated into concepts for subsystems, components and elements.

The sixth and final characteristic of the systems approach is its emphasis on test and evaluation. Throughout the application of the systems approach, developed concepts and criteria are constantly subjected to a rigorous evaluation process, such that the system which finally does emerge is truly tested and proven. Specific test and evaluation criteria vary as a function of the objectives of an individual test effort, but the major system level criteria areas must include: affordability - are system costs affordable; effectiveness - can the system meet its requirements; availability - is the system sufficiently on-line to meet its objectives; supportability - can the system be supported, supplied, and manned; and reliability can it be counted on.

System acquisition also requires that the systems approach is applied in discrete phases. The system acquisition process usually relegates certain specific activities to each phase. The usual phases, and major requirements of each, are as follows:

- Program initiation phase - selection of general design directions
- Conceptual phase - development of system concepts
- Preliminary design - development of subsystem concepts
- Detail design - development of system, subsystem and component level design criteria
- Production - development of system prototypes and final products
- Deployment - bringing the new system to market
- Operation - conduct of intended system operations in the real environment, until replacement by a new system

Milestones to be achieved at the end of each system acquisition phase are implied in the general requirements associated with each phase.

OVERVIEW OF THE ROLE OF HUMAN FACTORS IN SYSTEM ACQUISITION

Designers of systems must consider system personnel as a major component, equal in importance to system hardware, software, logistics, and procedures. When requirements of personnel who operate, control, maintain, and manage systems are not considered in system design, system effectiveness nosedives while system costs soar. The GAO recently reported that over half of the failures in military systems were due to human errors. Human errors are more likely when limitations of people have not been considered in the design of systems. More than half of the military's annual budget goes to manpower, personnel and training costs, which costs represent, in part, the cost of failing to consider the requirements of system personnel in system design. A number of catastrophic, and almost catastrophic events, such as the Three Mile Island accident, the misnavigation of KAL 707, the loss of the Apollo I astronauts, and the thousands of deaths that occur on our nation's highways every year, point to the proven fact that, if you fail to design for people in the system, the people will fail when you need them.

Human factors is the discipline that seeks to integrate people into systems. Human factors views people who operate, maintain, and manage systems as critical components of the system. As such, other components of the system (hardware, software, procedures, environments, information) must be designed to accommodate the unique capabilities and limitations of people, or the people must be changed, by changing their capabilities and limitations, through training. The major elements of human factors are manpower and training, human engineering, and life support and safety.

Human factors is applied to a system to influence the indicator of the extent to which system personnel can perform assigned functions and tasks. Readiness is a more generic construct than effectiveness since it demands not only effective performance when needed, but also the potential, the capability, for effective performance whenever it is needed.

Personnel readiness has a number of components, which in their totality, serve to define and delimit the term. It is composed of personnel availability, the availability of personnel, in required numbers, to meet the system workload. Personnel availability is a direct function of system manning, and varies as a function of workload.

Personnel readiness also includes personnel capability, or skills possessed by available system personnel. Personnel capability is directly related to training of system personnel.

Personnel readiness includes personnel performance, the ability of system personnel to perform at required levels, across all conditions and missions. Personnel performance depends in large part, on human engineering design of equipment, environments, information, procedures, software, and computer interfaces.

Closely aligned to personnel performance, personnnel readiness also includes personnel productivity. Productivity includes the sustained capability of personnel to meet production standards, to achieve required levels of throughput. Personnel productivity is a direct function of human engineering design of jobs, procedures, job aids, decision aids, work sites, environments, and information displays.

Finally, personnel readiness includes personnel safety and health. Included here are concerns for environmental design, protection systems, biomedical factors, and safety systems. Personnel safety and health is a direct function of system life support and safety.

THE HUMAN FACTORS DESIGN PROCESS IN SYSTEM ACQUISITION

Human factors ensures the integration of people into systems through the application of human factors analysis, design, and evaluation methods. As the application of these methods is standardized and formalized, and within the context of the phased system acquisition process, a human factors design process is developed. The essential features of a human factors design process are that it: 1) is standardized and formalized, 2) is interdisciplinary, 3) is geared to requirements, 4) emphasize interfaces, 5) is integrative, and 6) influences system design.

1. Standardized and Formalized Human Factors Process

The thrust of the human factors design process is to present an approach which is generic to all system acquisition efforts while at the same time, being able to be tailored to the specific requirements of each individual program. The characteristics of a human factors design process which impart its ability to be standardized and formalized include the extent to which it is: complete, correct, current, consistent within itself, compatible with the capabilities of its users, in compliance with governing specifications and standards, usable for its intended purpose, and effective in enabling human factors personnel to develop design concepts and criteria, conduct analyses, and perform evaluations.

2. Interdisciplinary Nature of the Human Factors Process

The human factors design process is made up of methods, procedures, techniques, principles and data from a number of disciplines, including human engineering, manning analysis, training, environmental design, biomedicine, and safety.

3. The Human Factors Design Process is Geared to Requirements

Throughout the human factors design process, design decisions are made on the basis of requirements. The primary requirements in a human factors effort are task requirements, including the tasks themselves, constraints on task performance, conditions which affect performance, and manning levels and skills necessary to complete the array of tasks. Task requirements lead to system requirements, which lead to design requirements for design of man-machine interfaces, environmental factors, and software. Finally, system and design requirements are used to develop test and evaluation criteria against which the system is assessed.

4. The Human Factors Design Process Emphasizes Interfaces

Since the objective of human factors is to integrate the personnel component with other elements of a system, it follows that the human factors process would place primary emphasis on the interfaces between personnel and these other elements.

The types of interfaces to be addressed in this integration of personnel into systems include:

- Functional interfaces
  - roles of personnel versus automation in system operation, control, maintenance and management
  - human functions and tasks
  - roles of system personnel in automated processes (e.g., monitoring, management, supervision, intervention, etc.)
- Information interfaces
  - information required by personnel from the computer
  - information required by the computer from personnel
  - protocols and dialogues for information access, entry, update, verification, dissemination and storage
- Environmental interfaces
  - physical environment (light, noise, temperature, etc.)
  - operational environment (operational conditions)
- Operational interfaces
  - procedures
  - workloads
  - personnel skill requirements

- personnel manning levels
- system response time constraints
- decision rules

- Organizational interfaces
  - management organization
  - job organization
  - data organization
  - command modes

- Cooperational interfaces
  - person-person interaction (communication, team performance)
  - person-computer interaction (cognitive maps, interactive dialogue, machine communication)

- Physical interfaces
  - input/output devices
  - maintainability design features
  - human-machine interfaces
  - safety features/protective devices

5. The Human Factors Design Process is Integrative

The human factors design process comprises a marriage of the human engineering process, manpower and training aspects of the logistic support analysis approach, and the life support and safety approach inherent in design for habitability and safety. This integration of human factors elements is achieved through: a) dependence on the results of a common front-end analysis which identifies requirements for personnel in the system. These requirements are equally important in guiding human engineering design decisions, in establishing manning and training approaches, and in delineating life support and safety concepts; b) reliance on a common data base of problems in predecessor systems, and performance capabilities for the new system; and, c) emphasis on test and evaluation of human factors aspects of the system as a whole, rather than in terms of human factors elements.

The human factors design process is also integrative in that it integrates with the system development process. Specific inputs and outputs between the two processes establish the importance of the human factors process to system design.

6. The Human Factors Design Process Should Influence System Design

The way in which the human factors design process seeks to ensure that human factors influences system design is to provide an approach for getting human factors concerns in system development early in the development process. This approach places emphasis on the use of information from existing systems to guide the selection of design directions in the new system. The human factors process requires evaluation of human factors aspects of predecessor systems prior to formal system initiation as an input to early system design concepts, and as an input to the system initiation products.

It is precisely in the program initiation phase of system development that human factors can exert its maximum influence in system design. It accomplishes this influence on design through two related efforts:

1) the identification of human factors problems and positive aspects in existing systems, which serve as lessons learned for the new systems;

2) the development of requirements for defining the role-of-man (versus automation) in the new system, as inputs to the establishment of alternate system concepts, and as issues to be considered in already established alternative concepts. Thus, design options will have received the benefit of human factors considerations since they should seek to avoid problems and deficiencies in existing systems while at the same time, employing established roles for system personnel which have proven especially effective.

ORGANIZATION AND MANAGEMENT OF HUMAN FACTORS IN SYSTEM ACQUISITION

The mechanism by which the human factors methods and techniques inherent in the design process get applied in system acquisition is through the human factors program. The human factors program includes the personnel, procedures, processes, data bases, and resources necessary for implementing human factors efforts. The three phases of the life of a human factors program are: program development, program implementation, and program evaluation.

Program development is largely a planning effort. In this effort, the standard human factors design process is tailored to the specific requirements of the system acquisition effort, and resources are identified to conduct each activity identified in the tailored design process.

Program implementation is primarily an implementation of the program plan at each stage of system acquisition. The major human factors concerns by system acquisition phase are:

- Program initiation - development of lessons learned from predecessor systems
- Conceptual phase - determination of the role-of-man in the system; allocation of functions
- Preliminary design phase - conduct of human factors studies to develop concepts and criteria
- Detail design phase - design of man-machine interfaces, training programs, determination of manning levels, etc.
- Production - verification of compliance
- Deployment - conduct of training
- Operation - acquisition of feedback data

Program evaluation seeks to satisfy two objectives. First of all, it attempts to determine the effectiveness of the program in meeting its objectives. This objective will serve to refine and improve program organization and management practices in later system development efforts. The second objective of program evaluation is to quantify the benefits of applying human factors to system development. The field of human factors is in dire need of a data base of such data; data which can be used to justify the expenditure of funds to develop and implement a human factors program in the next system development effort.

BENEFITS OF HUMAN FACTORS APPLICATION IN SYSTEM ACQUISITION

The primary benefit of applying human factors to system acquisition is that personnel readiness will be enhanced, leading to an increase in system effectiveness, availability, reliability and supportability. Human factors application will also significantly reduce system costs. Through the application of human factors, costly errors and accidents will be avoided, costs of training will be reduced, reduced manning will lead to decreased personnel costs, and costs of system redesign, to correct operability and maintainability problems, will be reduced.

Human Factors in Organizational Design and Management
H.W. Hendrick and O. Brown, Jr. (Editors)

THE SITUATIONAL APPROACH TO SYSTEMS IMPLEMENTATION

Jack T. Brimer

Institute of Safety and Systems Management
University of Southern California
Los Angeles, California
U.S.A.

Summary

This paper presents a proposed algorithm for systems implementation. The algorithm is driven by an initial definition of the system situation. The main features of the algorithm are that it focuses upon a step-wise procedure for the implementation process which explicitly takes account of human factors, and includes a documented output at each step in the process.

The algorithm is based upon standard; well documented theory and techniques which can be found in any standard management textbook. (see 2 for example)

I. The Situational Approach

A. Nature of The Approach

1. Definition. A situation is defined relative to a bounded system. It refers to the structure, conduct and performance of the system.

2. Focus. This paper attempts to identify an orderly scheme of situational definition, and presents an algorithm for system implementation based upon the situation definition.

3. Relationship of Situational Approach with System Concept. A system consists of the elements: Inputs, transformation, goal, process, feedforward, feedback, history and control. The relationship of these elements are shown in Figure 1. Structure refers to how the transformation element is built. Conduct refers to how the transformation element converts input to output. Performance deals with how feedback or feed forward are utilized to optimize the production of outputs relative to the goal.

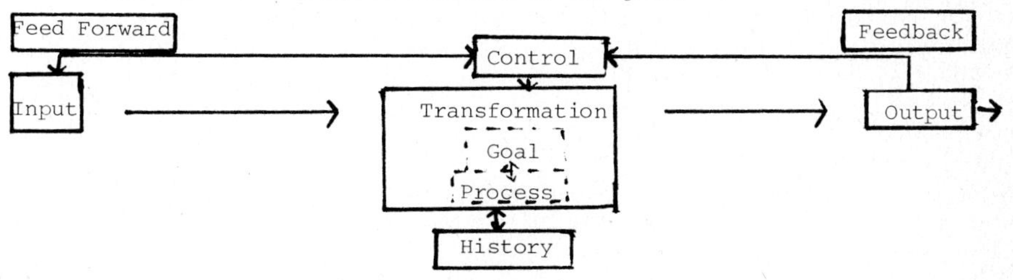

Figure 1 System Elements

4. Overview of the Structural Approach. Beginning with a description of the structural, conduct and performance characteristics of a system, then the implementation process is prescribed as a function of the situation in the form of a step-wise-algorithm.

5. Implication for Implementation. The situational approach does not treat all situations as homogeneous, but recognizes their heterogeneous dimensions, then uses this information to prescribe a systematic implementation path which minimizes user resistance, maximizes user acceptance, and optimizes system performance. This implies a rational, structured, iterative, and economical approach.

B. Situational Types

1. Ergonomic Grid. Figure 2 shows the Ergonomic Grid describing various situational types in terms of interactions.

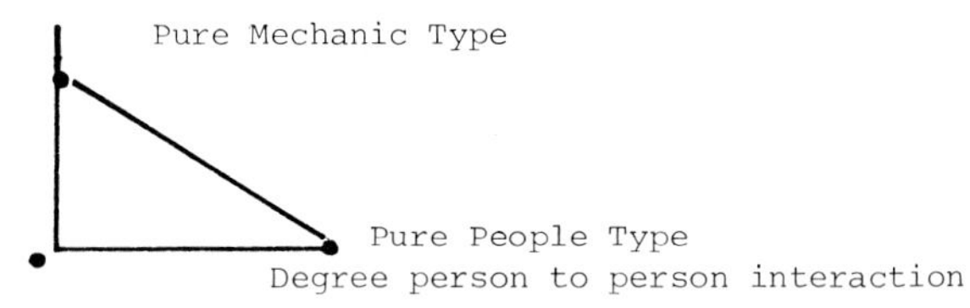

Figure 2 The Ergonomic Grid

The two extremes types of situations are:

People People Type: All man to man interaction is the dominant characteristics. Example: A management meeting.

Machine Machine Type: All machine to machine interaction is the dominant characteristic. Example: A robot assembly line.

All situations fall between these two extremes, and within the boundaries noted. The boundaries represent interaction limitations along both dimensions. The trade off curve indicates the strong possibilities that in order to achieve more machine interaction, a certain amount of human interaction must be sacrificed.

2. Implication of the Grid for Implementation. The implementation path is determined by the present situation's location on the ergonomic grid. Change depends upon the structure, conduct and performance of the interaction characteristics (system nature of relationships) of the situation.

C. Situational Parameters

1. Necessity to Balance Number of Parameters. The usefulness of situational approach must include enough information to differentiate situations yet not be so exhaustive as to defy understanding and analysis. Seven parameters seems about right.

2. List of Parameters

a. System size. The number of distinct system elements such as machine or people. Element definition can also be a grouping such as work group or machine pod.

b. Communication. The flow of information between system elements, the exchange of ideas and the level of formal informal contacts.

c. Personality. The psychological attributes and expectations of people in the situation.

d. Control. The process by which a congruence of individual and system goals is achieved.

e. Technology. The methodology for accomplishing a goal and the level of complexity of the machines used in the process.

f. Stage. The growth stage of the system with regard to the categories of expanding, stable, or declining.

g. Decision level. The location of decision power with respect to the system which executes the decision.

3. Relation of Parameters and Characteristics. Figure 3 summarizes the influence each parameter has on a particular characteristic. The relationship is conjectural and subject to empirical verification by future research. The +, ✓, and - indicates the influence of the parameters. This rating scheme recognizes the overlapping nature of the characteristic - parameters relationship.

Situational Parameters	Situational Characteristic		
	Structure	Conduct	Performance
Size	+	✓	-
Communication	-	+	✓
Personality	-	+	✓
Control	-	✓	+
Technology	+	✓	-
Stage	+	+	+
Decision level	-	+	✓

Figure 3 Situation Parameter - Characteristic Relationship

+ indicates primary influence category
✓ indicates secondary influence category
- indicates tersery influence category

D. The Situational Matrix

1. Primacy of the Growth Stage. Stage of development is a convenient initial organization factor for other parameters given the belief that it is equal in influence across all situational characteristics.

Ergonomic Type	Parameters	Growth Stage		
		Expanding	Stable	Declining
Machine Dominant	Size			
	Communication			
	Personality			
	Control	Value of Parameters		
	Technology			
	Decision level			
Man Dominant	Size			
	Communication			
	Personality	for particular situation		
	Control			
	Technology			
	Decision level			

Figure 4 The Situation Matrix

2. Application of the Situation Matrix. The following steps describe the use of the matrix to identify the situation.

Step I. Determine the Ergonomic type

Step II. Determine the Growth Stage
Step III. Data collection and analysis
Step IV. Perform evaluation of parameters to determine proper square which fits the situation
Step V. Utilize information for all subsequent planning

3. Measurement. Some form of measurement scale must be utilized for each parameter. A test, survey, or delphi are possibilities.

II The Implementation Process

A. Goal of the Implementation Process. The dictionary defines implementation as process of enacting a change. The goal of this process is to enact a change which enhances the performance of the system. The process of implementation will involve change in both the structure and conduct of the system.

B. Stages in the Implementation Process

1. Nature of the Implementation Process. Figure 5 depicts the idea that implementation proceeds, includes and proceeds system analysis. It literally envelopes the system analysis.

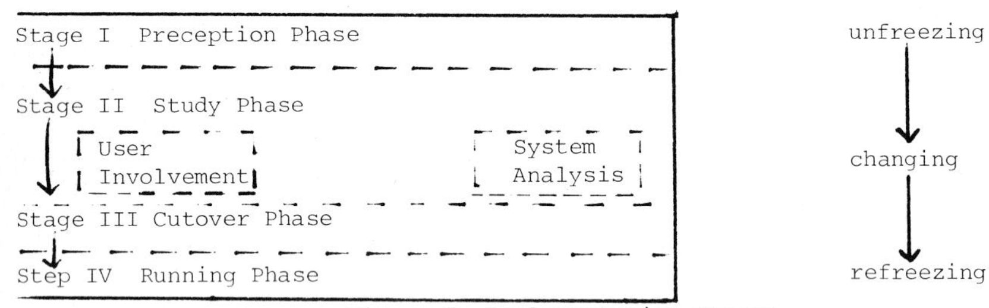

Figure 5 Implementation Stages

The categories of unfreezing, changing and refreezing are noted on the side of the diagram to denote the scope of human factors activity involved in each step.

2. Description of Stages

Stage I The Preception Phase

A need for change is preceived as a result of poor system performance, and it is generally acknowledged that something should be done. A steering committee or individual is appointed to see what can be done. This person lays the groundwork for the implementation process. Some sort of analysis is conducted to initially define the source of poor system performance and a proposal to make a detailed study of the problem is produced.

Stage II The Study Phase

Based upon the proposal in Stage I a detailed system analysis is conducted. The standard parts of such a study are: Problem Definition, Setting Objectives, Creating Alternatives, Modelling Alternatives, Evaluating Alternatives, Decision Making, Planning for Implementation. (3)

The situational approach emphasizes the fact that implementation is involved in all stages of the systems analysis process through user involvement, collaboration and participation. The scope of involvement in the systems analysis is shown in Figure 9. Involvement can take many forms of active consultation and advice.

Note that the systems analysis study is produced during this phase.

Stage III Cutover Phase

During this phase users are trained in the application of the new system, and human or technical design problems are addressed through operational tests and evaluations.

Stage IV Running Phase

The change is brought on line. The system is made a part of routine operations which includes continued evaluation of performance.

C. Characteristics of the Change Effort. Planning for change involves some mix of top down and bottom up approaches. Figure 6 shows how various planning areas are affected by the nature of the plan.

Planning Area	Type of Change	
	Bottom up	Top Down
Degree of Planning	Little	Total
Size and Scope of Change	Incremental	Chunk
Initiation Point	System Members	Top Management
Pace of Change	Gradual	Rapid
Elements to be changed	Structure/Conduct Performance	Whole system
User Involvement	High	High

Figure 6 The Character of the Change Effort

Implementation Stage	Involvement Activity	Scope of Involvement	Degree of Involvement
Stage I	Initial Diagnosis Data analysis	Inquiry Group	Selective Involvement of whole system
	Feedback Confrontation		
Stage II	Participation Planning	Advisory Group	
Stage III	Team building intergroup development subenrichment	Involvement Group	
Stage IV	Follow-up Evaluation	Work group	Total system involvement

Figure 8 Involvement Strategy

1. The scope of group envolvement runs from a minimal level. At the beginning with a small study group or steering committee, to the entire work group at the end.

2. Team building and job enrichment are two methods employed in Stages III and IV. A substantial body of literature exists on these two subjects. (2)

Implementation Process	Implementation Phases			
	Stage I	Stage II	Stage III	Stage IV
Main Process	Unfreezing	Study	Training/ Testing	Refreezing
Subprocess I (objectives)	General Change Objectives Emerge	Objectives Refined Subgoals Emerge	Objectives Integrated into behavior/ interaction	Achievement of objectives
Subprocess 2 (Interactions)	Tension in Interactions is recognized	Expectation of change established	New Attitudes centered around new activities	New attitudes are operational
Subprocess 3 (Enrichment)	Lower moral is felt	moral enhanced by participation involvement in change	moral enhanced by mastering new task	moral enhanced by task accomplishment
Subprocess 4 (motivation)	Scheme for change emerges (external motive)	New scheme for change conceptualized	internalization of new scheme by reality testing and improvising	New scheme drive behavior

Figure 7 Model of the Acceptance Process

3. The evaluation activity for involvement revolves around the following criteria. (a) some decision making is shared by group members, (b) there is a good repoire evident in man-man interaction, (c) individuals express a reasonable level of job satisfaction.

4. User involvement in the system study is depicted in Figure 9.

Systems Analysis	User Role	Degress of Involvement Low High
Problem Definition	Elaborate and Clearify	
Objectives	Specify	
Evaluation	Advise on Criteria	
Alternatives	Help to Elaborate	
Model	Assist in Formulation and Testing	
Evalution and Decision	Observe Selection of Alternative	

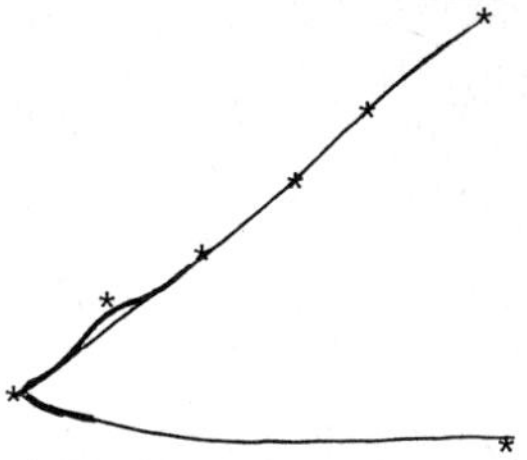

Figure 9 User Involvement in Systems Analysis

D. Human Factors in the Implementation Process.

1. User Resistance

a) Reason for Resistance. Employees usually resist social change not technical change. Social change is a change in human relationships which accompany technical change. Resistance arises with a preoccupation with technical aspects of change and forcing such change on the user.

b) Sources of Resistance. Insecurity of now knowing what to expect in terms of position after a change. Social loss of friends and coworkers. Economic loss in terms of job or wage losses. Change history in terms of time since the last change and success of prior change. Unanticipated System Impacts in terms of effects which were not predicted. External Forces such as a union or political body with conflicting interests.

c) Signals of Resistance. Denial of change by insistance on doing things as they have always been done. Sabotage of the change effort. Ignoring the change by procrastination or postponing deadlines.

2. User Acceptance

a) Focus. Acceptance is the process of anticipating change and planning for it through the active involvement and participation of users.

b) Acceptance Process. Dalton (1) has proposed a bottom up model of acceptance which urges a four pronged effort to produce positive results. The following four subprocesses are identified.

Subprocess 1 Making explicit the goals and objectives of the change.
Subprocess 2 Building and replacing interaction disrupted by change.
Subprocess 3 Job enrichment as a result of the change.
Subprocess 4 Maintaining or increasing motivational level.

Figure 7 shows these processes and relates them to each of the implementation phases.

3. Involvement Strategies. Figure 7 depicts the relationship of implementation stages, acceptance activities, and scope of involvement. Several points to note are:

4. Asymmetry of Resistance and Acceptance. Resistance and acceptance are not two sides of the same coin. Resistance is a signal that the acceptance process is not in control, and that corrective action is necessary.

III Implementation Task Algorithm

a. General Considerations

1. The situational parameters drive the implement action strategy.

2. Change involves people and technology. The focus below is on changing people, since changing technology has been overemphasized.

3. Duplication of extensive literature on specific techniques is avoided given space limitations.

b. Task Algorithm. Figure 10 presents the task algorithm.

References

(1) Dalton, Gene, "Influence and Organization Change", in Changing Organizational Behavior, Ed., Alton C. Bartlett and Thomas A. Kayser (Englewood Cliffs, N.J. , Prentice-Hall, 1973).

(2) Koontz, Harold, O'Donnell, Cyril and Weihrich, Heinz, Management (New York, N.Y., McGraw-Hill Book Co., 1980).

(3) Mosard, Gil, "A Generalized Methodology for Systems Analysis," IEEE Transaction on Engineering Management, Volume EN-29, Number 3, August, 1982, p. 81-87.

Implementation Stage	Step Name	Task Name	Disription of Activities	Documentation
I	Situation Review	Enquiry	.Appoint and charge inquiry group	Mission statement Change statement
		Present situation review	.Review present situation to determine structure, conduct, and performance characteristics. Figure 4	Situation diagram Interaction charts Personnel summary Operations summary
I	Change Assessment		.Proscribe direction, scope, extent and pace of change effort and likely scenerio. Figure 6 and 7	Change statement
			.Conduct economic study (preliminary cost/benefit analysis)	Economic balance sheet
			.Decide on Go-no go	Decision statement
I	Requirement's Identification		.Gather data to determine specifically what is to be changed and how change can be effected	Data records Data summary
			.Develop a detailed set of requirements	Requirement's statement
I	Checkpoint one		.Review all documentation	Recommendation
	Change planning	Project Planning	.Activity identification and sequencing	Project Plan
		Acceptance Strategy	.Determine scope and character of user involvement based upon values of structural parameters and the acceptance model. Figure 8 and 9	Involvement Program
			.Set up control procedures and criteria	Control Statement
I	Initialize System Study	Advisor Team	.Select advisers from in-house users	Change statement
		Study Team	.Select study team from in-house staff or outside consultant	Change statement

Implementation Stage	Step Name	Task Name	Discription of Activities	Documentation
II	Conduct System Study	Study Planning	.Activity identification and sequencing	Study Plan
		Analysis	.Identification and specification of new system	Detailed Goal Statement
			.Development of new system design	Blueprints
			.Development of instructions for users	User manual operating procedures
			.Maintain laision with user body by briefing and feedback sessions	Periodic Newsletters
		Education	.Devise an education program	Curriculum lesson plans instruction materials
II	Conversion Plan	Cutover Plan	.Scheme for parallel or direct cutover	Conversion plan
		Advisory Role	.Determine low advisory group will assist/ supervise team building effort	Team building plan
II	Checkpoint two		.Review all Documentation .Decide on continue-not continue	Recommendation
III	Conversion	Training	.Conduct training program .Evaluate training Accomplishment .Conduct job enrichment	Progress summaries Student grade reports
IV	Checkpoint three		.Review all documentation .Decide on continue-not continue .Run system	Operation reports
IV	Evaluation		.Appraisal of structure, conduct, and performance of emergent system. .Fine tuning and minor adjustments.	Control reports

Figure 10 Implementation Task Algorithm

Human Factors in Organizational Design and Management
H.W. Hendrick and O. Brown, Jr. (Editors)

FROM ERGONOMIC THEORY TO PRACTICE: ORGANIZATIONAL FACTORS AFFECTING THE UTILIZATION OF ERGONOMIC KNOWLEDGE

JEFFREY K. LIKER, BRADLEY S. JOSEPH and THOMAS J. ARMSTRONG

Industrial and Operations Engineering
University of Michigan, Ann Arbor, Michigan 48109

Despite decades of accumulating scientific knowledge of ergonomic principles, many workplace designs fail to incorporate human factors considerations. This paper proposes a general model of the organizational process which influences the degree to which the human side of human-machine systems is considered in workplace design. A case example is described which illustrates an innovative approach to workplace design and redesign at an automotive plant.

INTRODUCTION

Human factors specialists in Universities teach their students that any complete human-machine design should consider the capabilities of both the human and the machine in the original system design. Humans should not be fit into the system in a post-hoc fashion. The technology for human-machine design has become quite sophisticated and a well trained specialist can make very specific recommendations on ways to make maximum use of human capabilities without placing unhealthy strains on the human. Despite the level of technical sophistication in the theory of human factors design, the practice of work design generally fails to incorporate these principles.

We begin with the observation that ergonomic theory is all too rarely used in practice despite its tremendous utility for firms and employees and seek to identify organizational factors that explain this lack of application. The general problem of managing change has been the topic of much research (Coch and French,1948; Alderfer,1976; Conlon,1983; Kantor,1983; Kotter and Schlesinger,1979). Some general issues from this broader area are described below as applied specifically to human factors design. We conclude by briefly describing a case example which illustrates an innovative approach to workplace design and redesign at an automotive plant.

Three recommendations commonly appear in the literature on managing change: 1) Top management committment is a necessary prerequisite to successful change efforts; 2) all persons affected by the change should participate in the change; and 3) all persons affected should be trained for their roles in bringing about the change and (if applicable) the new procedures they will be using. To understand why these steps might be necessary for successful human factors engineering, it is useful to think through the reasons why human factors engineering is not more successfully utilized in industrial operations. We identified six general "obstacles to utilization of ergonomic knowledge."

OBSTACLES TO UTILIZATION OF ERGONOMIC KNOWLEDGE

1. Lack of General Ergonomic Knowledge - Human factors engineering is a multi-disciplinary science. Yet our universities and industry are organized around disciplinary specialization. The general knowledge base of ergonomics includes an understanding of human capabilities and limitations, physiology, psychology, manufacturing processes, machine design and process design. Manufacturing processes are generally designed by process engineers (e.g. mechanical or chemical) who are trained primarily in the machine side of the human-machine system.

2. Lack of Specific Job Knowledge - Beyond knowledge of general principles, a good workplace design requires understanding of the specific jobs done by people, and the specific technical demands of the system. This comes from experience with the machinery and knowledge of how operators do their job, knowledge which designers of manufacturing processes often lack.

3. Poor Inter-Departmental Communication - If one searches through a plant, it is generally possible to find a group of persons who collectively have knowledge of all the areas that comprise ergonomics. These may include a machine operator, a plant physician, a personnel administrator trained in psychology, a process engineer, and an industrial engineer. The problem is getting them to communicate with each other on a collaborative workplace design. As discussed below, organizational reward systems seldom promote inter-departmental collaboration.

4. Perceived Cost-Benefits - People tend to act in their own self-interest, in part as defined by their position in the organization, and tend to have a short-term focus. This suggests that an understanding of why ergonomics is not more readily utilized must take into account the interests of relevant actors in industrial organizations.

At a minimum, key actors include the process engineer, production supervisor, and machine operator. Other relevant actors may include the plant controller whose concern is cost justification purchasing people who often select suppliers based on price and delivery records, plant medical personnel trained primarily in disease detection, not prevention, a safety engineer who monitors accidents, and a union representative concerned with serving his constituency to get votes. For each of these industrial positions, human factors engineering is not a high priority. Moreover, performance is generally judged on doing well at one's specialty, not collaborating with other functions.

An excellent discussion of the motives of human factors engineers, process engineers, and machine operators and how their behavior is shaped by the plant social system can be found in Perrow (1983). If there is a plant human factors engineer, that individual has a stake in applying ergonomic principles to workplace design. Process engineers, however, do not necessarily have such a stake. In fact, sharing the design with a human factors engineer may have a number of costs from the viewpoint of the process engineer, including: sharing credit, taking additional time to coordinate efforts with others, admitting past designs that did not incorporate ergonomic considerations were inadequate, and possibly increasing the purchase price of the system.

It might appear that the worker has an interest in a workplace design that minimizes muscle strain and fatigue; however, such designs may appear to be in conflict with incentive systems that link pay to productivity. While the costs of changing the workplace or work methods can be felt immediately by the worker as a slowdown in production rates, the benefits are invisible. If the worker avoids getting carpal tunnel syndrome, for example, how can he know that he would have gotten it had his workplace or work methods been different?

5. Organizational Politics - If people in different positions in the organization have interests that are in conflict, who wins out? This brings in the dynamics of power as a consideration in workplace design. Perrow (1983) argues that human factors engineers are far less likely to be powerful actors in industrial organizations compared to process engineers. The former are more apt to be identified as "do-gooders" concerned with worker's welfare, while the latter are viewed as pragmatists with a concern for cost and efficiency -- the interests of management.

Human factors engineers are part of a staff function in industrial organizations. They have no direct authority over production. Yet, once they propose changes in workplaces and jobs, they are tredding on the turf of production people -- those with line authority. The conflicts between staff and line are well documented (Dalton, 1950). Professional staff get their rewards not only from the organization in which they work, but also from the professional community of which they are a part. They attend conferences and their reputation among peers comes from applying the latest technological developments. In contrast, line supervisors are judged by output, unit cost and quality. Until a new system or redesigned workplace is functioning up to speed, the line supervisor must eat the costs of part shortages or high defect rates.

6. Emotions/Attitudes - For whatever reason, people have a bias towards the status quo. Such terms as habit, tradition, or one's "niche" refer to the human tendency to prefer old ways to new. Operator's tend to prefer established work methods to new ones, purchasing agents tend to prefer long-time vendors to new ones even if the latter make better designed tools from an ergonomic standpoint, and process engineers tend to design jigs and fixtures in a particular way that may not be best for the operator. Why? Because "we've always done it that way." In changing machinery and tooling it is easy to forget the operator's emotional attachment to his or her workplace. The implicit message to the operator is, "this is not your workplace, this is simply a place where you come to work for us."

IMPLEMENTATION STRATEGIES

How can these obstacles to application of ergonomic theory be overcome? If no one person has all of the knowledge, authority, and access to resources to ensure that workplaces are well designed ergonomically, we are left with the question: How can we mobilize existing plant resources -- material, financial, and human -- and gain the committment of all parties involved in an ergonomic program. Unfortunately, there is no simple answer to this question. Different strategies will work best under different circumstances. An approach currently being used at one site (described below) combines three elements -- top management committment, ergonomic training, and a participative task force structure. In the best of circumstances, these three factors provide the knowledge base for workplace design, foster communication and cooperation between departments who normally do not work together, provide incentives to individuals involved in the program, and help to gain the committment of all persons involved (see Figure 1).

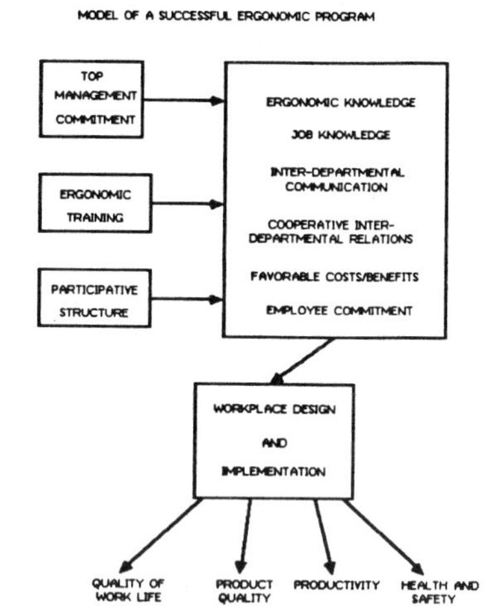

Figure 1

1. Top Management Committment - Top management committment is a minimum requirement for the success of a new program. Upper management directly controls the salaries and career prospects of their immediate subordinates and directly or indirectly control the money and other resources (e.g. manpower) needed to effectively mount a new program. Unfortunately, outside consultants have little influence over top management, apart from their powers of persuasion. Fortunately, in the U.S. auto industry there are indications that ergonomics is receiving much higher priority at the corporate level than was ever true in the past and this is being communicated to plant managers. As a result, some plant managers are willing to invest in ergonomically designed workplaces, even if tangible, short-term benefits are not apparent. Assuming top management is committed, what is the next step?

2. Participative Structure - One justification for meetings is that they permit the combined insight and judgement of several people. A participative or 'cooperative' meeting is one where a number of individuals work together to make a decision. Group decision-making brings a greater sum of knowledge to bear on the problem than any individual in the group possessed, and through group "synergy" can generally arrive at a superior quality solution. However, high quality solutions are not enough. Solutions are only effective if they are implemented as designed. This involves the acceptance of the decision by the group (Maier, 1963). People who participate in making a change tend to become committed to the change. They become allies instead of obstacles to be removed or manipulated.

If we stop here, we are left with a universal truth: Participation is better than non-participation. However, if not effectively managed, participation may lead to poorer quality solutions than the best individual solution, waste a great deal of time, cost money (at minimum employee salary), and even create tensions between individuals or groups. If participation is not always the answer, how does one know when it is the preferred approach? The approach by Vroom and Yetton (1973) answers this question using a decision tree. Managers answer a number of questions and each answer helps isolate the approaches to change that are viable. Of these viable approaches, the manager might use other criteria not in the model, such as the style that suits his personality, or economic considerations.

AUTOMOTIVE CASE STUDY

An innovative approach to applying ergonomic principles is being used at an automotive plant. Top management is highly committed, all participants have gone through 4 or 8 hour training sessions (on upper extremity cumulative trauma disorders, the focus of the program), and a participatory task force structure (similar to quality circles) is in the process of being developed.

The plant organization is based on a decentralized area management concept which is increasingly being used at auto plants. Two area managers report directly to the plant manager, and each area functions quite autonomously. The two production areas (Areas A and B) make suspension and drive train components for several car sizes. Plant wide administrative units support both areas and include safety, cost analysis, and personnel.

As structured, there was no obvious "home" for an ergonomic program at the plant. Hence, a task force structure was set up as shown in Figure 2. Persons from each production area of the plant participate in separate task forces. Top managers from plant wide administrative units were pulled together into an advisory committee. Workplace designs and redesigns will be developed by the task forces and proposals for expenditures greater than the sum area managers can authorize ($10,000) must be approved by the advisory committee.

The task forces from each area are composed of the area manager, a quality circle facilitator, the maintenance superintendent, the engineering superintendent, an industrial engineer and the local committeeman from the union. In Area A, which will focus on redesigning existing jobs, a separate production supervisor/ worker group has been established which will interact with the management level task force. In Area B, which will focus on design of new jobs coming on line, several production workers are included in the main task force. The advisory committee is composed of the plant manager, union chairman, industrial relations manager, plant engineering manager, a quality circles facilitator, the plant nurse, and the union safety representative.

Since the plant has an active quality circle program, our first thought was to combine the ergonomic program with QC circles. However, the circles choose their own agenda, which was heavily filled for months to come. Since the ergonomics program requires special training and a specific set of tasks, the plant chose to set up a separate task force structure. The benefit of the quality circles program is that managers and workers are trained in problem solving methods and are experienced at working in participative groups. It is too early to determine the success of the program, however, we are closely monitoring progress through questionnaires and systematic observation.

CONCLUSION

In general, human factors engineering would appear to be an excellent candidate for a participatory approach. Knowledge relevant to workplace design is widely distributed throughout the plant, there is often sufficient time to plan the design, workers identify with the goals of improving workplace safety and health, and it is important to gain the worker's acceptance of the design. After all, it is their job. Moreover, evidence from a national survey (Seashore,1980) suggests that 76% of all adult employees would like more influence over safety and health at work. A substantially smaller portion (41%) want say over "job tasks," though we believe this is partly because workers often are not aware of the close connection between job tasks and health. A number of experimental studies of workplace design have demonstrated the superiority of a participative approach over traditional autocratic approaches (Chaney,1969; Pasmore and Friedlander, 1982). The automotive case study, in progress, will add to the growing body of research on the organizational context of workplace design.

REFERENCES

1. Alderfer, C. "Change Processes in Organizations." Handbook of Industrial and Social Psychology, M. Dunnette, ed. Chicago: Rand McNally.
2. Chaney, F. (1969) "Employee Participation in Manufacturing Job Design." Human Factors 11 (2) 101-106.
3. Coch, L. and French, J. (1948) "Overcoming Resistance to Change". Human Relations, 512-532.
4. Conlon, E.J. (1983) "Managing Organizational Change." in Scientists, Engineers and Organizations by T. Connolly, California: Brooks/Cole, 363-378.
5. Dalton, M. (1950) "Conflicts between Staff and Line Managerial Officers." American Sociological Review, Vol. 15: 342-351.
6. Kantor, R.M. (1983) The Change Masters. NY: Simon and Schuster.
7. Kotter, J.P. and Schlesinger (1979) "Choosing Strategies for Change." Harvard Business Review, March-April, 106-113.
8. Maier, N. (1963) Problem-Solving Discussions and Conferences, NY: McGraw Hill.
9. Pasmore, W. and Friedlander, F. (1982) "An Action Research Program for Increasing Employee Involvement in Problem Solving." Administrative Science Quarterly 27, 343-362.
10. Perrow, C. (1983) "The Organizational Context of Human Factors Engineering." Administrative Science Quarterly 28: 4, 521-541.
11. Seashore, S. (1980) "The Social Psychology of Worker Participation." in Extending Workplace Democracy by Nickelhoff (ed.) University of Michigan ILIR.

Human Factors in Organizational Design and Management
H.W. Hendrick and O. Brown, Jr. (Editors)

Managing Ergonomic and Process Design Changes

Andrew S. Imada
University of Southern California
Los Angeles, California
U.S.A.

Debora Shoquist
Hewlett Packard
Santa Clara Division
Santa Clara, California
U.S.A.

This paper describes problems, solutions, and outcomes associated with a process design change. While correct from engineering and human factors perspectives, the change had unanticipated effects on both worker and system effectiveness. This case study points to the value of effective management of technical, ergonomic, and human resource systems. Evidence suggests that there may be important considerations when designing systems across cultures.

Introduction

In recent years there has been an increased awareness of the human factors contributions to improving productivity, system effectiveness, reducing accidents, and improving working conditions (e.g., Corden, 1983). One such example is Hewlett Packard's assembly of printed circuit boards for electronic test equipment. Six loading and one preforming lines produce circuit boards for approximately 100 different products. There are about 800 different board types; 80% of these products are high volume, easily assembled boards; 20% are specialized boards and are more difficult to build.

The previous production configuration was organized by product grouping. Workers were responsible for assembling different products and had opportunities to work on a variety of boards. This process design created several problems. First, responsibility for completing different products or problem boards was not specific. This led to unfinished boards left on shelves thereby cluttering the work area. Second, because the worker processed the entire product there were excessive walking requirements. A link analysis for the centermost work station revealed that for routine task completion a minimum of 1,030 ft. was required. Moreover, four of these 10 links retraced other links. Third, when workers took the products to a process station they may have had to wait while another worker's batch was being processed. This led to waiting and talking time. Together these problems contributed to lower system effectiveness. It was clear that an alternative straight through assembly process would be more effective from a production standpoint.

THE NEW SYSTEM

Analysis of the existing system and system requirements made it clear that conversion of the entire system to a straight through assembly process was not feasible. The high volume boards that are easier to assemble could be processed in a straight through assembly configuration. However, the more complicated specialized boards were left in a "deli" operation where the workers continued to work by product grouping. This led to a mixed design where the straight through line was functionally grouped while the deli operation was left in the original product grouping.

Ergonomic consultants contributed to the design of the straight through line. This work station design was self-contained and nearly eliminated traveling requirements. Parts, tools, and equipment were located within the worker's reach envelope thereby minimizing reaching requirements. This design change eliminated may of the problems inherent to the prior layout.

To familiarize the workers with the design changes, a scale model of the new straight through configuration was displayed in a central area for several months. Discussions with workers and weekly meetings with supervisors were held for several months before the system changeover.

PROBLEMS

Once the hybrid (straight through and deli) system had been installed a number of unanticipated problems emerged.

Identifiable groups of workers and low morale. Lower performers were put on the straight through line. These jobs involved simpler work and fewer decisions on high volume boards. The new work station design facilitated this simplicity. The higher performers continued working in the product-grouped configuration and produced the more complex specialized boards. Workers on the straight through line recognized that their group members were the lower performers. Consequently morale went down, self-esteem was damaged, and perhaps some self-fulfilling prophesy about their performance capabilities had an effect.

Variety ⟶ Specialization. Workers on the straight through line were now responsible for a function within an area as opposed to a product. Individual workers now controlled single operations. While these workers were loading an entire board and had specific responsibility for each board, they felt they were on an assembly line. This perception was enhanced by the work station design (e.g., in-line positions, moving conveyor).

Reduced Freedom. While the new system reduced movement requirements, it also restricted individual freedom to walk and talk. Because all tools and equipment were now within the reach envelope there was little need to walk to another station. This increased the visibility of a worker when he or she was not at the work station. It probably affected interpersonal relationships since worker communication links were greatly reduced. This perceived loss of freedom was compounded by the fact that the higher performers in the deli operation continued to walk in their work area.

A second restriction was the variety of work performed. Workers on the straight through line perceived themselves as having fewer alternatives. The move to a functional grouping not only caused a change from variety to specialization, it also reduced the selection of work to be performed.

Individuality of Work Stations. While ergonomically correct the new straight through line allowed little room for individuality. The convenient work envelopes, better lighting, and appropriate anthropometric dimensions were trade offs for the individuality common to the old system (e.g., room for radios, stuffed animals, pictures, coffee cups).

EFFECTS

System variance was adversely affected by these problems. Variance is a control measure that compares the predicted (weighted) costs with actual costs. Therefore, a variance of zero means the predicted and actual costs were the same; a postive variance represents actual costs in excess of projected costs; and a negative variance represents actual costs below projected costs. Data indicate significant positive variances after the initial system changeover. At least part of this increase can be attributed to the problems cited above. Additionally error rates were somewhat higher and more erratic.

SOLUTIONS

At least five important measures were taken to overcome these problems and their effects.

1. Recognition. Because the workers on the straight through line had identified themselves as the lower performers morale and self-esteem decreased. These groups were given several forms of recognition and attention: 1) recognition that their lines accounted for the highest volume; 2) automated devices to help them accomplish their tasks easier; 3) engineering support to solve their technical problems. Together these actions helped to increase their feelings of importance and contribution.

2. Training. Workers were given training in relevant skill areas. For example, loading and racking classes were given to help them do their jobs. The fact that management was spending time to train them may have increased the workers' sense of importance and esteem. The training reinforced the importance of quality in their work.

3. Engineering Support. In training sessions and weekly meet-

ings technical staff supported a bottom-up problem solving process. This bottom-up orientation encouraged workers to tell supervisors about the problem; supervisors explain the problem to the engineer; the engineer gives technical advice on how to solve the problem; and workers help solve the problem. Further, workers made suggestions and supervisors discussed management related problems. Engineers' responsibilities were arranged by function rather than by product. The technical support configuration is therefore consistent with the straight through production design.

4. Vertical Job Loading. Supervisory responsibilties were expanded from handling routine matters (e.g., health benefits, standard times, etc.) to problem solving. Instead of simply reporting a problem the supervisor investigates the problem and explores solutions with subordinates and engineers. Workers became involved in implementing ideas. This increased vertical responsibility provided a different kind of job variety. For example, workers now order fixtures from vendors directly. This increased responsibility leaves management out of this external interface and delegates responsibility to the workers and supervisors.

5. Responsive Management. Ideas from weekly meetings or suggestions are implemented soon after they are proposed. Examples include lighting changes, less control over documentation, and changes in racking documentation. For example, one complaint was that workers' fingers hurt from the finger-tightened thumb nuts used to secure boards to the rails. They suggested a fastener that could be tightened with a special wrench. The idea was implemented in 24 hours.

LEARNING POINTS

The design changes and solutions point to several valuable learning points.

1. Active involvement by system users is crucial for transitioning into a new system. It is not enough to passively display a model, talk with supervisors, and informally discuss design changes. Even after a new system is installed user suggestions can be invaluable to improving overall effectiveness. Imada (1982,1983) points out that these user inputs can be valuable, if not essential, contributions in coordinating technological changes.

2. Worker skills and technical support enable workers to deal with their new environment and equipment layout. If engineering does all the problem analysis and implementation, workers have no ownership in the system. Technical support from production engineers, human factors specialists, and other technically oriented personnel can provide skills and analytical tools to workers to solve problems and implement solutions.

3. Impersonal human factors can leave the human out of the

design process. Design efficiency at the expense of human dynamics (e.g., attitudes, self-esteem, and perceptions) is not necessarily an effective system design. Peoples' needs must be considered in a tradeoff analysis of system design.

Whether the need for "personal space" (i.e., room for pictures, radios, cups) is culturally-based or company specific is unknown. These needs, however, should be considered when developing like systems. This is particularly relevant to designers who develop systems to be used by different cultures.

4. The <u>problem solving orientation</u> may be more important than the problem solving technique or procedure. For example, over the past few years Quality Control Circles have gained wide attention. QCCs may not be effective for groups unfamiliar with the techniques and analyses; especially in short range or temporary problem areas. Instead, brainstorming in task circles may be better accepted. The process may draw people together who have common interests, particular skills, or vested interest. The group can be dissolved once the particular problem is solved. This flexibility approaches a matrix organization concept.

5.Perhaps most critical to the entire process is an <u>organizational "culture"</u> oriented toward problem solving, innovation, or quality. Peters and Waterman (1983) point out that this organization's orientation toward people and quality make it well suited to take on new challenges.
The design changes in this case may have altered jobs against the trend toward higher task identity and job variety (e.g., Hackman, Oldham, Janson & Purdy, 1975). The potentially negative effects of such a move may be offset by meeting individual growth needs, vertical job loading, open feedback channels, and a supportive organizational climate.

The present example illustrates how management, engineering, human resources, and ergonomic inputs can be integrated to overcome technological change and resistance to such changes. The total organizational climate must encourage and nurture this approach. Without this management and technical integration it is possible to factor out the human with too much human factors.

REFERENCES

(1) Corden, C. Human factors and nuclear safety: Grudging re-respect for a growing field. Am. Psychological Assn. Monitor 14 (1983).

(2) Imada, A.S. Productivity and worker involvement in the Japanese organization. Proceedings of the Human Factors Society Meeting. Seattle, Washington, U.S.A. (1982).

(3) Imada, A.S. Productivity through participation in large organizations. Proceedings of the Human Factors Society Meeting. Norfolk, Virginia, U.S.A. (1983).

(4) Peters, T.J. & Waterman, R.H.,Jr. In search of excellence. (Harper & Row, New York, 1982).

Human Factors in Organizational Design and Management
H.W. Hendrick and O. Brown, Jr. (Editors)

"ORGANIZATIONAL HUMAN ENGINEERING": THE DESIGN AND IMPLEMENTATION OF COMPUTERIZED SYSTEMS

Dan Zakay

Department of Psychology
Tel-Aviv University

Computerized systems are occasionally rejected or enjoyed questionable acceptance among their intended users. One possible cause for this state of affairs is the methodology utilized by designers of the system. An argument is made for a methodology called: "Organizational Human Engineering" (OHE) which is a combination of Human Engineering and Organizational Development. OHE deals with all the organizational problems caused by the implementation of a computerized system and takes them into account in the system's design.

INTRODUCTION

The implementation of computerized systems into organizations triggers a process of organizational change, which has been shown to produce considerable turbulence. This turbulence has in many cases the unhappy ending in which the intended system is rejected and discarded completely. This view is supported by Mackie (1980) who explored the fate of new computerized systems introduced into the U.S. navy. He concluded that despite an evident need many systems "have enjoyed questionable acceptance among their intended users, occasionally ending in outright rejection" (p. 80). The phenomenon of evidently needed but yet rejected computerized systems has been examined by many investigators. The following factors have been put forward as potential explanations:

1) Rogers and Shoemaker (1971) claim that all innovations carry some degree of subjective risk to the individual. This same line of reasoning was adopted by Zakay (1982), who has argued that the changes in methods of work, organizational norms and skills needed in order to be a successful worker or manager vis-a-vis new computerized systems create a process which is accompanied by an atmosphere of threat. This threat in turn produces negative attitudes toward the system. Biman (1982) has also argued that the tendency to resist change is natural and reflects fear of the unknown and of being inadequate in the new work situation.

2) Rogers and Shoemaker (1971) think that some computerized systems pose a specific threat to self-esteem since they appear to involve an erosion of responsibility in the intended user's traditional area of expertise. Guarnieri and Guarnieri (1982) claim that the meeting of a first time user with a computer is accompanied by general feelings of emotional and intellectual insecurity.

3) Rogers and Shoemaker (1971) suggest that a strong "not-invented-here"

syndrome might be developed among potential users who feel that their expertise in the subject was not utilized properly during the system's design. This becomes severe when Decision Support Systems (DSS) are introduced, since they are dealing in an area traditionally viewed as the exclusive domain of expert judgment (Mackie, 1980).

4) Biman (1982) points out that workers resent the idea of discarding acquired skills and learning new ones in face of the new system.

The whole pattern of reaction in face of a new computerized system is called by Guarnieri and Guarnieri (1982) The Psycho-Computer Syndrome. They describe it as a recognizable motivational pattern which includes stages of personal and of social reactions.

All the above factors produce a vicious circle. A lack of understanding of the operation and purpose of the system creates initial biases against it, which in turn, produce inadequate use and operation of the system. This reduces operating effectiveness, while further strengthening biases, and so on. Many investigators tried to identify factors in the design and development processes of computerized systems which facilitate and strengthen the creation of such a vicious circle. Carlson (1979) argues that the main problem is a mismatch between system's design and performance on the one hand and the requirements of users on the others. In regard to DSS, Levit et al. (1975) claim that a mismatch between the user's cognitive style and the mode of information presentation by the system is another barrier to its acceptability. These examples exemplifies that both individual level's factors (e.g. cognitive style) and organizational level's factors (e.g. users' requirements) are involved in the problem of computerized systems acceptability. It is claimed here that many of the above mentioned problems can be attributed to the nature of the basic approach adopted by the designers and developers of systems. An approach which can be called "Individual Human Engineering" (IHE) is identified as the source of the problem while another, called "Organizational Human Engineering" (OHE) is suggested as more suitable.

THE "ORGANIZATIONAL" VERSUS "INDIVIDUAL" HUMAN ENGINEERING APPROACHES

The individual (IHE) approach mainly considers the characteristics of the individual user or team of users in a static way while ignoring the dynamic emotional and motivational processes which take place in the organizational environment. The "IHE" approach focuses mainly on permanent characteristics of the individual like cognitive styles, information processing capabilities, perceptual and motor skills, anthropometric considerations etc. In the case of a team IHE looks at team's characteristics like formal structure, division of work, etc. What the IHE does not consider sufficiently is the emotional dynamic process which accompanies the design and implementation of a system. Thus, a practitioner following IHE methodology may end up with a well designed system which might be rejected.

"Organizational Human Engineering" (OHE), considers all the characteristics of individuals and teams but within the frame of reference of the dynamic processes of the whole organization. OHE is a combination of two disciplines: Human Engineering (e.g. Bailey, 1982) and Organizational Development (e.g. House, 1975).

THE PROBLEM OF IHE IN THE CASE OF COMPUTERIZED SYSTEMS

The traditional method for system design is hierarchical and composed of the following phases: (Huchingson, 1981)

Phase 1.	(a) Statement of objectives (b) Separation of functions (c) Allocation of functions		
Phase 2.	Development of equipment	Man Machine interface design	Development of Personnel
Phase 3.	System Integration		

Both the IHE and the OHE approaches can be represented within the framework of this methodology. When IHE is utilized, however, designers are usually interacting with potential users during phase 1 in general and during phase 1(a) in particular. In phase 3 designers are again interacting with the users and it is in this stage that resistance is most prominent. The big problem seems to develop in phase 2, where a crucial gap between designers and users might be developed. This methodology separates the equipment and human development. Once the system has been defined the equipment is developed separately by the design team. The human operators are being developed in order to fit the ongoing system definition. There is little interaction between these two independent lines of action, during the development phase. The main problem neglected by IHE is the duration of phase 2, usually measured by years. During this extended phase inevitable changes take place within any organization. Beside potential changes in work methods and organizational goals and objectives, people are moving in both horizontal and vertical directions. The result of this mobility is that those potential users who were involved in phase 1 are not necessarily the users of the system in phase 3. Often the system is accepted and operated by a whole new team of operators. The result of this process of change is that during implementation of the system, new operators and users do not feel involved and the "psycho-computer" syndrome and emotional factors mentioned earlier like the "not-invented-here" or "we-know-better" syndromes emerge. Another problem among operators is that designers and developers are more acquainted with those people with whom they worked together during phase 1, and continue to see them as their relevant reference group and relate mainly to them. The involvement of these people who have moved into the upper organizational level cannot meet the expectations of the designers. This produces a communication gap, which is perceived and interpreted by new operators as a devaluation of the system.

When the system enters the operational phase usually some need for change is identified. The designers, who are continuing to deal with such changes according to the IHE approach tend to work out their solutions independent of the organization. However, once the system is operating, direct users are adopting their own intuitive solutions and are acquiring habits which are subsequently difficult to change. Thus, when the designers are returning with solutions, again there is a high probability that it will not be accepted. This state of affairs may serve as an explanation to situations in which systems are used for routine report generation rather than for the more sophisticated objectives for the sake of which it was designed (Carlson, 1979).

ORGANIZATIONAL HUMAN ENGINEERING IN THE CASE OF COMPUTERIZED SYSTEMS

By this approach, while still following the basic logical phases outlined in figure 1, the interaction with the organization during all phases is continuous and the dynamic changes occurring within the organization are therefore accounted for. The following are guidelines which underly that approach.

A) The phase of system definition should be accompanied by a stage of organizational diagnosis. The nature of organizational climate, bases of power, information network, attitudes toward change in general and computers in particular should be identified.

B) System definition should be made on the basis of anticipated situations and conditions and not on the basis of the present situation.

C) The system should be adjusted to the nature of the organization in such a way that it will cause minimal threat.

D) Timely advance planning should be made by involving all the people directly affected by the anticipated change so as to develop a "consensus" toward a common goal (Bimon, 1982).

E) An Organizational Development process should take place while the system is being developed to ensure that the organization will be ready for the implementation phase.

F) The system should be developed in such a way that it will not be perceived as a product of outsiders but as a product of the organization itself.

G) A high managerial level figure should be assigned on the part of the organization as responsible for the system development and this figure should be selected so that he/she will be able to maintain his position throughout the whole development period.

H) The development team should be combined of designers and users and the development should be done as much as possible within the organization.

I) The system should not be declared prematurely as operational. The phase of changes and adaptations should be considered and perceived as part of the developmental phase. This should be done despite possible pressures on behalf of the organization itself.

J) The system should be accepted within the organization by both high level team and operators' team in collaboration and then transferred gradually into the operational level.

By following these guidelines, time and cost of the development might seem to be higher as compared to the IHE approach. This is only true when considering short term factors. In the long run, due to the increasing acceptability and decreased resistance level, costs might be much lower. This should be particularly true regarding hierarchical organizations like army and police as Rogers & Shoemaker (1971) suggest that changes brought about by an authoritative approach are more likely to be discontinued than those brought about by more participative approaches.

Human Engineering and Organizational Development were so far considered to be distinct disciplines. Herzog (1980) illustrated how the basic "action research" model so common in organizational development could be applied to productivity improvement efforts in the manufacturing setting. It seems that the combination of the two disciplines is possible. This can be done either by enriching the scope of skills of practitioners of both disciplines or by forming inter-disciplinary teams for coping with the complex problems posed by the implementation of computerized systems into organizations.

REFERENCES

(1) Bailey, R.W., Human Performance Engineering. (Prentice Hall, Englewood Cliffes, 1982).

(2) Biman, D., Consideration of Economical and Socio-psychological factors in the implementation of computer aided manufacturing, in: Edwards, R.E., Talin, P. (eds.). Proceedings of the Human Factors Society. 26th Annual Meeting. (Seattle, 1982).

(3) Carlson, E.D., An approach for designing support systems. Data base. Winter (1979) 3-12.

(4) Guarnieri, H., & Guarnieri, E., The Psycho-Computer Syndrome. Computer World. 16 (1982) 12-15.

(5) Herzog, E.L., Improving Productivity via organizational development. Training and Development. April (1980) 36-39.

(6) Huchingson, R.D., New Horizons For Human Factors in Design. (McGraw Hill, New York, 1981).

(7) Levit, R.A., Heaton, B.J., Alden, D.G., & Granada, T.M., Elements of decision support in a simulated tactical operation system, in: Proceedings of the 19th Congress of Human Factors (1975).

(8) Mackie, R.A., Design criteria for decision aids - The user perspective, in: Corrick, G.E., Haseltine, E.C., Durst, R.T. (eds.). Proceedings of the Human Factors Society. 24th Annual Meeting (Los Angeles, 1980).

(9) Rogers, E.M., & Shoemaker, F.F., Communication of innovations. (Free Press, New York, 1971).

(10) Zakay, D., Reliability of information as a potential threat to the acceptability of decision support systems. IEEE Transactions on systems, man and cybernetics. SMC-12 (1982) 518-520.

Human Factors in Organizational Design and Management
H.W. Hendrick and O. Brown, Jr. (Editors)

CASE STUDY: A METHOD OF COPING WITH AGGRESSIVE DEVELOPMENT CYCLES

W. B. Giesecke, K. R. Banning
R. E. Herder, W. S. James

IBM
Information Design and Development
Austin, Texas

A special work group has been formed at the IBM site in Austin, Texas. Its mission is to ensure the quality of the user interface of a new product that has a compressed development schedule. The group produced guidelines for use in the writing of specifications and reviewed these specifications to ensure compliance. Additionally, the group constructed and reviewed scenarios of user tasks in relation to the interface design.

DEVELOPMENT SCHEDULING

A special work group has been formed at the IBM site in Austin, Texas, to ensure the quality of the user interface of a new product that has a compressed development schedule. The topic of this paper is how non-development groups have become actively involved with development personnel in the design of a user interface system. This situation is the result of industry pressures working to shorten development schedules.

Traditional Scheduling

The traditional approach in setting development schedules is to establish a date sufficiently in the future by which time a product can be developed, produced and shipped. With an end date established, critical dates enroute to the end date are established. The bulk of this time is allotted to development work by architecture and programming departments which typically design the user interface for new products.

There is now a trend away from this form of scheduling. The shift is partly due to competitive pressures and partly due to the inherent weaknesses of the traditional approach to product scheduling. One of the major weaknesses in this approach is that the ship dates can be established inaccurately, leading to unnecessarily long development cycles. With increased competitive pressures, the time factor becomes important. Secondly, dates in the future become based on assumptions, on what we plan to have by a certain date, rather than being based on what we have or are relatively certain we can have.

Also under the traditional scheme, interested groups outside architecture and programming such as Human Factors, Product Assurance and Publications have had input primarily as reviewers of work already done. The long development cycle leads to a long review cycle, only adding to the above mentioned time problem inherent in this approach. To further complicate matters, this lengthy review cycle permits the detection of previously unnoticed weaknesses or flaws in published designs. While we want to detect any and all problems, development is put in the awkward position of being left open to undesirable exposures. Of course, another cycle to remedy any detected problems becomes necessary, consuming more time.

Compressed Scheduling

As the above weaknesses are realized, the traditional approach to product scheduling is being abandoned. Rather than making a number of assumptions, scheduling can be based on the technology and resources we have now or can reasonably expect to have. Thus, newer scheduling starts with the present and assesses what can be produced with current resources and how quickly it can be produced. In contrast to traditional scheduling, which is essentially a backward-looking process, scheduling under time constraints is a forward-looking process. The steps in the development cycle do not necessarily change; simply, the time element becomes shortened. The shortening of time allotted to the phases of development, while not eliminating the standard phases, does alter standard review procedures. The obvious problem is that time for extensive review and review meetings does not exist. As we all know, testing, reviewing, and documenting the testing and reviewing can be very time consuming. Facing the absence of adequate review or testing time under compressed scheduling, a new procedure for ensuring quality had to be found.

As a result, human factors and development have found the need to alter their traditional relationship. The need for a cooperative effort became obvious early in the development cycle. Faced with unusual time constraints and with the problem of having to incorporate the results of reviews and testing by Human Factors, Systems Assurance and Publications, a product development group asked these functional groups to participate in an interface review group. Human Factors was thereby given the opportunity to act in its traditional role of being development's conscience while participating as a partner. Human Factors, Systems Assurance, Publications, and Architecture each contributed individuals to form this group.

THE INTERFACE REVIEW GROUP

Purpose, Composition, and Duties

The specific goal of the interface review group (hereafter "the group") is to reduce the exposures of late changes resulting from reviews by the non-development groups, eliminating potential problems early in the development cycle. The group is to ensure the timely development of a usable, consistent, appropriate and complete interface for the product. The group is to make sure that everything the end user sees on the display screen, in the publications and on the keyboard meets his needs and expectations.

The User Interface Design Architecture group invited the interested functions to join the group and organized the effort. The architecture manager saw the work group as a way to involve these groups at an early date and to cut down review time. The ultimate responsibility for the group remained with the architecture department.

The group includes two members from Information Design and Development (one writer and one editor), two members from Assurance Human Factors, two members from Development Human Factors and one member from the product's User Interface Design Architecture group. The representative from Architecture acts as the group coordinator. Time commitments of these members varied. The members from IDD participated in the group, but dedicated 25% of their time to the group. The members from Assurance and Development Human Factors generally worked full-time on the group. The coordinator from Architecture worked half-time with the group. No members from Programming Development were members fo the group.

The specific duties of the group are to:

-- Produce guidelines for the development of menus, messages, helps and all other dialogues displayed on the screen;

-- Review published specifications of components of the interface soon after they appear, to ensure compliance with the guidelines and to review layout and terminology used in menus;

-- By using the system viewpoint, construct and review scenarios of user tasks to make sure that the interface design is coherent and consistent for the whole of the product, not just for particular components of the design;

-- Consult with developers of the interface, helping developers resolve problems uncovered by the scenario walkthroughs.

The group has published the guidelines and the scenarios for internal use by the development community working on the product. Both sets of publications include detailed information documenting the product interface.

Initial Challenges

Certain challenges existed in the creation of such a group. Difficulties in organizing the group stemmed from the fact that we had no precedent in guiding and coordinating work of this nature among functions and from the need for a high degree of clarity of purpose.

Initially, considerable time had to be spent on the fundamental issues of defining ourselves as a group, our purpose, our goals and our procedures. We had to begin thinking of ourselves as a viable working group, distinct from the functions we represent. The first concern was the time commitment involved. Our time commitments as noted above were agreed to.

With time commitments made, we began to think as a group rather than as individuals representing separate functional concerns. The group's purpose, goals and functions were established at this point. The group determined that guidelines for the writing of all aspects fo the interface system had to be established. We received a commitment from Programming that the guidelines would be adhered to.

Scenarios: An Afterthought

The group became busy reading the numerous specifications being written with the aid of the guidelines. As we read the specifications, we began to question the coherence and consistency of the interface design for the whole of the product. We found ourselves raising more and more questions for which we did not have the answers. We found that in some cases no answers existed and in other cases answers existed that were not consistent with other authors. The need for broad consistency checking became evident. The group redefined its mission to concentrate on this issue. We decided to construct and review scenarios of user tasks.

THE GROUP'S WORK

Guidelines

The following discussion of guidelines will attempt to outline some of the most pressing theoretical issues that our group had to deal with in the process of establishing guidelins for specification authors. The discussion will cover the design of menus in general and the design of helps.

Menus

One of the first questions to arise is the question of purpose. In the design of menus, do you want to try to teach a command interface via the interface menus, or do you simply want to make the interface simple? This has bearing on the conceptualization of all menu designs.

Next, you face mechanical questions. What size best suits your needs? Decisions of width and height must be made considering material that will be covered. In conjunction with size, you will want to consider placement.

Next you must consider how menus are invoked and made to go away. These questions are compounded by optional parameters or choices that may key off of selections made in a main menu. In these cases, does the main menu go away and yield to a submenu returning when the submenu is completed?

How are fields selected? Do you select a field before entering the selection, or do you join these processes? If you first select a field prior to entering its value, does the field become highlighted or in some other way indicate to the user that a field is selected? What happens to the cursor when a field is selected?

Once all selections on a menu are made, what course of action is taken? When entered, does the menu immediately go away? Does it remain on the screen until an initiated activity is running?

Last of all, do you provide helps? If so, for what?

Helps

Helps are simply a praticular type of dialogue; many of the above comments apply.

First, do you have helps online or in your publications? If you opt for publications, your user must be aware of where they are and how to use them. The help dialogues must be worked into a comprehensive indexing scheme and must be easy to find and use.

Ascertain the goal of your helps. Anticipate your audience needs. Do you have professionals in mind who will be familiar with jargon and procedures? If so, helps need not be as comprehensive and hand-holding as for new users.

Helps are only a type of dialogue, yet they are unique enough to present a few special problems, and they are growing in importance.

Scenarios

As mentioned above, the group found the need to construct and review scenarios of user tasks. The purpose was to obtain a system-wide viewpoint across all functions of the product. The group chose to concentrate on a set of typical user tasks including: customer set up, software installation, problem determination, text editing and system utilities. In reviewing for consistency across all applications or similar functions, we were able to suggest changes for the proper integration of their parts into the whole.

Group members participating in this work were from: System Design Externals, Development Human Factors and Assurance Human Factors. By working together as a single group, these individuals could accomplish their respective missions in a quicker and more efficient manner.

A list of proposed scenarios was developed. The members of the group to be responsible for creating each scenario were selected based on their background. In this way a minimum amount of learning time was required before effective work could be accomplished. The proposed scenario list was then reviewed by all interested parties.

Each member constructed a set of related scenarios. This began with a preliminary first pass at documenting the flow of information to the user. The scenario author depended upon development's documentation, the individual specification authors, and the software architects for the needed information.

Questions and problems relating to specific scenarios were documented and tracked. The scenario author would then discuss the problem and propose resolutions with the architect and the developer. The problems were tagged as: 1)resolved, 2) future review of proposed correction required, 3) acceptable work plan in place, or 4) rejected by Development as having no adequate solution or as being an insignificant problem.

When the majority of the questions and problems were resolved, the scenario was revised and the detail needed to fully document the user's view of the task was added. The scenario was again reviewed by the entire group. At that time, a sample of users viewed a simulation of the interface.

The simulation was performed using a general user interface simulator developed internally at IBM Austin. The simulator allows screens to be dynamically presented based on a predefined relationship. In this way, all potential dialogues for an individual scenario may be displayed in the correct sequence regardless of the specific user inputs. This served two purposes.

First, it was a method by which we couls "see" the flow of information as described in the scenario. This was useful in finding potentially misleading branches in the user decision process during the early phases of development. Second, it allowed us to look at the potential for visual complexity on the display. As separate components came together into an integrated package, the scenarios were expanded and viewed in simulation to ensure the ease-of-use for the total product. As a result of this work, a completed scenario, including proposed user interface changes, was presented to Development following this final refiew.

The contribution of the scenario development and review process was the uncovering of holes or inconsistencies across the product's specifications. It served to sensitize Development to the concern and awareness for potential user interface problems while proposing solutions to identified problems.

CONCLUSION

The members of the interface review group discussed in this paper realize the future possiblity of facing compressed development schedules. In such a case, the normal development and review processes have to be altered, and our group's formation is a result of such circumstances.

Our initial objective, that of creating guidelines for a user interface design, changed and became amplified as time passed and work progressed. In addition to drafting guidelines for the design of an interface, we constructed and reviewed scenarios of user tasks and reviewed, in detail, the final design of the interface.

Parallel with the group's expansion of responsiblities was a growing awareness of and interest in the group's activities. As a result, the group was able to have more of an impact than otherwise might have been the case. The group members and management responsible for the group feel that a significant contribution to the development and consistency of a user interface design has been made.

TRAINING SYSTEMS

Human Factors in Organizational Design and Management
H.W. Hendrick and O. Brown, Jr. (Editors)

TRAINING EFFECTIVENESS ANALYSIS SYSTEM
"An Organizational Capability"

Richard B. Wessling

Dynamics Research Corporation
60 Concord Street
Wilmington, Massachusetts 01887

and

Gilbert L. Neal and Thomas L. Paris

U.S. Army TRADOC Systems Analysis Activity
White Sands Missile Range, New Mexico 88002

ABSTRACT

The establishment of the Training Effectiveness Analysis System was an organizational response to a recognized need. This paper will explore the need and the actions taken to satisfy a structured and disciplined evaluation approach to training analysis and management.

INTRODUCTION

How does a large military organization with a world wide commitment provide for the training needs of over 785,000 soldiers, hundreds of hardware systems, and many different types of units and organizational levels? These training needs include training development requirements, training program delivery, training product support, and training evaluation and feedback mechanisms for both new and currently deployed systems. This paper will address just one of the organizational approaches used by the U.S. Army Training and Doctrine Command (TRADOC) to assess the validity and reliability of the Army Training System (Department of the Army, 1981). The approach to assess the effectiveness and efficiency of the Army Training System from a training point of view is accomplished through the use of the Training Effectiveness Analysis System (USA TRADOC, 1984b).

THE ARMY TRAINING SYSTEM

The Army Training System (Department of the Army, 1981) was established to provide coherent directions and guidelines for the management and implementation of military training for the Total Army. This system is codified in Army Regulations. It has only one organizational training goal and that is to develop a combat ready force which is physically and psychologically prepared to fight. The Army Training System carries this mission out through the establishment of objectives, training strategies, assignment of responsibilities, and allocation of resources. Major components of this system are individual training, unit training, and training support. TRADOC is responsible for the establishment, implementation and evaluation of these components. Both the training institutions (i.e., schools) and field commands are to insure that individual and collective training and the required training and training support requirements are achieved. TRADOC has further defined its mission in terms of training functions. These functions are to analyze, design, develop, implement and evaluate its responsibilities in support of the Army Training System and the Army training goal. To satisfy these functions does require a structured and disciplined effort. This structuring began with the "Model 76" reorganization of TRADOC. It continued through the associated establishment of procedures, the focus on criterion referenced and performance oriented training based on critical tasks to establish proficiency, the highlighting the need for evaluation and feedback and the impact of several associated management events which acted as a catalyst for training effectiveness analysis.

TRAINING STRUCTURE

The "Model 76" reorganization of TRADOC set the stage for the redefinition of the TEA System to meet new challenges. This reorganization recognized the need for a structured approach to training. It provided for a more analytical and disciplined approach to satisfy the training functions of analysis, design, development, implement and evaluation. The initial procedures used for these functions were governed by the Instructional System Development Procedures (USA TRADOC, 1975).

Currently these procedures have been somewhat amplified in the guidelines provided by the new TRADOC Regulation titled a "Systems Approach to Training", (USA TRADOC, 1984a). The ISD procedures were focused to institutional training while the "System Approach To Training" procedures embraced the training needs of the soldier for a particular Military Occupational Specialty (MOS) at all skill levels, with the attending emphasis on achieving soldier proficiency in the institution and field.

PROFICIENCY

A prime consideration that contributes to the attainment of organizational goals and objectives is a well-trained proficient work force in both static and dynamic conditions. To maintain organizational momentum, work force lost through attrition, turnover, and transfer must be replaced. These characteristics are especially true of the Army. To insure that replacement personnel are proficient, the organization must provide for an efficient training function. Furthermore, the proficiency of the work force must be maintained. Organization missions, goals and objectives change. Thus, an organization must have the means to upgrade and redirect proficiencies of the work force to meet the change. This is accomplished by the organization's training function. Most organizations have resource constraints. This means that execution of the training functions must be effected in terms of mediating proficiency, and this must be accomplished within (i.e., budgetary) constraints.

In the 1975 timeframe, the Army faced with: integrating increasingly complex hardware systems into the inventory; the need to train soldiers to effectively operate, maintain and support these systems; limited resources; and a diminishing manpower pool, launched a series of events that led to the evolution of the Training Effectiveness Analysis System (Neal, 1982).

EVALUATION

During the period 1976 to early 1980, feedback and evaluation took a back seat to other training functions within TRADOC. The service schools were not staffed to conduct meaningful feedback and evaluation programs because the resources were needed to produce training programs and products and to train soldiers. It was during this timeframe that the TEA System contributed significantly to the identification of training inadequacies in a broad variety of military occupational specialties (MOS) and hardware systems. This resulted in a gradual shift in the scope of the TEA System. Initially all Training Effectiveness Analyses were performed to provide an input to the Cost and Operational Effectiveness Analysis (USA TRADOC, 1977). It is a good shift in scope from the early phases of life cycle system management model (LCSMM) (Department of the Army, 1975) to the later phases of LCSMM and into the sustainment phase of a hardware system life.

CATALYST

Four additional events occurred which acted as a catalyst to influence training evaluation and feedback which contributed to the reassessment of "how to" obtain better feedback and training analysis. First, was the establishment of the Branch Training Teams by each service school to go to the field units and to collect data on the training programs, policies, and products. Second, was the emphasis on the standardization of drills, loading plans, and other training outputs, which highlighted that standardization seldom went below the "form: level at the school or in the field. Third, during the development of the Army Training 1990 Action Plan (USA TRADOC, 1982) greater clarity was provided to the importance of the Military Occupational Specialty, System and Unit interfaces. The collection of feedback and the prompt and accurate assessment of data with emphasis on individual and collective proficiency gave evaluation equal recognition with the other training functions. Fourth, the establishment of the Department of the Army, Force Modernization Office gave added clarity to the importance of the change in scope. Emerging from the force modernization efforts was a requirement to assess the training impacts of the introduction of new hardware into the field units. This training assessment needed to expand into doctrine, logistics, organization, and facilities. This assessment could provide for a quick correction to the next unit achieving initial operational capability. Out of these activities emerged the redefinition of the TEA System into the Developmental TEA Category and the Post Fielding Category.

CONTEXT

Life Cycle System Management Model

The U.S. Army Materiel Acquisition Process follows a model called the Life Cycle System Management Model (LCSMM) (Department of the Army, 1975). The LCSMM is outlined in Figure 1.

Army weapon systems, in general, are developed and fielded in accordance with this process and model. The model has four phases. Also indicated is a sustainment phase, which is a notional phase, but illustrates graphically that another set of conditions might exist after deployment of new systems to the field. Sustainment is normally addressed as a subset of deployment. The LCSMM prescribes the actions, events, and decisions that must occur as a system progresses from requirement through concept to production and issue to the Field Army. During the process many analyses and studies are carried out. The results of these studies and analyses impact design, system support, and management. Certain types of data are needed in each LCSMM phase to provide decision makers information needed relevant to system design, support, doctrine, performance, cost, etc. Based on this information, decisions are made concerning continued development--and, ultimately production--at the end of each phase of the LCSMM.

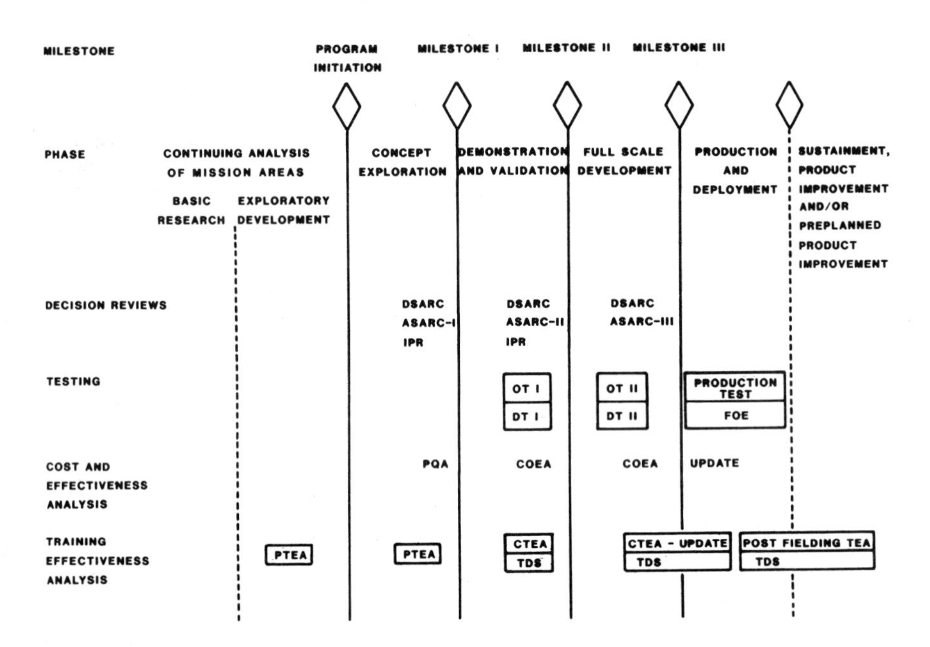

FIGURE 1. SYSTEM ACQUISITION PROCESS AND THE TRAINING EFFECTIVENESS ANALYSIS SYSTEM

The TRADOC training development process for new systems is carried out against the background of the LCSMM to insure that an effective training subsystem is available when the weapon system is fielded. The TRADOC training developments process produces products such as training programs, training devices, training manuals, training guidance, etc., for both institutional (i.e., Army School) and unit (i.e., Field Army) training.

Training Effectiveness Analysis (TEA) (USA TRADOC, 1984b) begins in the concept and validation phase--and sometimes before--and continues beyond the production and deployment phase, usually as long as the system is in the Army inventory. TEA provides data on the proficiency of individuals and/or proficiency-related effectiveness of systems, teams, and units and on the adequacy of TRADOC training development products as they evolve during the LCSMM. Prior to production and deployment, this type information serves as inputs to either the Preliminary Qualitative Analysis (PQA) or TRADOC Cost and Operational Effectiveness Analysis (COEA) studies. This information is also an input to the TRADOC training development actions and processes. After deployment, TEA is used to evaluate the effectiveness of institutional and unit training and as the basis for improving training where needed.

Since 1975 TEA has been an evolving concept. Currently, TEAs fall into two basic categories--Developmental Training Effectiveness Analysis (DTEA) and Post Fielding Training Effectiveness Analysis (PFTEA). This categorizing has resulted from the need to better align the TEA System with the LCSMM. In general, DTEAs are conducted prior to production and deployment of a weapon system and PFTEAs are conducted after the system is in the hands of the soldiers or units.

DTEAs, in general, assess training supportability and/or identify training requirements for a hardware system. There are four types of DTEAs, the

Preliminary Training Effectiveness Analysis (PTEA), the Cost and Training Effectiveness Analysis (CTEA), CTEA update, and the Training Development Study (TDS). Each DTEA type has a specific set of purposes and is intended to provide decision data at the completion of an associated phase of the LCSMM.

·· Preliminary Training Effectiveness Analysis (PTEA). This analysis provides for the identification of training issues with the anticipated target population prior to the concept definition for the early-on Cost and Operational Effectiveness Analysis. The PTEA can support training requirements, constraints, and resources decisions at Milestones 0 and I of the acquisition process. At these points, the design of the hardware may still be influenced by the findings of the preliminary training effectiveness analyses. The PTEA will be an estimate of training strategy and concept needs, focused to--who will be trained, what should be trained, and when, where, and how training should be accomplished. The output of this analysis will contribute to the Initial Outline Individual and Collective Training Plan (USA TRADOC, 1982b) for the materiel system under analysis. Generally, the costs are estimated and the effectiveness goals of training are stated but not measured or collected unless a predecessor system exists.

·· Cost and Training Effectiveness Analysis (CTEA). This analysis is a detailed training assessment or analysis of the comparative effectiveness and costs of the predecessor system with the proposed alternative hardware systems at Milestone II. This analysis shall continue to assess the output of the PTEA and the Initial Outline Individual and Collective Training Plan. The use of the Integrated Logistics Support elements, LSA, and LSAR (Department of the Army, 1983) and program management documents such as the Integrated Logistics Support Plan, Materiel Fielding Plan, Basis of Issue Plan, Qualitative and Quantitative Personnel Requirements Information and the available Reliability, Availability, and Maintainability Data will enhance training estimates. The wartime and peacetime mission profile can greatly contribute to a comprehensive analysis of all factors which relate to or interact with training issues. The CTEA will add muscle to the Initial Outline Individual and Collective Training Plan. It will also provide more objective data concerning manpower needs, personnel requirements, Military Occupational Specialty requirements by skill level, and control specialty requirements. Both individual and collective training requirements, organizational structuring, doctrine implications, and the need for institutional and/or field training devices or simulators are addressed. The results of this needs assessment will be the recommendation of a preferred training alternative to include costs which has been subjected to sensitivity analysis or a tradeoff analysis among the key factors. The recommendations of this analysis can be used to verify or amend previous training strategy and concept decisions contained in the Outline Individual and Collective Training Plan for the hardware system and individual training plan(s) for specific Military Occupational Specialties or Officer's Specialty Skill Identifiers. This analysis is normally conducted at Milestone II of the acquisition strategy.

·· Cost and Training Effectiveness Analysis Update. This analysis should occur during the Operational Testing II or III or even during a follow-on evaluation for a developing system. This analysis generally measures the worth, value, effectiveness, and efficiency of all or part of the implemented portions of the Outline Individual and Collective Training Plan to include New Equipment Training for individuals, crew or units. Generally, this analysis is focused to the initial transition and sustainment training as it is actually conducted and codified in the New Equipment Training Plan. The purpose is to measure the design effectiveness of the man-machine combination prior to, during, and after training. The actual effectiveness is then determined and if actual effectiveness is between the required effectiveness and the design effectiveness, the training goals have been achieved. Costs are still of concern to establish the optimum effectiveness and reduce costs for each task taught. If the measured results are below the required effectiveness (training standard), the corrective measures are recommended to the training development process, training implementation and/or training evaluation process. It is important to note that this classification is heavily dependent on the actual measurement of the soldier's and crew's proficiency. This analysis verifies and validates the recommendations provided by the CTEA which is used to derive the Individual and Collective Training Plan (USA TRADOC, 1982b).

.. Training Development Study (TDS). The TDS is normally embedded in a CTEA Update for a major system acquisition when a training device or simulator is a part of the training program. However, during the performance of one of the above or in response to an approved training device requirement, and/or training device letter requirement, a training development study can be conducted independently. The TDS evaluates the effectiveness and efficiency of a training device or simulator. This analysis should recognize that simulators requirements, generally scenario dependent, is individual and collective task-oriented since the fidelity of the device or simulator must be equal to performance required on the actual equipment. Additionally, the training device or simulator must satisfy the hardware, doctrine, and software tasks when the device or simulator is to provide meaningful operator (student) feedback relative to anticipated engagement criterion and parameters.

Post-fielding Training Effectiveness Analysis (PFTEA). The PFTEA is a major category of the TEA. It is conducted after the Initial Operational Capability date for an emerging system. It can be conducted on a current system as a part of a training assessment to support a product improvement program and/or to evaluate compliance with training standardization goals. The analysis includes the examination of the soldier, system, and unit level training requirements and actual training practices in the unit using standardized products, programs, and procedures developed by the Training and Doctrine Command. It is a systematic, structured, and specialized methodology to gather data and conduct hands-on performance evaluation to verify the quality of products and programs in the field. The post-fielding procedures are capable of examining the eight training elements: initial; integration; cross; train-up; transition; mobilization; sustainment; and mobilization at the institution and in the field. This analysis can examine doctrine, materiel, facilities, and organizational considerations to capitalize on quality and proven field practices that can be transferred to other units. It also identified deficiencies in unsatisfactory training products, programs, and procedures that were found in the field and institution. A graphic illustration of the Training Effectiveness Analysis System is provided in Figure 2.

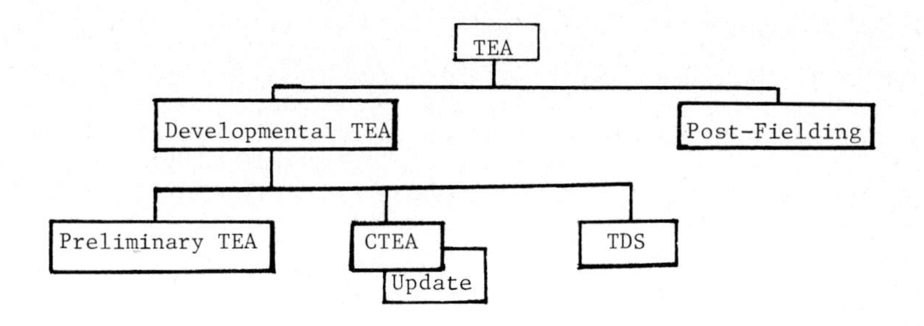

Figure 2
Training Effectiveness Analysis System

TRAINING EFFECTIVENESS ANALYSIS SYSTEM

ORGANIZATION

U.S. Army Training and Doctrine Command

The TEA System serves the United States Army Training and Doctrine Command. This command is composed of three Integrating Centers, the Combined Arms Center, the Logistics Center and the Soldiers Support Center. The Command is also composed of twenty-one branch schools. The Training and Doctrine Command exercises its mission through the Integrating Centers, Army Training Support Center, and the Schools. The authority and functions are governed by regulation (Department of the Army, 1981). The Training and Doctrine Command directs mission and function tasks to subordinate elements through a regulation (USA TRADOC, 1980). The Deputy Chief of Staff for Combat Developments, TRADOC has staff responsibility for all Studies and Analyses, and exercises management over this area. However, the Deputy Chief of Staff for Training, TRADOC has staff responsibility for training management.

Army Training Support Center

The mission of the Army Training Support Center includes monitorships of the TEA System. The results of the TEA System applications contribute to the overall knowledge and direction of the individual and collective training plan, evaluation and feedback networks, training management concerns, and training device and simulator requirements.

Army Service Schools/Army Training Centers

The benefactors of the TEA System are the Army Service Schools and Army Training Centers. It is the Commandant of the Army Service School who is the proponent for performance of a TEA. It is by his/her authority that a study is initiated. It is for the following benefits in terms of individual and/or collective proficiency resource conservation and effectiveness and efficiency that a TEA is performed. So the control of the uses of the TEA System rest with each Service School or Army Training Center. Within the school the staff supervision of the TEA System is exercised by the Directorate of Evaluation and Standardization yet all school activities participate in this system.

TORA/TRASANA

The TRADOC Systems Analysis Activity (TRASANA), an element of the TRADOC Operations Research Activity (TORA), is responsible to Headquarters, TRADOC for the performance of training effectiveness analyses, development of the methodology, maintenance of the data bases, and to provide training and consultant services concerning the TEA System to the Service Schools and/or Army Training Centers.

TEA Division

TRASANA's Training Effectiveness Analysis Division performs the above functions and is composed of four branches. Three of these branches are located at TRASANA Headquarters at White Sands Missile Range, New Mexico. The fourth branch is located at the Seventh Army Training Center in Germany. The TEA Division performs all types of TEAs for the proponent schools and centers.

SCOPE

A very brief description of the organizational scope of the training effectiveness analysis system includes the following:

- The training effectiveness analysis system is an evaluation tool of products, processes, programs, and procedures, which interacts with training at Army schools or field units for the soldiers, system, and unit.

- It is an independent assessment of the relationship of materiel, organization, and doctrine viewed from a trainers and human factors point of view to determine the individual and collective "fit" of unit readiness requirements.

- It uses an interdisciplinary approach combining operations research, system analysis, educational research, training technology research, military operational skills, and human factors engineering to estimate, assess or evaluate training products and programs at the school and in field units.

- It is designed to meet Training Effectiveness Analysis requirements during each phase of life cycle system management.

- It provides timely training considerations to the combat and materiel developer prior to finalizing of the hardware design.

- It provides managers with logistics, organizational, manpower, personnel, and human factors considerations which impact on training and associated benefits trade-offs.

- It provides for training feedback to the appropriate organizational levels to manage change in support of force modernization actions and the system approach to training.

- It is normally conducted on one or more of the following organizational levels: the soldier, the system [crew], and the unit [organization].

METHODOLOGY

It is appropriate to highlight the factors which form the methodology of the Training Effectiveness Analysis. Underlying each study is a central concept which provides direction to the analysis and is applied in whole, or part, to each analysis. This central concept is that soldier proficiency must be achieved prior to the examination of system (crew) and unit proficiency. Supporting this assumption is the idea that the trainer, training subsystem, hardware, and training environment also contribute to the soldier's proficiency and each factor must operate within some recognizable tolerance of effectiveness or efficiency which we will also address as proficiency.

The methodology recognizes that the ultimate concern of training is to provide for successful mission accomplishment by soldiers, systems (crews) and units. Therefore, the methodology includes an examination of individual training and, at times, collective training at the school and in the field units. It begins with the soldier and continues through the system or crew and ends with a unit training assessment in either a pure unit setting or task force organization.

By definition, the analysis addresses the area of cost analysis. It is unfortunate that this area of costing can be viewed differently at different organizational levels and the total implication of costing methodology are seldom understood by the generalist. Training Program costs are generally compared in the following manner:

- Equal effectiveness/variable costs.
- Followed by a sensitivity analysis.

DEFINITION OF THE FIVE PROFICIENCY FACTORS

It is essential at this point to say more about the five proficiency factors which are the centerpiece of the Training Effectiveness Analysis. The central concept is that proficiency, how well someone performs his or her job, is determined by five proficiency factors and their relationships. It is the measurement of the proficiency of the soldier-trainee, the trainer, the training subsystem, the design effectiveness of the hardware, and training environment factors to support training that provides the objective data for analysis (USA TRADOC, 1982a).

The first factor is the soldier. The soldier, alone or in combination with materiel, must be able to perform assigned military duties and survive. The examination of the soldier includes several dimensions. The soldier's mental ability, aptitudes, physical ability, previous training and experience, motivation, attitudes and perceptions contributed to achieving proficiency. Successful completion of training can be predicted through the analysis of these dimensions. Further, training can alter most of these dimensions. The need to measure the soldiers-trainers proficiency is central to the purposes of the evaluation of training.

The second factor is the trainer or instructor. Training can only be provided by knowledgeable, effective trainers. The success of training is related to the trainer's technical and tactical knowledge of the subject matter, and the trainer's skill in communicating or transferring their knowledge to the soldier-trainee. This is true of the trainer no matter what the training environment. So it is equally important to determine if the trainer has the prerequisite skills and knowledges and can motivate the soldier-trainee to seek improvement in his or her military job skills. The maturity and judgment of the trainer must be displayed in the proper attitudes toward his soldier-trainee and the subject matter to insure that the correct perception of the goals of training are understood by all.

The third factor is the training subsystem. The assessment of the training subsystem is generally related to the Individual and Collective Training Plan for systems and the Individual Training Plan for each applicable Military Occupational Specialty (MOS) for the enlisted and warranted personnel and the skill specialty indicator for the commissioned officer. Each of these documents contains the training processes, training programs, training support materials, and training devices and simulators required for each skill level for enlisted personnel or the military quality standards for officer training. The training subsystem includes all training activities conducted at the institution or in the field for indivdual proficiency and/or collective effectiveness in meeting operational mission requirements. It is important for the training subsystem assessment to include consideration of effectiveness and efficiency in meeting analysis, design, development, implementation, and evaluation requirements. The examination of the training subsystem components also includes a detailed look at formative, summative, and ultimate evaluation programs, policies, and procedures.

The fourth factor is the hardware system. The basic question to be answered is whether the soldier is capable of doing what the hardware demands. Is there a match between soldier capabilities and hardware demands? Does this match provide for optimum proficiency of the man-machine combination? Does the man-machine combination properly respond to the doctrine and operational environment tasks when the total system is placed in action? This part of the analysis is human factors engineering oriented and includes the requirement to determine the required effectiveness of the man-machine combination if none exists. This required effectiveness becomes the benchmark to determine the hardware system proficiency. It may include a collective or crew analysis to determine the placing of "steel" on the target or live firing, or simulation of live firing, or a combination of both!

The fifth factor is the training environment. Analysis of the training environment either in the institution or the field units includes an assessment of how training is managed, to what extent training is emphasized, to what extent it is funded and supported, the manner in which operational, logistics and administrative functions are enhanced by training or vice versa, the availability of training time, the availability of facilities, resources, funds, and supplies, the frequency of training, both individual and collective, and the command <u>emphasis</u> and <u>actions</u> relating to training. Also examined are the training distractors, such as work details, operational commitments in excess of the system's mission profile, and peacetime status, equipment malfunctions, special projects, and lack of knowledgeable trainers to implement and sustain standardized training programs using available training support materials.

All of these five factors contribute to constructing an objective set of data for each. This data is examined as to its impact on each of the other five proficiency factor areas. Any degradation in any of these factors can be expected to produce a concomitant degradation in training effectiveness and soldier proficiency and unit readiness.

Training effectiveness is assessed using standardized tests, special purpose tests, cost analysis, special purpose surveys, and demographic surveys. Each of these areas are addressed below (USA TRADOC, 1982a).

Standardized Tests

Standardized tests are tests intended for widespread use to assess general attributes, i.e., abilities or traits possessed to some degree by virtually everyone. Standardized tests are used in TEA to determine sample representativeness, to develop selection criteria, to assess the appropriateness of training materials, and to assess soldier characteristics.

Special Purpose Tests

In addition to standardized tests, it is frequently necessary to develop special purpose tests and surveys to answer specific questions posed in a particular study. Two types of special purpose tests are used frequently in the TEA process: paper and pencil proficiency tests, also called skills and knowledge (S/K) tests; and performance proficiency tests, also known as hands-on tests (HOT). Written proficiency tests are used to assess the soldier's knowledge and understanding of what he has been taught. Performance proficiency tests require a soldier to actually perform tasks he was trained to do. Written and hands-on proficiency tests provide the most direct and objective data by which a training program can be evaluated and serve as the primary measures of training effectiveness (MOTE). For the most part, proficiency tests are developed (with the assistance of subject matter experts), administered, scored, and interpreted by a team of analysts.

Special Purpose Surveys

In addition to performance tests, soldier and trainer attitudes and perceptions also provide useful information in the TEA process. Typically, these data are obtained by various types of surveys. These surveys address general attitudes and/or specific attitudes toward hardware and training. General attitudes are assessed for a number of reasons. They serve as useful indicators of general motivation and can be used to control for differences in training and proficiency resulting from motivation and general outlook on training, the Army, and particular weapon systems. While attitudes do not necessarily correlate directly with measured proficiency, they do offer important and useful insights into the interpretation of proficiency scores and results of training. When carefully measured and interpreted, they serve as a valuable "temperature check" on morale and motivation.

Another frequently used survey is an assessment of soldier perceptions of the tasks related to his job. It is important to determine which tasks are the most important to the successful completion of a job or mission and whether these critical tasks receive corresponding emphasis in training. To determine the extent to which training addresses actual job requirements, several TEA's have included a "Task Frequency, Criticality, and Performance Survey" (which may be expanded to include perceptions of task difficulty and task training requirements). Constructing these surveys may be difficult for a developing system because task lists often are not available.

Demographic Survey

Another source of information in TEA is a demographic survey which yields background and descriptive information on the study participants, e.g., age, rank, educational level, and MOS. This information is used to describe the test subjects, determine if the sample is representative, and determine if demographic information is related to performance measures. Another use of demographic information is to compare two or more study groups to insure their approximate equivalence in terms of performance related variables.

Cost Analysis

The cost analysis associated with a TEA is aimed at the evaluation of economic considerations on which decisions will be made. It typically follows the same analytical process that is inherent in any cost analysis. The costing approach is conceptually and methodologically similar to the approach normally undertaken for Cost and Operational Effectiveness Analysis (COEA). The procedure for a cost analysis is as follows:

- Define objective
- Define alternatives
- Formulate assumptions
- Develop methodology for estimating cost
- Determine quantities cost
- Perform a detailed analysis
- Analyze uncertainties
- Write report

Tailoring

This methodology is tailored to meet the objectives and goals for each type analysis depending on which phase of the system acquisition process is being addressed. Of course, the Post Fielding Training Effectiveness Analysis takes the system into the operational or sustainment training world.

CONCLUSIONS

The Training Effectiveness Analysis System is not a concept, it has been working since 1975 and includes 60 completed studies. It has grown to meet the expanding needs of front-end analysis to estimate the training issues, program and resources for a developing system. It has also extended its scope into the area of post-fielding analysis at three organizational levels - the soldier - the system (crew) - and the unit.

AREAS OF APPLICATION

The track record of the Training Effectiveness Analysis System can be divided into four major areas of application.

- The evaluation of training which includes training on new equipment, institutional training, and field training.

- The evaluation of training devices which includes simulators for operators, maintainers, and leaders.

- The evaluation of soldier selection criteria for new systems or in response to a "red flag" such as high attrition rates and/or low enlistment rates.

The evaluation of soldier/hardware interface for developing systems or systems undergoing a product improvement program with emphasis on design or human factors engineering which adversely affect training and operational performance.

The scope of these applications contributed to real-world changes to training, to soldier proficiency and to unit readiness. It is apparent that the Training Effectiveness Analysis System is not only a control or evaluation tool, but it also contributes directly to organizational development to include the soldier, system (crew), and various organizational levels.

REFERENCES

Neal, G.L. Overview of training effectiveness analysis. In R. E. Edwards and T. Tolin (Eds.) Proceedings of the Human Factors Society, 26th Annual Meeting. Santa Monica, CA: Human Factors Society, 1982.

Department of the Army (DA) Life Cycle System Management Model for Army Systems (DA PAM 11-25). Washington, D.C.: Headquarters, May 1975.

Department of the Army (DA) Training-Army Training (Army Regulation 350-1). Washington, D.C.: Headquarters, 1 August 1981.

Department of the Army (DA) Military Standard, Logistics Support Analysis (Military Standard 1388-1A) (Draft). Washington, D.C.: Headquarters, 11 April 1983.

U.S. Army Training and Doctrine Command (USA TRADOC) Interservice Procedures for Instructional Development (TRADOC Pamphlet 350-30). Fort Monroe, VA: Headquarters, 1 August 1975.

U.S. Army Training and Doctrine Command (USA TRADOC) Cost and Training Effectiveness Analysis [TRADOC Pamphlet 71-10 (Draft)]. Fort Monroe, VA: HQ, TRADOC, 1 November 1976.

U.S. Army Training and Doctrine Command (USA, TRADOC) Combat Development Studies (TRADOC Regulation 11-8). Fort Monroe, VA: HQ, TRADOC, 18 Mar 1977.

U.S. Army Training and Doctrine Command (USA, TRADOC) Mission Assignments (TRADOC Regulation 10-41). Ft. Monroe, VA: HQ, TRADOC, 1 May 1980.

U.S. Army Training and Doctrine Command (USA, TRADOC) Army Training 1990 Action Plan, Draft II. Ft. Monroe, VA: HQ, TRADOC, 25 February 1983.

U.S. Army Training and Doctrine Command (USA, TRADOC) Training Effectiveness Analysis, A Process in Evolution (TRASANA Technical Report 49-82 (Revision 1)). White Sands Missile Range, NM: HQ, TRADOC Systems Analysis Activity, October 1982a.

U.S. Army Training and Doctrine Command (USA, TRADOC) Schools - Individual and Collective Training Plan for Developing Systems; Policy and Procedures (TRADOC Regulation 351-9). Ft. Monroe, VA: HQ, TRADOC, 8 November 1982b.

U.S. Army Training and Doctrine Command (USA, TRADOC) System Approach to Training [TRADOC Regulation 350-7 (Draft)]. Fort Monroe, VA: HQ, TRADOC, 11 January 1984a.

U.S. Army Training and Doctrine Command (USA, TRADOC) Training Effectiveness Analysis System, [TRADOC Reg 350-4 (Draft)]. Ft. Monroe, VA: HQ, TRADOC, February 1984b.

DISCLAIMER

Opinions and conclusions of this paper are those of the authors and do not necessarily reflect the view or endorsement of the Department of the Army and Dynamics Research Corporation.

ACKNOWLEDGEMENTS

The constructive criticism of Mr. Vernon Wilson and Mr. Terry Palmer is acknowledged. A special thanks is in order to MS Norma McGill who carried out the typing and format structuring of this paper.

Human Factors in Organizational Design and Management
H.W. Hendrick and O. Brown, Jr. (Editors)

DATA AGGREGATION PERCEIVED BEHAVIORAL CHANGE, AND ORGANIZATIONAL TRAINING PROGRAM VALIDATION

George M. Porchelli
Massachusetts Mutual Life Insurance Company
Springfield, Massachusetts 01111
U.S.A.

Various validity models have been proposed for assessing Corporate Management Training Programs. The majority of these models espouse a Pre-Post Design. This research used a Post-, Pre-Post Design. The attempt was made to validate manager "back on-the-job behavior change" via a post program questionnaire that asked "OTHERS around the manager" to rate what the manager's behavior "was" vs. what the behavior is "now". Two added dimensions of this research were (1) the emphasis on "perceived behavioral shift" as the truer measure of program validity, and (2) the method of "Aggregation of Data".

Introduction

Organizational training programs are coming under increasing pressure from top management to not only evaluate a costly training program, but to validate the program in some meaningful quantitative manner. Classical psychological research methods have limited usefulness, cost effectiveness, and practical value. The psychological emphasis on "pre-post" testing and skill measurement is appropriate to "skill acquisition" training, but has obvious limitations when applied to the validation of a management training program.

An alternate approach to management training program validation is to utilize the measurement process of "aggregating" behavior responses over situations, occasions, and training program areas to diminish the effects of fortuitous/uncontrollable factors effecting the research design and outcome. Secondary to this approach and specific to management training is the shift in experimental emphasis from measuring "skill acquisition" to the measuring of "perceived behavior change" as the critical dependent/outcome variable: such "shift" being tested using a post-, pre-post design. Such an overall approach was deployed to evaluate and validate a five-day management training program that was costing the company approximately $20,000 per year.

Organizational Realities and Science

To conduct research in an organizational setting is not an easy task. What this researcher believes is needed is a research model that can "flow with organizational realities", not interfere with them.

What Organizational Realities? Can science find a methodology that best fits the situation, is cost effective, timely and statistically sound both in theory and application? The basic realities that the researcher must deal with are: (1) turnover, (2) production demands versus research "time-to-complete", (3) sampling procedure control, (4) research grouping and control, (5) managers as difficult to control and monitor in the standard scientific way, (6) manager skills take longer to measure and evaluate their impact on the organization, (7) political factors often play roles that are uncontrollable, (8) statistical and computer capabilities for data handling and processing are not widely available, (9) the demand by top management for immediate results in ever present, (10) minimal appreciation for standard research requirements by management as a group, (11) managers are transferred, rotated, and fired, etc., (12) the organization is a fluid environment, (13) managers enjoy feedback but do not like to be studied, (14) validity is often "what works" and not what was efficient and scientifically sound, (15) organizations are continually in a state of change, and finally, (16) the human factor in organizational research is highly unpredictable.

To cope with the above difficulties, three research strategies were used. They were selected because of their inherent possibility of "best fit" to organizational realities and the approximate necessities of science.

The three stratgies used were:

1. Aggregation of data to handle organizational change and the power of uncontrollable influences.

2. Perception of human behavior versus skill measurement as the key criterion of valid human factor outcomes of manager training.

3. Post-, pre-post design as the best design for meeting the criterion of timeliness, effectiveness, and validity.

Despite the reality of a changing organization, there is certain stability of structure and behavior within the organization. Thus, scientific natural observation of human behavior in its "natural setting" is possible without the unfortunate emphasis on scientific experimental control. In fact, the artificiality of scientific control may in many respects be suspect as a means to valid and reliable outcomes for prediction and wider generalization.

Organizational Realities and Management Training

In the effort to validate a management training program, the widest scope of methodology and expertise must be applied. It is not enough to isolate a group of managers under experimental conditions and study outcomes. When studying the validity of a management training program, the researcher must look at the following brief list of overall considerations in the realities of the organization: (1) the total population called "management" must be determined to evaluate whether a given program is at the right level and meeting the needs of that level; (2) the program must be in line with preparing for such organizational concerns as rotation, succession planning, and providing a ready pool of well trained managers for the next higher level; (3) the skill of "managing people" is more complex to measure than the measurement of "skill acquisition". Thus, different measurement techniques are called for from the researcher; (4) managers are rarely brought in for training because they are deficient in a specific area. In many cases the programs have an inherent "reward the successful" tone to them. This makes validation "shift scores" difficult to demonstrate, (5) some organizational validation approaches administer tests of progress during the actual training sessions. This upsets the spontaneity of the program and often is self-defeating to the researchers purpose, (6) often the measure of whether a program is valid or not is determined by an obscure telephone call or participant verbal remark which keeps the originator socially and politically safe but fails to scientifically further the cause of scientific program validation; and finally, (7) often a program is measured based upon its political or smile factor rather than whether it met the marketed and purchased goals.

In conclusion, the training programs that train managers present a unique situation. What is called for is a unique methodology for validating these programs.

The Solution to Evaluating and Validating A Management Training Program (Perception and Aggregation of Data)

In discussing the methodology used in this validation project, I would like to reference an article in the September 1980 issue of the American Psychologist.[1] The article, "The Stability of Behavior: Implications for Psychological Research", by Dr. Seymour Epstein, makes specific reference to two (2) principles that this project utilized. They are the finding that (1) "personality scales and other self-report procedures are stable, but that objective measures are not" when it comes to estimating the stability of personality variables, and (2) we can utilize the method of aggregating data of behavior "over a wide range of variables so long as the behavior in question is averaged over a sufficient number of occurrences." A corollary to the second principle is "reliable relationships can be demonstrated between ratings by others and self-ratings...so long as the objective behavior is sampled over an appropriate level of generality and averaged over sufficient number of occurrences."

According to Epstein, "single observations on a single item of behavior" in a laboratory experiment suffer the problem of poor experimental "replication." Further, while laboratory experiments afford maximum control possibilities for independent variables, etc., the dependent variable (results) have low (poor) generalization to a wider context. Yet, as scientists, we continue to generalize rat bar pressing to child rearing. A further vulnerability of our present scientific situation is that our experiments lack a method for screening out uncontrollable influences (experimental noise) when applied to human subjects and research. In short, multiple observation leads to a more stable measure of behavior than single observation measures. In view of these findings, I decided to use post-, pre-post design and the method of aggregating data. The post program questionnaire used requested a post-, pre-post decision by employees working around or for the manager. Inherent within this rating was the perceptual judgment of change in behavior over six (6) managerial tasks (below). This one rating required the rater to rate in one sitting a perceived change in behavior, before and after training. These "temporal" judgements, summated over a six-month period after training, constitute the measure of "cross situational stability" as described by Epstein. The method of averaging constituted the main methodology about which (1) we control the influence of uncontrollable variables, and (2) have the greatest hope for replicability and transfer of research findings to the human condition in the organization.

If responses to a single event produce low stability and responses over many events gain stability, then multiple observations over many perceptually generalized events should stabilize "What's Going On" and whether our validity data is a reliable index of a change in managerial behavior. I equated the respondees (observers) to "occasions" and summated and averaged their perceptual ratings over all managerial factors (perceived behavior on the tasks of planning and organizing, monitoring and controlling, leadership and motivation, interpersonal competence, problem solving and decision making, and time management). This, I felt, would give greater data stability, relibility, replicability, and transferable generalizability to the real world of management training outcomes within the organization. There is no way to control all "live" variables in the management situation. To compensate for this, the method of aggregating data is an excellent tool for validating management training programs. Aggregating data "peaks" the factor(s) we're measuring and diminishes the "uncontrollable" variable influence factor(s).

Epstein makes note of two types of stability: (1) temporal, and (2) cross-situational. This research utilized both but focused, specifically, on managerial human factors across situations on an everyday managerial life cycle basis. For example, how does the manager do on planning? The questionnaire respondees' and manager's response is a forced global perception over many reflected (summated) "temporal" perceptions regarding the manager's planning performance, i.e., my boss (on planning) plans at a "3 level". The combined perceptions of multiply-perceived situational behaviors of the manager yield the data we're interested in. When aggregated over all respondees, we have a stable picture of the manager's ability to plan.

The next question is whether any change in that planning ability (human factor) has occurred as a result of a management training program. The post-, pre-post decision is an assessment of perceived change or "shift" by those most affected by the manager's ability to plan effectively. Training program evaluators tend to assume that skill change, practice and reinforcement are casually related. In skill acquisition level type training programs, this is logical and experimentally supported.

However, in management training programs, the higher order skills are more like the "hall of mirrors" (they are very complex and subtle). The single observation is possible, but a "red herring", i.e., one observed manager decision, may appear an effective short-term decision but the longer-range consequences may go totally unnoticed and be totally ineffective. A more appropriate methodology is to average data across situations, data that is derived from those who have an on-site perception of it over time and can make an on-site one-sitting decision. In short, I know if my boss plans well or not.

Six months after training, the researches asks for a post-, pre-post decision on the manager's planning ability. After six months, the new post-, pre-post aggregated perception is 4.5. From this, we see that the data indicates a perceived "shift" in the manager's planning ability. Thus, the research tells the organization that the training program is valid and works. "Maybe I'll send someone else to the program" is top managements reaction to the program. Organizations can't afford pre-planned buildings with no toilets, cars that explode, and products that no one knows how to service or use.

From this point in the research, the researcher then aggregated average scores over all sub-tests (stimuli). Each sub-test had 8 questions specific to the above-mentioned human managerial factor. Collectively, these items yield a relatively stable measure of the specific factor. We then averaged together, for each sub-test, all questionnaire respondees who observed the manager useage of the human factor. This led to aggregated data on each factor from all respondees who directly observed the manager. Aggregating sub-test scores for all respondees and all managers, I believe, decreases the "error of measurement" by averaging out uncontrollable influence and still lets the organization alone to "get the product out!"

This post-, pre-post design was a "single occasion" rating. Perhaps a more applied example might make the point. When you pick up your car from the garage after the fix-up work is done and it doesn't work, you make a post-, pre-post decision. Based on the results, you either "go back" to that garage for further service on your car (buy another program) or you never "use that garage and mechanic again" (cut the program). It's a post, one-occasion decision, and it's final. Your decision is ultimately based on your aggregated temporal experience with that garage (training program). Finally, according to Epstein and regarding single ratings following multiple or extended observations, "although the ratings consist of a single response, they represent an intuitive averaging of many observations. As a result, such ratings have the potential for producing highly replicable and valid results." Additionally, "such ratings have a very important advantage in that they need not be limited by insufficient opportunity for observation." In conclusion, this researcher feels that this total method and research system best matches organizational realities of measuring training program validity in the highly volatile and changing world of the manager and business.

[1] Epstein, S., The Stability of Behavior: Implications for Psychological Research, American Psychologist, Vol. 35, No. 9 (1980) 790-806.

Human Factors in Organizational Design and Management
H.W. Hendrick and O. Brown, Jr. (Editors)

ARTIFICIAL INTELLIGENCE AND AN ADAPTIVE TRAINING CONCEPT

Leighton L. Smith, Ph.D., and Roy E. Connally, Ph.D.

University of Central Florida
Orlando Florida
U.S.A.

Adaptive training is where trainees are instructed on a one-on-one basis. This method provides better training than the traditional training systems with high trainee to trainer ratios. An artificially intelligent training system can provide better adaptive training than can a human because it can better adjust to individual needs.

INTRODUCTION

The traditional method of training is for one person to teach many students over the course of a structured time period. This method is designed to accomodate the ubiquitous paucity of qualified teachers.

Throughput (the percentage of starting students who finish) has generally been high, and therefore training administrators have maintained that this method continues to be effective. But the designated training period is, as often as not, subjected to logistic influences rather than to training quality considerations.

In effect, the training system (syllabus, media, and duration) is not designed with any specific type of student in mind at all. Further, there is no concrete evidence that the resulting quality of training achieved via this traditional method is anywhere near an optimal level.

An additional problem is that testing techniques in the traditional method have evolved to a form which provide the trainers with only marginal feedback as to whether the trainees have adequately mastered the course objectives. Unfortunately the current form of tests is one which lacks flexibility and is generally easy for the trainer to correct.

ADAPTIVE TRAINING

The concept of adaptive training is remarkably simple, and as a matter of fact, not at all original. Essentially, the concept describes a training system where there is a one-to-one trainer to trainee ratio. Tutors have been teaching one-to-one for centuries. The 'adaptive' descriptor is used because a truly effective one-on-one training system is one which conveys the course objectives to the trainee as effectively as possible for each trainee. In other words, each trainee is subjected to a training scenario catered to his/her specific needs. Hence, the course is generally completed no sooner than the trainee is ready (i.e., has an adequate mastery of the course objectives). In addition, the duration of the course is no longer than is necessary for the trainee to achieve this level of mastery. Thus, adaptive training is a training system which is capable of catering to the individual characteristics of each trainee.

The adaptive training concept, as embodied by a human trainer one-on-one with a trainee has not achieved the degree of success that might be expected. There are two major reasons for this. First, and most obvious, there have not been, and are not, enough qualified trainers available for one-on-one training to be a viable alternative to traditional training systems. And second, human trainers (or tutors in this context) have not been adequately prepared (nor perhaps aware of the need) to evaluate and assess those individual characteristics of each trainee which correspond to successful training. They have merely provided reinforcement of those course objectives on which trainees were weak. And therefore, that is why traditionally tutors have only trained those individuals who have failed to succeed in traditional training systems. It is noteworthy at this juncture that traditionally tutors have not trained talented trainees.

ARTIFICIAL INTELLIGENCE

Artificial Intelligence (AI) is a concept describing machines which have one or more intelligent characteristics or capabilities. AI appears to be one of the most desirable goals of current computer design research and development efforts.

There has been interest in using machines to aid in training for may years. The terms CAI (computer aided instruction), CBI (computer based instruction), CII (computer integrated instruction), etc., have been used extensively to describe various approaches to the use of machines in the training environment. The incorporation of AI into a training device is an appropriate application of the technology.

To date there have been a number of successful developments of AI methodologies and techniques. Most of these have embraced intelligence only in a superficial manner. But there are numerous talanted researchers currently developing new approaches and methodologies and it will not be long before there are thinking machines performing all sorts of sundry tasks.

Basically a machine which is intelligent is one which on its own (i.e., without a human operator), can either make decisions or judgements based on entered information. The logic is to model the machine's operating system (or computer) after a human being's decision/judgement making process. The major difficulty with this approach is that the current level of understanding of how a human decision/judgement making process works is far below where it needs to be for adequate AI model development, except for very structured cases, such as chess playing and simple medical diagnoses.

ARTIFICIAL INTELLIGENCE AND ADAPTIVE TRAINING

The goal is to develop a thinking machine which would serve as a perceptive and innovative tutor in a one-on-one training system, i.e., to achieve symbiotic synergism.

Once AI development achieves sufficient sophistication it would be possible for a training device to work interactively with any individual trainee whether he/she be very talented or marginally so, and acheive adequate mastery of course objectives in nearly every case. Thus, the throughput theoretically should be very close to 100%. It is not conceivable that the best of AI based adaptive training systems can achieve 100% throughput, simply because there is too large a variance of human individual differences. Consequently, it is anticipated that non-symbiotic match-ups will occur from time to time.

Therefore, an AI based adaptive training system would accomplish what is unattainable today. Namely, a nearly ideal throughput commensurate with a maximization of course objective mastery on an individual basis would be achieved without requiring a greater availability nor competancy of the existing population of human trainers.

In addition, the AI training system would be most conducive to technical updating and field system reconfigurations. Further, the hardware would usually be compatible with new developments in AI methodologies and techniques.

Finally, as the technology of microcomputers advances it appears that the AI adaptive training device may become small enough for the trainee to take it into the field with him/her. The ramifications of this concept are extensive and make the applications of AI adaptive training nearly boundless.

RECOMMENDATIONS

New developments in AI techniques and methodologies are needed forthwith.

In addition, new technologies of human/computer interfacing are necessary to enhance the efficiency and effectiveness of human to computer interaction. In this same vein, much more needs to be quantitatively known about how humans process coded and reproduced stimuli such as computer graphics on CRTs.

Such technologies as natural language processing and semantic information processing have natural application here due to the need for the human to comfortably communicate with the training machine. New innovations in natural/semantic language processing will be needed to support AI based adaptive training systems. Further, the technology of voice synthesization (both speech and understanding) can significantly support a successful AI training system.

It should also be mentioned that in order to be effective, an AI adaptive training system would have to process data acquired in a training scenario in real-time, i.e., at the time of acquisition - not at some later time. This poses a burden on the data processing capability of the hardware. New developments in high speed data processing are currently approaching the needed capabilities in this area.

Finally, there are a number of theorists, including the authors, who advocate the hypothesis that human physiological parameters such as heart rate pattern, respiration rate pattern, electromyographic data, electroencephalographic data, galvanic skin response pattern, etc., can be shown to correlate to human task performance. If this hypothesis can be empirically supported, then an AI adaptive training system can tap into these physiological parameters for the purpose of either predicting task performance and/or enabling the human to monitor some parameter via feedback from the system so as to try to bring it into a range suitable for optimal performance.

REFERENCES

(1) Barr, A., and Feigenbaum, E.A., The handbook of artificial intelligence (William Kaufmann, Inc., Los Altos, CA, 1981.

(2) Carbonell, J.R., AI in CAI: An artificial-intelligence approach to computer assisted instruction, IEEE Transactions on Man-Machine Systems. MMS-11 (1970) 190-202.

(3) Kelly, C.R., What is adaptive training? Human Factors. 11 (1969) 547-556.

(4) Lintern, G., and Gopher, D., Adaptive training of perceptual-motor skills: Issues, results, and future directions, Journal of Man Machine Studies. 10 (1978) 521-551.

(5) Lumsdane, A.A., and Glaser, R. (Eds.), Teaching machines and programmed learning, a source book (National Education Association, Washington, D.C., 1960).

(6) Minsky, M. (Ed.), Semantic information processing (The MIT Press: Cambridge, MA, 1968).

(7) Williges, R.C., The tide of computer technology, Human Factors. 26(1) (1984) 109-114.

LATE PAPERS*

*These papers arrived too late to be included under their appropriate session headings. The paper by K. Morooka et al. belongs to the "Human-Computer Interaction" section and that of W. Rohmert to the "Organizational Design and Organizational Development" section.

Human Factors in Organizational Design and Management
H.W. Hendrick and O. Brown, Jr. (Editors)

AN ERGONOMICAL STUDY OF VISUAL DISPLAY TERMINALS

K. Morooka and S. Yamada

Department of Management Engineering
Tokai University, Japan

S. Fujimura

F&F Design Office, Japan

Eleven male subjects performed one hour test under several environment using "KANGI" vocablaries. Concerning illumination, heart-rate was the lowest around 320 Lux with high significance and failing eye-sight was found with significance. There was not significant difference between each color and each CRT size regarding heart-rate, but failing eye-sight was found on 14-inch CRT test. However, CFF changing-rate of 12-inch test was higher than 14-inch, further more, there was high significance on 12-inch CRT test.

Introduction

It is a great mistake for us to neglect ergonomical consideration of visual desplay terminals (VDTs). Various troubles on the body have springed up and various cases were reported in the some field[1,2]. Almost of these research were concerned with alphabetical display, however, KANGI and HIRAGANA characters are used in Japanese VDTs. Studies of using Japanese characters are only a few.

The field research was studied about actual VDT operators. In the field two women resigned their post as VDT operator in the cause of failing eye-sight for two years. These situation was thought to be caused a lack of ergonomical consideration on VDT work.

Ergonomical factor of VDT work was classified into machines, source document, facilities and working environment. Still more machines were classified into CRT and key-board, facilities were classified into desk, chair and source document keeper.

While on the field research, some subjects were picked up, for example illumination (working environment), color, cathode-ray tube (CRT) size (machine) and so on.

Method

A simulated experiment using VDT was made. Some targets were gathered up with the results of field research. The targets were as follows;

1) Illumination (10, 32, 100, 320, 1000 Lux)
2) Color (white, green, red, blue, yellow)
3) CRT size (12-inch, 14-inch)

Apparatus

Working environment

An experimental booth which was furnished with adjustable illumination (0-1000 Lux), set in the laboratory. The environmental condition of experimental booth was controled temperature (24± 1°C) and humidity. Color and CRT size examination were carried out under 320 Lux.

Display

The CRT display of terminal device was used in this study. 14-inch and 12-inch displays were prepared, and a character of displaies was formed by dot matrices which were 24X24 dot and 16X 16 dot. Each CRT was a color display.

Visual load measurement divice

The visual load was investigated using CFF(TAKEI), near point (TAKEI) and eye-sight(TOPCON) value, and heart-rate(SANEI-SOOKI).

Procedure

Figure 1 shows the simple procedure of this experiment. At the first step, the incorrect word and the correct one were displayed on the CRT. Words were "KANGI" vocabularies and thirty two vocabularies were prepared (Table 1).

At the second step, the word was displayed on the CRT one after another from left top and displayed row-wise. Each displayed word was put out after about two seconds.

The words per one CRT were fifty and twenty-five incorrect words were hidden in the CRT. The subjects realized the word if it was right or not. Whenever the subjects realized the incorrect word, they pressed down the zero key of the 10-key numerics in the key-board. One vocabulary test was completed about two minutes, and the positions of the incorrect and the correct word were changed at random. This test was repeated for about one hour.

Measurement

Figure 2 shows the simple procedure of measurement. Befor and after one hour test eye-sight, CFF and near point value were measured and fatigue was investigated. Still more heart-rate was recorded during one hour test.

Subjects

Eleven males of university students between the age of 22-25 were selected as subjects for this study. They were all in good physical conditions.

Results

1) Illumination

On right eye, failing eye-sight was found on 10, 32, 100 and 320 Lux with significance and changing-rate was shown in Figure 3. On left eye, failing eye-sight was found on 100 and 1000 Lux with high significance and found on 32 and 100 Lux with significance.

Decline of CFF value was found on 10 and 1000 Lux with high significance and found on 32 and 100 Lux with significance, and changing-rate was shown in Figure 4.

Extension of near point value was found on 10 and 1000 Lux with significance and found on 32 and 100 Lux with high significance, and changing-rate was shown in Figure 5.

On the heart-rate, the effect of illumination was found on

Variance analysis with high significance (Figure 6).

2) Color and CRT size

On right eye, failing eye-sight was found on Green(14-inch) with significance. On left eye, failing eye-sight was found on Green(14-inch), Blue and White with significance, and changing-rate was shown in Figure 7.

Decline of CFF value was found on Green(12-inch), and Red with high significance and found on Yellow, Blue and White with significance, and changing-rate was shown in Figure 8.

Extension of near point value was found on Blue with high significance and found on Red and White with significance. and changing-rate was shown in Figure 9.

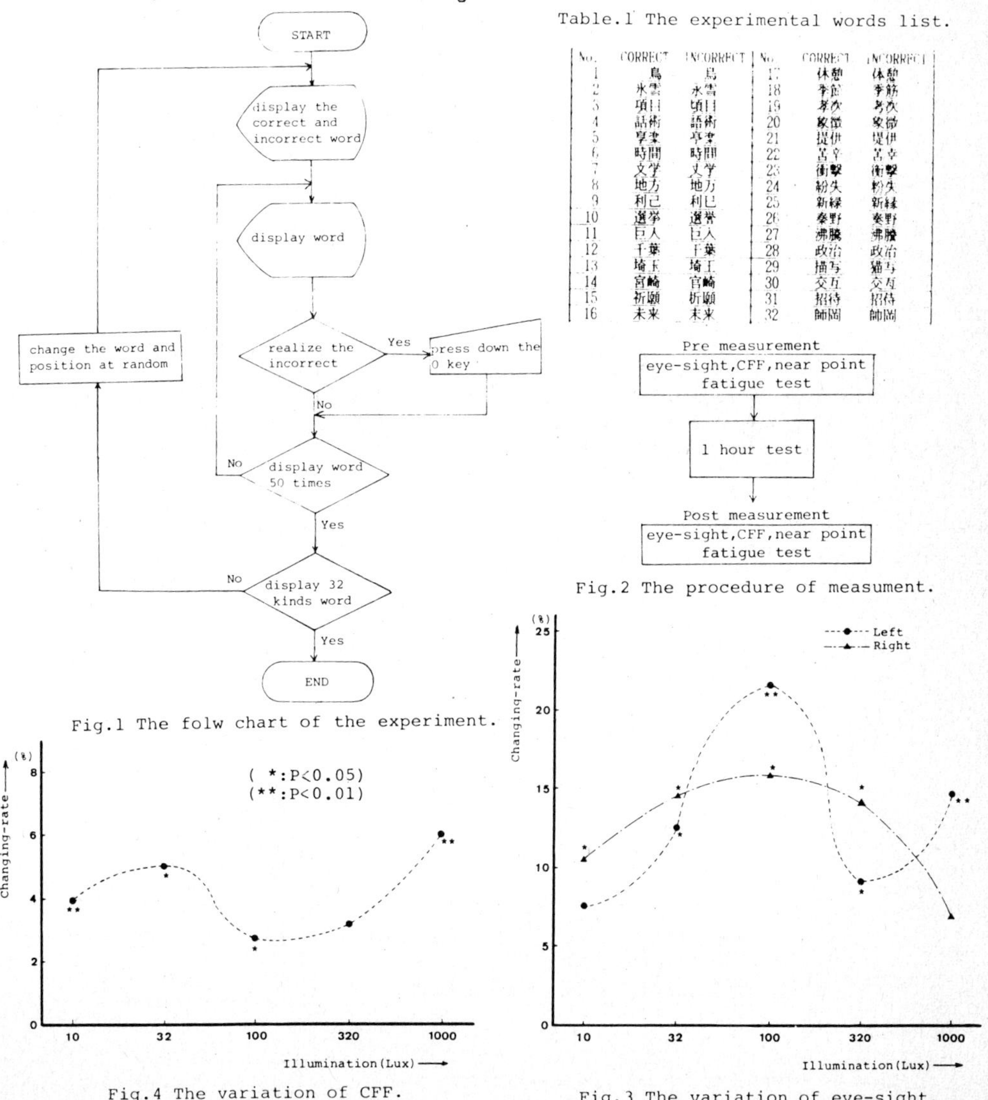

Table.1 The experimental words list.

No.	CORRECT	INCORRECT	No.	CORRECT	INCORRECT
1	鳥	烏	17	休憩	体憩
2	氷雲	水雲	18	季節	季筋
3	項目	頃目	19	孝次	考次
4	話術	語術	20	象徴	象微
5	享楽	亭楽	21	提供	堤供
6	時間	時閒	22	苦辛	苦幸
7	文学	丈学	23	衝撃	衡撃
8	地方	地万	24	紛失	粉失
9	利己	利巳	25	新緑	新縁
10	選挙	選誉	26	秦野	奏野
11	巨人	巨入	27	沸騰	沸謄
12	千葉	干葉	28	政治	政冶
13	埼玉	埼王	29	描写	猫写
14	宮崎	官崎	30	交互	交亙
15	祈願	折願	31	招待	招侍
16	未来	末来	32	師岡	帥岡

Fig.1 The folw chart of the experiment.

Fig.2 The procedure of measument.

Fig.3 The variation of eye-sight.

Fig.4 The variation of CFF.

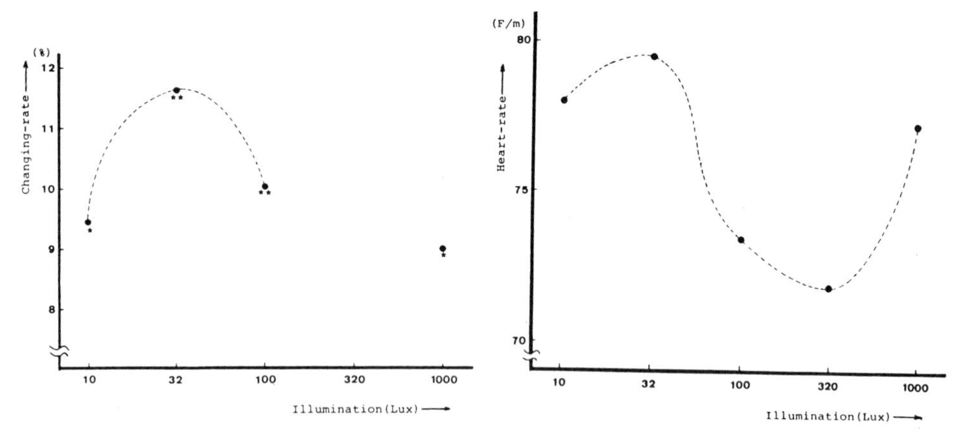

Fig.5 The variation of near point.

Fig.6 The variation of heart-rate.

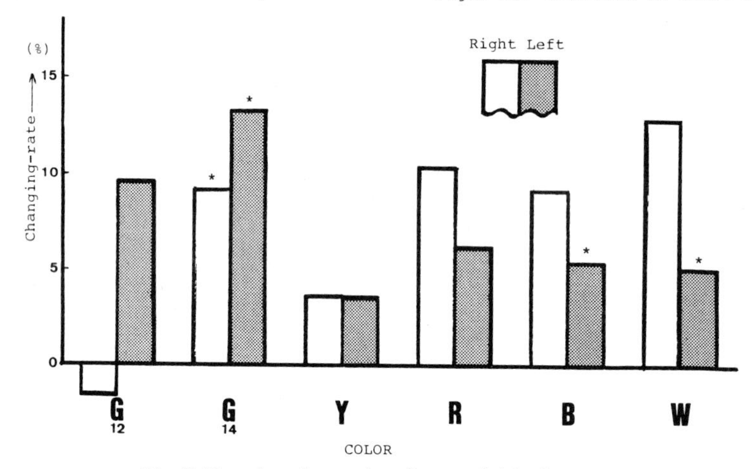

Fig.7 The changing-rate of eye-sight.

Fig.8 The changing-rate of CFF.

Fig.9 The changing-rate of near point.

Discussion

Under controled illumination, variation of heart-rate was the most notable and high significance ($P<0.01$: variance analysis). According to this result, it was clear that factor of illumination influenced VDT operator and heart-rate was the highest around 32 Lux and the lowest around 320 Lux. This tendency fitted with the accomplished result concerning refering work using hard source document [3,4]. Regarding CFF, the curve was drawn as well as heart-rate. As CFF means awakening level, it was thought that visual load was the lowest around 200 Lux and the highest around 32 and 1000 Lux. Furhter more, concerning to eye-sight especially left eye, changing-rate was the lowest among significant value, accordingly, it was considered that visual load was the lowest around 320 Lux.

Only Green CFF value of CRT colors was not significant between before and after test. It was regaded that green did not lower awakening level, namely, Green could produce low visual load. However, failing eye-sight was found on Green with significance and changing-rate was high, too. Changing-rate of Yellow was the lowest on each eye concerning eye-sight and CFF changing-rate was the lowest among significant color, too. It was considered that this result fitted the recent state of things. Namely, Yellow is thought of the recommendable color. However, it is needful to consider uses of each color, for example, a certain color suits entry operation and another suits dialog operation.

Concerning CRT size, failing eye-sight was found with significance on each eye of 14-inch and was not found on 12-inch. As changing-rate of 12-inch was relatively high on left eye, it could not judged that 12-inch was better than 14-inch. On the other hand, decline of CFF value was found on 12-inch with high significance and changing-rate was higher than 14-inch. Further more, 14-inch CRT(24X24 dot) is necessary for "KANGI" character to ensure certain operation.

Conclusion

1) Illumination was recommended nearly 300 Lux for VDT operator.
2) Green and Yellow were more suitable than other colors. But the close examination was necessary how to use colors.
3) It was not clear whether CRT size influenced visual load. Still more consideration is neccessary for using "KANGI" chracter.

Acknowledgment

We acknowledge the suggestions and advice of Prof. T. Michishita and Prof. T. Awano of the Department of Management Engineering, Tokai University.

Reference

1) E.Grandjean, E.Vigliani, Ergonomic Aspects of Visual Display Terminals (Taylor & Francis Ltd, London, 1980).
2) E.Grandjean, k.Nishiyama, A laboratory study on preferred and imposed settings of a VDT workstation, Behaviour and Inforamation Technology 3(1982) 289-304.
3) S.Sugimoto, Physiological Effects of Environmental Lighting conditions (Part 1), J.Illum. Engng. Inst. Jpn. 4(1980) 26-30.
4) S.Sugimoto, Physiological Effects of Environmental Lighting conditions (Part 2), J.Illum. Engng. Inst. Jpn. 4(1981) 41-45.

Human Factors in Organizational Design and Management
H.W. Hendrick and O. Brown, Jr. (Editors)

DIFFERENT STEPS IN ORGANIZATIONAL WORK DESIGN

Walter Rohmert

Institut für Arbeitswissenschaft, University
of Technologie
Darmstadt, West-Germany

State of work design results from different steps of design and their limiting conditions. But there is a feedback from the first organizational step of the design process to the prior selections of design-options dependent on the criteria of evaluation of work (related to a four-level hierarchy for assessment of work design) and the limiting factors involved.

INTRODUCTION

Over a long period of development, man was so superior to the machine, that he could always adapt himself to its demands as well as to his own needs. This has been changed by progress in technology and by the division of tasks. Now the content and environment of work are changed to such an extent, that they can be inappropriate for the natural conditions of human life. Therefore every work design is obliged to consider all the needs of men at their work. There is an increasing need to consider the quality of working life. Work design should meet these requirements.

STEPS OF WORK DESIGN AND THEIR LIMITING CONDITIONS

Job demands are influenced not only by technical sophistication of man-at-work systems. Different design areas are important. Technological design first of all selects the type of technique and fixes the functions, which have to be executed at all. Technical design is related to the utilization of technical equipment based on applied technology. The result of this design is the functional work partition between man and machine. In consequence to this we know all functions which may be executed by man. Ergonomic design covers all aspects of adapting work to man. By ergonomic design the working conditions for men will be fixed. The main objectives of ergonomic design are increased output by man, and improved working conditions. As the result of organizational design the total working sphere for the individual will be predetermined.

Work design is a cyclic process dependent on criteria which are fixed and accepted. This paper is going into details as far as organizational criteria are included. The organizational design process is a series of steps within a cycle. After analyzing the design problem the step of prognosis follows; the stress which will be imposed on the workers and their needs can be predicted. If the result of such an evaluation procedure satisfies the fixed criteria which are accepted by all institutions and people concerned with work conditions, the design loop can be completed and the design can be realized. More often, however, the result of evaluation requires modifications which have to be reanalyzed. Then after making corrections in the design the design loop starts again. In general, the result of the cyclic process of work design is negative. Corrections are necessary. The design loop will be traversed several times. This means that the total procedure will be carried out often. This procedure shows the state of each work design option which comes about as a result of the different steps of design and their limiting conditions. In general, each purpose and aim of a man-at-work system allows a choice among different kinds of technology. By selecting one, the first condition of work design is fixed. The next decision means the selection of the technique of producing goods and services; at this stage of the design process the limiting conditions are determined by economic and technological factors as well as by regulations and laws of safety at work and in the community, and by the applications of ergonomics. Then the degree of mechanization has to be fixed. Among different alternatives a particular structure of work has to be chosen. This selection is influenced by a choice of the degree of functional division between man and man (job partition) or man and machine (allocation of functions), and by economic as well as by social or sociopolitical factors.

After fixing these general conditions the process of work design in the very detailed sense can start. At first we can design specific working tools, machines, places and environment in the course of which limiting conditions of ergonomics and industrial hygiene have to be taken into account, as well as legal postulates and the restrictions of labour law and negotiation agreements.

In the final step of the design, work has to be organized with respect to the working time (e.g. shift schedules, duration of shifts, location and length of rest pauses, etc.). In this area, also, the same limiting conditions of design have to be respected. The result of the total design procedure emerges as a specific state of work design.

There is a feedback from the final steps of the design process to the prior selections of design-options dependent on the criteria of evaluation of work and the limiting factors involved.

Looking to the process of work design one could see that the procedure of ranking the different hierarchical design areas runs in an order. It can be seen that organization work design has two positions within this order. The first step of organizational work design includes job partition between men and fixing the distinguished functions either to man or to machine and computer. Within this first step are included all more or less new techniques of work structuring (e.g. job rotation, job enlargement, job enrichment, work-simplification, deverticalisation, work- structuring and job-redisign respecticely, socio-technical approach, autonomous work-groups; Rohmert and Weg, 1976). The second step of organizational work design includes problems like working time and rest pauses, the daily position of work and rest, problems of shift work, additional holidays, earlier retirements, etc.

HUMANITARIAN CONSIDERATIONS AND PROFITABILITY AS MAIN DESIGN AIMS

When planning and designing work systems, the designer must have some knowledge of the various alternative approaches to design (mentioned earlier), their effects on the goals and objectives of the work system and their effects on, or demands they make of, the work personnel. In planning and designing work systems, the designer has to evaluate and select from various alternative designs. Therefore it is necessary to formulate design aims.

The interpretation of these aims is important as regards the proposed preparation and interpretation of design knowledge. The user of this knowledge must be aware of this in order that he can form his own opinion about the interpretation. The general aims of work design in the context of designing the man-at-work system will therefore be discussed briefly. Work system design should, first and foremost, achieve the desired purpose of the work system (e.g. "object of work system"). However, other criteria must be used to decide the best way to realize the aims. These design aims may vary considerably from case to case, depending on the particular set of design problems. However, main aims can be defined, and the specific, detailed aims should contribute to these.

The efficiency of work systems under design or evaluation to fulfil a specific purpose, is always a main aim. There should be an optimum ratio of outlay to return, as this means that the most can be achieved with, or the greatest benefit derived from, the minimum of resources. However, as far as the benefit is concerned and how to achive it, we find that it always comes back to people: as profit makers and profit losers. Consequently, work system design must always take into account the human contribution to doing the work and human needs with respect to what is produced by the work. Work systems designing must, therefore, include human criteria (that is to say, relate to human beings). Profitability as a work output must also serve humanity.

In our competitive economic and social system, profitability of individual economic units or organizations is a prime determinant. However, most research and experience has shown that man and his contribution to work can greatly affect the achievement of the economic purpose in a work system. Here we would simply remind you of the importance of motivation in output and production and keeping down costs, and the long-term advantage to be gained from human capabilities given a suitable work-load as compared with the disadvantages of short-term exploitation of human working power. Therefore, while pursuing economic aims, we also have to consider humanitarian aims, lest we reduce the chances of achieving the economic aim in the long term.

Therefore, humanitarian considerations and profitability are always regarded as the joint leaders in the league of design aims. Of course, specific design measures will tend to promote one or the other more, but within the frame-work of a design project they must complement each other in a sensible way.

CONSTRUCTION OF A HIERARCHICAL SYSTEM OF DESIGN AIMS

The both primary design aims (humanitarian considerations and profitability) are too general, and cannot reasonably be used as a basis for stipulating the concrete factors and measures governing design. By looking for secondary aims which may contribute to the particular primary aims, we can construct a hierarchical system of aims. Formulating secondary aims leads to a list of aims (Rohmert and Weg, 1976). These aims are formulated as desired changes in a situation, since the aim of designing is always to alter a given situation defined as unsatisfactory. This can apply to an existing work system which is to be improved, or to the original designing of a work system from scratch. The aims are formulated by active "change" words, e.g. increase, reduce, ensure, facilitate, etc.

The aims mentioned in both the main areas of aims are equivalent. The evaluation of the aims is a subjective one and is related to decisions based not only on ergonomics but on other disciplines as well as on sociopolitical agreements. Often, achieving one aim contributes to achieving another. On the other hand, two aims can also conflict with each other. Therefore, the aims in the list of aims should be arranged as an hierarchical aims system (Rohmert and Weg, 1976). This aims system is offered as a stimulus and an aid to the specific designing of organizational problems. Of course, we cannot claim that this system has absolute validity, since it must necessarily include judgements which can never have absolute validity but can only find general acceptance. Thus, the aims system proposed for the organizational and ergonomical designing of human work can be supplemented or modified.

After having formulated the hierarchical system of design aims we are able to set up guide lines. For each guide line the aims will be mentioned which should be archieved by the measure.

A FOUR-LEVEL HIERARCHY FOR ASSESSMENT OF WORK DESIGN

To assess the results of design of human work in the context of an overall man-at-work system design, apart from the primary aims of humanity and profitability, there are four cardinal criteria, the fulfilment of which contributes to the primary aims. Therefore the evaluation of any concrete work design has to consider and to satisfy the four-level hierarchy:

ability
tolerability
acceptability
job satisfaction.

The ability of man to do the job is a basic prerequisite for profitability. Human capabilities and characteristics must enable man to do the work before the work can be carried out at all. This limit of man's ability means a level of practicability of using human input. If the work design does not consider these maximal capacities, even if only required for a limited time by men at their work, then the designer is obliged to give up all claim to human power; the designer has to design a completely automatic work system. Therefore the level of practicability of human work sets imperative claims to cater for human functions (e.g. body dimensions, muscular strength, speed and accuracy of movements, sensory- organ functions, etc.).

In most cases, we are not only interested in the "one off" completion of a job (as in sports), but we require a daily repetition in the work during a normal shift's duration and over the total working life's period. Under these longer lasting conditions the demands placed on men have to be lower. The highest level of demand which can be endured to the end of the normal working life without any work-related damage to health or normal human functional abilities is called the level of tolerability. The qualitative and quantitative output of a man affects profitability. Then there is effectiveness in the sense of general efficiency. Not all working conditions are equally conductive to quality and quantity of output or to incurring minimum costs; this, in turn, affects profitability.

The cardinal criteria for assessing the humanity of work and working conditions are health, well-being and job satisfaction. However, it is important to see that well-being and satisfaction always include long-term tolerability of work in health terms. Therefore the second level of tolerability in assessing work systems also makes imperative claims on the designer as well as the worker and the trade unions, or the employers and the employers' federations.

The third criterion level means that working conditions should be accepted by those who are social partners on the shop floor or at the labour agreement level. Unfortunately conditions can be accepted by local agreement as well as imposed by labour laws which are not tolerable over the total life's period. Therefore the third level of acceptability of human work-design needs must respect the hierarchical level of tolerability first of all, but only after ensuring tolerable working conditions. Understandably, social and perhaps individual values have a large part to play here. Also, there are likely to be changes in attitudes in the course of time. This is shown clearly when regarding the development of the modern slogan of "improving quality of life". There are, of course, human working conditions which were acceptable in the past or which will be accepted in times of high unemployment, which in normal conditions today are assessed to be inhumane. However, the more important postulate for humanity is to respect the conditions of long-term tolerability.

The hierarchy of assessment of human working conditions or design claims which starts with the individual level of man's ability ends at an individual level: the highest level of satisfaction. But the both levels in between are mainly oriented to more or less homogeneous groups and not necessarily to the individual. Striving for conditions of highest job satisfaction does not necessarily ensure tolerable conditions. For example, if someone is very content with his working conditions it might be, nevertheless, that his work load is intolerably high due to the fact that one stress factor, not necessarily its intensity but its duration, is too high. This can occur if a man works too much overtime or omits rest pauses, recovery periods and holidays under the illusion that he is doing a very satisfactory job. Also, the long distance jumper at the Olympic games of Mexico City will be highly content with his performance although he may never succeed in doing this maximal performance again. The two examples given show the importance of the basic levels in the hierarchy, as well as the necessity of setting and controlling design aims merely at these basic levels.

There are very critical limits to the self-regulating of human working conditions. That means that there is a real need in disciplines, methods, techniques and knowledge to adapt work to man and man to work to ensure "tolerable" working conditions. Here is a real challenge for the human sciences, especially for techniques of organizational work design. By these techniques standards of organizational work design may be fixed primarily of the basic levels of design assessment.

REFERENCES

Rohmert, W. and Weg, J.: Organisation teilautonomer Gruppenarbeit. Betriebliche Projekte - Leitregeln zur Gestaltung (Carl Hanser, München, Wien, 1976).

AUTHOR INDEX